IEE CIRCUITS, DEVICES AND SYSTEMS SERIES 9

Series Editors: Dr D. G. Haigh
　　　　　　　　Dr R. S. Soin
　　　　　　　　Dr J. Wood

VLSI TESTING

digital and mixed analogue/digital techniques

Other volumes in this series:

Volume 1	**GaAs technology and its impact on circuits and systems** D. G. Haigh and J. Everard (Editors)
Volume 2	**Analogue IC design: the current-mode approach** C. Toumazou, F. J. Lidgey and D. G. Haigh (Editors)
Volume 3	**Analogue-digital ASICs** R. S. Soin, F. Maloberti and J. Franca (Editors)
Volume 4	**Algorithmic and knowledge-based CAD for VLSI** G. E. Taylor and G. Russell (Editors)
Volume 5	**Switched-currents: an analogue technique for digital technology** C. Toumazou, J. B. Hughes and N. C. Battersby (Editors)
Volume 6	**High frequency circuits** F. Nibler and co-authors
Volume 7	**MMIC design** I. D. Robertson (Editor)
Volume 8	**Low-power HF microelectronics** G. A. S. Machado (Editor)

VLSI TESTING

digital and mixed analogue/digital techniques

STANLEY L. HURST

The Institution of Electrical Engineers

Published by: The Institution of Electrical Engineers, London,
United Kingdom

© 1998: The Institution of Electrical Engineers
Reprinted with corrections 1998

This publication is copyright under the Berne Convention and the
Universal Copyright Convention. All rights reserved. Apart from any fair
dealing for the purposes of research or private study, or criticism or
review, as permitted under the Copyright, Designs and Patents Act, 1988,
this publication may be reproduced, stored or transmitted, in any forms or
by any means, only with the prior permission in writing of the publishers,
or in the case of reprographic reproduction in accordance with the terms
of licences issued by the Copyright Licensing Agency. Inquiries
concerning reproduction outside those terms should be sent to the
publishers at the undermentioned address:

The Institution of Electrical Engineers,
Michael Faraday House,
Six Hills Way, Stevenage,
Herts. SG1 2AY, United Kingdom

While the author and the publishers believe that the information and
guidance given in this work is correct, all parties must rely upon their own
skill and judgment when making use of it. Neither the author nor the
publishers assume any liability to anyone for any loss or damage caused
by any error or omission in the work, whether such error or omission is
the result of negligence or any other cause. Any and all such liability is
disclaimed.

The moral right of the author to be identified as author of this work has
been asserted by him/her in accordance with the Copyright, Designs and
Patents Act 1988.

British Library Cataloguing in Publication Data

A CIP catalogue record for this book
is available from the British Library

ISBN 0 85296 901 5

Typeset by Euroset, Alresford

Printed in England by Short Run Press Ltd., Exeter

If you truly want to understand something,
try changing it
Kurt Lewin

Contents

Preface	xi
Acknowledgments	xiii
List of symbols and abbreviations	xv

1 Introduction   1
   1.1 The need for testing   1
   1.2 The problems of digital testing   7
   1.3 The problems of analogue testing   9
   1.4 The problems of mixed analogue/digital testing   11
   1.5 Design for test   11
   1.6 Printed-circuit board (PCB) testing   13
   1.7 Software testing   15
   1.8 Chapter summary   16
   1.9 Further reading   16

2 Faults in digital circuits   19
   2.1 General introduction   19
   2.2 Controllability and observability   20
   2.3 Fault models   25
      2.3.1 Stuck-at faults   26
      2.3.2 Bridging faults   30
      2.3.3 CMOS technology considerations   31
   2.4 Intermittent faults   35
   2.5 Chapter summary   38
   2.6 References   40

3 Digital test pattern generation   43
   3.1 General introduction   43
   3.2 Test pattern generation for combinational logic circuits   47
      3.2.1 Manual test pattern generation   48
      3.2.2 Automatic test pattern generation   48

viii *Contents*

		3.2.2.1	Boolean difference method	51
		3.2.2.2	Roth's D-algorithm	55
		3.2.2.3	Developments following Roth's D-algorithm	63
	3.2.3		Pseudorandom test pattern generation	67
3.3	Test pattern generation for sequential circuits			70
3.4	Exhaustive, nonexhaustive and pseudorandom test pattern generation			73
	3.4.1		Exhaustive test pattern generators	73
	3.4.2		Nonexhaustive test pattern generators	74
	3.4.3		Pseudorandom test pattern generators	75
		3.4.3.1	Linear feedback shift registers (LFSRs)	76
		3.4.3.2	Cellular automata (CAs)	90
3.5	I_{DDQ} and CMOS testing			95
3.6	Delay fault testing			105
3.7	Fault diagnosis			108
3.8	Chapter summary			110
3.9	References			112

4 Signatures and self test — 119

- 4.1 General introduction — 119
- 4.2 Input compression — 120
- 4.3 Output compression — 124
  - 4.3.1 Syndrome (ones-count) testing — 125
  - 4.3.2 Accumulator-syndrome testing — 128
  - 4.3.3 Transition count testing — 129
- 4.4 Arithmetic, Reed-Muller and spectral coefficients — 130
  - 4.4.1 Arithmetic and Reed-Muller coefficients — 131
  - 4.4.2 The spectral coefficients — 133
  - 4.4.3 Coefficient test signatures — 138
- 4.5 Signature analysis — 141
- 4.6 Online self test — 153
  - 4.6.1 Information redundancy — 153
    - 4.6.1.1 Hamming codes — 157
    - 4.6.1.2 Two-dimensional codes — 160
    - 4.6.1.3 Berger codes — 161
    - 4.6.1.4 Residue codes — 163
  - 4.6.2 Hardware redundancy — 169
  - 4.6.3 Circuit considerations — 175
- 4.7 Self test of sequential circuits — 188
- 4.8 Multiple-output overview — 191
- 4.9 Chapter summary — 193
- 4.10 References — 195

5 Structured design for testability (DFT) techniques — 201

- 5.1 General introduction — 201
- 5.2 Partitioning and *ad hoc* methods — 201
- 5.3 Scan-path testing — 206
  - 5.3.1 Individual I/O testing — 211
  - 5.3.2 Level-sensitive scan design (LSSD) — 211
  - 5.3.3 Further commercial variants — 218
  - 5.3.4 Partial scan — 225
- 5.4 Boundary scan and IEEE standard 1149.1 — 230
- 5.5 Offline built-in self test — 241
  - 5.5.1 Built-in logic block observation (BILBO) — 242
  - 5.5.2 Cellular automata logic block observation (CALBO) — 264

	5.5.3	Other BIST techniques	274
5.6		Hardware description languages and test	286
5.7		Chapter summary	289
5.8		References	291

6 Testing of structured digital circuits and microprocessors — 297

- 6.1 Introduction — 297
- 6.2 Programmable logic devices — 298
  - 6.2.1 Programmable logic arrays — 298
    - 6.2.1.1 Offline testable PLA designs — 306
    - 6.2.1.2 Online testable PLA designs — 311
    - 6.2.1.3 Built-in PLA self test — 314
  - 6.2.2 Programmable gate arrays — 320
  - 6.2.3 Other programmable logic devices — 323
- 6.3 Read-only memory testing — 329
- 6.4 Random-access memory testing — 331
  - 6.4.1 Offline RAM testing — 333
  - 6.4.2 Online RAM testing — 337
  - 6.4.3 Buried RAMs and self test — 345
- 6.5 Microprocessors and other processor testing — 347
  - 6.5.1 General processor test strategies — 350
  - 6.5.2 The central processing unit — 352
  - 6.5.3 Fault-tolerant processors — 355
- 6.6 Cellular arrays — 362
- 6.7 Chapter summary — 371
- 6.8 References — 372

7 Analogue testing — 381

- 7.1 Introduction — 381
- 7.2 Controllability and observability — 382
- 7.3 Fault models — 385
- 7.4 Parametric tests and instrumentation — 386
- 7.5 Special signal processing test strategies — 392
  - 7.5.1 DSP test data — 392
  - 7.5.2 Pseudorandom d.c. test data — 395
- 7.6 The testing of particular analogue circuits — 398
  - 7.6.1 Filters — 398
  - 7.6.2 Analogue to digital converters — 403
  - 7.6.3 Digital to analogue converters — 414
  - 7.6.4 Complete A/D subsystems — 419
- 7.7 Chapter summary — 421
- 7.8 References — 422

8 Mixed analogue/digital system test — 427

- 8.1 Introduction — 427
- 8.2 Mixed-signal user-specific ICs — 430
- 8.3 Partitioning, controllability and observability — 436
- 8.4 Offline mixed-signal test strategies — 441
  - 8.4.1 Scan-path testing — 441
  - 8.4.2 Built-in self-test strategies — 446
  - 8.4.3 Built-in analogue test generators — 450
- 8.5 Boundary scan and IEEE standard 1149.4 — 457
- 8.6 Chapter summary — 470
- 8.7 References — 471

x   Contents

9   The economics of test and final overall summary   **477**
   9.1   Introduction   477
   9.2   A comparison of design methodologies and costs   478
   9.3   The issues in test economics   484
   9.4   Economic models   489
       9.4.1   Custom ICs   489
       9.4.2   The general VLSI case   490
       9.4.3   Test strategy planning resources   496
   9.5   Final overall summary   499
       9.5.1   The past and present   499
       9.5.2   The future   503
   9.6   References   510

Appendix A   Primitive polynomials for $n \leq 100$ **515**

Appendix B   Minimum cost maximum length cellular automata for $n \leq 100$ **519**

Appendix C   Fabrication and yield   **523**

Index   **527**

Preface

Historically, the subject of testing has not been one which has fired the imagination of many electronic design engineers. It was a subject rarely considered in academic courses except, perhaps as the final part of some laboratory experiment or project, and then only to confirm the correct (or incorrect!) working of some already-designed circuit. However, the vast increase in on-chip circuit complexity arising from the evolution of LSI and VLSI technologies has brought this subject into rightful prominence, making it an essential part of the overall design procedure for any complex circuit or system.

The theory and practice of microelectronic testing has now become a necessary part of both IC design and system design using ICs. It is a subject area which must be considered in academic courses in microelectronic design, and which should be understood by all practising design engineers who are involved with increasingly complex and compact system requirements.

Written as a self-contained text to introduce all aspects of the subject, this book is designed as a text for students studying the subject in a formal taught course, but contains sufficient material on more advanced concepts in order to be suitable as a reference text for postgraduates involved in design and test research. Current industrial practice and the economics of testing are also covered, so that designers in industry who may not have previously encountered this area may also find information of relevance to their work.

The book is divided into nine principal chapters, plus three appendices. Chapter 1 is an introductory chapter, which explains the problems of microelectronic testing and the increasing need for design for test (DFT) techniques. Chapter 2 then continues with a consideration of the faults which may arise in digital circuits, and introduces the fundamentals of controllability, observability, fault models and exhaustive versus non-exhaustive test. This consideration of digital circuit testing continues in

Chapter 3, where circuit simulation, automatic test pattern generation (ATPG), fault coverage, the particular test requirements of CMOS technology and other fundamentals are covered.

Chapter 4 surveys some of the techniques which have been proposed to simplify the testing procedures for combinational logic networks, largely with the objective of providing a simple final faulty/fault-free answer from the test. These include signature analysis and other methods which usually involve some form of counting of the logic signals which are generated during the test procedure. The following chapter, Chapter 5, continues these considerations, introducing structured DFT techniques such as scan testing which forms the basis of most of the currently-employed VLSI test techniques.

Chapter 6 is a survey of the means of testing very specific digital architectures, including microprocessors, ROM, RAM and PLA circuits, and also cellular arrays.

Chapters 7 and 8 move on from the considerable volume of development for testing digital circuits to analogue and mixed analogue/digital circuit test. This is a more recent area of concern than the digital area, but one of rapidly increasing importance in view of the growing mixed analogue/digital signal processing interests — 'systems on a chip'. Chapter 7, therefore, considers analogue testing fundamentals, including instrumentation and how much parametric testing the system designer may need to do on bought-in microelectronic parts. Chapter 8 then considers mixed analogue/digital testing, including partitioning and the possibility of applying some of the established digital DFT techniques to the analogue area. Other concepts still being researched will also be mentioned.

Finally, Chapter 9 looks at a wider scene rather than at specific test methodologies. This includes a consideration of the economics of test and design for test, and future possible expectations.

The book assumes that readers are already familiar with basic electronic circuits, both digital and analogue, and with microelectronic fabrication technologies, both bipolar and MOS. The importance of computer-aided design (CAD) in circuit and system synthesis is also assumed to be known. The mathematics involved in some sections should be well within the knowledge of the majority of readers, and therefore the text as a whole should be straightforward to follow, and hopefully helpful to many.

Stanley L. Hurst

Acknowledgments

I would like to acknowledge the work of the many pioneers who have contributed to our present state of knowledge about and application of testing methodologies for microelectronic circuits and systems, oftimes with belated encouragement from the manufacturing side of industry. Conversations with many of these people over the years have added greatly to my awareness of this subject area, and the many References given in this text are therefore dedicated to them.

On a more personal note, may I thank former research colleagues at the University of Bath, England, and academic colleagues in electronic systems engineering at the Open University who have been instrumental in helping my appreciation of the subject. Also very many years of extremely pleasant co-operation with the VLSI design and test group of the Department of Computer Science of the University of Victoria, Canada, must be acknowledged. To them and very many others, not excluding the students who have rightfully queried many of my loose statements, specious arguments or convoluted mathematical justifications, my grateful thanks.

Finally, I must acknowlege the cheerful help provided by the administrative and secretarial staff of the Faculty of Technology of the Open University, who succeeded in translating my original scribbles into recognisable text. In particular, may I thank Carol Birkett, Angie Swain and Lesley Green for their invaluable assistance. Any errors remaining in the text are entirely mine!

S.L.H.

List of symbols and abbreviations

The following are the symbols and abbreviations that may be encountered in VLSI testing literature. Most but not necessarily all will be found in this text.

ABSC	analogue boundary scan cell
ADC	analogue to digital converter
AI	artificial intelligence
ALU	arithmetic logic unit
ASIC	application-specific intregrated circuit (see also CSIC and USIC)
ASM	algorithmic state machine
ASR	analogue shift register
ASSP	application-specific standard part
A-to-D	analogue to digital
ATE	automatic test equipment
ATPG	automatic test pattern generation
BIC	built-in current (testing)
BiCMOS	bipolar/CMOS technology
BDD	binary decision diagram
BILBO	built-in logic block observation
BIST	built-in self test
BJT	bipolar junction transistor
BSC	boundary scan cell
BSDL	boundary scan description language
BSR	boundary scan register
C	capacitor
CA	cellular automata
CAD	computer-aided design

CALBO	cellular automata logic block observation
CAE	computer-aided engineering
CAM	computer-aided manufacture
CAT	computer-aided test
CCD	charge-coupled device
CFR	constant failure rate
CML	current-mode logic
CMOS	complementary MOS
CPU	central processing unit
CSIC	custom or customer-specific IC
CTF	controllability transfer factor
CUT	circuit under test
DA	design automation
DAC	digital to analogue converter
DBSC	digital boundary scan cell
DDD	defect density distribution
DEC	double error correction
DED	double error detection
DFM	design for manufacturing
DFR	decreasing failure rate
DFT	design for test, or design for testability
DIL or DIP	dual-inline package, or dual-inline plastic package
DL	default level
DMOS	dynamic MOS
DR	data register
DRC	design rule checking
DSP	digital signal processing
D-to-A	digital to analogue
DUT	device under test
EAROM	electrically alterable ROM
ECAD	electronic computer-aided design
ECL	emitter-coupled logic
EEPROM	electrically erasable PROM
EPLD	electrically programmable logic device
EPROM	erasable programmable ROM
E^2PROM	electrically erasable PROM
ERC	electrical rules checking
eV	electron volt
FC	fault cover or coverage
FET	field effect transistor
FFT	fast Fourier transform
FPGA	field-programmable gate array

FPLA	field-programmable logic array
FPLS	field-programmable logic sequencer
FRAM	ferroelectric random-access memory
FRU	field replacable unit
FTD	full technical data (cf. LTD)
GaAs	gallium arsenide
GDS II	graphics design style, version II (a mask pattern standard)
Ge	germanium
GHz	gigahertz
GPIB	general purpose instrumentation bus, electrically identical to HPIB bus
HDL	hardware description language
HILO	a commercial simulator for digital circuits
IC	integrated circuit
I_{DDQ}	quiescent current in CMOS
IEE	Institution of Electrical Engineers
IEEE	Institute of Electrical and Electronics Engineers
IFA	inductive fault analysis
IFR	increasing failure rate
IGFET	insulated gate FET
I/O	input/output
IR	instruction register
ISL	integrated Schottky logic
I_{SSQ}	quiescent current in CMOS
JEDEC	Joint Electron Device Engineering Council
JFET	junction FET
JIT	just in time (manufacture)
JTAG	Joint Test Action Group
k	1000
K	1024
kHz	kilohertz
L	inductor, or length in MOS technology
λ	lambda, a unit of length in MOS layout design
LAN	local area network
LASAR	logic automated stimulus and response
LCA	logic cell array
LCC	leaded ceramic chip carrier
LCCC	leadless ceramic chip carrier

LFSR	linear feedback shift register	
LSB	least significant digit	
LSI	large scale integration	
LSSD	level-sensitive scan design	
LSTTL	low-power Schottky TTL	
LTD	limited technical data (cf. FTD)	
MCM	multichip module	
MHz	megahertz	
mil	one thousandth of an inch	
MIPS	million instructions per second	
MISR	multiple input signature register	
MOS	metal-oxide-semiconductor	
MOST	MOS transistor	
MPGA	mask-programmable gate array	
MPLD	mask-programmable logic device	
ms	millisecond	
MSB	most significant bit	
MSD	most significant digit	
MSI	medium scale integration	
MTBF	mean time between failure	
MTTR	mean time to repair	
NFF	no fault found	
nMOS	n-channel MOS	
NRE	nonrecurring engineering	
ns	nanosecond (10^{-9})	
OEM	original equipment manufacturer	
op. amp.	operational amplifier	
OTA	operational transconductance amplifier	
OTF	observability transfer function	
PAL	programmable array logic	
PCB	printed-circuit board	
PDM	process device monitor (see also PED and PMC)	
PED	process evaluation device (see also PDM and PMC)	
PG	pattern generation	
PGA	pin grid array, or programmable gate array	
PLA	programmable logic array	
PLCC	plastic leaded chip carrier	
PLD	programmable logic device	
PLS	programmable logic sequencer	
PMC	process monitoring circuit (see also PDM and PED)	

List of symbols and abbreviations xix

pMOS	p-channel MOS
PODEM	path-oriented decision making (algorithm)
POST	power-on self test
PRPG	pseudorandom pattern generation
PROM	programmable read-only memory
ps	picosecond (10^{-12} s)
PSA	parallel signature analyser
PUC	per-unit cost
QCM	quiescent current monitoring
QIL	quad inline (package)
QTAG	Quality Test Action Group
R	resistor
RAM	random-access memory
RAPS	random path sensitising
RAS	random access scan
ROM	read-only memory
RTL	register transfer language
RTOK	retest OK
S-A or s-a	stuck-at
s-a-0	stuck-at 0
s-a-1	stuck-at 1
SAFM	stuck-at fault model
SASP	signature analysis and scan path
SDI	scan data in
SDO	scan data out
SEC	single error correction
SED	single error detection
SEM	scanning electron microscope
SIA	Semiconductor Industry Association
Si_3N_4	silicon nitride
SiO_2	silicon dioxide
SMD	surface mount device
SNR	signal to noise ratio
SOI	silicon-on-insulator
SOS	silicon-on-sapphire
SPICE	simulation program, integrated circuit emphasis
SRAM	static read-only memory
SSI	small scale integration
SSR	serial shadow register
STL	Schottky transistor logic
STR	structured test register
STTL	Schottky TTL

TC	test coverage
TAP	test access port
TCK, TMS, TDI, TDO	boundary-scan terminals of IEEE standard 1149.1
TDD	test directed diagnosis
TDL	test description language
TMR	triple modular redundancy
TPG	test pattern generation (see also ATPG)
TPL	test programming language
TQM	total quality management
TTL	transistor-transistor logic
UCA	uncommitted component array
ULSI	ultra large scale integration
µP	microprocessor
USIC	user-specific IC (preferable term to ASIC)
UUT	unit under test
V_{BB}	d.c. base supply voltage
V_{CC}	d.c. collector supply voltage
V_{DD}	d.c. drain supply voltage
V_{EE}	d.c. emitter supply voltage
V_{GG}	d.c. gate supply voltage
V_{SS}	d.c. source supply voltage
VDU	visual display unit
VHDL	very high speed hardware description language
VHSIC	very high speed IC
VLSI	very large scale integration
VMOS	a MOS fabrication technology
W	width in MOS technology
WSI	wafer scale integration
Y	production or wafer yield

Chapter 1
Introduction

1.1 The need for testing

The design and production of the majority of manufactured products usually involves a rigorous design activity followed by detailed testing and evaluation of prototypes before volume production is begun. Once in production, continuous quality control of components and assembly is desirable, but comprehensive testing of each finished item is not usually affordable. However, the more sophisticated the product, or the components used within the product, then the more need there will be to try to ensure that the end product is fully functional. The ability to test end products as comprehensively as possible within a given budget or time constraint is therefore a universal problem.

Electronics, and here more specifically microelectronics, has the fundamental feature that, in contrast to mechanical and electromechanical devices, visual inspection is of very little use. Parametric or functional testing must be used. The range of testing involves both the IC manufacturer and the original equipment manufacturer (OEM), in total including the following principal activities:

(i)   tests by the IC manufacturer (vendor) to ensure that all the fabrication steps have been correctly implemented during wafer manufacture (IC fabrication checks);
(ii) tests to ensure that prototype ICs perform correctly in all respects (IC design checks);
(iii) tests to ensure that subsequent production ICs are defect free (IC production checks);
(iv) possibly tests by the OEM of incoming ICs to confirm their functionality (acceptance tests);
(v)   tests by the OEM of final manufactured products (product tests).

The first of these five activities is the sole province of the vendor, and does not involve the OEM in any way. The vendor normally incorporates 'drop-ins' located at random positions on the wafer, these being small circuits or geometric structures from which the correct resitivity and other parameters of the wafer fabrication can be verified before any functional checks are undertaken on the surrounding circuits. This is illustrated in Figure 1.1. We will have no occasion to consider wafer fabrication tests any further in this text.

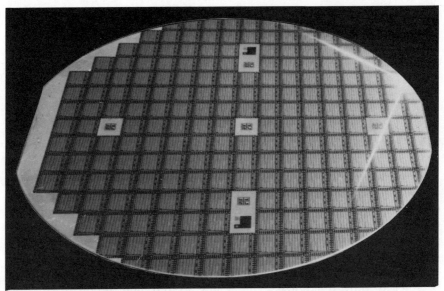

Figure 1.1 *The vendor's check on correct wafer fabrication, using drop-in test circuits at selected points on the wafer. These drop-ins may be alternatively known as process evaluation devices (PEDs), process device monitors (PDMs), process monitoring circuits (PMCs), or similar terms by different IC manufacturers (Photo courtesy Micro Circuit Engineering, UK)*

In the case of standard off the shelf ICs, available for use by any OEM, the next two test activities are also the sole responsibility of the vendor. However, in the case of customer-specific ICs, usually termed ASICs (application-specific ICs) or more precisely USICs (user-specific ICs), the OEM is also involved in phase (ii) in order to define the functionality and acceptance details of the custom circuit. The overall vendor/OEM interface details therefore will generally be as shown in Figure 1.2.

The final two phases of activity are clearly the province of the OEM, not involving the vendor unless problems arise with incoming production ICs. Such testing will be unique to a specific product and its components, although in its design the requirements of testing, and possibly the incorporation of design for test (DFT) features such as will be covered later in this text, must be considered.

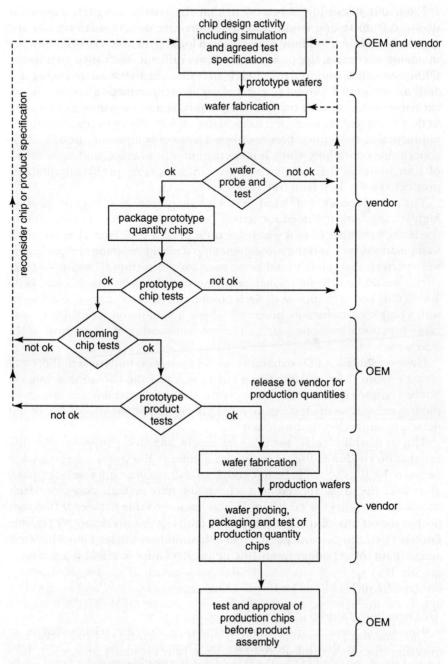

Figure 1.2   The IC vendor/OEM test procedures for a custom IC before assembly and test in final production systems. Ideally, the OEM should do a 100 % functional test of incoming ICs in a working equipment or equivalent test rig, although this may be impractical for VLSI circuits

From this discussion it is evident that the system designers cannot be divorced from testing considerations. In the case of very small circuits and systems containing, say, only a few hundred logic gates or a very small number of analogue circuits, the problem is not too difficult. With such circuits the OEM can often undertake a 100 % fully-functional test of incoming ICs and/or completed products, particularly if the production volume is not excessive, but as complexity increases or production rises then the pressure to do a restricted number of tests escalates rapidly. For example, it would be impractical to test a microprocessor-based piece of equipment through all the conceivable conditions which it may encounter in practice, and some subset of tests must be defined that will give an acceptable probability that the product as a whole is fault free.

The development of VLSI has clearly played the biggest part in highlighting both the need for testing, and the difficulties of doing so with the limited number of pins which are present on a packaged VLSI circuit. If wafer manufacture and the subsequent chip scribing, bonding and packaging were perfect, then there would be no need for production IC testing — every circuit would be fully functional, assuming that the prototype ICs had been 100 % checked and approved. Such absolute perfection cannot be achieved with complex production processes where a submicron sized defect may cause functional misbehaviour, and hence some testing of all production ICs is necessary.

However, unless a fully-exhaustive test of each IC is undertaken, there will always remain the possibility that a circuit will pass the chosen tests but will not be completely fault free. The lower the yield of the production process or the less exhaustive the testing, then the greater will be the probability of not detecting faulty circuits during test.

This probability may be mathematically analysed. Suppose that the production yield of fault-free circuits on a wafer is Y, where Y has some value between $Y = 0$ (all circuits faulty) and $Y = 1$ (all circuits fault free). Suppose, also, that the tests applied to each circuit have a fault coverage (fault detection efficiency) of FC, where FC also has some value between 0 (the tests do not detect any of the possible faults) and 1 (the tests detect all possible faults). Then the percentage of circuits that will pass the tests but which will still contain some fault or faults (the defect level after test, DL) is given by:

$$DL = \left\{1 - Y^{(1-FC)}\right\} \times 100 \ \%$$

This equation is shown in Figure 1.3.

The significance of this very important probability relationship is as follows. Suppose that the production yield was 50 % ($Y = 0.5$). Then if no testing was done at all ($FC = 0$), 50 % of the ICs would clearly be faulty when they came to be used. If testing is now done and the efficiency of the tests is only 80 % ($FC = 0.8$), then the percentage of faulty circuits when used would now be about 15 % (85 % fault free, 15 % faulty). This is still about one in

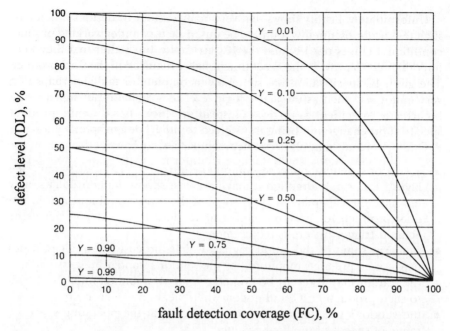

Figure 1.3 The defect level, DL, of a circuit after test with different manufacturing yields Y and fault detection coverage, FC. Only $Y = 1.0$ or $FC = 1.0$ will guarantee 100 % fault-free circuits (no defects) after test

seven ICs faulty, which is far too high for most customers. Hence to ensure a high percentage of fault-free circuits after test, either the manufacturing yield Y or the fault coverage FC, or both, must be high.

This, therefore, is the dilemma of testing complex integrated circuits, or indeed any complex system. If testing efficiency is low then faulty circuits may escape through the test procedure. On the other hand, the achievement of near 100 % fault detection ($FC = 1.0$) may require such extensive testing as to be prohibitively costly unless measures are taken at the design stage to facilitate such a level of test.

Before we continue with the main principles and practice of integrated circuit testing, let us consider a little further Figure 1.2 and problems which can specifically arise with OEM use of custom ICs. During the design phase, simulation of the IC will have been undertaken and approved before fabrication was commenced. This invariably involves the vendor's CAD resources for the final post-layout simulation check, and from this simulation a set of test vectors for the chip may be automatically generated which can be downloaded into the vendor's sophisticated general-purpose test equipment. The vendor's tests on prototype custom ICs may therefore be based upon this simulation data, and if the circuits pass this means that they conform to the simulation results which will have been approved by the OEM.

Unfortunately, history shows that very many custom IC designs which pass such tests are subsequently found to be unsatisfactory under working product conditions. This is not because the ICs are faulty, but rather that they were not designed to provide the exact functionality required in the final product. The given IC specification was somehow incomplete or faulty, perhaps in a very minor way such as the active logic level of a digital input signal being incorrectly specified as logic 1 instead of logic 0, or some product specification change not being transferred to the IC design specification.

Other problems may also arise between parties, such as:

- a vendor's general-purpose computer-controlled VLSI test system, see Figure 1.4, which although it may have cost several million dollars, may not have the capability to apply some of the input conditions met in the final product, for example nonstandard digital or analogue signals or Schmitt trigger hysteresis requirements;
- similarly, some of the output signals which the custom circuit provides may not be precisely monitored by the vendor's standard test system;
- the vendor may also only be prepared to apply a limited number of tests to each production IC, and not anywhere near an exhaustive test;
- the vendor's test system may not test the IC at the operating speed or range of speeds of the final product.

The main point we need to make is that in the world of custom microelectronics the OEM and the vendor must co-operate closely and intelligently to determine acceptable test procedures. When the OEM is using standard off the shelf ICs then the responsibility for incoming component and product testing is entirely his. However, in both cases the concepts and

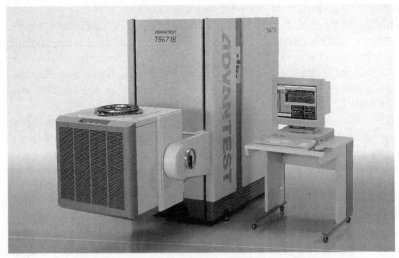

Figure 1.4 A typical general-purpose VLSI tester as used by IC vendors and test houses (Photo courtesy Avantest Corporation, Japan)

the theory and practice of microelectronic circuit testing such as we will be considering later must be understood.

1.2 The problems of digital testing

Before continuing in the following detailed chapters with the theory and practice of IC testing, let us broadly consider the problems which are present.

First, it may be appropriate to define three basic terms which arise in digital circuit testing. These are:

(i) *Input test vector* (or input vector or test vector): this is a combination of logic 0 and 1 signals applied in parallel to the accessible (primary) inputs of the circuit under test. For example, if eight primary inputs are present then one test vector may be 0 1 1 1 0 0 1 1. (A test vector is the same as a word, but the latter term is not commonly used in connection with testing.)
(ii) *Test pattern*: a test pattern is an input test vector plus the addition of the fault-free output response of the circuit under test to the given test vector. For example, if there are four accessible (primary) outputs, then with the above test vector the expected outputs might be 0, 0, 0, 1.
(iii) *Test set*: a test set is a sequence of test patterns, which ideally should determine whether the circuit under test is fault free or not. As will be seen later, a test set may be a fully-exhaustive, a reduced or minimum, or a pseudorandom test set. Continuing from the above example, a test set may be as shown in Table 1.1.

Table 1.1 An example of a test set for a combinational network with eight inputs and four outputs

	Test vectors								Output response			
	x_1	x_2	x_3	x_4	x_5	x_6	x_7	x_8	y_1	y_2	y_3	y_4
First test pattern	0	1	1	1	0	0	1	1	0	0	0	1
Next test pattern	0	1	1	1	0	1	0	0	0	1	0	0
Next test pattern	1	1	1	1	0	1	0	0	0	1	1	0
.												
.												
.												

Unfortunately, the terms 'vectors', 'patterns' and 'sets' are sometimes loosely used, and care is therefore necessary when reading literature from different sources. It should also be noted that these terms apply directly to purely combinational networks; where sequential networks are involved then clearly some additional information is needed, for example the number of clock pulses and the initial state of the circuit at the beginning of the test.

There is nothing fundamentally difficult in the functional testing of digital networks; no parametric testing such as may be involved in analogue testing (see later) is normally required. All that is necessary is to apply some input test patterns and observe the resulting digital output signals. The problem, basically, is the volume of data and the resultant time to test.

Take a simple combinational network with n primary inputs and m primary outputs. Then to test this network exhaustively with every possible input condition requires 2^n input test vectors, that is the complete truthtable of the network. If n is, say, 32 and the test vectors are applied at a frequency of 10 MHz, this would take $2^{32} \times 10^{-7}$ seconds, which is about seven minutes, to complete this simple exhaustive test. It would also require the $m\, 2^n$ output signals to be checked for correct response during this test against some record of the expected (fault free) output signals.

The presence of memory (storage) which is inherent in sequential networks greatly increases this problem. For a network containing s storage circuits (latches or flip-flops) there are 2^s possible internal states of the network. For a fully-exhaustive test, every possible input combination to the primary inputs should be applied for each of these 2^s internal states, which with n primary inputs gives an exhaustive test set of $2^n \times 2^s = 2^{n+s}$ test vectors.

An often-quoted example of this is that of the arithmetic 16-bit accumulator IC, which can add or subtract a 16-bit input number to/from any 16-bit number already stored in the accumulator. This relatively simple circuit has 19 inputs (16 bits plus three control signals) plus the 16 internal latches. Hence $n = 19$ and $s = 16$, giving $2^{19} \times 2^{16} = 2^{35} = 34\,359\,738\,368$ possible different logic conditions for this circuit. If test vectors were applied at 10 MHz, it would therefore take $2^{35} \times 10^{-7}$ seconds, which is about an hour, to test this MSI circuit exhaustively, and of course the 2^{35} output signals on each of the 16 output pins would need to be checked.

Normally all testing of ICs has to be performed through the input and output terminals of the circuit, either the bonding pads around the perimeter of each chip during the wafer stage testing by the vendor, or the pins on the packaged IC during subsequent testing by vendor or OEM. Hence, IC test procedures are often termed 'pin-limited' because of this accessibility problem, all the test input signals and their response having to be applied and detected at these relatively few points in comparison with the very large numbers of gates and macros within the circuit.* The order of application of test vectors to the circuit may also be important, which is an aspect that will be considered later particularly in connection with CMOS circuits where certain open-circuit faults may cause unexpected circuit action.

In summary, therefore, it is basically the number of tests which it may be necessary to apply to test a digital circuit, together with the limited number of accessible input and output pins, which complicates digital testing, rather

---

* It may be physically possible for the vendor to probe internal points on an IC before packaging, particularly to look for inexplicable faults, but this is increasingly difficult as device geometry shrinks in size, and can cause scar damage on the surface making the chip subsequently imperfect. See Chapter 3, Section 3.7, for additional information.

than any fundamental difficulty in the testing requirements. For circuits with a small number of logic gates and internal storage circuits, the problem is not acute; fully-exhaustive functional testing may be possible. As circuits grow to LSI and VLSI complexity, then techniques to ease this problem such as will be considered later become increasingly necessary.

1.3 The problems of analogue testing

In contrast to the very large number of logic gates and storage circuits encountered in digital networks, purely analogue networks are usually characterised by having relatively few circuit primitives such as operational amplifiers, etc. The numbers complexity is replaced by the increased complexity of each building block, and the need to test a range of parameters such as gain, bandwidth, signal to noise ratio, common-mode rejection (CMR), offset voltage and other factors. Although faults in digital ICs are usually catastrophic in that incorrect 0 or 1 bits are involved, in analogue circuits degraded circuit performance as well as nonfunctional operation has to be checked.

Prototype analogue ICs are subject to comprehensive testing by the vendor before production circuits are released. Such tests will involve wafer fabrication parameters as well as circuit parameters, and the vendor must ensure that these fabrication parameters remain constant for subsequent production circuits. The vendor and the OEM, however, still need to test production ICs, since surface defects or mask misalignments or other production factors may still cause unacceptable performance, and therefore some subset of the full prototype test procedures may need to be determined. Completed analogue systems obviously have to be tested by the OEM after manufacture, but this is unique to the product and will not be considered here.

The actual testing of analogue ICs involves standard test instrumentation such as waveform generators, signal analysers, programmable supply units, voltmeters, etc., which may be used in the test of any type of analogue system. Comprehensive general-purpose test systems are frequently made as an assembly of rack-mounted instruments, under the control of a dedicated microcontroller or processor. Such an assembly is known as a rack-and-stack test resource, as illustrated in Figure 1.5. The inputs and outputs of the individual instruments are connected to a backplane general purpose instrumentation bus (GPIB), which is a standard for microprocessor or computer-controlled commercial instruments. (The HP instrumentation bus HPIB is electrically identical.)

In the case of custom ICs it is essential for the OEM to discuss the test requirements very closely with the vendor. With very complex analogue circuits a high degree of specialist ability may be involved, which can only be acquired through considerable experience in circuit design.

10 VLSI testing: digital and mixed analogue/digital techniques

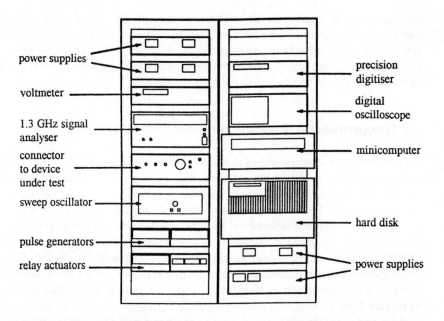

Figure 1.5   A rack-and-stack facility for the testing of analogue circuits, involving a range of instrumentation with computer control of their operation (Photo courtesy IEEE, © 1987 IEEE)

In summary, in analogue test it is circuit complexity rather than volume complexity which dominates the problem. However, as will be covered in later pages, the need for rapid acceptance testing of analogue ICs is often an important factor, and concepts towards this objective will be considered.

1.4 The problems of mixed analogue/digital testing

Testing of the analogue part and testing of the digital part of a combined analogue/digital circuit or system each require their own distinct forms of test, as introduced in the two preceding sections.* Hence, it is usually necessary to have the interface between the two brought out to accessible test points so that the analogue tests and the digital tests may be performed separately.

In the case of relatively unsophisticated mixed circuits containing, say, a simple input A-to-D converter, some internal digital processing, and an output D-to-A converter, it may be possible to define an acceptable test procedure without requiring access to the internal interfaces. All such cases must be individually considered, and so no hard and fast rules can be laid down.

There are on-going developments which seek to combine both analogue and digital testing by using multiple discrete voltage levels or serial bit streams to drive the analogue circuits, or voltage-limited analogue signals for the digital circuits. However, this work is still largely theoretical; more successful so far in the commercial market has been the incorporation of both analogue test instrumentation and digital test instrumentation within one test assembly, as illustrated in Figure 1.6. Here the separate resources are linked by an instrumentation bus, with the dedicated host processor or computer being programmed to undertake the necessary test procedures. When such resources are used it is essential to consider their capabilities, and perhaps constraints, during the design phase of a mixed analogue/digital IC, particularly as appropriate interface points between the two types of circuit may still be required.

1.5 Design for test

From the previous sections it will be clear that the problems of test escalate as the complexity of the IC or system increases. If no thought is given to 'how-

---

* We exclude here in our considerations the testing of individual A-to-D and D-to-A converters, particularly high speed flash converters. The vendor testing of these and similar mass-production circuits usually involves extremely expensive test resources especially designed for this purpose, similar in appearance to the VLSI test equipment shown in Figure 1.4.

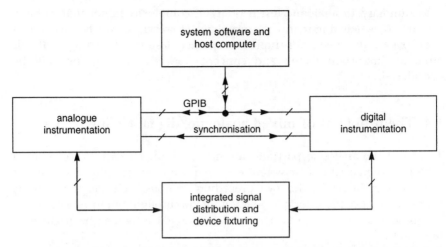

Figure 1.6 *A mixed analogue/digital test resource, with the analogue and the digital tests synchronised under host computer control. This can be a rack-and-stack assembly, as in Figure 1.5 for the analogue-only case*

shall-we-test-it' at the design stage then a product may result which cannot be adequately tested within an acceptable time scale or cost of test.

Design for test (DFT) is therefore an essential part of the design phase of a complex circuit. As we will see later, DFT involves building into the circuit or system some additional feature or features which would not otherwise be required. These may be simple features, such as:

- the provision of additional input/output (I/O) pins on an IC or system, which will give direct access to some internal points in the circuit for signal injection or signal monitoring;
- provision to break certain internal interconnections during the test procedure, for example feedback loops;
- provision to partition the complete circuit into smaller parts which may be individually tested;

or more complicated features such as reconfigurable circuit elements which have a normal mode of operation and a test mode of operation.

As will be seen later, one of the most powerful test techniques for digital VLSI circuits is to feed a serial bit stream into a circuit under test to load test signals into the circuit. The resulting signals from the circuit are then fed out in an output serial bit stream, and checked for correct response. Such techniques are scan test techniques; they become essential when, for example, the IC under test is severely pin-limited, precluding parallel feeding in of all the desired test signals in a set of wide test vectors. One penalty for having to adopt scan test methods is an increase in circuit complexity, and hence chip size and cost. We will be considering all these factors in detail in later chapters of this text.

1.6 Printed-circuit board (PCB) testing

All OEM systems employ some form of printed-circuit board (PCB) for the assembly of ICs and other components. PCB complexity ranges from very simple one- or two-sided boards to extremely complex multilayer boards containing ten or more layers of interconnect which may be necessary for avionics or similar areas.

PCB testing falls into three categories, namely:

(i) bare-board testing, which seeks to check the continuity of all tracks on the board before any component assembly is begun;
(ii) in-circuit testing, which seeks to check the individual components, including ICs, which are assembled on the PCB;
(iii) functional testing, which is a check on the correct functionality of the completed PCB.

Bare-board testing

Simple one- or two-sided bare boards may be checked by visual inspection. However, as layout size and complexity increases, then expensive PCB continuity testers become necessary. Connections to the board under test are made by a comprehensive 'bed-of-nails' fixture, which is unique for every PCB layout, test signals being applied and monitored by a programmed sequence from the tester's dedicated processor or computer. One hundred per cent correct continuity is required from such tests.

In-circuit testing

In-circuit testing, the aim of which is to find gross circuit faults before commencing any fully-detailed functional testing, may or may not be done. If it is, then electrical connections to the individual components are again made via a bed-of-nails fixture, the processor of the in-circuit tester being programmed to supply and monitor all the required test signals.

The fundamental problem with in-circuit passive component measurement is that the component being measured is not isolated from preceding and/or following components on the board. For discrete components a measurement technique as shown in Figure 1.7 is necessary. Here Z_1 and Z_2 are the impedance paths either side of the component Z_x being measured. By connecting both of these paths to ground, virtually all the current flowing into the in-circuit measuring operational amplifier comes from the test source v_S via the component Z_x. Current flowing through Z_2 is negligible because of the virtual earth condition on the inverting input of the operational amplifier. Hence:

$$v_{OUT} = \frac{-v_S}{Z_x} \times R_{fb}$$

whence

$$Z_x = -\frac{v_S R_{fb}}{v_{OUT}}$$

The PCB is not powered up for these component value tests.

To test active devices on the PCB, the board is powered up with its normal supply or supplies. For operational amplifiers and other analogue devices, test signals are applied to inputs, and output voltages measured. However, to test the operation of onboard digital devices, it is necessary to impose 0 or 1 logic signals on gate inputs which may otherwise have the opposite logic value on them. This is done by applying low source impedance logic 0 or logic 1 pulses, which are long enough to force the required value on inputs and measure the resulting logic outputs, but not long enough to cause damage to any preceding logic gates by what is effectively a short circuit of their output to supply or ground potential.

In-circuit PCB testing matured in the days of analogue and simple digital circuits, particularly SSI and MSI TTL circuits which could withstand test pulses of about 10 μs duration to drive gate inputs to opposite logic levels. However, it becomes impractical for PCBs with a high population of digital ICs, particularly LSI and VLSI packages, and with present-day small geometry, low-power gates and macros. Functional PCB testing therefore is currently prevalent, following initial bare-board inspection and/or test.

Functional testing

In contrast to the bed of nails fixtures noted above, PCB functional testing must access the circuit through its normal edge connector(s) or other I/O

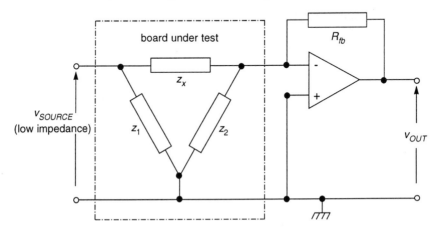

Figure 1.7 The technique used for the in-circuit testing of passive component values. In practice additional complexity may be present to overcome errors due to source impedance, lead impedance, offset voltages, etc.

terminals. Probing of internal tracks on the PCB may be done for specific fault-finding purposes, but not during any automated test procedure.

With fully-assembled PCBs we are effectively doing a systems test. This will be unique for every OEM product, and may involve an even greater complexity of test than individual VLSI circuits. We will not pursue functional PCB testing any further in this text, except to note that all the theory and techniques which will be discussed in the following chapters apply equally to complex PCB assemblies; design for test (DFT) techniques must be considered at the design stage, and PCB layouts must incorporate the necessary provisions for the final functional tests. Scan testing (see Chapter 5) in particular may need to be incorporated to give test access from PCB I/Os to internal IC packages.

1.7 Software testing

Software is an essential element in perhaps the majority of present-day systems, and can account for a greater design time and cost than the hardware costs. Examples of software problems in complex systems such as airport baggage handling, combined police, fire and ambulance control systems and other similar distributed systems are increasingly known, and have demonstrated that software design and test is critical to final system performance.

The principal subject of this text, namely IC testing, has a long history based upon research into electrical and electronic reliability and failure mechanisms; software reliability, testing and failure, however, is a more recent subject area. Indeed, until recently any system fault in service was assumed to be a hardware failure, the software program(s) being considered fault free. However, with increasing system complexity, larger distributed systems and parallel processing, this is no longer the case.

The objectives of software testing are to ensure correct system operation under:

- extreme conditions of input parameter values, timing and memory utilisation;
- ranges of input sequences;
- fault conditions on input data;
- other abnormal as well as normal operating conditions.

The greatest problems are software sneak conditions, namely:

- sneak outputs, where the wrong output code is generated;
- sneak inhibits, where an input or output code is incorrectly inhibited;
- sneak timings, where an incorrect output code is generated due to some timing irregularity;
- sneak messages, where a program message incorrectly reports a system status.

As with VLSI hardware testing, system complexity may make it impossible to undertake a 100 % fully-functional test of the system under all possible system conditions.*

Computer science is tackling this problem, and developing reliability prediction and measurement modelling for software programs. It remains, however, a complex and specialist area, but one about which designers of very large software-based systems must be aware.

1.8 Chapter summary

This first chapter has been a broad overview of the problems of testing circuits of VLSI complexity and systems into which they may be assembled. As will be appreciated, the testing problem is not usually one of fundamental technical difficulty, but much more one of the time and/or the cost necessary to undertake a procedure which would guarantee 100 % correct functionality.

The subsequent chapters of this text will therefore consider the types of failures which may be encountered in microelectronic circuits, fault models for digital circuits, the problems of observability and controllability and the various techniques that are available to ease the testing of both digital and mixed analogue/digital circuits. Finally, the financial aspects of testing which reflect back upon the initial design of the circuit or system will be considered, as well as the production quantities that may be involved.

We will conclude this chapter with a list of publications which may be relevant for further general or specific reading. The more specialised ones may be referenced again in the following chapters.

1.9 Further reading

1. ABRAMOVICI, M., BREUER, M.A., and FRIEDMAN, A.D.: 'Digital systems testing and testable design' (Computer Science Press, 1990)
2. LALA, P.K.: 'Fault tolerant and fault testable hardware design' (Prentice Hall, 1985)
3. WILKINS, B.R.: 'Testing digital circuits: an introduction' (Van Nostrand Reinhold, 1986)
4. FUJIWARA, H.: 'Logic testing and design for testability' (MIT Press, 1985)
5. MICZO, A.: 'Digital logic testing and simulation' (Harper and Row, 1986)
6. PYNN, C.: 'Strategies for electronic test' (McGraw-Hill, 1986)
7. STEVENS, A.K.: 'Introduction to component testing' (Addison-Wesley, 1986)
8. PARKER, K.P.: 'Integrating design and test: using CAE tools for ATE programming' (IEEE Computer Society Press, 1987)
9. YARMOLIK, V.N.: 'Fault diagnosis of digital circuits' (Wiley, 1990)

---

* The extreme case of this is possibly the Star Wars research and development programme, which would have been impossible to test under operational conditions.

10 BARDELL, P.H., McCANNEY, W.H., and SAVIR, J.: 'Built-in test for VLSI: pseudorandom techniques' (Wiley, 1987)
11 BENNETTS, R.G.: 'Design of testable logic circuits' (Addison-Wesley, 1984)
12 BATESON, J.: 'In-circuit testing' (Van Nostrand Reinholt, 1985)
13 GULATI, R.K., and HAWKINS, C.F.: 'I_{DDQ} testing of VLSI circuits' (Kluwer, 1993)
14 BLEEKER, H., Van den EIJNDEN, P., and De JONG, F.: 'Boundary scan test: a practical approach' (Kluwer, 1993)
15 RAJSUMAN, R.: 'Digital hardware testing: transistor-level fault-modeling and testing' (Artech House, 1992)
16 MILLER, D.M. (Ed.): 'Developments in integrated circuit testing' (Academic Press, 1987)
17 WILLIAMS, T.W. (Ed.): 'VLSI testing' (North-Holland, 1986)
18 RUSSELL, G., and SAYERS, I.L.: 'Advanced simulation and test methodologies for VLSI design' (Van Nostrand Reinhold, 1989)
19 RUSSELL, G. (Ed.): 'Computer aided tools for VLSI system design' (Peter Peregrinus, 1987)
20 MASSARA, R.E. (Ed.): 'Design and test techniques for VLSI and WSI circuits' (Peter Peregrinus, 1989)
21 SOIN, R., MALOBERT, F., and FRANCA, J. (Eds.): 'Analogue digital ASICs: circuit techniques, design tools and applications' (IEE Peter Peregrinus, 1991)
22 TRONTELJ, J., TRONTELJ, L., and SHENTON, G.: 'Analogue digital ASIC design' (McGraw-Hill, 1989)
23 ROBERTS, G.W., and LU, A.K.: 'Analogue signal generation for the built-in-self-test of mixed signal ICs' (Kluwer, 1995)
24 NAISH, P., and BISHOP, P.: 'Designing ASICs' (Wiley, 1988)
25 'Design for testability'. Open University microelectronics for industry publication PT505DFT, 1988
26 HURST, S.L.: 'Custom VLSI microelectronics' (Prentice Hall, 1992)
27 NEEDHAM, W.M.: 'Designer's guide to testable ASIC devices' (Van Nostrand Reinhold, 1991)
28 DI GIACOMO, J.: 'Designing with high performance ASICs' (Prentice Hall, 1992)
29 WHITE, D.E.: 'Logic design for array-based circuits: a structure design methodology' (Academic Press, 1992)
30 BENNETTS, R.G.: 'Introduction to digital board testing' (Edward Arnold, 1982)
31 MAUNDER, C.: 'The board designer's guide to testable logic circuits' (Addison-Wesley, 1992)
32 O'CONNOR, P.D.T.: 'Practical reliability engineering' (Wiley, 1991)
33 CHRISTOU, A.: 'Integrating reliability into microelectronics manufacture' (Wiley, 1994)
34 MYERS, G.J.: 'Software reliability: principles and practice' (Wiley, 1976)
35 MYERS, G.J.: 'The art of software testing' (Wiley, 1979)
36 SMITH, D.J., and WOOD, K.B.: 'Engineering quality software' (Elsevier, 1989)
37 MITCHELL, R.J. (Ed.), 'Managing complexity in software engineering' (Institution of Electrical Engineers, 1990)
38 SOMMERVILLE, I.: 'Software engineering' (Addison-Wesley, 1992)
39 SIMPSON, W.R., and SHEPPARD, J.W.: 'System test and diagnosis' (Kluwer, 1994)
40 ARSENAULT, J.E., and ROBERTS, J.A. (Eds.): 'Reliability and maintainability of electronic systems' (Computer Science Press, 1980)
41 KLINGER, D.J., NAKADA, Y., and MENENDIZ, M.A. (Eds.): 'AT+T reliability manual' (Van Nostrand Reinhold, 1990)

Chapter 2
Faults in digital circuits

2.1 General introduction

In considering the techniques that may be used for digital circuit testing, two distinct philosophies may be found, namely:

(*a*) to undertake a series of functional tests and check for the correct (fault-free) 0 or 1 output response(s);
(*b*) to consider the possible faults that may occur within the circuit, and then to apply a series of tests which are specifically formulated to check whether each of these faults is present or not.

The first of the above techniques is conventionally known as functional testing. It does not consider how the circuit is designed, but only that it gives the correct outputs during test. This is the only type of test which an OEM can do on a packaged IC when no details of the circuit design and silicon layout are known.

The second of the above techniques relies upon fault modelling. The procedure now is to consider faults which are likely to occur on the wafer during the manufacture of the ICs, and compute the result on the circuit output(s) with and without each fault present. Each of the final series of tests is then designed to show that a particular fault is or is not present. If none of the chosen set of faults is detected then the IC is considered to be fault free.

Fault modelling relies upon a choice of the types of fault(s) to consider in the digital circuit. It is clearly impossible to consider every conceivable imperfection which may be present, and therefore only one or two types of fault are normally considered. These are commonly stuck-at faults, where a particular node in the circuit is always at logic 0 or at logic 1, and bridging faults, where adjacent nodes or tracks are considered to be shorted together. We will consider these faults in detail in the following sections.

The potential advantage of using fault modelling for test purposes over functional testing is that a smaller set of tests is necessary to test the circuit.

This is aided by the fact that a test for one potential fault will often also test for other faults, and hence the determination of a minimum test set to cover all the faults being modelled is a powerful objective. However, in theory a digital circuit which passes all its fault modelling tests may still not be fully functional if some other, possibly obscure, fault is present, but the probability of this is usually considered to be acceptably small.

2.2 Controllability and observability

Two terms need to be considered before discussing further aspects of digital circuit testing. These are controllability and observability. These terms were first introduced in the 1970s in an attempt to quantify the ease (or difficulty) of testing a digital circuit, with the aim of bringing to the attention of the circuit designer during the design phase potentially difficult to test circuit arrangements so that circuit modifications could be considered.

The basic concept of controllability is simple: it is a measure of how easily a node in a digital circuit can be set to logic 0 or to logic 1 by signals applied to the accessible (primary) inputs. Similarly, the concept of observability is simple: it is a measure of how easily the state of a given node (logic 0 or logic 1) can be determined from the logic signals available at the accessible (primary) outputs.

Consider the small circuit in Figure 2.1. Nodes 1, 2, and 3 are immediately controllable, since they are connected directly to primary inputs. Node 7, on the other hand, is clearly not so readily controllable; it requires either node 6 to be held at logic 0 and node 5 switched from 0 to 1 (toggled), or vice versa, which in turn must be traced back to the primary inputs. One test set for node 7 would be to set inputs F and G to logic 1 and toggle input E, giving the two test vectors:

			input test vector						
			E	F	G				node 7
•	•	•	0	1	1	•	•	•	1
•	•	•	1	1	1	•	•	•	0

Clearly, the further a node is from the primary inputs the more indirect it is to control the logic value of that node.

Some nodes in a working circuit may not be controllable from the primary inputs. Consider the monitoring circuit shown in Figure 2.2a. With a healthy circuit the outputs from the two subcircuits will always be identical, and no means exists of influencing the output of the monitoring circuit. An addition such as shown in Figure 2.2b is necessary in order to provide controllability of the monitoring circuit response.

Faults in digital circuits 21

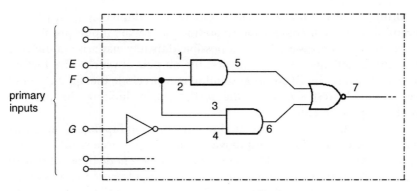

Figure 2.1 A simple circuit to illustrate controllability

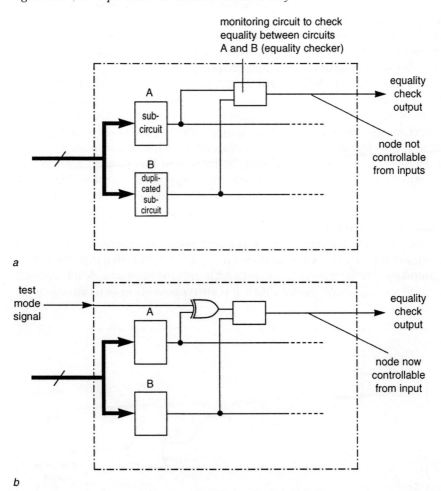

Figure 2.2 An example of a circuit node that is not controllable from a primary input
  a basic system with duplicated circuits A and B
  b an addition necessary to give controllability

Monitoring circuits and also circuits containing redundancy both present inherent difficulties in controllability; additional primary inputs just for test purposes become necessary. Another possible difficulty may arise in the case where the IC contains a ROM structure, the outputs of which drive further logic circuits. Here the ROM programming may preclude the application of certain desirable test signals to the further logic, limiting the possible controllability of the latter.

The controllability of circuits containing latches and flip-flops (sequential circuits) is also often difficult or impossible, since a very large number of test vectors may have to be applied to change the value of a node in or controlled by the sequential circuits. Additionally there will often be certain states of the circuit which are never used, for example in a four-bit BCD counter which only cycles through ten of the possible 16 states. It is, therefore, frequently necessary to partition or reconfigure counter assemblies into a series of smaller blocks, each of which can be individually addressed under test conditions. (This is one of the design for test philosophies that we will consider in greater detail in a later chapter.)

Turning to observability, consider the simple circuit shown in Figure 2.3. Suppose it is necessary to observe (monitor) the logic value on node 2. In order that this logic value propagates to the primary output Z, to give a different logic value at Z depending upon whether the node is at logic 0 or logic 1, it is clear that nodes 1 and 4 must be set to logic 1 and node 6 to logic 0. Hence, the primary signals must be chosen so that these conditions are present on nodes 1, 4 and 6, in which case output Z will be solely dependent upon node 2. Node 2 will then be observable. This procedure is sometimes termed sensitising, forward driving or forward propagating the path from a node to an observable output.

The general characteristics of controllability and observability for any given network are therefore as shown in Figure 2.4. Provided that there is no redundancy in the network, that is all paths must at sometime switch in order to produce a fault-free output, then it is always possible to determine two (or

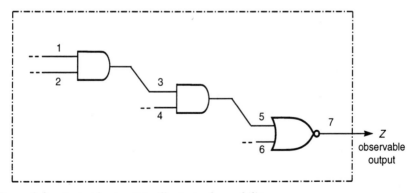

Figure 2.3   A simple circuit to illustrate observability

more) input test vectors which will check that each internal node of the circuit correctly switches from 0 to 1 or from 1 to 0, or fails to do so if there is a fault at the node. However, the complexity of determining the smallest set of such vectors to test all internal nodes is high, way beyond the bounds of computation by hand except in the case of extremely simple circuits. If sequential circuits are also present, then there is the additional complexity of ensuring that the sequential circuits are in their required states as well.

Many attempts have been made to quantify the controllability and observability of a given circuit, to allow difficult to test circuits to be identified and hopefully modified during the design phase[1-10]. The software packages which were developed include TMEAS, 1979[1] TEST/80, 1979[3], SCOAP, 1980[4], CAMELOT, 1981[6], VICTOR, 1982[7], and COMET, 1982[8]. These are discussed in Bennetts[11], Bardell et al.[12] and Abramovici et al.[13]. The basic concepts used in the majority of these developments involve (i) the computation of a number intended to represent the degree of difficulty in setting each internal node of the circuit under test to 0 and 1 (0-controllability and 1-controllability) and (ii) the computation of another number intended to represent the difficulty of forward propagating the logic value on each node to an observable output. The difficulty of testing the circuit is then related to some overall consideration of these individual numerical values. Further developments in the use of this data so as to ease the testing difficulty were also pursued[14].

In the majority of these developments, controllability was normalised to the range 0 to 1, with 0 representing a node which was completely uncontrollable from a primary input, to 1 representing a node with direct controllability. Typically, a controllability transfer factor, CTF, for every type of combinational logic gate or macro is derived from the expression:

$$CTF = 1 - \left| \frac{N(0) - N(1)}{N(0) + N(1)} \right| \qquad (2.1)$$

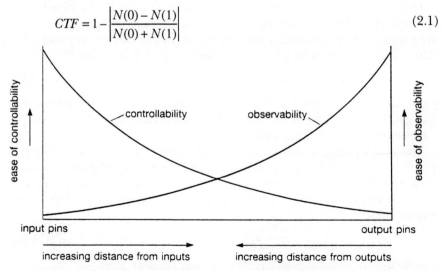

Figure 2.4 *The general characteristic of both controllability and observability*

where $N(0)$ and $N(1)$ are the number of input patterns for which the component output is logic 0 and logic 1, respectively, in other words $N(0)$ is the number of 0s in the truthtable or Karnaugh map for the component, and $N(1)$ is the number of 1s. Components such as an inverter gate or an exclusive-OR gate have a value of $CTF = 1$, since they have an equal number of 0s and 1s in their truthtable; n-input AND, NAND, OR and NOR gates, however, have a controllability transfer factor value of $1/2^{n-1}$, as may readily be shown by evaluating the above equation.

The controllability factor, CY, of a component output with inputs which are not directly accessible from primary inputs is then computed by the equation:

$$CY_{output} = (CTF \times CY_{inputs}) \qquad (2.2)$$

where CTF is the controllability transfer factor for the component and CY_{inputs} is the average of the CTFs on the component input lines. (For components with inputs which are directly accessible $CY_{inputs} = 1$, and hence $CY_{output} = CTF$ for this special case.) Hence, working through the circuit the controllability value of every node from primary input to primary output can be given a numerical value.

In a similar manner to determining the controllability transfer factor value for each type of gate or macro, in considering observability an observability transfer factor, OTF, is determined for each component. This factor is the relative difficulty of determining from a component output value whether there is a faulty input value (error) on an input to the component. An inverter gate clearly has an observability transfer factor value of 1; for n-input AND, NAND, OR and NOR gates the value is $1/2^{n-1}$, the same as the CTF value. The exact equation for OFT may be found in the literature[11-13], and does not necessarily have the same value as CTF for all logic macros.

The observability factor, OY, for each node in a circuit is next determined by working backwards from primary outputs through components towards the primary inputs, generating the observability value OY for every node. The value of OY is given by an equation similar to eqn. 2.2, namely:

$$OY_{inputs} = (OTF \times OY_{outputs}) \qquad (2.3)$$

where OTF is the observability transfer factor for each component, and $OY_{outputs}$ is the average of the observability values for the individual output nodes of the component or macro.

This computation of controllability and observability values, however, is greatly complicated by circuit factors such as reconvergent fan-out, feedback paths and the presence of sequential circuits. Combining the two values so as to give a single measure of testability, TY, is also problematic. The simple relationship:

$$TY_{node} = (CY_{node} \times OY_{node}) \qquad (2.4)$$

may be used, with an overall circuit testability measure given by:

$$TY_{overall} = \frac{\sum_{all\ nodes}(TY_{node})}{number\ of\ nodes} \tag{2.5}$$

but this in turn is not completely satisfactory since it does not, for example, show any node(s) which are not controllable or observable, i.e. nodes which have a value $TY_{node} = 0$. Although these and other numerical values for controllability and observability generally follow the relationships shown in Figure 2.4, experience has shown that quantification has relatively little use. In particular:

- the analysis does not always give an accurate measure of the ease or otherwise of testing;
- it is applied after the circuit has been designed, and does not give any help at the initial design stage or direct guidance on how to improve testability;
- it does not give any help in formulating a minimum set of test vectors for the circuit.

Hence, although controllability and observability are central to digital testing, programs which merely compute testability values have little present-day interest, particularly with VLSI circuits where their computational complexity may become greater than that necessary to determine a set of acceptable test vectors for the circuit, see later. SCOAP[4], CAMELOT[6] and VICTOR[7] possibly represent the most successful programs which were developed. For further comments see Savir[15], Russell[16], and Agrawal and Mercer[17].

We will have no need to refer again to testability quantification in this text, but the concept of forward propagating test signals from primary inputs towards the primary outputs and backward tracing towards the inputs will arise in our further discussions.

2.3 Fault models

The consideration of possible faults in a digital circuit is undertaken in order to establish a minimum set of test vectors which collectively will test that the faults are or are not present. If none of the predefined faults are detected, then the circuit is considered to be fault free. This procedure is sometimes termed structurally-based test pattern generation; when the test vectors are automatically generated from the circuit layout (or from any other data) the term automatic test pattern generation (ATPG) is commonly used.

The faults in digital circuits which are usually considered are:

(i)   stuck-at faults;
(ii)   bridging faults;

(iii) stuck-open faults;
(iv) pattern sensitive faults.

The first two of these categories apply to any fabrication technology, and will be considered in the following two sections. The third category is normally considered in the context of CMOS technology, where it results in a potentially unique fault condition as will be considered in section 2.3.3. The fourth category is normally considered in connection with specific devices with a regular silicon layout, such as ROMs and RAMs, and consideration of this will therefore be deferred until Chapter 6.

2.3.1 Stuck-at faults

The most common model that as been used for faults in digital circuits is the single stuck-at fault model. This assumes that any physical defect in a digital circuit results in a node in the circuit (a gate input or output) being fixed at logic 0, stuck-at 0, or fixed at logic 1, stuck-at 1. This fault may be in the logic gate or macro itself, or some open circuit or short circuit in the interconnections such that the node can no longer switch between 0 and 1. The abbreviations s-a-0 and s-a-1 are usually used for these two stuck-at conditions, and will be adopted from here on.

Consideration of the possible faults in a digital circuit will show that the stuck-at model is very powerful. Even if a particular node is stuck outside its normal 0 or 1 voltage tolerance, subsequent logic gates will generally interpret this value as either logic 0 or 1. It has also been proved by Kohavi and Kohavi[18] that if a test set which detects all possible single stuck-at faults in any two-level combinational circuit is determined, then this exhaustive test set will also detect all possible multiple stuck-at faults. In other words, it is not possible for two or more stuck-at faults in a two-level circuit to mask each other and give a fault-free output for all the input test vectors. However, for circuits with more than two levels of combinational logic this is no longer necessarily true, although in practice it is found that the single stuck-at test vectors do always test for a high percentage of possible multiple stuck-at faults, the precise coverage being circuit and layout dependent.*

For a three-input NAND gate there are eight possible single stuck-at faults, as shown in Table 2.1. However, it will be seen that four of these are indistinguishable at the gate output, each giving a s-a-1 condition on the output node. A total of four test input vectors as shown in Table 2.2 will detect (but not identify) the presence of all eight possible stuck-at faults. Similar developments can be shown for all other combinational gates and macros, each requiring less than 2^n input test vectors to test for all possible stuck-at faults on a n-input component.

---

* For a function with N nodes a total of $2N$ single stuck-at faults have to be considered, but the theoretical number of possible multiple stuck-at faults is 3^N-2N-1. This is clearly a very large number to consider.

Faults in digital circuits 27

Table 2.1 All possible stuck-at faults on a three-input NAND gate; the wrong outputs have been circled

Inputs A B C	Fault-free	Output Z							
		A s-a-0	A s-a-1	B s-a-0	B s-a-1	C s-a-0	C s-a-1	Z s-a-0	Z s-a-1
0 0 0	1	1	1	1	1	1	1	⓪	1
0 0 1	1	1	1	1	1	1	1	⓪	1
0 1 0	1	1	1	1	1	1	1	⓪	1
0 1 1	1	1	⓪	1	1	1	1	⓪	1
1 0 0	1	1	1	1	1	1	1	⓪	1
1 0 1	1	1	1	1	⓪	1	1	⓪	1
1 1 0	1	1	1	1	1	1	⓪	⓪	1
1 1 1	0	①	0	①	0	①	0	0	①

Table 2.2 The minimum test set for detecting all possible stuck-at faults in a three-input NAND gate

Input test vector A B C	Healthy output	Wrong output	Faults detected by test vector
0 1 1	1	0	A s-a-1 or Z s-a-0
1 0 1	1	0	B s-a-1 or Z s-a-0
1 1 0	1	0	C s-a-1 or Z s-a-0
1 1 1	0	1	A or B or C s-a-0 or Z s-a-1

Automatic test pattern generation (ATPG) programs for determining a minimum set of test vectors based upon this single stuck-at model have been extensively developed. These programs, see Section 3.2 later, generate appropriate input vectors which:

(a) attempt to drive the node under consideration to logic 1 when a s-a-0 check is being made;
(b) attempt to drive the node to logic 0 when a s-a-1 check is being made;
(c) propagate the signal on this node to a primary output so that its logic state can be checked.

For example, in the circuit shown in Figure 2.5, when it is required to check that the output of the inverter gate G1 is not stuck-at 0, the input test vector shown must be applied. This vector will attempt to set this output node to logic 1, and whatever logic signal is at the node will be propagated to the primary output, giving 0 in the presence of the stuck-at fault and 1 if the node is not stuck-at 0. Note that this fault is indistinguishable from the Inverter input node stuck-at 1, or the lower input of the NAND gate G4 stuck-at 0, or

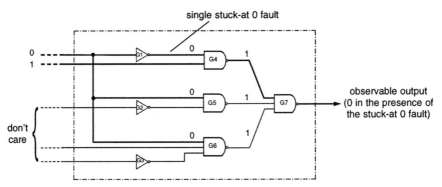

Figure 2.5   *A simple example showing the test vector required to test for and propagate the output of gate G1 stuck-at 0 to the primary output, the three lower inputs being all don't cares*

its output stuck-at 1. Hence, in total there is a great deal of commonality in the set of test vectors for checking the individual nodes s-a-0 or s-a-1, from which a minimum test set can be compiled by the ATPG program.

A more complex example is shown in Figure 2.6. Here 15 input test vectors are sufficient to check for all possible s-a-0 and s-a-1 faults in the circuit. The amount of work involved in deriving a minimum test set clearly increases sharply when there are more than two levels of logic (excluding input inverters) in the circuit, and when there is a high fan-out from the logic gates.

A further problem that we have not considered is where there is feedback in the circuit and where sequential circuits (latches and/or flip-flops) are present. This greatly complicates the problem, although it is always possible to find a test for a node s-a-0 or s-a-1 provided that the node at some time controls the value of a primary output.

If it is not possible to find any test for a node s-a-0, then we can apply a steady logic 0 at this node without affecting the output function; similarly if no test for s-a-1 can be found then we can apply a steady logic 1 at the node without affecting the output function. In both cases there must be some redundancy in the existing circuit at this point; Figure 2.7 is an illustration of this. It should, therefore, be appreciated that if redundancy is deliberately introduced into a circuit this will inevitably complicate the subsequent circuit testing; for example, if in an asynchronous logic design an additional prime implicant term is introduced specifically to eliminate a potential race hazard in the circuit, then either no s-a-0 or no s-a-1 tests will be possible for this additional logic gate.

In summary, the use of stuck-at fault modelling is a powerful tool, but is not necessarily the best approach for the testing of present-day very complex VLSI circuits. For further details, see existing literature[13,16,18–25].

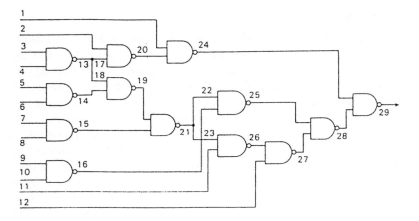

Inputs	Faults covered
10×0×0× × ×01×	1/0 2/1 20/0 24/1 29/0
00×0×0× × ×01×	1/1 24/0 29/1
11×0×0× × ×01×	2/0 17/0 20/1 24/0 29/1
0×11×0×0× × 11	3/0 4/0 13/1 18/1 19/0 21/1 23/1 26/0 27/1 28/0 29/1
0101×0×01111	3/1 13/0 18/0 19/1 21/0 23/0 26/1 27/0 28/1 29/0
01×001×01111	5/1 14/0 19/1 21/0 23/0 26/1 27/0 28/1 29/0
0× × ×11111111	7/0 8/0 15/1 21/0 23/0 26/1 27/0 28/1 29/0
0× × ×1101× × 11	7/1 15/0 21/1 23/1 26/0 27/1 28/0 29/1
01×0×0× ×011×	9/1 16/0 25/1 28/0 29/1
01×0×0× ×101×	10/1 16/0 22/0 25/1 28/0 29/1
01×0×0× ×1101	11/1 26/0 27/1 28/0 29/1
1111× × ×0× × ×1	17/1 20/0 24/1 29/0
0× × ×11×0×0×0	12/1 22/1 25/0 27/0 28/1 29/0
011010×01111	4/1 6/1 9/0 10/0 11/0 13/0 14/0 16/1 18/0 19/1 21/0 23/0 25/0 26/1 27/0 28/1 29/0
01×01110× × 11	5/0 6/0 8/1 12/0 14/1 15/0 19/0 21/1 23/1 26/0 27/1 28/0 29/1

Figure 2.6 *A further example giving a minimum test set of 15 test vectors which test all the circuit nodes for possible s-a-0 and s-a-1 faults. The ×s in the input vectors are don't cares; 1/0, 2/1, etc. in the faults-covered listing means node 1 s-a-0 tested, node 2 s-a-1 tested, and so on (Acknowledgement, Oxford University, UK)*

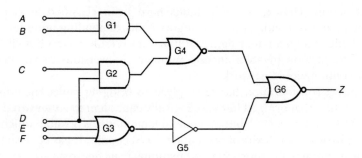

Figure 2.7 *An example of overlooked redundancy in a combinational network. It is left as an exercise for the reader to try to establish stuck-at tests for the nodes around gate G2, and the reason for any failure to establish such tests*

2.3.2 Bridging faults

Unlike stuck-at faults, where a node is considered to be stuck-at one or other logic value, bridging faults may result in a node being driven to 0 or 1 by the action at some other node(s). Since this depends upon the driving impedance of the logic signals on the lines which are bridged, no precise decisions can be made on what the effect of bridging faults will be without some consideration of (i) the technology, (ii) where the faults are in the circuit layout and (iii) how extensive is the bridging. In TTL, for example, a logic 0 signal on two nodes which are shorted will normally dominate; in ECL a logic 1 signal will dominate, although in CMOS an intermediate logic value can result. Also, when shorted nodes individually have the same logic value (0 or 1) then the circuit output(s) will be correct and no fault can be observed; it is only when they differ that some logic conflict has to be resolved.

The number of bridging faults which are theoretically possible between any two lines in a circuit of n lines is:

$$\frac{n(n-1)}{2}$$

This assumes that a short may occur between any two lines, but in practice shorts between physically adjacent lines are clearly more realistic. However, if bridging between more than two lines is considered, then the number of theoretically possible bridging faults escalates rapidly[13]. In general, the number of feasible bridging faults between two or more lines is usually greater than the theoretical number of possible stuck-at faults in a circuit, and although it is straightforward to derive a single test vector that will cover several stuck-at faults in the circuit, this is not so for bridging faults.

A further difficulty with bridging faults is that a fault may result in a feedback path being established, which will then cause some sequential circuit action. Hence, bridging faults have been classified as sequential bridging faults or combinational bridging faults, depending upon the nature of the fault. A sequential bridging fault may also result in circuit oscillation if the feedback loop involves an odd number of inversions in an otherwise purely combinational network.

Extensive consideration has been given to bridging faults, including the consideration that bridged lines are logically equivalent to either wired-OR or wired-AND functions[25-33], but no general fault model is possible which caters for the physical complexity of present-day VLSI circuits. It has been suggested that most (but not all) shorts in combinational logic networks are detected by a test set based upon the stuck-at model, provided that the latter is designed to cover 99 % or more of the possible stuck-at circuit faults[12]. This statement, however, is increasingly debatable as chip size increases and where CMOS technology is involved, see the following discussions.

2.3.3 CMOS technology considerations

CMOS technology is the most pervasive of present-day microelectronic technologies, and it is therefore very relevant to consider specifically the faults which may be present in CMOS gates and interconnections.

The stuck-at fault model will clearly be appropriate should, say, a gate input be bridged to V_{CC} (logic 1) or ground (logic 0). However, it has been estimated that over one-third of the faults which occur in CMOS circuits are not covered by this classical stuck-at fault model[34], and, therefore, a consideration of other possible failure modes is essential.

The circuits of typical CMOS NAND and NOR logic gates are shown in Figure 2.8. The circuit arrangement of the p-channel field-effect transistors (FETs) is always the dual of the circuit arrangement of the n-channel FETs, that is a parallel connection of p-channel FETs is mirrored by a series connection of n-channel FETs, and vice versa.* When the circuit is healthy, there is always a conducting path from either V_{DD} or from ground to the gate output, but no conducting path from V_{DD} to ground except partially during output transitions. Under steady-state conditions, the current taken from the supply is virtually zero, since gate input resistance is extremely high (tens or hundreds of megohms), and the inherent circuit capacitances at gate output connections are either charged to V_{DD} via the conducting p-channel FETs or discharged to 0 V via the conducting n-channel FETs.

However, consider a fault in the p-channel transistor T1 in Figure 2.8a such that it becomes open circuit. With inputs AB = 0 0, 1 0 and 1 1 the gate output will be unaffected since there is always a healthy conducting path from either V_{DD} or 0 V to the gate output. With input 0 1, however, there will be no conducting path through T1, and hence no path to the output from either supply rail. The gate output will therefore be floating.

If under test conditions the input test vectors are applied in the order 0 0, 0 1, 1 0 and 1 1, the first vector 0 0 will establish logic 1 at the gate output through the working transistor T2. When the input changes to 0 1 there will be no immediate change in the gate output, since the circuit capacitance will maintain the output voltage at the logic 1 level. Should the input 0 1 be held for a long time (possibly minutes), then eventually the output voltage will decay sufficiently so that it is no longer an effective logic 1 signal to the following gates, but if (as is invariably the case) the input vector is quickly changed to 1 0 then no fault will have been detected. The output has remembered the correct logic value, and the fault is said to be a memory fault.

However, if the input test vector 0 1 had been preceded by the test vector 1 1, then the gate output would not have been precharged to logic 1, and on input 0 1 would remain at logic 0. Therefore, to detect this particular

---

* Because the resistance of p-type FETs is higher than that of similar dimension n-type FETs, it is preferable to series the n-channel FETs and parallel the p-channel FETs. Hence, NAND gates rather than NOR gates become the preferred basic logic element.

32 VLSI testing: digital and mixed analogue/digital techniques

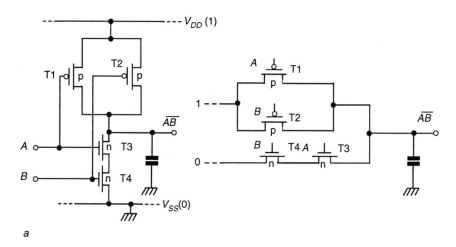

a

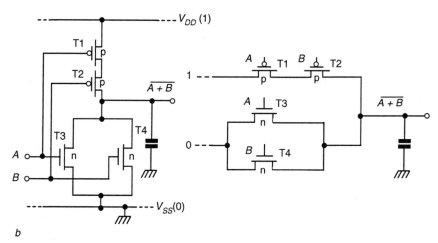

b

Inputs		State of the transistors				Output state
A	B	T1	T2	T3	T4	
0	0	on	on	off	off	1
0	1	on	off	off	on	1
1	0	off	on	on	off	1
1	1	off	off	on	on	0

c

Figure 2.8   CMOS technology

    a a two-input NAND gate drawn conventionally and in an alternative way which emphasises the paths from V_{DD} and ground to the gate output;
    b two-input NOR gate, again drawn conventionally and in an alternative form;
    c the fault-free action of the 2-input NAND gate.

    Note that the capacitor shown in *a* and *b* represents the lumped capacitance at the gate output

open-circuit fault it is necessary for the test vector sequence to be 1 1 followed by 0 1. Should T2 rather than T1 be the open-circuit path, then it requires 1 1 followed by 1 0 to detect this fault.

A different problem can arise with a fault in the series connection of the n-channel FETs in Figure 2.8a. Here an open-circuit fault in either transistor will prevent the gate output from being discharged to 0 V, but a short circuit on a transistor can prevent the output from being fully charged to V_{DD} via the conducting p-channel transistor(s). However, when the n-channel transistors are in parallel, as in the NOR gate of Figure 2.8b, open-circuit faults are now memory faults, with short-circuit faults in the p-channel transistors preventing the output from being fully discharged to 0 V via the conducting n-channel transistor(s).

CMOS short-circuit faults, therefore, do not have the memory characteristic of open-circuit faults, but instead allow a conducting path to be established between V_{DD} and ground on certain input vectors. For example, in Figure 2.8a should T3 be short circuit then on the test vector 0 1 there will be a path through T1, T3 and T4, giving rise to an ambiguous output voltage and an excessive current flow from V_{DD} to ground. A check for this condition is now usually made by I_{DDQ} testing, which will be covered in Chapter 3.

A fully-exhaustive test set for single open-circuit and short-circuit faults in a two-input NAND gate is therefore as shown in Table 2.3, where pairs of test vectors may be regarded as an initialisation vector followed by a test vector.

Table 2.3 An exhaustive test set for single open-circuit (O/C) and short-circuits (S/C) faults in a two-input CMOS NAND gate

Input test vector AB	Healthy gate output Z	Check
0 0	1	
0 1	1	T3 S/C check*
1 1	0	T1 or T2 S/C check*; T3 or T4 O/C check
1 0	1	T2 O/C check
0 0	1	
1 0	1	T4 S/C check*
1 1	0	(as test vector 1 1 above)
0 1	1	T1 O/C check
0 0	1	

* excessive current if transistor short circuit

Bridging faults within CMOS gates also cause failures which may not be modelled by the stuck-at fault model, particularly where more complex CMOS structures are present. Consider the circuit shown in Figure 2.9. The bridging fault shown will connect the gate output to ground under the input conditions of $AB + CD + AD + BC = 1$ instead of the normal conditions of $AB + CD = 1$. However when input conditions AD or $BC = 1$ are present, there

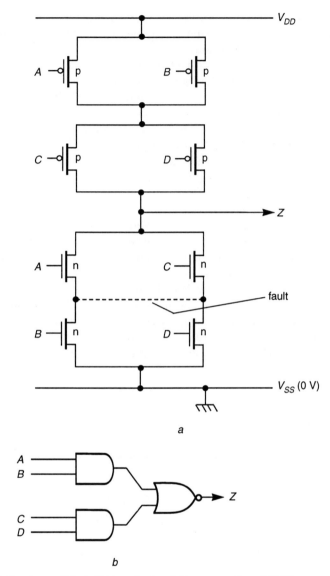

Figure 2.9 A possible bridging fault within a complex CMOS gate, the fault-free output being $Z = \overline{(AB + CD)}$
a the circuit topology
b the equivalent Boolean circuit

will be a conducting path from V_{DD} to ground, and therefore excessive dissipation will be present under these conditions. Notice that the equivalent Boolean circuit for this complex gate, shown in Figure 2.9b, cannot model this internal bridging fault, either as a stuck-at or bridging failure, and hence fault modelling is generally inappropriate.

The faults peculiar to CMOS structures were first described by Wadsack in 1978[35] and have been intensively considered since that date[36–40]. As will be appreciated, the problems with CMOS testing are inherently related to the fact that every CMOS logic gate consists of two dual parts, the p-channel and the n-channel logic, and a fault in one half but not in the other half gives rise to problems such as those which we have just considered.

The further problem of open or short circuits in FET transmission gates, such as shown in Figure 2.10, does not appear to have received a great deal of specific attention[40,41], but the prevailing approach is that all CMOS faults are best tested by functional testing and I_{DDQ} current testing, rather than by the theory and application of any form of fault modelling.

We will return to this important subject of CMOS testing again in the following chapter.

2.4 Intermittent faults

Nonpermanent faults in a circuit or system are faults that appear and disappear in a random way. There is, therefore, no predetermined set of tests which can be formulated to detect such malfunctioning. Unfortunately,

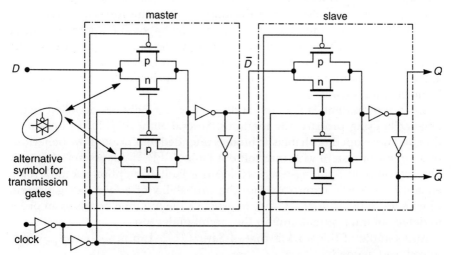

Figure 2.10 A CMOS D-type master-slave circuit using tranmission gate configurations, where each p-channel or n-channel FET may be considered open circuit or short-circuit under fault conditions. Note, if entirely separate clock signals are supplied to the master and slave circuits, it is possible to cause the complete circuit from D to Q to be conducting (transparent) in test mode conditions, which may be useful for test purposes but at the expense of doubling the clock lines in the routed IC

it has been reported that a major portion of digital system faults when in service are intermittent (temporary) faults, and that the investigation of such faults accounts for more than 90 % of total maintenance expenditure[42,43].

Nonpermanent faults may be divided into two categories, namely:

(i) transient faults, which are nonrecurring faults generally caused by some extraneous influence such as cosmic radiation, power supply surges or electromagnetic interference;
(ii) intermittent faults, which are caused by some imperfection within the circuit or system and which appear at random intervals.

In practice it may not always be clear from the circuit malfunctioning which of these categories is present, particularly if the occurrence is infrequent.

By definition it is not possible to test for transient faults. In VLSI memory circuits α-particle radiation can cause wrong bit patterns in the memory arrays, and in other latch and flip-flop circuits it is possible to encounter latch-up due to some strong interference. However, once the memory or storage circuits have been returned to normal, no formal analysis of the exact cause of the malfunctioning is possible. Experience and intuition are the principal tools in this area.

Intermittent faults, however, may possibly be detected if tests are repeated enough times. This may involve repeating again what the system was doing and the state of its memory at the time of the transitory fault, if known. Alternatively, some abnormal condition may be imposed upon the circuit or system, such as slightly increasing or decreasing the supply voltage, raising the ambient temperature or applying some vibration, with a view to trying to make the intermittent fault a permanent one which can then be investigated by normal test means.

Since intermittent faults are random, they can only be modelled mathematically using probabilistic methods. Several authorities have considered this, and developed probabilistic models intended to represent the behaviour of a circuit or system with intermittent faults[12,34,42–49]. All assume that only one intermittent fault is present (the single fault assumption), and develop equations which attempt to relate the probability of a fault being present when a test is being applied, and/or estimate the test time necessary in order to detect an intermittent fault in the circuit under test.

An example of this work is that of Savir[12,44,45]. Two probabilities are first introduced, namely:

(i) the probability of the presence of a fault, f_i, in the circuit, expressed as PF_i = probability (fault f_i is present);
(ii) the probability of the activity of the fault, f_i, expressed as PA_i = probability (fault f_i is active, fault f_i being present).

The circuit is assumed to have a fault set $f_1, f_2, ..., f_m$, with only one fault, f_i, being present at a time. It should be appreciated that fault f_i can be present but not affect the circuit output because that part of the circuit containing the fault is not actively controlling the present output; probability PA_i therefore has a (usually very considerable) lower value than PF_i. Also PF_i, $i = 1$ to m, forms the fault probability distribution

$$\sum_{i=1}^{m} PF_i = 1 \qquad (2.6)$$

but this does not hold for PA_i.

A further probability PE_i, termed the error probability is then introduced, PE_i being the conditional probability of an incorrect output being given by the application of a randomly chosen input test vector, given that the fault f_i is active. Thus the probability, PD_i, of detecting the fault f_i by one randomly chosen input test vector is given by:

$$PD_i = PA_i \cdot PE_i \qquad (2.7)$$

This mathematical development continues with an aim of determining expressions for the mean time, MT_i, which may be necessary to detect fault f_i with a given confidence limit and for other factors, for example:

$$MT_i = k\left(\frac{1}{PD_i}\right) \qquad (2.8)$$

where k is the time between test vectors, from which the lowest upper bound on the number of tests, T, necessary to detect fault f_i with a given confidence limit, c_d, is expressed by:

$$T = \left(\frac{\log_e c_d}{\log_e(1 - PD_i)}\right) \qquad (2.9)$$

Further developments consider the effect of intermittent faults in disjoint subsets, that is groups of faults which may be considered separately from other groups of faults. However, although these mathematical considerations throw light on, and to some extent quantify, the difficulties of testing for intermittent faults, it is generally impossible to estimate real values for the probability parameters PF_i and PA_i[12]. Thus it is not possible to compute any meaningful results that will directly help the test or maintenance engineer in the detection of such faults in a given situation.

Alternative developments using Marcov models and probability matrices[50] have been pursued by other authorities[34,51-53], but again no real values are generally available to use in the algebraic developments. We are, therefore, left with the following strategies for intermittent fault detection which do not involve any probabilistic modelling:

(i) repeatedly apply the tests which have been derived for the circuit or system to detect permanent faults;

(ii) build into the circuit or system some self-checking monitoring such as parity checks (see Chapter 4), which will detect some if not all of the possibly intermittent faults and ideally provide a clue to their location in the circuit;
(iii) in the case of computer systems, continuously run a series of check tests when the computer is idle, and from these hopefully build up information pointing to the intermittent fault source when particular check tests fail;
(iv) as previously mentioned, subject the circuit or system under test to some abnormal working condition in an attempt to make the intermittent fault permanent.

In the case of circuits used in safety critical applications, redundancy techniques which will mask the effect of intermittent faults must be used. If it is then found that, say, one of three systems operating in parallel is occasionally out of step with the other two, this is an indication that there is an intermittent fault in the first circuit, with possibly some clue to its source.

2.5 Chapter summary

This chapter has introduced the basic concepts of controllability and observability of nodes within a digital circuit, but has shown that quantification of these two parameters does not have a great practical significance, particularly for VLSI circuits. Modelling the faults that can arise in digital circuits, with a view to establishing tests which specifically detect whether such faults are or are not present, has a longer record of success, the stuck-at fault model being particularly significant.

Other fault models to cover bridging faults and intermittent faults have not achieved the same success as the stuck-at model, largely because of the complexities of modelling these other fault conditions. However, practical experience, particularly with bipolar technology, has shown that the stuck-at model will also detect most bridging faults and may detect intermittent faults if applied a sufficient number of times. On the other hand, CMOS technology exhibits some unique fault characteristics which are not amenable to simple fault modelling.

In looking at all these fault models, it will be seen that they concentrate on the gate level of the circuit design, and do not directly consider the overall primary input/primary output functionality of the circuit. But as more and more complexity is built into circuits, it has become no longer feasible to start at gate level, or use gate level simulation. Instead, present-day VLSI test philosophies concentrate on the functional or behavioural level, with functional testing being adopted without specific reference to internal circuit design and layout details[54–56]. Later chapters will, therefore, be largely concerned with functional testing techniques, including self test, and little reference will then be made to fault modelling except in very specific cases such as memory or array structures.

Finally, although this chapter has been particularly concerned with faults in and fault models for digital circuits, no mention has yet been made of the possible physical causes of such failures. A major problem here is that IC manufacturers are reluctant to reveal exact details of their fabrication defects; those statistics which are available are usually obsolete due to the rapid and continual developments in fabrication technology and expertise.

There is a considerable volume of information available on the potential causes of IC failure[58–64]. We must, however, distinguish between:

(a) some major failure during manufacture, such as incorrect mask alignment or a process step incorrectly carried out, which it is the province of the professional production engineer to detect and correct before chips are bonded and packaged;
(b) the situation where wafer processing is within specification, but circuits still need to be tested for individual failings.

The latter category of random scattered failures is our concern.

Considering the possible failure mechanisms which can occur in VLSI circuits, these may be chip related, that is some fault within the circuit itself, or assembly related, that is some fault in scribing, bonding and encapsulating the chip, or operationally related, for example caused by interference or α-particle radiation. Further details of these categories may be found discussed in Rajsuman[58] and in Prince[64] for memory circuits, but no up to date global information on the frequency of occurrence of these failures is available. An early failure mode statistic quoted by Glazer and Subak-Sharpe[59] gives the following breakdown:

- metalisation failures, 26 % of all failures;
- bonding failures, 33 %;
- photolithography defects, 18 %;
- surface defects, 7 %;
- others, 16 %.

Metalisation defects still seem to be a prominent category of defect, caused particularly by microcracks in metal tracks where they have to descend into steep narrow vias to make contact with an underlying layer of the planar process, together with the difficulties of cleaning out etched vias before metalisation. Dielectric breakdown also remains a problem should a SiO_2 insulation layer be of imperfect thickness*, and chip bonding is still at times a known cause of failure.

These failings may readily be related to the stuck-at fault model and open-circuit faults, but from the test engineer's point of view (as distinct from the IC manufacturer's point of view) the precise cause of a functional fault is of academic interest only. We shall, therefore, have no occasion to look deeper

---

* The dielectric breakdown strength of SiO_2 is about 8×16^6 V/cm, and the usual thickness of a SiO_2 layer is about 200 $A^0 = 2 \times 10^{-6}$ cm. There is, therefore, not a very great safety factor if the SiO_2 is too thin.

into failure mechanisms in this text, but the reader is referred to the references cited above for more in-depth information if required.

2.6 References

1. GRASON, J.: 'TMEAS—a testability measurement program'. Proceedings of 16th IEEE conference on *Design automation*, 1979, pp. 156–161
2. STEVENSON, J.E., and GRASON, J.: 'A testability measure for register transfer level digital circuits'. Proceedings of IEEE international symposium on *Fault tolerant computing*, 1976, pp. 101–107
3. BREUER, M.A., and FRIEDMAN, A.D.: 'TEST/80—a proposal for an advanced automatic test generation system'. Proceedings of IEEE *Autatestcon*, 1979, pp. 205–312
4. GOLDSTEIN, L.M., and THIGAN, E.L.: 'SCOAP: Sandia controllability and observability analysis program'. Proceedings of 17th IEEE conference on *Design automation*, 1980, pp. 190–196
5. KOVIJANIC, P.G.: 'Single testability figure of merit'. Proceedings of IEEE international conference on *Test*, 1981, pp. 521–529
6. BENNETTS, R.G., MAUNDER, C.M., and ROBINSON, G.D.: 'CAMELOT: a computer-aided measure for logic testability', *IEE Proc. E.*, 1981, **128**, pp. 177–189
7. RATIU, I.M., SANGIOVANNI-VINCENTELLI, A. and PETERSON, D.O.: 'VICTOR: a fast VLSI testability analysis programme'. Proceedings of IEEE international conference on *Test*, 1982, pp. 397–401
8. BERG, W.C., and HESS, R.D.: 'COMET: a testability analysis and design modification package'. Proceedings of IEEE international conference on *Test*, 1982, pp. 364-378
9. DEJKA, W.J.: 'Measure of testability in device and system design'. Proceedings of IEEE Midwest symposium on *Circuits and systems*, 1977, pp. 39–52
10. FONG, J.Y.O.: 'A generalised testability analysis algorithm for digital logic circuits'. Proceedings of IEEE symposium on *Circuits and systems*, 1982, pp. 1160–1163
11. BENNETTS, R.G.: 'Design of testable logic circuits' (Addison-Wesley, 1984)
12. BARDELL, P.H., McANNEY, W.H., and SAVIR, J., 'Built-in test for VLSI: pseudorandom techniques' (Wiley, 1987)
13. ABRAMOVICI, M., BREUER, M.A. and FRIEDMAN, A.D.: 'Digital system testing and testable design' (Computer Science Press, 1990)
14. CHEN, T-H., and BREUER, M.A.: 'Automatic design for testability via testability measures', *IEEE Trans.*, 1985, **CAD-4**, pp. 3–11
15. SAVIR, J.: 'Good controllability and observability do not guarantee good testability', *IEEE Trans.*, 1983, **C-32**, pp. 1198–1200
16. RUSSELL, G. (Ed.): 'Computer aided tools for VLSI system design' (Peter Peregrinus, 1987)
17. AGRAWAL, V.D., and MERCER, M.R.: 'Testability measures—what do they tell us?'. Proceedings of IEEE international conference on *Test*, 1982, pp. 391–396
18. KOHAVI, I., and KOHAVI, Z.: 'Detection of multiple faults in combinational networks', *IEEE Trans.*, 1972, **C-21**, pp. 556–568
19. HLAVICKA, J., and KOLTECK, E.: 'Fault model for TTL circuits', *Digit. Process.*, 1976, **2**, pp. 160–180
20. HUGHES, J.L., and McCLUSKEY, E.J.: 'An analysis of the multiple fault detection capabilities of single stuck-at fault test sets'. Proceedings of IEEE international conference on *Test*, 1984, pp. 52–58
21. NICKEL, V.V.: 'VLSI—the inadequacy of the stuck-at fault model'. Proceedings of IEEE international conference on *Test*, 1980, pp.378–381

22 KARPOVSKI, M., and SU, S.Y.H.: 'Detecting bridging and stuck-at faults at the input and output pins of standard digital computers'. Proceedings of IEEE international conference on *Design automation*, 1980, pp. 494–505
23 SCHERTZ, D.R., and METZE, G.: 'A new representation of faults in combinational logic circuits', *IEEE Trans.*, 1972, **C-21**, pp. 858–866
24 BATTACHARYA, B.B., and GUPTA, B.: 'Anomalous effect of a stuck-at fault in a combinational circuit', *Proc. IEEE*, 1983, **71**, pp. 779–780
25 TIMOC, C., BUEHLER, M., GRISWOLD, T., PINA, C., SCOTT, F., and HESS, L.: 'Logical models of physical failures'. Proceedings of IEEE international conference on *Test*, 1983, pp. 546–553
26 KARPOVSKI, M., and SU, S.Y.H.: 'Detection and location of input and feedback bridging faults among input and output lines', *IEEE Trans.*, 1980, **C-29**, pp. 523–527
27 ABRAHAM, J.A., and FUCHS, W.K.: 'Fault and error models for VLSI', *Proc. IEEE*, 1986, **75**, pp. 639–654
28 MEI, K.C.Y.: 'Bridging and stuck-at faults', *IEEE Trans.*, 1974, **C-23**, pp. 720–727
29 FRIEDMAN, A.D.: 'Diagnosis of short-circuit faults in combinational circuits', *IEEE Trans.*, 1974, **C-23**, pp. 746–752
30 ABRAMOVICI, M., and MENON, P.R.: 'A practical approach to fault simulation and test generation for bridging faults', *IEEE Trans.*, 1985, **C-34**, pp. 658–663
31 XU, S., and SU, S.Y.H.: 'Detecting I/O and internal feedback bridging faults', *IEEE Trans.*, 1985, **C-34**, p. 553–557
32 KODANDAPANI, K.L., and PRADHAM, D.K.: 'Undetectability of bridging faults and validity of stuck-at fault tests', *IEEE Trans.*, 1980, **C-29**, pp. 55–59
33 MALAIYA, Y.K.: 'A detailed examination of bridging faults'. Proceedings of IEEE international conference on *Computer design*, 1986, pp. 78–81
34 LALA, P.K.: 'Fault tolerant and fault testable hardware design' (Prentice Hall, 1985)
35 WADSACK, R.L.: 'Fault modelling and logic simulation of CMOS and MOS integrated circuits', *Bell Sys. Tech. J.*, 1978, **57**, pp. 1449–1474
36 GALIAY, J., CROUZET, Y., and VERGNIAULT, M.: 'Physical versus logic fault models in MOS LSI circuits', *IEEE Trans.*, 1980, **C-29**, pp. 527–531
37 EL-ZIQ, Y.M., and CLOUTIER, R.J.: 'Functional level test generation for stuck-open faults in CMOS VLSI'. Proceedings of IEEE international conference on *Test*, 1981, pp. 536–546
38 CHIANG, K.W., and VRANESIC, Z.G.: 'Test generation for complex MOS gate networks'. Proceedings of IEEE international symposium on *Fault tolerant computing*, 1982, pp. 149–157
39 RENOVELL, M., and CAMBON, G.: 'Topology dependence of floating gate faults in MOS integrated circuits', *Electron. Lett.*, 1986, **22**, pp. 152–157
40 JAIN, S.K., and AGRAWAL, V.D.: 'Modelling and test generation algorithms for MOS circuits', *IEEE Trans.*, 1985, **C-34**, pp. 426–433
41 REDDY, M.K., and REDDY, S.M.: 'On FET stuck-open fault detectable CMOS memory elements'. Proceedings of IEEE international conference on *Test*, 1985, pp. 424–429
42 LALA, P. K., and MISSEN, J.I.: 'Method for the diagnosis of a single intermittent fault in combinational logic circuits', *Proc. IEE*, 1979, **2**, pp. 187–190
43 CLARY, J.B., and SACANE, R.A.: 'Self-testing computers', *IEEE Trans.*, 1979, **C-28**, pp.49–59
44 SAVIR, J.: 'Detection of single intermittent faults in sequential circuits', *IEEE Trans.*, 1980, **C-29**, pp. 673–678
45 SAVIR, J.: 'Testing for single intermittent failures in combinational circuits by maximizing the probability of fault detection', *IEEE Trans.*, 1980, **C-29**, pp. 410–416

46 TASAR. O., and TASAR, V.: 'A study of intermittent faults in digital computers'. Proceedings of *AFIPS* conference, 1977, pp. 807–811
47 McCLUSKEY, E.J., and WAKERLY, J.F.: 'A circuit for detecting and analysing temporary failures'. Proceedings of IEEE *COMCON*, 1981, pp. 317–321
48 MALAIYA, Y.K., and SU, S.Y.H.: 'A survey of methods for intermittent fault analysis'. Proceedings of national *Computer* conference, 1979, pp.577–584
49 STIFLER, J.I.: 'Robust detection of intermittent faults'. Proceedings of IEEE international symposium on *Fault tolerant computing*, 1980, pp. 216–218
50 SHOOMAN, M.L., 'Probabilistic reliability: an engineering approach' (McGraw-Hill, 1968)
51 KOREN, I., and KOHAVI, Z.: 'Diagnosis of intermittent faults in combinational networks', *IEEE Trans.*, 1977, **C-26**, pp. 1154–1158
52 SU, S.Y.H., KOREN, I., and MALAIYA, Y.K.: 'A continuous parameter Marcov model and detection procedure for intermittent faults', *IEEE Trans.*, 1978, **C-27**, pp. 567–569
53 KRISHNAMURTHY, B., and TALLIS, I.G.: 'Improved techniques for estimating signal probabilities', *IEEE Trans.*, **C-38**, pp. 1041–1045
54 BREUER, M.A., and PARKER, A.C.: 'Digital circuit simulation: current states and future trends'. Proceedings of IEEE conference on *Design automation*, 1981, pp. 269–275
55 RENESEGERS, M.T.M.: 'The impact of testing on VLSI design methods', *IEEE J.*, 1982, **SC-17**, pp. 481–486
56 'Test synthesis seminar digest of papers'. IEEE international conference on *Test*, Washington, DC, USA, October 1994
57 BERTRAM, W. J.: 'Yield and reliability', *in* SZE, S.M. (Ed.): 'VLSI technology' (McGraw-Hill, 1983)
58 RAJSUMAN, R.: 'Digital hardware testing: transistor level fault modelling and testing' (Artech House, 1992)
59 GLASER, A.B., and SUBAK-SHARPE, G.E.: 'Integrated circuit engineering: design fabrication and applications' (Addison-Wesley, 1979)
60 GALLACE, L.J.: 'Reliability', *in* Di GIACOMO, J. (Ed.): 'VLSI handbook' (McGraw-Hill, 1989)
61 GULATI, R. K., and HAWKINS, C.F. (Eds.): 'I_{DDQ} testing of VLSI circuits' (Kluwer, 1993)
62 CHRISTOU, A.: 'Integrating reliability into microelectronics and packaging' (Wiley, 1994)
63 SABNIS, A.G.: 'VLSI reliability' (Academic Press, 1990)
64 PRINCE, B.: 'Semiconductor memories' (Wiley, 1995, 2nd edn.)

Chapter 3
Digital test pattern generation

3.1 General introduction

The final objective of testing is to prevent faulty circuits from being assembled into equipment, or to detect circuits which have developed faults subsequent to their commitment. Our discussions here are largely in the context of LSI/VLSI circuits, but apply equally to digital systems which have comparable controllability and observability limitations.

Digital testing may be considered to have three purposes, namely:

(i) fault detection, which is to discover something wrong in a circuit or system, ideally before it has caused any trouble;
(ii) physical fault location, which is the location of the source of a fault within an integrated circuit;
(iii) component fault location, which is the location of a faulty component or connection within a completed system.

This is illustrated in Figure 3.1. It is the top part of this diagram with which we will be largely concerned; fault location is IC or system specific, requiring intimate knowledge and expertise covering the particular component or system.

Every digital test involves a situation such as shown in Figure 3.2a. At each step every output test response has to be checked, which requires prior knowledge of what the fault-free responses should be. For very simple circuits, particularly of SSI and MSI complexity, the procedure shown in Figure 3.2b may be used; for more complex circuits the healthy output responses may be held in memory as shown in Figure 3.2c.

In the situation shown in Figure 3.2b the input test set is usually fully exhaustive, testing the circuit through all its possible input/output conditions. However, as shown in Chapter 1, this becomes increasingly impractical as circuit size escalates, and some more restricted test set to save testing time is then required.

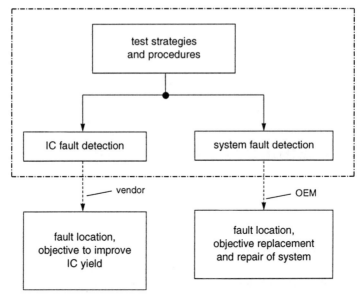

Figure 3.1   *The heirarchy of digital testing objectives. Many test procedures are equally applicable to IC and system test, but some may be IC or system specific. Analogue testing, see later, has an identical heirarchy*

One objective of digital testing, therefore, is to determine a nonexhaustive test set for the circuit under test which will adequately test the circuit within an acceptable time and cost. This reduced test set may be based upon fault models, particularly the stuck-at model, or upon other procedures such as we will consider later. However, it is often required to estimate how effective this nonexhaustive test set is in detecting possible faults in the circuit under test. This requires a fault simulation procedure, which is a parallel activity to test pattern generation, and which may involve a very great deal of additional computational effort[1-6].

Early methods of fault simulation considered single stuck-at faults one at a time. If there were f possible single stuck-at faults to be considered, then f computer models of the circuit under test were generated, each containing one fault source. The proposed test set was applied to each fault model in turn, and a count, t, made of the number of faulty circuits which were not detected by the proposed test set. From this procedure a value for the fault coverage, usually known as the test coverage, TC, of the proposed test set was calculated using the equation:

$$TC = \left(\frac{f-t}{f}\right) \times 100\% \qquad (3.1)$$

If the value for TC was unacceptably low, additional test vectors were sought specifically to cover some of the faults that had escaped detection. This is illustrated in Figure 3.3.

Digital test pattern generation 45

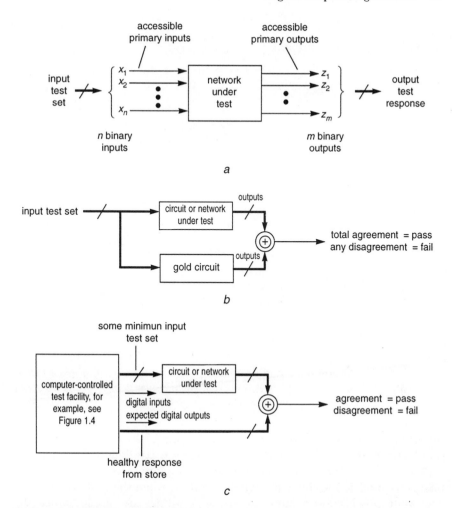

Figure 3.2 Digital testing procedures
   a the basic procedure
   b output response on every input test vector checked against a known-good circuit, known as a gold circuit
   c output response on every input test vector checked against the stored healthy values held in the computer memory

This serial fault simulation procedure rapidly became very inefficient as circuit size increased, and was superseded by three alternative techniques[4] namely:

- parallel fault simulation;
- deductive fault simulation;
- concurrent fault simulation.

46 VLSI testing: digital and mixed analogue/digital techniques

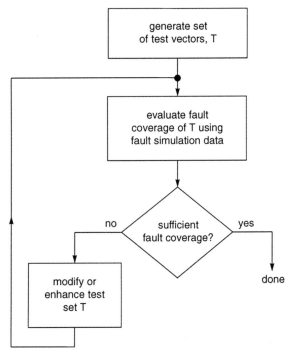

Figure 3.3 The general concept of determining the fault coverage of a given test set from fault simulation. An eventual modification to increase FC may be by interactive intervention by the test designer

All of these differ from the serial fault simulation method by simultaneously simulating a set of faults rather than just one fault at a time. In parallel fault simulation, one n-bit word in the simulation program (where $n = 8$, 16 or 32 bits) is used to define the logic signal on a node when the node is fault free and when $n-1$ chosen faults are present within the circuit. Since the computer operates on words rather than bits, logical operations between words corresponding to the logic of the circuit between the nodes (i.e. AND, NAND, OR, NOR, etc.), allows the simultaneous simulation of the n copies of the circuit to be implemented on each input test vector, thus speeding up the simulation by a factor of about n compared with the one at a time serial fault simulation. The main problem is that the circuit being simulated must be expressed in Boolean terms, which means that memory and large sequential circuit blocks are impractical or impossible to handle.

Deductive fault simulation, however, relies upon fault list data, that is the input/output relationship of logic gates and macros under healthy and chosen fault conditions. A fault list is associated with every signal line within the circuit, including fault data flow through storage and memory elements. For each input test vector fault lists are serially propagated through the circuit to the primary output(s), all the faults covered by the test vector being noted after each pass. The time taken for one pass through the simulator is much greater than the time for one pass through a parallel simulator, but a large number of circuit faults will be covered on each pass. Dynamic memory capacity, however, has to be very large in order to handle all the continuously changing data in the propagation of the fault lists.

Concurrent fault simulation is the preferred present method of fault simulation. It avoids the complexity of implementing deductive fault simulation, yet retains a speed advantage over series and parallel fault simulation. Here a comprehensive concurrent fault list for each line is compiled for each test vector, which includes a chosen fault on the line plus all preceding faults which propagate to this line; if a preceding fault produces a response which is the same as the healthy response on this line to the test vector, the former is deleted from this concurrent fault list. By completing this procedure from primary inputs to primary outputs, a record is built up of the number of the chosen faults detected by the given set of test vectors, and hence the fault coverage; only those paths in the circuit which differ between the faulty and the fault-free state need be considered in each simulation.

Further details of fault simulation procedures may be found in the literature[3,7-13]. It must, however, be appreciated that when using any fault model simulation to determine a fault model coverage value, *FMC*, for a given set of test vectors, the resultant value only relates to the set of faults which have been chosen in the simulation. Thus, *FMC* = 100 % only indicates that all the faults introduced in the fault simulation will be detected, which is not the same as saying that the circuit is completely fault free. The implication of this is that the value of *FMC* is not necessarily the same as the value of fault coverage, *FC*, introduced in Chapter 1, see Figure 1.3, and in theory should not be used in the defect level equation $DL = (1 - Y^{1-FC})$. However, if *FMC* is nearly 100 %, then it is often assumed that *FMC* = *FC*, and this value of *FMC* may be used in the equation for *DL*. A more direct use for *FMC* is as a useful parameter in the development and grading of automatic test pattern generation (ATPG) algorithms, see later. We will use the designation *FC* rather than *FMC* subsequently, but the distinction when used in the equation for *DL* should not be forgotten.

3.2 Test pattern generation for combinational logic circuits

Test pattern generation is the design process of generating appropriate input vectors to test a given digital design, whether it is a printed-circuit board (PCB) or an individual IC. As previously noted, exhaustive testing is usually prohibitively long unless steps are taken to partition the circuit or system into smaller parts, in which case exhaustive testing of each partition may become acceptable. However, assuming that this cannot be or is not done, then some method of generating a reduced (nonexhaustive) test set has to be undertaken.

The generation of an acceptable reduced set of test vectors may be done in the following ways:

(i) manual generation;
(ii) algorithmic generation;
(iii) pseudorandom generation.

3.2.1 Manual test pattern generation

Manual TPG is a method which the original circuit designer or OEM may adopt, knowing in detail the functionality of the circuit or system involved. The test patterns may be specified by considering a range of functional conditions, and listing the input test vectors and healthy output responses involved in these situations. Alternatively, the input vectors that will cause all the gates to switch at least once may be considered. If a working breadboard prototype has been made, then this may be used to assist the OEM in this manual compilation, or alternatively the CAD program used in the design may assist.

This strategy of relying upon the circuit designer to propose some minimum set of test vectors can be a reasonable procedure to undertake for circuits containing, say, a thousand but not tens of thousands of gates. It is a procedure which has been and still is widely used in custom microelectronics (ASICs or USICs), where the IC vendor is manufacturing a specific circuit to meet the OEM's requirements[14]. The vendor will take the OEM's suggested test vectors, and check that they are acceptable to both parties by performing some CAD simulation; this may take the form of checking how many of the internal nodes of the circuit are toggled by the suggested test vectors rather than by considering any functionality of the tests. If this toggle coverage is considered to be inadequate (or incomplete), additional test vectors will be requested from the OEM to remedy the shortfall. It should be appreciated that the computer processing time for this procedure is relatively small, certainly far less than the excessive times which can build up when automatic test pattern generation is attempted, see the following section.

3.2.2 Automatic test pattern generation

Automatic (or algorithmic) test pattern generation, usually abbreviated to ATPG, becomes increasingly necessary as the gate count in the circuit increases to the thousands upwards. ATPG programs normally use a gate-level representation of the circuit, with all nodes or paths enumerated.

Figure 3.3 illustrated the concept of determining the fault coverage of a given test set for a given circuit; here we will be considering starting with a proposed fault in the circuit and finding a test for this fault, this procedure being repeated for all the chosen set of faults in the circuit, see Figure 3.4.

The first test vectors that are determined tend to cover a number of faults, see Figure 2.6 for example and Figure 3.13 later, and hence the initial fault

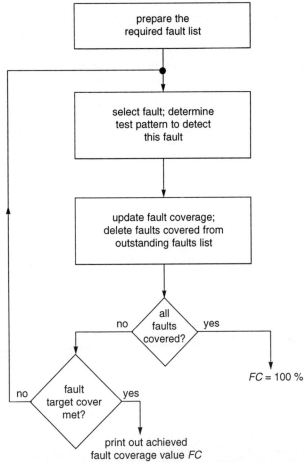

Figure 3.4 The ATPG procedure starting with a given fault list for the circuit and finishing with a set of test patterns and their fault coverage value, FC

coverage can be rapid. However, for complex circuits the processing time to find the tests for the final remaining faults may become unacceptably long, particularly if feedback loops or other circuit complications are present. If any redundancy is present then the ATPG program will, of course, never succeed in finding a test for certain nodes. Hence, economics may dictate the termination of an ATPG program when *FC* has reached an acceptable level, say 99.5 %, or has run for a given time, leaving the outstanding faults to be considered by the circuit designer if necessary. For very complex VLSI circuits even ATPG programs are now proving to be inadequate, being replaced by self test and other test strategies such as will be considered in later chapters.

All ATPG programs based upon fault models assume that a single fault is present when determining the test vectors. The usual fault model is the stuck-at model, which as we have seen does in practice cover a considerable number of other types of faults, but not all. The results of an ATPG program cannot, therefore, guarantee a defect-free circuit.

A basic requirement in test pattern generation is to propagate a fault at a given node in the circuit to an observable output, such that the output is the opposite value in the presence of the fault compared with the fault-free output under the same input test vector. This procedure may be termed path sensitising or forward driving. A second requirement is that the test input vector shall establish a logic value on the node in question which is opposite to the stuck-at condition under consideration, i.e. to test for a s-a-0 fault at node x the test vector will give a logic 1 at the node under fault-free conditions, and vice versa.

The principle of propagating a stuck-at fault condition on a node to an observable output has been illustrated in Chapter 2, Figure 2.5. In this earlier example a single path was sensitised to the primary output, but in more complex circuits it may be necessary to consider more than single-path sensitisation. Consider, for example, a simple part of a larger circuit as shown in Figure 3.5, and let us consider the signals required to drive a stuck-at 0 fault on line Q through G1 to the observable output node. Clearly we require $P = 1$, $Q = 1$ and $R = 0$ to establish this single path. However, if due to the preceding logic it is not possible to have $R = 0$ when $Q = 1$, then this single path sensitisation is not possible. But making $P = 1$, $Q = 1$, $R = 1$ will allow the fault to be detected at the output, the parallel paths through G1 and G2 both being sensitised. This is known as parallel reconvergence with dual-path sensitisation.

In Figure 3.5 the two paths which reconverge always took the same logic value when testing for stuck-at 0. However, it is possible to encounter reconvergence where the two converging signals are always opposite to each other under test conditions. (This is sometimes termed negative reconvergence, as distinct from positive reconvergence where the signals are the same.) This form of reconvergence is not testable, and indicates some local redundancy

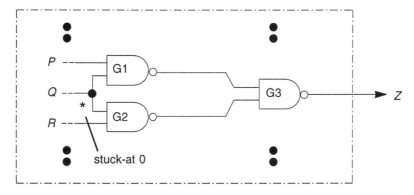

Figure 3.5 *A fragment of a combinational logic to illustrate dual-path sensitization and positive reconvergence*

left in the circuit design. The more complex the combinational logic network, the more likely it will be that reconvergence is present in the circuit. Hence, the need to sensitise more than one path from a stuck-at node is often encountered, which necessitates effective algorithms that can handle this situation.

Most test pattern generation algorithms but not all have as their underlying basis the procedure which we have now indicated, namely:

(i) choose a faulty node in the circuit;
(ii) propagate the signal on this node to an observable output;
(iii) backward trace to the primary inputs in order to determine the logic signals on the primary inputs which correctly propagate this fault signal to the observable output.

Additionally the procedure should:

(iv) ensure that the derived test vectors cater for all possible fan-out and reconvergence effects in the circuit, with the possibility of multiple-path sensitisation;
(v) keep a record of what additional faulty nodes are covered by each test vector when generating a test for a chosen node (i.e. the fault cover of each test vector), so that duplication of effort is minimised.

Here we will consider two methods which have been used for test pattern generation, the first of which does not in fact use the above signal propagation procedure, and the second which does use such a procedure.

3.2.2.1 Boolean difference method

The Boolean difference method is one test pattern generation technique which does not rely upon path sensitisation, but instead is functionally based using Boolean algebraic relationships to determine the test vectors. The method is based upon the principle of using two Boolean expressions for the combinational network, one of which represents the fault-free behaviour of the circuit and the other the behaviour under a single complementary fault condition. If these two functions are then exclusive-ORed together and the result is not logic 0 this fault can be detected; if the result is logic 0, which means that the two functions are identical under this fault condition, then this fault cannot be detected[1,2,15,16].

The formal definition of the Boolean difference of a given function $f(X)$ of n independent input variables x_1 to x_n with respect to one of its input variables x_i is defined as:

$$\frac{d}{dx_i} f(X) = f(x_1,...,x_i,...,x_n) \oplus f(x_1,...,\bar{x}_i,...,x_n) \qquad (3.2)$$

If $(d/dx_i)f(X) = 1$ then the fault on input line x_i can be detected, if 0 it cannot be detected since $f(X)$ is then independent of x_i. Notice that this operator is not a true differential operator in the full mathematical sense, since it does not distinguish between a change of $f(X)$ from 0 to 1 or vice versa, and hence it is defined as a difference operator rather than a differential operator. Also notice that the functions being exclusive-ORed together in eqn. (3.2) represent the complete truthtable of $f(X)$, and are not concerned with just one input combination.

Further properties of the Boolean difference operator are as follows[15]:

- function complementation:

$$\frac{d}{dx_i} f(X) = \frac{d}{dx_i} \overline{f(X)} \qquad (3.3)$$

- literal complementation:

$$\frac{d}{dx_i} f(X) = \frac{d}{d\overline{x}_i} f(X) \qquad (3.4)$$

- difference of Boolean difference:

$$\frac{d}{dx_i}\left\{\frac{d}{dx_j} f(X)\right\} = \frac{d}{dx_j}\left\{\frac{d}{dx_i} f(X)\right\} \qquad (3.5)$$

- difference of Boolean product:

$$\frac{d}{dx_i}\{f(X).g(X)\} = \left(f(X).\frac{d}{dx_i} g(X)\right) \oplus \left(g(X).\frac{d}{dx_i} f(X)\right) \oplus \left(\frac{d}{dx_i} f(X).\frac{d}{dx_i} g(X)\right) \qquad (3.6)$$

- difference of a Boolean sum:

$$\frac{d}{dx_i}\{f(X)+g(X)\} = \left(\overline{f(X)}.\frac{d}{dx_i} g(X)\right) \oplus \left(\overline{g(X)}.\frac{d}{dx_i} f(X)\right) \oplus \left(\frac{d}{dx_i} f(X).\frac{d}{dx_i} g(X)\right) \qquad (3.7)$$

- if the function $g(X)$ is independent of x_i, then eqns. (3.6) and (3.7) collapse to:

$$\frac{d}{dx_i}\{f(X).g(X)\} = g(X).\frac{d}{dx_i} f(X) \qquad (3.8)$$

and

$$\frac{d}{dx_i}\{f(X)+g(X)\} = \overline{g(X)}.\frac{d}{dx_i} f(X) \qquad (3.9)$$

and similarly should $f(X)$ be independent of x_i.

The test vector for primary inputs stuck-at 0 or stuck-at 1 are given by the following:

$$x_i \cdot \frac{d}{dx_i} f(X), \text{ for } x_i \text{ s-a-0} \qquad (3.10)$$

and

$$\bar{x}_i \cdot \frac{d}{dx_i} f(X) \text{ for } x_i \text{ s-a-1} \qquad (3.11)$$

For example, consider the circuit shown in Figure 3.6a. The Boolean difference with respect to x_1 is given by the following algebraic development:

$$\frac{d}{dx_1} f(X) = \frac{d}{dx_1}(x_1 x_2 x_3 + x_2 x_3 x_4 + \bar{x}_2 \bar{x}_4)$$

$$= \left\{ (\overline{x_1 x_2 x_3 + x_2 x_3 x_4}) \cdot \frac{d}{dx_1}(\bar{x}_2 \bar{x}_4) \right\} \oplus \left\{ (\overline{\bar{x}_2 \bar{x}_4}) \frac{d}{dx_1}(x_1 x_2 x_3 + x_2 x_3 x_4) \right\}$$

$$\oplus \left\{ \frac{d}{dx_1}(\bar{x}_2 \bar{x}_4) \cdot \frac{d}{dx_1}(x_1 x_2 x_3 + x_2 x_3 x_4) \right\}$$

$$= (\overline{\bar{x}_2 \bar{x}_4}) \frac{d}{dx_1}(x_1 x_2 x_3 + x_2 x_3 x_4) \text{ since } \frac{d}{dx_1}(\bar{x}_2 \bar{x}_4) \text{ is zero}$$

$$= (\overline{\bar{x}_2 \bar{x}_4}) \left(\left\{ (\overline{x_1 x_2 x_3}) \frac{d}{dx_1}(x_2 x_3 x_4) \right\} \oplus \left\{ \overline{x_2 x_3 x_4} \frac{d}{dx_1}(x_1 x_2 x_3) \right\} \right.$$

$$\left. \oplus \left\{ \frac{d}{dx_1}(x_2 x_3 x_4) \cdot \frac{d}{dx_1}(x_1 x_2 x_3) \right\} \right)$$

$$= (\overline{\bar{x}_2 \bar{x}_4}) \left((\overline{x_2 x_3 x_4}) \frac{d}{dx_1}(x_1 x_2 x_3) \right)$$

$$= (\overline{\bar{x}_2 \bar{x}_4}) \cdot (\overline{x_2 x_3 x_4}) \cdot (x_2 x_3)$$

$$= (x_2 + x_4)(\bar{x}_2 + \bar{x}_3 + \bar{x}_4)(x_2 x_3)$$

$$= x_2 x_3 \bar{x}_4 \qquad (3.12)$$

Thus, the test vector $x_1 x_2 x_3 \bar{x}_4$ will test for x_1 s-a-0, and $\bar{x}_1 x_2 x_3 \bar{x}_4$ will test for x_1 s-a-1.

This result is very easily confirmed by looking at the Karnaugh map of $f(X)$ given in Figure 3.6b, and considering the $x_1 = 0$ and $x_1 = 1$ halves of the map. These two decompositions of $f(X)$ differ only in the minterms $\bar{x}_1 x_2 x_3 \bar{x}_4$ and $x_1 x_2 x_3 \bar{x}_4$, being $f(X) = 0$ in the former and $f(X) = 1$ in the latter. Hence the Boolean difference $(d/dx_1)f(X)$ is $x_2 x_3 x_4$, giving the stuck-at test vectors for x_1 shown above.

If the Boolean difference method is used to generate test vectors for internal nodes of the circuit, the output function must be expressed in terms

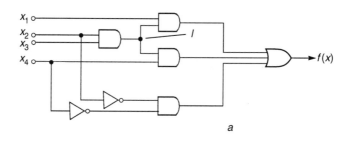

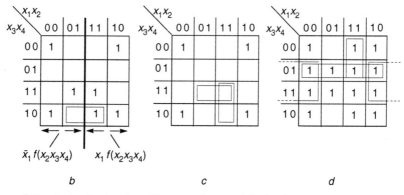

Figure 3.6 A simple circuit to illustrate the use of the Boolean difference
  a circuit, $f(X) = x_1 x_2 x_3 + x_2 x_3 x_4 + \bar{x}_2 \bar{x}_4$
  b Karnaugh map of $f(X)$, showing the decomposition about x_1
  c output function when internal node I is stuck at 0, the difference between this function and the fault-free output being as indicated
  d as c but I stuck at 1

of the internal node being considered. For example, should internal node I in Figure 3.6 be considered, the output function becomes

$$f(X) = x_1 I + x_4 I + \bar{x}_2 \bar{x}_4$$

whence

$$\frac{d}{dI} f(X) = \overline{(\bar{x}_2 \bar{x}_4)} \frac{d}{dI} \{ I(x_1 + x_4) \}$$
$$= (x_2 + x_4)(x_1 + x_4)$$
$$= x_1 x_2 + x_4 \qquad (3.13)$$

Therefore to test for I s-a-0 we have:

$$I \left(x_1 x_2 + x_4 \right)$$
$$= x_2 x_3 \left(x_1 x_2 + x_4 \right)$$
$$= x_1 x_2 x_3 + x_2 x_3 x_4$$

and for I s-a-1 we have

$$\overline{(x_2 x_3)}(x_1 x_2 + x_4)$$
$$= \bar{x}_2 x_4 + x_1 x_2 \bar{x}_3 + \bar{x}_3 x_4$$

Figures 3.6c and d will confirm these results.

It will be seen that the Boolean difference method generates all the possible test vectors for each faulty node considered, covering both s-a-0 and s-a-1 faults. Should an algebraic expression reduce to 0 this indicates a redundant line in the circuit under consideration. However, because Boolean algebraic manipulations are tedious to handle with conventional computer programs, the method requires a excessive amount of computer time and memory to handle circuits of more than a few hundred logic gates, and as a result algorithms which generate tests for more than one node at a time and which we will consider shortly prove more satisfactory.

Further developments in the Boolean difference method may be found[16]. It is also possible to consider more than one node simultaneously faulty using higher-ordered Boolean difference equations, which if all possible single and multiple stuck-at nodes are considered requires the evaluation of 2^N Boolean difference equations, where N is the number of nodes or lines in the circuit. This is even more impractical than the single stuck-at fault condition which involves only N Boolean difference equations. A more powerful technique for handling this method has also been investigated; this uses the Gibbs dyadic differentiator and logic values of $-1, +1$ rather than 0, 1, and can incorporate matrix manipulations in the computational procedures[17–19]. However, no breakthrough in using these further developments has been reported, leaving the Boolean difference method as a good educational exercise but with little VLSI application.

3.2.2.2 Roth's D-algorithm

In contrast to the educationally-interesting Boolean difference method, Roth's D-algorithm[20] forms the underlying concept of many practical ATPG programs. It sensitises all paths from the site of a chosen fault to an observable output, and therefore inherently caters for reconvergent fan-out situations. It does, however, operate at the individual gate level, and requires knowledge of all the gates and their interconnection topology which functionally-based ATPG programs do not necessarily need.

The D-algorithm involves five logic states, namely

0 = normal logic zero
1 = normal logic one
D = a fault-sensitive state of a line or node, where $D = 1$ under fault-free
  conditions but is 0 under the particular fault condition being considered
$\bar{D}$ = a fault-sensitive state of a line or node, where $\bar{D} = 0$ under fault-free
  conditions but is 1 under the particular fault condition being considered
X = an unassigned logic value, which can take any value 0, 1, D or $\bar{D}$

Thus D identifies a line or node which is 1 under fault-free conditions and changes to 0 under the chosen fault condition, and $\bar{D}$ identifies a line or node which changes from 0 to 1 under fault conditions. D and $\bar{D}$ are always opposite logic values.

Using these five logic states, the primitive D-cubes of failure for any type of logic gate may be defined.[*] Table 3.1 gives the relationships for all three-input Boolean gates. The D-cube 1 1 1 D, see first line of Table 3.1, gives the input pattern necessary to prove whether the output line of an AND gate is fault free or stuck-at 0; $D = 1$ if no fault is present, and is 0 with the stuck-at fault. Similarly, X X 1 $\bar{D}$ checks for a s-a-1 fault at the output of a three-input NOR gate. Similar tables may be compiled for all other gates and combinational macros.

Table 3.1 *The primitive D-cubes of failure for three-input Boolean logic gates, giving the input signals necessary to distinguish the presence of the output line stuck-at*

	Required inputs to detect faulty output				Corresponding D-cubes of failure			
	inputs			output				
	A	B	C	Z	A	B	C	Z
AND gate	1	1	1	s-a-0	1	1	1	D
	0	X	X	s-a-1	0	X	X	$\bar{D}$
	X	0	X	s-a-1	X	0	X	$\bar{D}$
	X	X	0	s-a-1	X	X	0	$\bar{D}$
NAND gate	1	1	1	s-a-1	1	1	1	$\bar{D}$
	0	X	X	s-a-0	0	X	X	D
	X	0	X	s-a-0	X	0	X	D
	X	X	0	s-a-0	X	X	0	D
OR gate	0	0	0	s-a-1	0	0	0	$\bar{D}$
	1	X	X	s-a-0	1	X	X	D
	X	1	X	s-a-0	X	1	X	D
	X	X	1	s-a-0	X	X	1	D
NOR gate	0	0	0	s-a-0	0	0	0	D
	1	X	X	s-a-1	1	X	X	$\bar{D}$
	X	1	X	s-a-1	X	1	X	$\bar{D}$
	X	X	1	s-a-1	X	X	1	$\bar{D}$

The requirements for propagating (sometimes termed D-drive) any D value through following fault-free gates are given in Table 3.2. This information is known as the propagation (or propagating) D-cubes for each type of gate. For example, with a single stuck-at 0 signal on input x_2 of a three-input NAND

---

[*] In this context, a cube is an ordered set of symbols such that each symbol position defines a particular input or output node, and the value of the symbol identifies its logic state.

gate we have the D-cube 1 D 1 $\bar{D}$, see the top line of Table 3.2. Notice also that:

(i) the propagating D-cubes also define the propagation when more than one input is D or $\bar{D}$, which can arise in a circuit with reconvergent fan-out from the original D (or $\bar{D}$) source. However, should D and $\bar{D}$ both converge on a Boolean gate there will be no further propagation of the D or $\bar{D}$ value; D and $\bar{D}$ on an AND gate will always give an output 0 and on an OR gate an output 1, and hence the D and $\bar{D}$ values will be lost.

(ii) for the propagation of a D (or $\bar{D}$) value through a Boolean gate there is only one possible input condition; there is therefore no choice of logic 0 or 1 signals on the gate inputs to propagate the D signal(s).

Table 3.2 *The propagation D-cubes for (a) three-input AND and NAND gates, and (b) three-input OR and NOR gates, the gates being fault free*

a

Gate inputs, $x_1\ x_2\ x_3$					Gate output $f(X)$	
					AND gate	NAND gate
1 1 D	or	1 D 1	or	D 1 1	D	$\bar{D}$
1 1 $\bar{D}$	or	1 $\bar{D}$ 1	or	$\bar{D}$ 1 1	$\bar{D}$	D
1 D D	or	D 1 D	or	D D 1	D	$\bar{D}$
1 $\bar{D}$ $\bar{D}$	or	$\bar{D}$ 1 $\bar{D}$	or	$\bar{D}$ $\bar{D}$ 1	$\bar{D}$	D
		D D D			D	$\bar{D}$
		$\bar{D}$ $\bar{D}$ $\bar{D}$			$\bar{D}$	D

b

Gate inputs, $x_1\ x_2\ x_3$					Gate output $f(X)$	
					OR gate	NOR gate
0 0 D	or	0 D 0	or	D 0 0	D	$\bar{D}$
0 0 $\bar{D}$	or	0 $\bar{D}$ 0	or	$\bar{D}$ 0 0	$\bar{D}$	D
0 D D	or	D 0 D	or	D D 0	D	$\bar{D}$
0 $\bar{D}$ $\bar{D}$	or	$\bar{D}$ 0 $\bar{D}$	or	$\bar{D}$ $\bar{D}$ 0	$\bar{D}$	D
		D D D			D	$\bar{D}$
		$\bar{D}$ $\bar{D}$ $\bar{D}$			$\bar{D}$	D

The propagating D-cubes represent a subset of all the possible five-valued input/output relationships of combinational logic gates, being the subset in which D (or $\bar{D}$) is driven from gate input(s) to gate output. For two-input Boolean gates we have the full relationships shown in Figure 3.7, with six input conditions being fault propagating. For three-input Boolean gates there are $5 \times 5 \times 5 = 125$ possible input conditions, but only fourteen are fault propagating.

A simple example of D propagation is shown in Figure 3.8. The chosen fault source on the output line from G3 is driven to both primary (observable) outputs by the signals shown. Should G4, G5 and G6 be other than the two-input gates shown here, then the propagation D-cubes for these gates would define the required logic signals for forward driving the D or $\bar{D}$ conditions. All possible paths from the D or $\bar{D}$ source towards the primary outputs are normally considered[16,18], although only one primary output needs to be finally monitored for the stuck-at test.

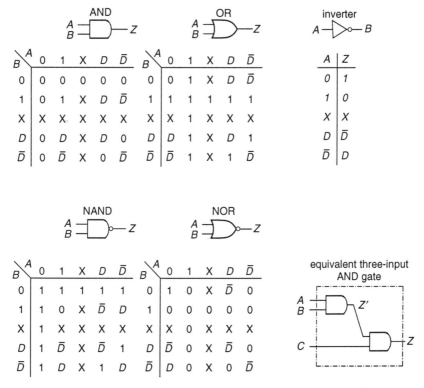

Figure 3.7 Roth's five-valued D-notation applied to two-input Boolean logic gates. The relationships for three (or more) input gates may be derived by considering a cascade of two-input gates, since commutative and associative relationships still hold

Roth's full algorithm for test pattern generation thus consists of three principal operations as shown in Figure 3.9, namely:

(i) choose a stuck-at fault source, and from the primitive D-cubes of failure data identify the signals necessary to detect this fault;

(ii) forward drive this fault D or $\bar{D}$ through all paths to at least one primary output, using the information contained in the propagation D-cubes;

(iii) perform a consistency operation, that is backward trace from a primary output to which D or $\bar{D}$ has been driven, to the primary inputs, allocating further logic 0 and 1 values as necessary to give the final test input vector.

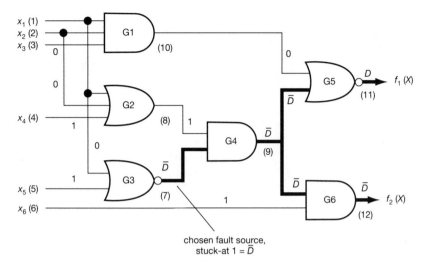

Figure 3.8 An example using Roth's notation, showing a stuck-at 1 fault being driven to both primary outputs. $\bar{D}$ represents the same logic value on all lines so marked, with the line marked D having the opposite logic value. The numbers in parentheses are used later in the text

This procedure is repeated until all the chosen stuck-at paths have been covered.

In undertaking the D-drive, the operation known as D-intersection is performed for each gate encountered from the source fault to the primary output(s). This is an algebraic procedure which formally matches the logic signals on the gates with the appropriate propagation D-cube data. Recall that the propagation data for any D or $\bar{D}$ gate input is unique. The D-drive procedure for the simple circuit shown in Figure 3.8 would therefore proceed as shown in Table 3.3, having first identified all the paths in the circuit.

Table 3.3 The D-drive conditions for Figure 3.8

Circuit path	1	2	3	4	5	6	7	8	9	10	11	12
Select (7) stuck-at 1	0	•	•	•	1	•	$\bar{D}$	•	•	•	•	•
Intersect with G4 propagation D-cube	0	•	•	•	1	•	$\bar{D}$	1	$\bar{D}$	•	•	•
Intersect with G5 propagation D-cube	0	•	•	•	•	•	$\bar{D}$	1	$\bar{D}$	0	D	•
Intersect with G6 propagation D-cube	0	•	•	•	1	1	$\bar{D}$	1	$\bar{D}$	0	D	$\bar{D}$

The consistency operation to follow this D-drive evaluation is shown in Table 3.4. No inconsistencies are found, and hence the stuck-at fault at (7) can be detected by the input test vector $\bar{x}_1\bar{x}_2\bar{x}_3x_4x_5x_6$, giving a logic 0 at the primary output $f_1(X)$ if the fault is present.

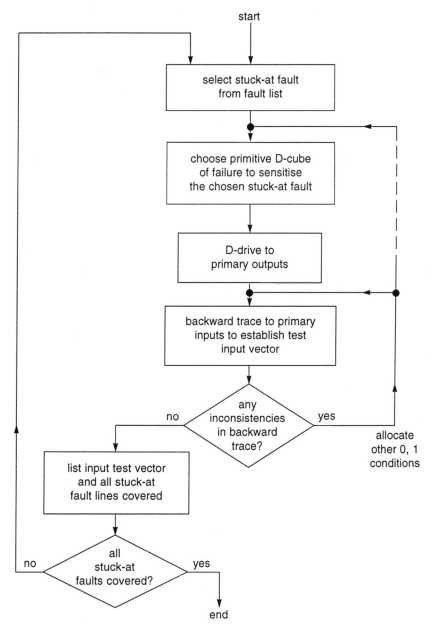

Figure 3.9 *The outline schematic of Roth's D-algorithm ATPG procedure*

Table 3.4 The backward tracing consistency operation for Table 3.3

Circuit path	1	2	3	4	5	6	7	8	9	10	11	12
End of D-drive	0	●	●	●	1	1	$\bar{D}$	1	$\bar{D}$	0	D	$\bar{D}$
Check (8) is at 1 from G2; OK if (4) = 1	0	●	●	1	1	1	$\bar{D}$	1	D	0	D	$\bar{D}$
Check (10) is at 0 from G1;												
OK if (1) (2) & (3) = 0	0	0	0	1	1	1	$\bar{D}$	1	D	0	D	$\bar{D}$

In this example no inconstancies are encountered in completing the backward tracing operation. However, if we had started with the equally valid primitive D-cube of failure for gate G3 of $x_1 \, x_5 = 1\,0$ or $1\,1$ instead of $0\,1$, then we would have encountered an inconstancy in gate G1 when backward tracing from node 10 to the inputs. Hence, in practice, we may have to recompute the D-drive conditions trying alternative primitive D-cubes as the starting point.

Further difficulties with the D-algorithm arise when exclusive-OR/NOR gates or local feedback conditions are encountered. The problem with exclusive gates is that there is not a unique input condition for propagating a D or $\bar{D}$ signal, see Table 3.5, and reconvergence of D or $\bar{D}$ or D and $\bar{D}$ will be nonpropagating.

Table 3.5 The propagation D-cubes for two-input exclusive gates

Gate inputs, $x_1 \, x_2$	Gate output $f(X)$	
	ex-OR	ex-NOR
0 D or D 0	D	$\bar{D}$
0 $\bar{D}$ or $\bar{D}$ 0	$\bar{D}$	D
1 D or D 1	$\bar{D}$	D
1 $\bar{D}$ or $\bar{D}$ 1	D	$\bar{D}$

To summarise the principal characteristics of Roth's algorithm:

(i) the algorithm guarantees to find a test for any stuck-at fault in a circuit if such a test exists; if a stuck-at 0 fault cannot be detected then a steady logic 0 signal could be substituted at this node without affecting any primary output; similarly, if a stuck-at 1 fault cannot be detected a steady logic 1 signal could be substituted; if neither a s-a-0 or a s-a-1 fault can be detected then this node is completely redundant and could be removed from the circuit.*

(ii) the test vector for the detection of any given stuck-at fault source is also the test vector for all the following stuck-at conditions encountered in the D-drive from the fault source to the observable output; for example the test vector $\bar{x}_1 \bar{x}_2 \bar{x}_3 x_4 x_5 x_6$ in the above example will detect s-a-1

---

* There are many published examples of circuits which purport to show why a stuck-at fault cannot be detected by fault modelling. What is not always clear is that these are usually examples where there is redundancy in the circuit which the author may not mention.

conditions on lines (9) and (12) and s-a-0 on line (11) as well as the fault source on line (7).

(iii) it is the difficulty of assigning input signals for a given gate output in the backward tracing operation which largely causes problems; unlike the forward D-drive operation there can be a choice of gate input signals for a given output, since $2^n - 1$ input combinations of any n-input Boolean gate give rise to the same gate output condition, and hence several retries of the algorithm to find a consistent backward tracing operation may be necessary.

A more detailed analysis and discussion of the D-algorithm may be found in the appendix of Bennetts[20c]*; other texts have worked examples[1,2,11,20,21] and software code fragments[11]. However, it remains a difficult topic to learn, partly because of the terminology which was introduced in the original disclosures, and because of the supporting mathematics based upon the so-called calculus of D-cubes. Nevertheless, it remains a foundation stone in ATPG theory.

Finally, for interest, the Boolean difference technique applied to the circuit of Figure 3.8 would give the Boolean difference functions:

$$\frac{d}{dF} f_1(X) = x_2\bar{x}_2 + x_2\bar{x}_3 + \bar{x}_1 x_2 + \bar{x}_1 x_4 \tag{3.14}$$

and

$$\frac{d}{dF} f_2(X) = x_1 x_6 + x_2 x_6 + x_4 x_6 + x_5 x_6 \tag{3.15}$$

for the two outputs $f_1(X)$ and $f_2(X)$ with respect to the fault source F on line (7).

Hence the tests for a stuck-at 1 fault on line (7) would be:

$$(x_1 + x_5)(x_1\bar{x}_2 + x_2\bar{x}_3 + \bar{x}_1 x_2 + \bar{x}_1 x_4)$$
$$= x_1\bar{x}_2 \text{ or } x_1 x_2\bar{x}_3 \text{ or } x_2\bar{x}_3 x_5 \text{ or } \bar{x}_1 x_2 x_5 \text{ or } \bar{x}_1 x_4 x_5$$

at output $f_1(X)$, and

$$(x_1 + x_5)(x_1 x_6 + x_2 x_6 + x_4 x_6 + x_5 x_6)$$
$$= x_1 x_6 \text{ or } x_5 x_6$$

at output $f_2(X)$. The tests for a stuck-at 0 fault on line (7) similarly would be:

$$(\overline{x_1 + x_5})(x_1\bar{x}_2 + x_2\bar{x}_3 + \bar{x}_1 x_2 + \bar{x}_1 x_4)$$

at $f_1(X)$ and

$$(\overline{x_1 + x_5})(x_1 x_6 + x_2 x_6 + x_4 x_6 + x_5 x_6)$$

at $f_2(X)$. Notice that our Roth's algorithm example merely identified one of the possible test vectors which detect the s-a-1 fault on line (7).

---

* The terminology 'dual' used in Reference 20c should be read with care. It is not the Boolean dual $f^D(X)$ of $f(X)$ where $f^D(X)$ is the complement of $f(X)$ with all gate inputs individually complemented, but is the changing of all Ds to $\bar{D}$ and all $\bar{D}$s to D in any given propagation D-cube, leaving the 0s and 1s unchanged.

3.2.2.3 Developments following Roth's D-algorithm

Although Roth's algorithm guarantees to find a test for any stuck-at fault if a test exists, and is found to be more computational efficient than the Boolean difference method, a very great deal of CPU time is still involved which precludes its use for circuits containing thousands of logic gates. A later modification by Roth *et al.*[20a] improved the situation to some degree by performing the consistency backward tracing as the D-drive progresses, rather than waiting until the D-drive has reached a primary output, but alternative developments have now built upon these pioneering approaches.

A modification of the *D*-algorithm which works from the primary outputs backwards to determine a sensitive path was the LASAR (logic automated stimulus and response) algorithm[22-24]. Essentially, this algorithm assigned a logic value to a primary output, and then followed paths back through the circuit assigning gate input values which gave the required output value. However, this algorithm in turn has been succeeded by the PODEM (path oriented decision making) algorithm[25-27], which has formed a basis for many still-current ATPG programs.

PODEM uses the basic concept of using five logic values 0, 1, X, D and $\bar{D}$ and a *D*-drive operation as introduced by Roth, but operates in a dissimilar manner by initially backward tracing from each chosen fault source (sometimes termed the target fault) towards the primary inputs before commencing a *D*-drive towards the primary outputs. The procedure is as follows.

(i) set all nodes (lines) in the circuit to X (unassigned);
(ii) choose a stuck-at 0 (D) or a stuck-at 1 ($\bar{D}$) gate output fault from the fault list; assign this node to 1 if s-a-0 is chosen and to 0 if s-a-1 is chosen (these are the logic levels necessary for the stuck-at test); leave all remaining nodes at X;
(iii) choose one of the gate input nodes, and assign a logic 0 or 1 to it which is appropriate for testing the chosen stuck-at gate output (see the primitive *D*-cubes of failure given in Table 3.1);
(iv) backward trace from this assigned gate input node to a primary input or inputs, allocating 0 or 1 as appropriate to the primary input(s) which give this gate input condition; leave all unassigned primary inputs and other circuit nodes at X;*
(v) simulate the circuit with these 0, 1, X values, and check whether the required logic 1 (now identified as D) or logic 0 (now identified as $\bar{D}$) is established at the chosen fault source;

---

* Care should be taken with the terminology that may be found in existing literature. The term 'backtrace' is what we are here calling backward trace, and is a consideration of the signals on certain nodes in the circuit working backwards towards the primary inputs. The term 'backtrack', however, which we will not use here, is a starting again (retry) of a procedure or part of a procedure, which may proceed backwards towards the primary inputs or forwards towards the primary outputs.

(vi) if it is established, which means that the primary input assignment of 0, 1, X is appropriate for the stuck-at test of the target fault, proceed to (viii) below;

(vii) if it is not established, repeat (iii), (iv) and (v) (possibly several times) with alternative assignments until the correct primary assignment for the chosen stuck-at condition is achieved (note, if no primary assignment can be found for the chosen fault, then this fault is untestable);

(viii) with this partial input test vector 0, 1, X, D-drive the chosen stuck-at gate output node to a primary output by randomly allocating 0 and 1 to the remaining X inputs, until D (or $\bar{D}$) is detected at an output by simulation of each situation;

(ix) when successful accept this primary input assignment of 0, 1, X as the test vector for the chosen stuck-at fault;

(x) repeat procedures (i) to(viii) until all stuck-at nodes in the fault list have been covered, returning all nodes to X at the beginning of each test.

This procedure is summarised in Figure 3.10.

Like Roth's original D-algorithm, PODEM guarantees to find a test vector for any stuck-at fault in the circuit, if such a test is possible. The initial choice of the partial primary input assignment 0, 1, X before undertaking the D-drive operation is found to be particularly effective in reducing the total computational effort for the complete circuit under test compared with Roth's original procedure.

There are additionally further mechanisms built into the full PODEM procedure which are not detailed in Figure 3.10. These include:

(a) recognition of AND, NAND, OR and NOR gates for both the backward tracing and subsequent D-drive operations;

(b) where one gate input in the backward tracing operation is sufficient, the gate input which has the easiest controllability from the primary inputs is first assigned; where all gate inputs have to be assigned then the hardest controllability paths to the primary inputs are chosen so as to cover them at this early stage of the procedure;

(c) the simulator used is a simple zero-delay, five-valued simulator, which can rapidly check the 0, 1, X, D, $\bar{D}$ values without involving any timing or possible circuit race conditions.

The full three-part schematic diagram for PODEM may be found in Reference 26, and republished in References 1 and 2. A more readily-followed three-part schematic and worked examples may be found in Bottorff[28]. A further detailed worked example may also be found in Reference 27, which includes a circuit known as the Schneider counter example shown in Figure 3.11; the stuck-at 0 fault on gate output G2 would be detected by first completing the backward trace from the fault source, giving:

$$x_1\ x_2\ x_3\ x_4 = X\ 0\ 0\ X$$

Digital test pattern generation 65

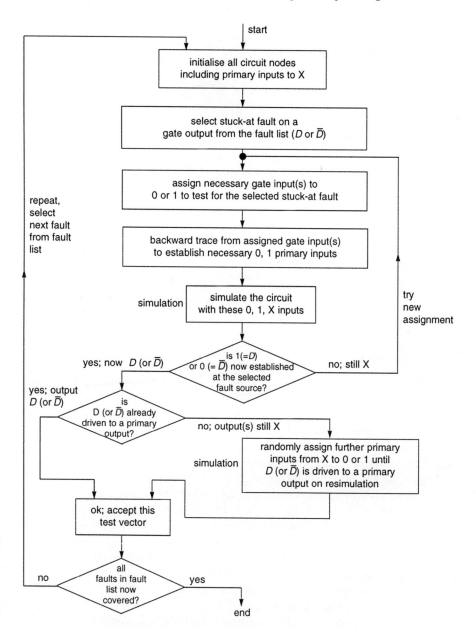

Figure 3.10 Outline schematic of the PODEM ATPG program, starting with all nodes and primary inputs at X. Exit paths (not shown) are present if no test for a given stuck-at fault is possible

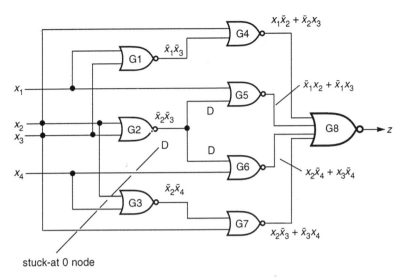

Figure 3.11 Schneider's symmetrical circuit example, which is frequently used to demonstrate test pattern generation aspects. See also the example given in Figure 3.8

Attempting a single-path *D*-drive from this fault source to the primary output via gate G5 or gate G6 would reveal an inconsistency; for example, driving through G5 only would result in 0 $\bar{D}$ 0 1 on gate G8 with the input test vector 0 0 0 1, or driving through gate G6 only would result in 1 0 $\bar{D}$ 0 with the input test vector 1 0 0 0, neither of which would give a *D* output. The only test possible is the test vector 0 0 0 0, which drives $\bar{D}$ through both G5 and G6 to G8. The PODEM algorithm, however, would have found this test almost immediately by trying this test vector from the given starting point of X 0 0 X. Notice that with this fairly trivial example there are only four possible test vectors to try, namely 0 0 0 0, 0 0 0 1, 1 0 0 0 and 1 0 0 1. Also the actual circuit is highly artificial, being merely $Z = \bar{x}_1 \bar{x}_2 \bar{x}_3 \bar{x}_4 + x_1 x_2 x_3 x_4$ which changes to $x_1 x_2 x_3 x_4$ under the given stuck-at fault condition.

A further development by IBM of PODEM-X has been used for the test pattern generation of circuits containing tens of thousands of gates[26]. PODEM-X incorporates an initial determination of a small set of test vectors which will cover a high percentage of faults in the fault list, leaving the PODEM procedure to cover the remainder. This will be considered further in section 3.2.3. However, a more distinct variation of the PODEM algorithm is the FAN (fan-out oriented) ATPG program of Fujiwara and Shimono[29], which specifically considered the fan-out nodes of a circuit, and uses multiple-path forward and backward tracing.

The major introduction in FAN is when considering a backward trace from a given *D* or $\bar{D}$ node. The procedure is broadly as follows:

(i) from a given fault source, backward trace appropriate 0 or 1 assigned logic values until a gate output is reached which has fan-out, i.e. one or more extra paths is fed from this node in addition to the path from the chosen fault source; stop the backward trace at this point if encountered; otherwise continue right back to the primary inputs (this is unlike PODEM, which always continues back to the primary inputs irrespective of fan-out conditions);

(ii) with this assigned set of 0, 1 and X signals forward trace to determine all the resulting 0, 1 and X signals, including the fan-out branches if the backward trace was halted at a fan-out node;

(iii) if D (or $\bar{D}$) is driven to a primary output continue with (v) below; if D (or $\bar{D}$) is not yet driven to a primary output, backward trace as in (i) above from all gates to which D or $\bar{D}$ has now been driven (the D-frontier), assigning 0 or 1 as appropriate to drive each D or $\bar{D}$ forward;

(iv) Repeat (ii) and (iii) until D (or $\bar{D}$) is driven by a forward trace to a primary output;

(v) with D (or $\bar{D}$) at a primary output, backward trace to assign all remaining necessary 0 and 1 signals from the fan-out nodes to the primary inputs, leaving any unrequired inputs at X.

It will therefore be seen that the FAN algorithm proceeds in steps of forward and backward tracing, attempting at each step to assign consistent 0 and 1 logic values to the nodes which are instrumental in propagating D or $\bar{D}$ towards a primary (observable) output. However, the steps are more complex than indicated here, since conflicts of assigned 0 and 1 logic values may arise which necessitate retries with different assigned values where a choice is possible, but unlike PODEM such conflicts are largely localised and resolved by retry at each step. It is therefore claimed that FAN is three to four times more efficient than PODEM for circuits containing thousands of logic gates.

The FAN algorithm is summarised in Figure 3.12. A full three-part schematic may be found in Reference 29 and reprinted in Reference 1. A comprehensive but simplified schematic may also be found in Reference 28.

3.2.3 Pseudorandom test pattern generation

The ATPG algorithms of the previous section are deterministic, being based upon the choice and detection of a single stuck-at fault by an appropriate input test vector. Although each test considers one fault as the starting point for the determination of a test vector, each test vector covers more than one stuck-at node, and a final consolidation to the minimum number of test vectors (the minimum test set) to cover the complete stuck-at fault list can be implemented. See Figure 2.6 for example.

Deterministic ATPG algorithms provide the smallest possible test set to cover the given fault list. The disadvantage is the complexity and cost of

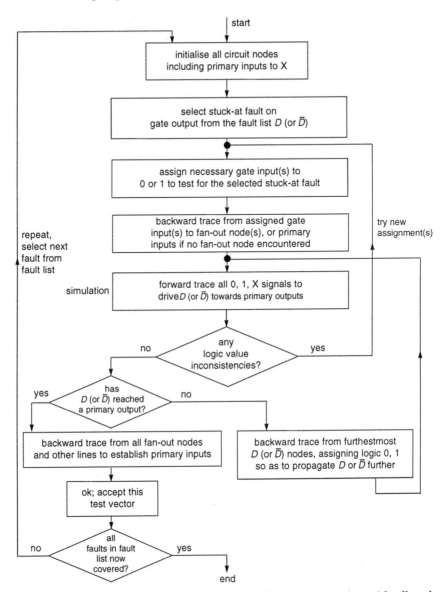

Figure 3.12 Outline schematic of the FAN ATPG program, starting with all nodes and primary inputs at X. Exit paths (not shown) are present if no test for a given stuck-at fault is possible

generating this minimum test set.* On the other hand a fully exhaustive test set will incur no ATPG costs, but will usually be too long to employ for large circuits. There is, however, an intermediate possibility which has been used.

---

* It has been reported[28] that millions of retries have been found necessary in some circuits of VLSI complexity before the test vectors to cover the complete set of faults were determined.

It is intuitively obvious that fault coverage increases with the number of input test patterns which are applied, up to the full fault coverage; a single randomly chosen input test vector is also likely to be a test for several faults in a complex circuit. Hence if a sequence of random or pseudorandom input test vectors (see later) is used, it is probable that a number of circuit faults will be covered without incurring any ATPG cost.

The relationship between the number of applied random test vectors and the resultant fault coverage has been studied[21,30–33]. Tests made by Eichelburger and Lindbloom[32] on two combinational circuits containing approximately 1000 gates are shown in Table 3.6. From this work the expected fault coverage, FC, achieved by the application of N random test vectors is given by the relationship:

$$FC = \left[1 - e^{\left(-\lambda \log_{10} N\right)}\right] \times 100\,\% \tag{3.14}$$

where λ is a constant for the particular circuit under test. The general characteristic of this relationship is shown in Figure 3.13, which confirms the intuitive concept that it is relatively easy to begin the fault coverage but becomes increasingly difficult to cover the more difficult remaining faults in the fault list.

Table 3.6 *The fault coverage obtained on two circuits by the application of random test vectors. Note, a fully-exhaustive functional test set would contain 2^{63} and 2^{54} test vectors, respectively*

Circuit	No. of primary inputs	No. of gates	% fault coverage obtained with N random test patterns		
			$N = 100$	1000	10000
(a)	63	926	86.1	94.1	96.3
(b)	54	1103	75.2	92.3	95.9

A truly random sequence of test vectors, which includes the possibility of the same vector occurring twice within a short sequence, is difficult to generate. Also, it is usually unproductive to apply the same test vector more than once in a test procedure. Hence it is more practical to consider the use of pseudorandom patterns, which are very easy to generate. The theory of pseudorandom pattern generation will be given in Section 3.4; suffice here to note that in a n-bit vector there can be a maximum of 2^n-1 sequences before the sequence repeats, with each bit in the sequence having the same number of changes between 0 and 1. Hence 0s and 1s are equally probable on each bit.

Therefore, for a circuit with n primary inputs, it is appropriate to take a very small subset of the 2^n-1 pseudorandom sequence to use as the random test set. The number of faults that are covered by this test is determined by normal simulation, leaving the small percentage of faults which have not been detected to be covered by using PODEM or FAN or some other

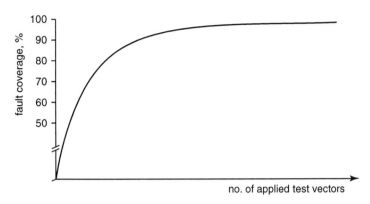

Figure 3.13 The general characteristics of fault coverage versus the number of random test vectors applied to the circuit

deterministic ATPG procedure. Prior analysis by probability theory[34-38] of the effectiveness of the pseudorandom vectors in testing a given circuit is not particularly useful, since with a complex circuit the computational effort involved could be comparable with performing a deterministic ATPG computation for the whole circuit, which is the thing we are trying to avoid having to do.

A slightly different approach has been attempted in the RAPS (random path sensitising) procedure, which is combined with PODEM in the PODEM-X procedure to complete the test coverage[26,39]. RAPS selects at random a primary output and allocates again at random a logic 0 or 1 to it. From this assignment a backward trace is made to determine the necessary primary inputs. This is repeated until a number of input test vectors are generated. The global fault coverage of this randomly generated input test set is then computed by simulation, leaving the remaining clean-up faults on the fault list to be covered by PODEM.

A particular characteristic of this approach is the very small number of random input test vectors used, being a very small multiple of the number of primary outputs irrespective of the complexity of the network. This is due to the complete sensitisation of multiple paths from outputs back to inputs during each random test vector determination. A worked example of this may be found in Reference 27.

3.3 Test pattern generation for sequential circuits

So far our considerations have involved combinational circuits only, which have been flattened to gate level with stuck-at conditions considered at every node. In general, the latch and flip-flop elements of a sequential network cannot be broken down to individual gate level without breaking all the inherent feedback connections—if they could then the test pattern generation problem would become purely combinational with perhaps limited controllability and observability—and therefore we have to consider the possible states of the circuit as well as the combinational aspects. It will also be recalled from earlier in this text that to consider the exhaustive test of a circuit containing n logic gates and s latches or flip-flops would theoretically require 2^{n+s} input test vectors, which is an impossibly high number to consider for LSI and VLSI circuits.

The classic model for any sequential network is shown in Figure 3.14. If the present primary outputs z_1 to z_m are a function of the present secondary inputs y_1 to y_s (the memory states) and the present primary inputs x_1 to x_n, then the model is a Mealy model; if the outputs are a function of y_1 to y_s only then the model is a Moore model. In both cases it is not possible to define the fault-free outputs resulting from a primary input test vector without knowledge of the internal states of the memory. Also it will be appreciated that the two halves shown in Figure 3.14 are in practice often inextricably mixed, with w_i and y_i lines buried within the circuit unless some specific silicon layout is present.

The basic difficulty with testing the circuit model of Figure 3.14 is that the logic values y_1 to y_s are generally unknown. If they were observable then testing of the top half of the circuit could be entirely combinational, and testing of the bottom half would be by monitoring y_1 to y_s during an appropriate set of primary input test vectors. However, assuming y_1 to y_s are inaccessible, developments as follows must be considered.

The first problem is to initialise the memory elements to a known state before any meaningful test procedure can continue. This may be done in the following ways:

(i) Apply a synchronising input sequence to the primary inputs which will force a fault-free circuit to one specific state irrespective of its initial starting state[1,40]. This is only possible in special cases where the states always home to a given state and remain there in the presence of a certain input vector—a counter which free runs but is stopped when it reaches 1 1 1 1... by a certain input vector is a trivial example.

(ii) Apply a homing input sequence to the primary inputs which finally gives a recognisable output sequence from a fault-free circuit irrespective of its initial starting state[1,40-43]. Depending upon which output sequence is recognised, the final state (but not the initial state) of the circuit is known.

(iii) A special case of the homing input sequence is the distinguishing input sequence, where the circuit output response is different for every possible starting state[41]. If this is possible, then both the initial and final states of the circuit are known. Most sequential circuits, however, do not have a distinguishing input sequence, although all do have one or more homing input sequences.

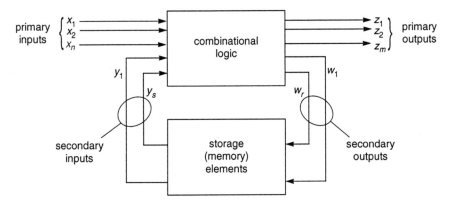

Figure 3.14 *The model for sequential logic networks, where all combinational logic gates are lumped into one half of the model and all memory elements are lumped into the other half*

(iv) Finally, a separate asynchronous reset (clear) may be applied to all internal latches and flip-flops to reset them to the 0 0 0 ... 0 0 state, or they may all be reset or set on the application of a specific input test vector which the circuit designer has built into the circuit design.*

A discussion of the procedures (i) to (iii) above may be found in Rajsuman[1] and in Lala[2]. It has been shown[42] that for any machine containing s memory (storage) elements and which therefore has S, $S \leq 2^s$, states, the upper bound on the length of a homing sequence is given by $S(S-1)/2$. This is a very high, possibly prohibitively high, number as s increases, and hence current design practice is to ensure that initialisation before test can be directly accomplished by means of specific test input signal(s).

With the circuit initialised to a known starting state the primary outputs may be checked for correctness, with appropriate primary test input vectors if necessary. If this test fails, the circuit is faulty and no further testing need be continued unless some diagnostic information is also sought. It has also been suggested that the successful completion of a homing sequence, particularly if a long sequence, is itself a good test for the complete circuit, any deviation indicating some failure in the combinational or sequential elements. However, it is difficult to determine the fault cover that a homing test provides, and additional testing from the starting point of the initialised circuit is usually required.

In spite of research and development activities[20d] it is here that viable ATPG programs are not available, use of the stuck-at model becoming difficult because of feedback complexities. Some efforts have been made to partition the sequential circuit of Figure 3.14 into an interactive cascade of one-state circuits[21], effectively spreading out the synchronous machine

---

* This is a design for test feature, one of many such considerations which we will consider in detail in Chapter 5.

linearly in time instead of going around the one circuit model on each clock pulse, but unfortunately this introduces the equally difficult problem of having to model multiple stuck-at combinational faults. From the dates of the references which we have cited it will be seen that there is very little new reported work in this area; the only realistic way of continuing from the initialisation stage is a functional approach, verifying the sequential circuit operation by consideration of its state table or state diagram or ASM (algorithmic state machine) chart, rather than by any computer modelling and simulation technique [2,43].

This functional approach in turn becomes impractical as circuit size increases to, say, 20 or more storage elements and possibly tens of thousands of used and unused circuit states. For VLSI it is now imperative to consider testing requirements at the circuit design stage, and build in appropriate means of testing large sequential circuits more easily than would otherwise be the case. This will be a major topic in Chapter 5; as will be seen partitioning, re-configuration and other techniques may be introduced, giving both a normal operating mode and a test mode for the complete circuit design.

3.4 Exhaustive, nonexhaustive and pseudorandom test pattern generation

An ATPG program that produces a set of test vectors which detects all the faults in a given fault list for a circuit has obvious advantages, since it provides a minimum length test vector sequence to test the circuit to a known standard of test. The difficulty and cost of generating this test set, which is a one-off operation at the design stage, must be set against the resulting minimum amount of data to be stored in the test system, see Figure 3.2c, and the minimum time to test each production circuit.

In general, the order of generating and applying the test vectors in a system such as in Figure 3.2c is fully flexible, the test vectors and expected (healthy) output responses being stored in ROM. However, deterministic test pattern generation based upon (usually) the stuck-at model does not generally require any specific ordering of the test vectors, each test being independent of the other tests. Unfortunately the difficulties of determining this test set for complex VLSI circuits has become too great to undertake, and therefore present test philosophies are moving away from the cost of ATP generation to design for test strategies with the use of exhaustive, nonexhaustive or pseudo-random test patterns. The cost of test pattern generation in the latter cases is now usually some relatively simple hardware circuitry, such as we shall consider below.

3.4.1 Exhaustive test pattern generators

If the circuit or system under test is sufficiently small or if it has been appropriately partitioned into blocks which can be independently tested (see

Chapter 5), then exhaustive testing applying every possible input combination is conceptually and physically the easiest means of test. A basic system such as that shown in Figure 3.15 would be appropriate. No ATPG costs are involved, the only test design activity being the determination of the healthy output responses to the test vectors.

The simplest hardware generator that can be used in this situation is a normal binary counter, which with n stages produces the usual 2^n output vectors. Other binary counter arrangements, such as a Gray cyclic binary counter, have little practical advantage over a simple binary counter, because we are aiming for the full 2^n possible output vectors.

The disadvantage of this simple exhaustive test pattern generation is that the order of application of the test vectors is fixed. Also with a natural binary sequence the most significant bit only changes from logic 0 to 1 once in the complete 2^n sequence. The questions that should be asked when considering the use of a hardware binary counter test generator are therefore:

(i) first of all, is it feasible to employ a fully exhaustive test set for the circuit (or partition of the circuit) under test, or will this be too long a test sequence?
(ii) is there any special circuit requirement which requires the test vectors to be applied in some specific order?
(iii) does it matter if the frequency of changing from logic 0 to 1 of the individual bits in the input test vector is dissimilar?
(iv) is there any special necessity to repeat short sequences of test vectors within the total test[40]?

It may well be that one or more of these special considerations rules out the use of a simple binary counter for a given circuit. However, in practice it is usually preferable to use pseudorandom test vector hardware generation, as will be considered in section 3.4.3 below, except perhaps for fairly trivial applications of MSI complexity.

3.4.2 Nonexhaustive test pattern generators

The above fully-exhaustive test set would check all possible input/output functional relationships of a combinational circuit.* This test set would undoubtedly toggle most of the internal nodes of the circuit more than once, and hence it may be acceptable to use a nonexhaustive test pattern sequence for the circuit test.

If BCD counters instead of normal binary counters are used, then the reduction in input test vectors is as shown in Table 3.7. The hardware to generate BCD coding is only marginally more complicated than a simple binary counter, and hence there is little hardware penalty in this alternative.

---

* We are ignoring in our present discussions the peculiar problems of CMOS testing, which will be considered further in Section 3.5.

Table 3.7 The number of test vectors available from binary and BCD counters

No. of input bits	No. of input test vectors	
	fully exhaustive binary	binary-coded decimal
$n = 4$	16	10 (1 decade)
$n = 8$	256	100 (2 decades)
$n = 16$	65536	10000 (4 decades)
$n = 32$	4.3×10^9	1×10^8 (8 decades)

However, there is not a very substantial saving in this alternative, and fault coverage is now unknown. If some other subset of a full binary sequence is considered then the greater the reduction in the number of test vectors the lower the potential fault coverage. A more satisfactory nonexhaustive test set strategy is that discussed in Section 3.2.1, where the circuit designer specifies from his or her knowledge of the circuit a set of vectors which will exercise the circuit with certain key or critical input/output functional requirements, or alternatively will cause all or most of the internal gates to change state. This procedure will produce a nonexhaustive set of test vectors. As covered in Section 3.2.1, the effectiveness of this suggested test set may be investigated by a computer check to determine how many of the internal nodes of the circuit are toggled by this set of vectors; if this coverage is near 100 % then the probability of passing faulty circuits when under test will be acceptably small.

The source of the test vectors for nonexhaustive tests such as above cannot be made using simple hardware in the form of binary or BCD counters. Instead we have to revert to supplying these vectors from a programmable source such as ROM. This is back to the test set arrangement illustrated in Figure 3.2c rather than the simple hardware generation of Figure 3.15.

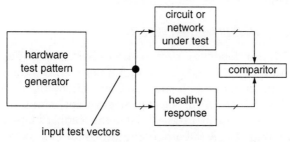

Figure 3.15   Hardware test pattern generation, similar to the general cases shown in Figure 3.2a and b. The input test sequence is usually exhaustive

3.4.3 Pseudorandom test pattern generators

By far the most common hardware generation is now one or other form of pseudorandom test pattern generator. Such generators may be standalone generators, as in Figure 3.16, or reconfigurations of flip-flops within the circuit under test such that they become in-circuit pseudorandom test generators when in test mode.

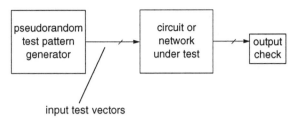

Figure 3.16 Test generation using a pseudorandom test pattern generator. The output check is often by signature analysis (see later) rather than by comparison against a healthy response

We will be considering the place of pseudorandom generators extensively in later chapters, particularly when considering signature analysis and built-in self test. At this stage we will merely consider the theory and circuit configurations involved, so that when we later use them in specific test strategies we will have the necessary knowledge of their action. It is possible to use a pseudorandom generator instead of a binary counter for simple exhaustive testing as in Figure 3.15, but there is no fundamental advantage in so doing compared with the use of a simple natural binary counter.

There are two principal forms of pseudorandom generators, namely:

(i) linear feedback shift registers, usually known as LFSRs;
(ii) one-dimensional linear cellular automata, usually known as CAs.

As will be seen, both have a basic shift register construction with exclusive-OR gates to derive their particular output sequences. LFSRs are more generally known and used than CAs, but the latter are becoming more widely appreciated and researched. We will look at both in turn.

3.4.3.1 Linear feedback shift registers (LFSRs)

The principle of linear feedback shift registers is shown in Figure 3.17. The LFSR is a normal shift register configuration, which when clocked progresses its stored pattern of 0s and 1s from left to right, but which has feedback exclusive-ORed from selected stages (taps) along the register to form the serial input to the first stage.* When the feedback taps are appropriately chosen a n-stage LFSR will autonomously count in a pseudorandom manner through $2^n - 1$ states before repeating its sequence; this is known as a maximum length pseudorandom sequence, or M-sequence, and contains all possible states of the register except for one forbidden state.

---

* The terminology linear is because the exclusive logic relationship realises the mod_2 addition of binary values, which is a linear relationship that preserves the principle of superposition. For example, $(0\ 0\ 1\ 1 \oplus 1\ 0\ 0\ 1) \oplus 0\ 0\ 1\ 0 = 1\ 0\ 0\ 0$; $1\ 0\ 0\ 0 \oplus (0\ 0\ 1\ 1 \oplus 1\ 0\ 0\ 1) = 0\ 0\ 1\ 0$, and so on. No information is lost going through an exclusive-OR (or exclusive-NOR) gate.

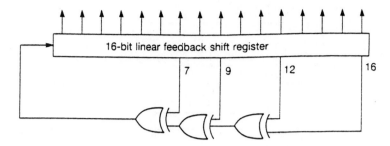

Figure 3.17 The linear feedback shift register consisting of n D-type flip-flops in cascade, with feedback arranged to generate a maximum length pseudo-random sequence when clocked. Sixteen stages are indicated here, which would give a maximum length sequence of $2^{16} - 1 = 65{,}535$ states before the sequence repeats. Alternative feedback connections to those shown here are also possible, see Appendix A

Figure 3.18 gives a small example of a maximum length pseudorandom sequence; for convenience only four stages are involved, giving a maximum length sequence of 15. It will be noted that the all-zeros state of the counter is absent, this being the forbidden state with this particular configuration as may readily be seen by noting that when all 0s are present in the register then the feedback signal to the first stage is also 0. It therefore requires an initialisation signal, a seed signal of ...1..., to initiate the pseudorandom sequence. Many texts imply that the forbidden output of an M-sequence must always be 0 0 0 ... 0, but this is only so if the LFSR outputs and the feedback taps are taken from output Q and not $\bar{Q}$ of each flip-flop, that all shift register interconnections are logically Q to D between consecutive stages, and that exclusive-OR and not exclusive-NOR relationships are present in the feedback circuits. For example, in Figure 3.18 if the second output had been taken from $\bar{Q}_2$ instead of Q_2, leaving the rest of the circuit unchanged, the forbidden vector output would clearly then have been 0 1 0 0 instead of 0 0 0 0. However, 0 0 0 ... 0 0 is the normally encountered forbidden state, and we will have no occasion to consider otherwise in the following discussions.

If we analyse the vectors produced by an autonomous maximum length LFSR generator, we can list the following properties:

(i) Starting from any nonforbidden state the LFSR generates $2^n - 1$ vectors before the sequence is repeated—this is obvious because the period of the M-sequence is $2^n - 1$.
(ii) In the complete M-sequence the number of 1s in any bit differs from the number of 0s by one, the 1s appearing 2^{n-1} times and the 0s $2^{n-1} - 1$ times—again, this is obvious from the period of the sequence.

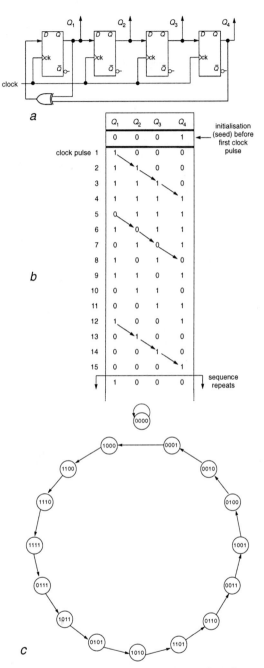

Figure 3.18 An example autonomous maximum length pseudorandom sequence generator using four D-type flip-flops. (Alternative taps are possible)
a circuit
b state table
c state diagram to emphasise the difference between the pseudo-random sequence and a simple binary sequence

Considering any one of the n bits and visualising its value written in a continuous circle of $2^n - 1$ points, we may consider the runs of consecutive 0s and 1 in the sequence, the length of a run being the number of 0s or 1s in a like-valued group. Then:

(iii) In the complete M-sequence there will be a total of 2^{n-1} runs in each bit; one half of the runs will have length 1, one quarter will have length 2, one eighth will have length 3, and so on as long as the fractions 1/2, 1/4, 1/8, ... are integer numbers, plus one additional run of n 1s. In the example of Figure 3.18 there are $2^3 = 8$ runs, four of length 1, two of length 2, one of length 3 plus one run of four 1s in each bit.

(iv) From (iii), it follows that the number of transitions between 0 and 1 and vice versa of each bit in a complete M-sequence is 2^{n-1}.

(v) Every M-sequence has a cyclic shift and add property such that if the given sequence is term-by-term added mod_2 to a shifted copy of itself, then a maximum length sequence results which is another shift of the given sequence. For example, if the M-sequence shown in Figure 3.18b is added mod_2 to the same sequence shifted up six places, it can easily be shown that this results in the sequence which is a cyclic shift of eight states from the given sequence.

(vi) Finally, if the autocorrelation of the M-sequence of 0s and 1s in each bit is considered, that is knowing a particular entry has the value 0 (or 1) how likely is any other entry in the same sequence to be 0 (or 1), we have the autocorrelation function:

$$C(\tau) = \frac{1}{p}\left\{\sum_{1}^{p}(a_\tau)\right\} \quad (3.15)$$

where τ is the shift between the entries in the same sequence being compared, $1 \leq \tau \leq 2^n - 2$, i.e. when $\tau = 1$ adjacent entries in the sequence are being compared:

$p = 2^n - 1$

and $a_\tau = 1$ if the two entries being compared have the same value 0 or 1, $= -1$ if the two entries being compared have differing values.*

The value of $C(\tau)$ for any M-sequence and any value of τ is:

$$C(\tau) = \frac{-1}{p} \quad (3.16)$$

This may be illustrated by taking the Q_1 sequence in Figure 3.18b and considering a shift of, say, three positions. This gives the results tabulated in

---

* We may express $C(\tau)$ more explicitly than above using logic values of +1 and −1 instead of 0 and 1. This will be introduced in Chapter 4, but we have no need to do so at the present.

Table 3.8, with the total agreements and disagreements being 7 and 8 respectively, giving the autocorrelation value of $-1/15$.

It will also be appreciated that the shift register action between stages of a LFSR means that all the n bits in the sequence have exactly the same properties.

Table 3.8 The agreements/disagreements between an M-sequence and a shifted copy of itself, giving the autocorrelation value $C(3) = (7 - 8)/p = -1/15$

Given M-sequence	Shifted 3 positions ($\tau = 3$)	Agreements (A) or disagreements (D)
1	1	A
1	0	D
1	1	A
1	0	D
0	1	D
1	1	A
0	0	A
1	0	D
1	1	A
0	0	A
0	0	A
1	0	D
0	1	D
0	1	D
0	1	D
		Total: A = 7, D = 8

The maximum length of pseudorandom sequence is therefore seen to have many interesting properties. The autocorrelation value has no specific application in digital testing, but it is an indicator of the randomness of the sequence. This is not a truly random sequence, however, since what the sequence will be with a given circuit configuration is fully deterministic, and it has a specific periodicity, but as we have seen it is more random in nature than any normal binary or BCD counter. The designation pseudorandom therefore is justified, and its ease of generation leads to its use in many applications.

However, we have not yet considered how the taps on a shift register are chosen so as to generate a maximum length sequence. Are only two taps as in Figure 3.18 always sufficient, and where should they be connected? Is there more than one possible choice for any n-stage LFSR? It may easily be shown that a random choice of taps on a shift register will not always generate a maximum length sequence—if we take a three-stage shift register and exclusive-OR all three flip-flop outputs back to the serial input, four distinct sequences can result, namely:

- stuck, circulating all 0s;
- stuck, circulating all 1s;
- a sequence of 0 1 0, 1 0 1, repeat...;
- a sequence of 0 0 1, 1 0 0, 1 1 0, 0 1 1, repeat....

Clearly, a theory to determine the appropriate taps to give an M-sequence is required.

The initial point to re-emphasise is that in the LFSR the serial input and hence the next state of the register is fully deterministic, being a function of the present state and of the exclusive-OR connections which are present. If at time t_i we know the 0 or 1 serial input signal, y_i, to the first flip-flop just before the shift register is clocked, then at this time the output from the first register will be the previous serial input y_{i-1}, the output from the second flip-flop will be y_{i-2} and so on. This is shown in the general schematic of Figure 3.19.

The value of the input bit, y_i, at any given time, t_i, is therefore the mod_2 sum of the feedback taps, which may be expressed by the recurrence relationship:

$$y_i = \sum_{j=1}^{\infty} c_j y_{i-j} \tag{3.17}$$

where $y_i, c_j \in \{0, 1\}$.

Now from the theory originally developed for cyclic error detecting codes[44,45], any binary sequence of 0s and 1s may be expressed by a polynomial:

$$G(x) = a_0 x^0 + a_1 x^1 + a_2 x^2 + \ldots + a_m x^m + \ldots$$

$$= \sum_{m=0}^{\infty} a_m x^m \tag{3.18}$$

where $x^0, x^1, x^2 \ldots$ represent the positions in the bit sequence with increasing time and a_i are the binary coefficients, $a_i \in \{0, 1\}$. For example, an output sequence ... 1 1 1 0 0 1 with the first received bit on the right is given by:

$$G(x) = 1x^0 + 0x^1 + 0x^2 + 1x^3 + 1x^4 + 1x^5 + \ldots$$

$$= x^0 + x^3 + x^4 + x^5 + \ldots$$

The algebraic manipulations of $G(x)$ are all in the Galois field of $GF(2)$, that is mod_2 addition, subtraction, multiplication and division of binary data. Recall also that mod_2 addition and subtraction are identical, being:

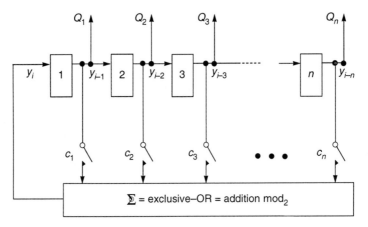

Figure 3.19 The general schematic of a LFSR with n stages, the taps $c_1, c_2, c_3, ..., c_n$ being open ($c_i = 0$) if no connection is present. y_i is the input signal at particular time t_i

$0 + 0 = 0$

$0 + 1 = 1$

$1 + 0 = 1$

$1 + 1 = 0$

$0 - 1 = 1$

$1 - 0 = 1$

$1 - 1 = 0$

and hence polynomial multiplication and division follow as illustrated below;*

$(x^3 + x^2 + x + 1) \times (x^2 + x + 1)$ is given by:

$$x^3 + x^2 + x + 1$$
$$\underline{x^2 + x + 1}$$
$$x^3 + x^2 + x + 1$$
$$x^4 + x^3 + x^2 + x$$
$$\underline{x^5 + x^4 + x^3 + x^2}$$
$$x^5 + 0\ + x^3 + x^2 + 0 + 1$$

---

* Note, as we are doing all the algebra here in GF(2), we use the conventional algebra addition sign + rather than the logical exclusive-OR symbol ⊕. The latter symbol may be found in some publications in this area, but not usually. Also, we will discontinue the circle in the symbol Σ used in eqn. 3.17 from here on.

since $x^i + x^i = 0$. Similarly, $(x^5 + x^3 + x^2 + 1) \div (x^2 + x + 1)$ is given by:

$$
\begin{array}{r}
x^3 + x^2 + x + 1 \\
x^2 + x^1 + 1 \overline{\smash{\big)} x^5 + 0 + x^3 + x^2 + 0 + 1} \\
\underline{x^5 + x^4 + x^3} \\
x^4 + 0 + x^2 \\
\underline{x^4 + x^3 + x^2} \\
x^3 + 0 + 0 \\
\underline{x^3 + x^2 + x} \\
x^2 + x + 1 \\
\underline{x^2 + x + 1} \\
0
\end{array}
$$

since $x^i - x^i = 0$. There is no remainder in this particular polynomial division, and therefore $x^5 + x^3 + x^2 + 1$ is said to be divisible by $x^2 + x + 1$. Division of say $x^5 + x^4 + x^2 + 1$ by $x^3 + x + 1$ will be found to leave a remainder of x^2.

Reverting to the autonomous LFSR circuit, each of the terms in the polynomial expansion of eqn. 3.18 represents the value of the input bit y_i, given by eqn. 3.17 at increasing time intervals (clock pulses). The first term $a_0 x^0$ in eqn. 3.18 is the present value of y_i, the next term $a_1 x^1$ is the next value of y_i and so on. The overall relationship is therefore as follows:

feedback taps

$\{0,1\}$ sequence $G(x)$ with increasing time $t_0, t_1, \ldots, t_m, \ldots$

$$a_0 x^0 = y_0 = c_1 y_{-1} + c_2 y_{-2} + c_3 y_{-3} + \ldots + c_n y_{-n}$$
$$a_1 x^1 = y_1 = c_1 y_0 + c_2 y_{-1} + c_3 y_{-2} + \ldots + c_n y_{1-n}$$
$$a_2 x^2 = y_2 = c_1 y_1 + c_2 y_0 + c_3 y_{-1} + \ldots + c_n y_{2-n}$$
$$a_3 x^3 = y_3 = c_1 y_2 + c_2 y_1 + c_3 y_0 + \ldots + c_n y_{3-n}$$
$$\vdots$$
$$a_m x^m = y_m = c_1 y_{m-1} + c_2 y_{m-2} + c_3 y_{m-3} + \ldots c_n y_{m-n}$$

Hence the input signal, y_i, of eqn. 3.17 may now be replaced by y_m, where m denotes the increasing time of the sequence defined by eqn. 3.18.*

We may therefore rewrite eqn. 3.17 as:

$$y_m = \sum_{j=1}^{n} c_j y_{m-j} \qquad (3.17a)$$

---

* Notice that the relationships shown above are in a matrix-like form. It is possible to use matrix operations for this subject area, see Yarmolik[21] for example, but we will not do so in this text.

Substituting for a_m in eqn. 3.18 now gives us:

$$G(x) = \sum_{m=0}^{\infty} \left\{ \sum_{j=1}^{n} c_j y_{m-j} \right\} x^m \qquad (3.19)$$

This may be rearranged as follows:

$$G(x) = \sum_{j=1}^{n} c_j x^j \left\{ \sum_{m=0}^{\infty} y_{m-j} x^{m-j} \right\}$$

$$= \sum_{j=1}^{n} c_j x^j \left\{ y_{-j} x^{-j} + \ldots + y_{-1} x^{-1} + \sum_{m=0}^{\infty} a_m x^m \right\}$$

$$= \sum_{j=1}^{n} c_j x^j \left\{ y_{-j} x^{-j} + \ldots + y_{-1} x^{-1} + G(x) \right\}$$

This has rearranged the terms in the brackets { } into negative powers of x (= past time), and positive powers of x (= present and future time), and has eliminated the summation to infinity. Collecting together terms:

$$G(x) \left\{ 1 - \sum_{j=1}^{n} c_j x^j \right\} = \sum_{j=1}^{n} c_j x^j \left\{ y_{-j} x^{-j} + \ldots + y_{-1} x^{-1} \right\}$$

$$G(x) = \frac{\sum_{j=1}^{n} c_j x^j \left\{ y_{-j} x^{-j} + \ldots + y_{-1} x^{-1} \right\}}{1 - \sum_{j=1}^{n} c_j x^j} \qquad (3.20)$$

Because addition and subtraction are the same in GF(2), we may replace the minus sign in the denominator of eqn. 3.20 with plus, giving the denominator:

$$1 + \sum_{j=1}^{n} c_j x^j \qquad (3.21)$$

$$= 1 + c_1 x + c_2 x^2 + \ldots + c_n x^n \qquad (3.21a)$$

Hence the sequence $G(x)$ is a function of:

(i) the initial conditions $y_{-1}, y_{-2}, \ldots, y_{-n}$ of the LFSR given by the numerator in eqn. 3.20;
(ii) the feedback coefficients $c_1, c_2, \ldots, c_n$ in the denominator.

Notice that the denominator is independent of the initial conditions and only involves the taps in the LFSR circuit, see Figure 3.19, c_j being 0 with the feedback connection open (not present) and 1 with feedback connection closed (present). For the four-stage LFSR shown in Figure 3.18 the denominator of $G(x)$ would therefore be:

$$1 + x^1 + 0 + 0 + x^4$$
$$= 1 + x^1 + x^4$$

This denominator which controls the sequence which the circuit generates from a given initial condition is known as the characteristic polynomial $P(x)$ for the sequence; the powers of x in the characteristic polynomial are the same as the stages in the shift register to which the feedback taps are connected. Two further points may be noted, namely:

(i) if the initial conditions $y_{-1}, y_{-2}, ..., y_{-n}$ were all zero, then the numerator of $G(x)$ would be zero, and the sequence would be 0 0 0 ... irrespective of the characteristic polynomial;
(ii) if the initial conditions were all zero except y_{-n} which was 1, then the numerator would become c_n, which if $c_n = 1$ gives:

$$G(x) = \frac{1}{P(x)} = \frac{1}{1 + c_1 x + c_2 x^2 + ... + c_n x^n} \tag{3.22}$$

Condition (i) above is a previously noted condition from a consideration of the working of the LFSR. Condition (ii) is the example initialisation (seed) signal used in Figure 3.18b.

It is straightforward to perform the division in $G(x)$ by hand, but in practice it is rarely necessary to do so except as an academic exercise. The simplest case to consider is eqn. 3.22. Taking the LFSR of Figure 3.18 with its characteristic polynomial $1 + x^1 + x^4$, we have the never-ending division as follows:

$$
\require{enclose}
\begin{array}{r}
1 + x^1 + x^2 + x^3 + x^5 + x^7 + ... \\
1 + x^1 + x^4 \enclose{longdiv}{1 } \\
\underline{1 + x^1 + 0 + 0 + x^4} \\
x^1 + 0 + 0 + x^4 \\
\underline{x^1 + x^2 + 0 + 0 + x^5} \\
x^2 + 0 + x^4 + x^5 \\
\underline{x^2 + x^3 + 0 + 0 + x^6} \\
x^3 + x^4 + x^5 + x^6 \\
\underline{x^3 + x^4 + 0 + 0 + x^7} \\
x^5 + x^6 + x^7 \\
\underline{x^5 + x^6 + 0 + 0 + x^9} \\
x^7 + 0 + x^9 \\
\end{array}
$$

•
•
•

The generated sequence is therefore:

$$1\;1\;1\;1\;0\;1\;0\;1\ldots$$

which if continued will be found to be the sequence detailed in Figure 3.18b. This pattern will start to repeat at x^{15} (the 16th bit) in the quotient, giving a maximum length sequence of $2^n - 1$. If we had begun with a different set of conditions for y_j then the same sequence would have been found but displaced (cyclically shifted) from the above calculation. This GF(2) division could of course equally be done by substituting 1 for each nonzero x^i, namely:

$$\begin{array}{r} 1\;1\;1\;1\;0\;1\;0\;1\ldots \\ 1\;1\;0\;0\;1\overline{)1\;0\;0\;0\;0\;0\;0\;0\ldots} \end{array}$$

See also and compare with the further development following, which involves a similar GF(2) division with the same characteristic polynomial.

It has been shown that for the pseudorandom sequence to be the maximum length, the characteristic polynomial $P(x)$ given by eqn. 3.21 must be a primitive polynomial, that is it cannot be factorised into two (or more) parts. For example $(1 + x^1 + x^3 + x^5)$ may be factorised into $(1 + x^1)(1 + x^3 + x^4)$, remembering that we are operating in GF(2), but the polynomial $(1 + x^2 + x^5)$ cannot be factorised. The latter but not the former, therefore, is a primitive polynomial. Additionally, the primitive polynomial must divide without remainder into the polynomial $1 + x^{2^n-1}$, or putting it the other way around the primitive polynomial must be a factor of $1 + x^{2^n-1}$, but not of $1 + x^{2^t-1}$ where $t < n$. For example, given the primitive polynomial $1 + x^1 + x^3$, which we may write as 1 1 0 1, we have:

$$\begin{array}{r}
1\;1\;1\;0\;1 \\
1\;1\;0\;1\overline{)1\;0\;0\;0\;0\;0\;0\;1} \\
\underline{1\;1\;0\;1} \\
1\;0\;1\;0 \\
\underline{1\;1\;0\;1} \\
1\;1\;1\;0 \\
\underline{1\;1\;0\;1} \\
1\;1\;0\;1 \\
\underline{1\;1\;0\;1} \\
0
\end{array}$$

This may be further illustrated by evaluating the previous LFSR example of Figure 3.18, dividing the primitive polynomial $1 + x^1 + x^4$ (1 1 0 0 1) into the polynomial $1 + x^{15}$. The result of this division is the same as in the previous worked example on page 85, except that we now have a 1 rather than a 0 in the 16th position of the numerator, which causes the division to terminate rather than continue onwards.

All the primitive polynomials for any n are therefore prime factors of the polynomial $1 + x^{2^n-1}$. Fortunately we do not have to calculate the primitive polynomials for our own use, since they have been extensively calculated and published. Appendix A at the end of this text gives the minimum primitive polynomials for $n \leq 100$, together with further comments and references to other published tabulations. The theory and developments of these polynomial relationships may be found in MacWilliams and Sloane[45], Brillhart et al.[46], Bardell et al.[47] and elsewhere, but some further comments may be appropriate to include here.

First, as n increases the number of possible primitive polynomials increases rapidly. The formula for this number may be found in the developments by Golomb[48] and listed in Bardell et al.[47], being for example 16 possibilities for $n = 8$, 2048 possibilities for $n = 16$ and 276 480 possibilities for $n = 24$. Not all are minimum, that is containing the fewest number of nonzero terms, but even so there are alternative possibilities with the fewest number of terms for $n \geq 3$. The listings given in Appendix A therefore are not the only possibilities. A complete listing of all the possible primitive polynomials for up to $n = 16$ is given in Peterson and Weldon[44].

Secondly, given any minimum primitive polynomial $P(x)$ such as listed in Appendix A, there is always the possibility of determining its reciprocal polynomial $P*(x)$, which is also a minimum primitive polynomial yielding a maximum length sequence[11,47,49]. The reciprocal of the polynomial is defined by:

$$P*(x) = x^n P\left(\frac{1}{x}\right) \tag{3.23}$$

that is given, say,

$$P(x) = 1 + x^1 + x^4$$

which is the generating polynomial in Figure 3.18a, the reciprocal polynomial is:

$$P*(x) = x^4 P\left(1 + \frac{1}{x} + \frac{1}{x^4}\right)$$
$$= x^4 + x^3 + 1$$
$$= 1 + x^3 + x^4$$

This polynomial yields the maximum length sequence of

1 0 0 1 1 0 1 0 1 1 1 1 0 0 0

which may be confirmed by evaluating:

$$G(x) = \frac{1}{1 + x^3 + x^4}$$

and which will be seen to be the reverse of the sequence generated by $P(x) = 1 + x^1 + x^4$. It is the property of every characteristic polynomial which generates a maximum length sequence that its reciprocal will also generate a maximum length sequence but in the reverse order with a cyclic shift. Hence, there are always at least two characteristic polynomials with the fewest number of terms for $n = 3$ upwards. (For $n = 2$, $P(x) = 1 + x^1 + x^2 = P*(x)$.) For $n = 15$ there are six possible polynomials (including the reciprocals) which have the minimum number of terms[50], as listed in Table 3.9.

Table 3.9 *The possible primitive polynomials for $n = 15$ with the least number of nonzero terms (trinomials)*

$P(x) = 1 + x^1 + x^{15}$	$P_1*(x) = 1 + x^{14} + x^{15}$
$P_2(x) = 1 + x^4 + x^{15}$	$P_2*(x) = 1 + x^{11} + x^{15}$
$P_3(x) = 1 + x^7 + x^{15}$	$P_3*(x) = 1 + x^8 + x^{15}$

Finally, the LFSR circuit configuration that we have considered so far has the feedback taps from the chosen LFSR stages all exclusive–ORed back to the first stage. However, for any given circuit and characteristic polynomial, an alternative circuit configuration with the same characteristic polynomial $P(x)$ and the same output sequence $G(x)$, see eqn. 3.22, is possible by including the exclusive–OR logic gates between appropriate LFSR stages. This is illustrated in Figure 3.20. For each nonzero c_i in Figure 3.20a there is an effective exclusive–OR gate between stages $n - i$ and $n - i + 1$ in Figure 3.20b; c_n is always nonzero, and therefore there is always a connection between Q_n and the first stage as shown. For example, the equivalent of the LFSR circuit shown in Figure 3.18 with the characteristic polynomial $P(x) 1 + x^1 + x^4$ would have one exclusive–OR gate between the third and final stages as shown in Figure 3.20c.

Notice that the same number of two-input exclusive–OR gates is necessary in both possible circuit configurations, which we have termed type A and type B in Figure 3.20. However, although the characteristic polynomial and hence the output sequence of both type A and type B can be the same, the precise n-bit data held in the n stages of the type A and type B LFSRs after each clock pulse will not always be exactly the same. It is left as an exercise for the reader to compile the state table for the type B LFSR with the characteristic polynomial $1 + x^1 + x^4$, and compare it with the state table given in Figure 3.18b.

In general the type A LFSR of Figure 3.20 is preferable to the type B from the manufacturing point of view, and most practical circuits show this configuration. However, we will briefly come back to the type B in Section 4.5 of the following chapter, since it has a certain mathematical advantage when the n-bit data in the LFSR rather than the $2^n - 1$ pseudorandom output sequence is of interest.

Further alternative circuit configurations have been investigated, particularly the hybrid Wang-McCluskey circuits which seek to minimise the number of exclusive-OR gates by a combination of the two circuit concepts

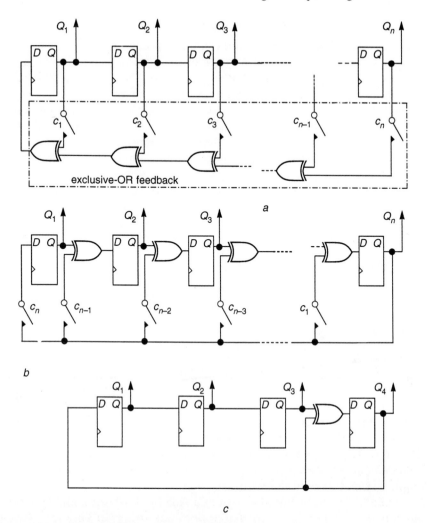

Figure 3.20 Two alternative circuit configurations for the same maximum length pseudorandom sequence generation
    a type A LFSR, which is the circuit so far considered
    b alternative type B circuit configuration
    c type B realisation of the maximum length LFSR of Figure 3.18 with the characteristic polynomial $P(x) = 1 + x^1 + x^4$. Note, other publications may refer to these two configurations as type 1 and type 2 LFSRs, but regretably there is a lack of consistancy whether *a* is type 1 and *b* is type 2, or vice versa

shown in Figure 3.20[51]. We will not pursue these and other alternatives such as partitioning a LFSR into smaller autonomous LFSRs, particularly as the cellular automata (CA) pseudorandom generators which we will introduce below have theoretical advantages over LFSR generators. Further reading may be found in References 11, 13, 44, 47 and 50.

3.4.3.2 Cellular automata (CAs)

Turning from LFSRs to autonomous cellular automata, again the basic circuit configuration is a shift register but now the connection between stages always involves some linear (exclusive–OR) relationship. The regular shift of 0s and 1s along the register, as illustrated in Figure 3.18b therefore is not present.

The CA circuits which have been developed to produce a maximum length pseudorandom sequence of $2^n - 1$ states from an n-stage register have the D input of each stage controlled by a linear relationship of the Q outputs of adjacent (near-neighbour) states. For example, input D_k of stage Q_k may involve the outputs Q_{k-1}, Q_k and Q_{k+1}, but no other global connection(s). The two functions $f(D_k)$ which are found to be relevant are as follows:

Q_{k-1}	Q_k	Q_{k+1}	$f_1(D_k) = Q_{k-1} \oplus Q_{k+1}$	$f_2(D_k) = Q_{k-1} \oplus Q_k \oplus Q_{k+1}$
0	0	0	0	0
0	0	1	1	1
0	1	0	0	1
0	1	1	1	0
1	0	0	1	1
1	0	1	0	0
1	1	0	1	0
1	1	1	0	1

Reading these two exclusive-OR functions as eight-bit binary numbers with the top entry as the least significant digit, the function $Q_{k-1} \oplus Q_{k+1}$ is termed function 90, and the function $Q_{k-1} \oplus Q_k \oplus Q_{k+1}$ is termed function 150, a terminology first introduced by Wolfram[52]. This gives us the two basic circuit blocks (cells) shown in Figure 3.21 from which autonomous maximum length n-stage CAs may be constructed.*

Originally, research work, largely in Canada by Hortensius, McLeod, Card and others[53–56], demonstrated that appropriate strings of 90 and 150 cells can produce maximum length pseudorandom sequences. The initial work of the early 1980s involved extensive computer simulations of different strings of 90 and 150 cells to determine the resulting sequences; it was found that 90 or 150 cells alone cannot produce an M-sequence. For small n a string of 90, 150, 90, 150, ..., 90, 150 was found to be successful, but in this early work no theory was yet developed to determine the arrangements of 90/150 circuits for maximum length generation. Further research has solved this problem (see later), and has also considered using the minimum number of

---

* Other functions of Q_{k-1}, Q_k and Q_{k+1} have been investigated[57,58], but it has now been formally proved that only the 90 and 150 functions are appropriate to generate maximum length sequences. We will, therefore, only consider the 90 and 150 functions in this text.

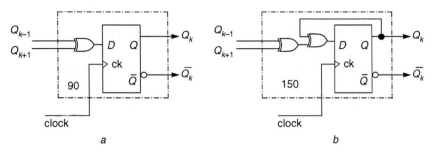

Figure 3.21 Two basic cells from which maximum length pseudorandom CA generators may be built
a type 90 cell
b type 150 cell

150 cells, the more expensive cell of the two, and maximising the use of the 90 cell. It has been shown[59] that for $n \leq 150$ at most two 150 cells are required, the remainder all being 90 cells. The list for $n \leq 150$ is given in Appendix B.

The circuit and resulting M-sequence for a four-stage autonomous CA generator with alternating 90 and 150 cells is given in Figure 3.22. As will be seen, the resulting maximum length sequence does not have the simple shift characteristic of the LFSR generator, but instead has a much more random-like relationship between the successive output vectors $Q_1 Q_2 Q_3 Q_4$. A forbidden state of 0 0 0 0 is present as in an autonomous LFSR generator, necessitating a seed of ...1... to be present to allow the M-sequence to proceed. The circuit for $n = 4$ using the data tabulated in Appendix B would be similar to Figure 3.22 but with the 90 and 150 cells interchanged—there is always a choice of circuit configurations for $n \geq 2$.

The analysis of the sequence generated by a given CA may easily be found by a matrix computation, where all multiplications and additions are in GF(2). For example, consider the string of 90 and 150 cells shown in Figure 3.23; then the state transition matrix [**T**] is given by:

$$\begin{bmatrix} 0 & 1 & 0 & 0 & 0 & 0 & 0 & 0 \\ 1 & 1 & 1 & 0 & 0 & 0 & 0 & 0 \\ 0 & 1 & 0 & 1 & 0 & 0 & 0 & 0 \\ 0 & 0 & 1 & 1 & 1 & 0 & 0 & 0 \\ 0 & 0 & 0 & 1 & 0 & 1 & 0 & 0 \\ 0 & 0 & 0 & 0 & 1 & 1 & 1 & 0 \\ 0 & 0 & 0 & 0 & 0 & 1 & 0 & 1 \\ 0 & 0 & 0 & 0 & 0 & 0 & 1 & 1 \end{bmatrix}$$

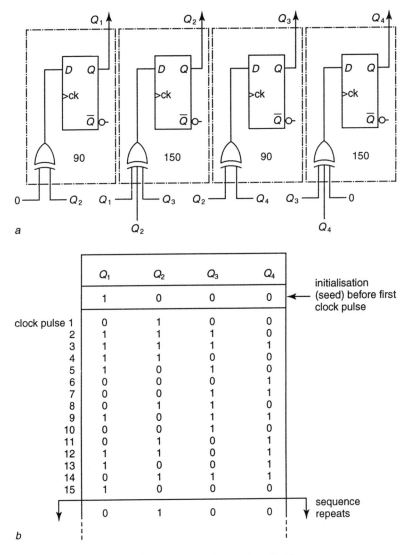

Figure 3.22 An autonomous four-stage maximum length CA generator
　　　　　　 a circuit
　　　　　　 b the resulting sequence, cf. Figure 3.18b

This is a tridiagonal matrix, that is all zeros except on three diagonals, the diagonal row entries being 1 1 1 for internal 150 cells and 1 0 1 for internal 90 cells. For any present-state output vector $Q]$, the next-state output vector $Q^+]$ of the CA is given by:

$$[T]Q] = Q^+] \tag{3.24}$$

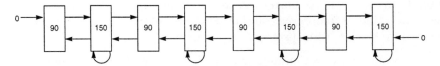

Figure 3.23 Schematic of a string of 90/150 CA cells

For example, taking the eighth vector in Figure 3.22b, namely 0 1 1 0, the next-state vector for this CA is given by:

$$\begin{bmatrix} 0 & 1 & 0 & 0 \\ 1 & 1 & 1 & 0 \\ 0 & 1 & 0 & 1 \\ 0 & 0 & 1 & 1 \end{bmatrix} \begin{bmatrix} 0 \\ 1 \\ 1 \\ 0 \end{bmatrix} = \begin{bmatrix} 0+1+0+0 \\ 0+1+1+0 \\ 0+1+0+0 \\ 0+0+1+0 \end{bmatrix} = \begin{bmatrix} 1 \\ 0 \\ 1 \\ 1 \end{bmatrix}$$

$\quad\quad$ [T] $\quad\quad\quad$ Q] $\quad\quad\quad\quad\quad\quad$ Q$^+$]

The following vector may be obtained by the transformation of 1 0 1 1, or by transforming the previous vector 0 1 1 0 by $[T]^2$, that is:

$$\begin{bmatrix} 0 & 1 & 0 & 0 \\ 1 & 1 & 1 & 0 \\ 0 & 1 & 0 & 1 \\ 0 & 0 & 1 & 1 \end{bmatrix} \begin{bmatrix} 0 & 1 & 0 & 0 \\ 1 & 1 & 1 & 0 \\ 0 & 1 & 0 & 1 \\ 0 & 0 & 1 & 1 \end{bmatrix} = \begin{bmatrix} 1 & 1 & 1 & 0 \\ 1 & 1 & 1 & 1 \\ 1 & 1 & 0 & 1 \\ 0 & 1 & 1 & 0 \end{bmatrix}$$

$\quad\quad$ [T] $\quad\quad\quad\quad$ [T] $\quad\quad\quad\quad$ $[T]^2$

In general, the kth vector Q^{+k}] after any given vector may be obtained by:

$$[T]^k Q] = Q^{+k}] \qquad (3.25)$$

where [T] is the given state transition matrix.

There is, therefore, relatively little difficulty in the analysis of a given cellular automaton.* The theory for the synthesis of a maximum length CA, however, is much more difficult; unlike the LFSR where polynomial division over GF(2) is involved, for the CA no polynomial operations are directly applicable and generally unfamiliar matrix operations are deeply involved.

It has been shown by Serra et al.[61–63] that autonomous LFSRs and CAs are isomorphic to each other, so that given any maximum length LFSR a corresponding maximum length CA may be determined; more precisely, given any characteristic polynomial for a LFSR a corresponding CA may be

---

* We could of course have performed a similar matrix operation to determine the next state of a LSFR; the state transition matrix of Figure 3.20a would be a tridiagonal of 1 0 0 except for the first row which must represent the generating polynomial[13]. However, the polynomial division that we have covered is generally quicker and gives the full sequence of each LFSR output.

found. In this context corresponding does not imply the same output sequence, but a reordered output sequence of the same length. The procedure involves three principal steps namely:

(i) compile the state transition matrix of the chosen LFSR generator;
(ii) determine the companion matrix[64] of this state transition matrix using similarity transformations, the former being a matrix which is isomorphic to the transition matrix;
(iii) tridiagonalise the companion matrix into a tridiagonal form, this being undertaken by a development of the Lanczos tridiagonalisation algorithm[64].

The last step generates the state transition matrix for the corresponding cellular automaton, which as we have seen must be in tridiagonal form because all interconnections involve only Q_{k-1}, Q_k and Q_{k+1} signals. Further details may be found in Serra *et al.*, particularly in References 61–63. Note, the type 1 and type 2 LFSRs in Serra is what we have termed type B and type A LFSRs, respectively.

However, this method of developing a maximum length autonomous pseudorandom CA generator is found to produce far from minimum CA assemblies, that is with the fewest number of the more expensive 150 cells, even when developed from minimum length primitive polynomials. No relationship exists between minimal LFSRs and minimal CAs, and hence the search for minimal cost CA realisations has been undertaken by a search procedure based upon the tridiagonal characteristics of the CA state transition matrix.

Looking back at the example state transition matrix of Figure 3.23 it will be seen that the main diagonal has the value 0 for a type 90 cell, and 1 for a type 150 cell; the two adjacent side diagonals are always all 1s. Also, if the transition matrix for any n produces the maximum length sequence of $2^n - 1$ states, then $[\mathbf{T}]^k$ will yield the identity matrix $[\mathbf{I}]$ for $k = 2^n$, but will not yield $[\mathbf{I}]$ for any $k < 2^n$. This is so because the maximum length sequence repeats after $2^n - 1$ different states to its starting state. The search procedure to identify the smallest number of type 150 cells for a maximum length sequence is therefore to set all the main diagonal entries initially to zero, and then progressive insert one 1, two 1s, ... in the main diagonal. The search is stopped for any given n when $[\mathbf{T}]^{2^n} = [\mathbf{I}]$. Further details may be found in Reference 59. This procedure has, as previously noted, shown that only two type 150 cells are necessary for $n \leq 150$, see details in Appendix B.

Looking at the autonomous LFSR circuits and the CA circuits which we have now examined, it will be appreciated that either could be used as a standalone hardware generator for supplying pseudorandom test vectors[50,57]. The total circuit requirements of a CA generator are more complex than that of a LFSR generator, since more exclusive–OR gates are required, but no interconnections running the full length of the shift register are ever required in the CA case as is always necessary in a LFSR generator. This may be advantageous in an integrated circuit layout. Also the test vector sequence produced by a CA generator has been shown to be more random in nature than that produced by a LFSR, having preferable testing properties. Indeed, as will be seen in the following section covering CMOS testing, the pseudorandom sequence produced by a LFSR generator fails to provide a test for certain CMOS conditions, whereas the CA generator will provide such a test.

However, the main present and future use of both LFSR and CA generators is not generally as standalone autonomous sources of pseudorandom test vectors. Instead both are intimately involved in built-in test methods for digital circuits. In particular the LFSR circuit will be encountered in the built-in logic block observation (BILBO) methodology for the self test of VLSI circuits, and the CA circuit will be found in the corresponding cellular automata logic block observation (CALBO) test methodology (see Chapter 5). The LFSR circuit will also be encountered when considering signature analysis in the following chapter. Having, therefore, now covered both means of pseudorandom generation, we will be able to appreciate their action when considering their utilisation in later pages.

3.5 I_{DDQ} and CMOS testing

The problem of CMOS testing with its dual p-channel and n-channel FET configurations was introduced in the preceding chapter. Functional testing and I_{DDQ} current tests were seen to be the most appropriate rather than test pattern generation based upon, say, the stuck-at fault model.

To cover open-circuit and short-circuit faults in a CMOS circuit we have seen that:

(i) for any open-circuit fault it is necessary to apply a specific pair of test vectors, the first being an initialisation vector to establish a logic 0(1) at the gate output, the second being the test vector which checks that the output will then switch to 1 (0);
(ii) for any short-circuit fault then on some test vectors there will be a conducting path from V_{DD} to ground through the gate, which may be detected by monitoring the supply current I_{DD} under quiescent (non-transitory) conductions, this being the I_{DDQ} measurement.

The latter feature is illustrated by Figure 3.24.

Let us consider first the functional tests for open-circuit faults. Table 2.3 in Chapter 2 illustrated the exhaustive test set for a simple two-input CMOS NAND gate; the pair of input test vectors 0 1, 1 1 checked for transistor T3 or T4 open circuit, the pair 1 1, 1 0 checked for T2 open circuit and the pair 1 1, 0 1 checked for T1 open circuit. The mathematics for determining an appropriate test vector sequence was not, however, considered.

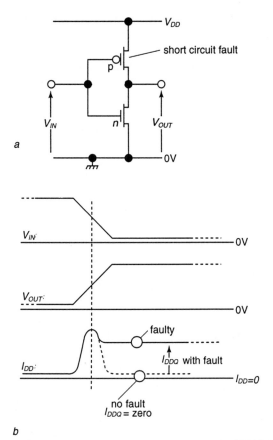

Figure 3.24 The voltage and current waveforms for a simple CMOS inverter gate with a short-circuit in the p-channel FET
a the circuit
b typical waveforms

The pseudorandom sequences produced by an autonomous LFSR generator can easily be shown to be unsatisfactory. Consider the situation shown in Figure 3.25 where two bits of the LFSR generator are used to test a two-input NAND gate. It is clear that the test vector sequence 1 1, 1 0 can never be applied to the NAND inputs due to the fixed correlation between bits k and $k+1$, bit $k+1$ always being the same as bit k one clock pulse earlier. This is still true if type B LFSRs, see Figure 3.20b, are used since the LFSR will still contain a certain number of adjacent bit patterns. The LFSR circuit is therefore a poor test vector generator for CMOS, even though it is frequently employed. The autonomous CA generator, however, does not have the adjacent bit correlation of the LFSR, and therefore is a much better pseudorandom generator to consider for CMOS testing.

A method of determining an appropriate exhaustive test vector sequence for all possible open-circuit faults in any n-input CMOS circuit has been

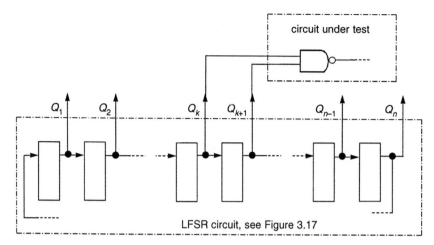

Figure 3.25 CMOS testing and inadequate LFSR test vector generation

published[65-67]. This gives a structured sequence rather than any pseudo-random sequence. If we plot, say, a four-variable function on a Karnaugh map, it may readily be found that it is impossible to find any simple path (input sequence) through the sixteen squares of the map which gives the required initialisation vector followed by the test vector for all possible functions and open-circuit faults. To construct such a universal sequence it is necessary to enter every map square once from each adjacent square. As shown in Reference 65, this is best illustrated by using a hypercube construction rather than a Karnaugh map; the construction for $n = 3$ is shown in Figure 3.26, from which it will be seen that all transitions involving one change of state of a variable are present. This is an Eulerian cycle, that is a cycle which begins at any vertex, passes along each edge in each direction exactly once, and returns to the initial vertex[66].

The algorithm that has been published for generating the required sequence takes as its starting point the sequence for n and from it builds the required sequence for $n + 1$. Since the cycle for $n = 2$ is easy to compile, $n = 3$, $n = 4$, etc., may be built up from this starting point. The required steps are:

(i)   take the Eulerian cycle for n with all zeros as the starting vector (note, more than one cycle is possible for $n \geq 2$);
(ii) add a leading zero to each vector in (i);
(iii) extend the sequence in (ii) by locating the first but not any subsequent occurrence of each nonzero vector 0**X** (where **X** denotes the following $n - 1$ bits), and inserting two additional vectors 1**X** and 0**X** immediately after each located vector;
(iv) add a further copy of the Eulerian cycle for n but with a leading 1 added to each vector;
(v)   finally, add the concluding all-zeros vector 0 0 ... 0.

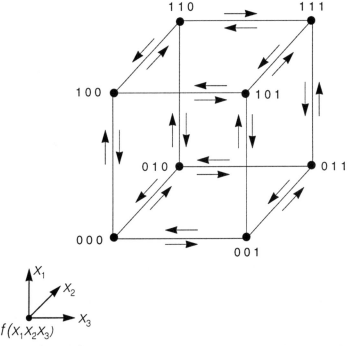

Figure 3.26 The directed hypercube for n = 3, where every node is entered once from all adjacent nodes

The procedure for generating the sequence for $n = 3$ from $n = 2$ is as shown in Table 3.10; the procedure which this algorithm follows may easily be illustrated and understood by plotting the steps in Table 3.10 on an $n = 3$ hypercube construction. Notice that a total of $n2^n + 1$ test vectors is present in this universal sequence for any n.

Such a test vector sequence will provide an exhaustive functional test for any n-input combinational circuit, since all possible input test vectors are present multiple times. However, four problems remain, namely:

(i) the length of this exhaustive test vector sequence has been increased from 2^n to $n2^n + 1$, which may be prohibitively long;
(ii) for a given combinational circuit not all the Eulerian cycle may be required to detect the possible open-circuit faults;
(iii) autonomous hardware generation of such a sequence is difficult;
(iv) applying the Eulerian test sequence to the primary inputs of a circuit under test does not ensure that an Eulerian sequence is being applied to all the internal logic gates if two or more levels of logic are present.

These features have been addressed by Bate and Miller[65], but it currently remains that nonEulerian test vectors are normally used for CMOS testing, even though they do not provide a good fault coverage for CMOS

Table 3.10 The generation of the Eulerian cycle for n = 3 from a given n = 2 sequence. Note, other sequences for n ≥ 2 are possible

Step (i) $n = 2$	Step(ii)	Step(iii)	Step (iv)	Step (v) $n = 3$
00	000	000	000	000
01	001	001	001	001
11	011	101	101	101
10	010	001	001	001
00	000	011	011	011
10	010	111	111	111
11	011	011	011	011
01	001	010	010	010
00	000	110	110	110
		010	010	010
		000	000	000
		010	010	010
		011	011	011
		001	001	001
		000	000	000
			100	100
			101	101
			111	111
			110	110
			100	100
			110	110
			111	111
			101	101
			100	100
				000

open-circuit defects. Further discussions on CMOS open-circuit faults may be found in the literature[68–71].

Other concepts for functionally testing CMOS circuits have been considered, based upon the duality of the p-channel and n-channel nets in any logic gate. These concepts include:

(a) isolating and testing the correct functionality of the p-channel nets separately from the n-channel nets, and vice versa;
(b) separating, recording and comparing the functionality of the p-channel nets against the dual functionality of the n-channel nets under the same test vectors;
(c) a variation of b, to self test the p-channel nets against the n-channel nets looking for any disagreement on each input test vector.

Figure 3.27 illustrates two of the above proposals. Signals T1, T2, T3 and T4 in Figure 3.27a are test mode signals which feed all the CMOS gates, 0 1 1 0 being the normal mode condition of these inputs, changing to 0 1 0 1 for the p-channel tests and to 1 0 1 0 for the n-channel tests. In Figure 3.27b only one test mode signal, T1, and its complement are required to switch from normal mode to test mode, but in both cases it will be appreciated that additional circuit complexity and possibly degradation of circuit performance has been built in.

Further details of these and other proposals may be found published[72–75], but none have yet found acceptance in the commercial field due to the circuit performance and area penalties involved. However, the final technique which we must consider, I_{DDQ} testing, has now considerable industrial significance.

The first proposal for CMOS current testing was in 1981 by Levi[76]. Since then a considerable amount of development has taken place to bring the methodology to its current status[1,77–81]; extensive references to these developments may be found in Reference 80.

Considering in more detail the fundamental concept, from Figure 3.24 the distinction between a fault-free circuit and one containing a short circuit is only apparent when switching transients have died away. Three problems, however, arise, namely:

(i) because the measurement of I_{DDQ} must be delayed until this transient current has decayed, slow rise and fall times t_r and t_f of the input vectors will delay the time at which the current measurements may be made;
(ii) the I_{DDQ} fault current is only present on the specific input test vector(s) which appropriately energise the faulty CMOS circuit;
(iii) in a very complex circuit there may be a great deal of continuous switching activity in the circuit, which means that a global quiescent condition is infrequent.

These factors were not too onerous when production testing of early SSI and MSI CMOS logic circuits was undertaken, and I_{DDQ} current checks could then be undertaken by vendors[81]. However, as complexity has increased to LSI and VLSI levels and with higher operating speeds, the time slots for appropriate measurements of I_{DDQ} have become less frequent and/or of shorter duration, unless specific means are incorporated to facilitate this test procedure.

The order of magnitude of the parameters involved is also significant. Typical values are as follows, although these may vary significantly with fabrication processing, gate dimensions (particularly transistor width, W, and transistor length, L), temperature, operating voltage and circuit loadings:

- average gate switching time $T_{av} \leq 5$ ns;
- average fault-free gate quiescent current < 5 pA;
- ohmic resistance of typical short circuits 100 Ω to 20 kΩ;
- required I_{DDQ} measurement sensitivity 1–50 µA, but one or two orders of magnitude lower for certain static RAM production tests;
- typical sample rate for I_{DDQ} measurements 10–100 kHz, say 50 µs per measurement.

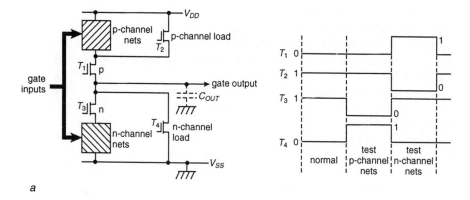

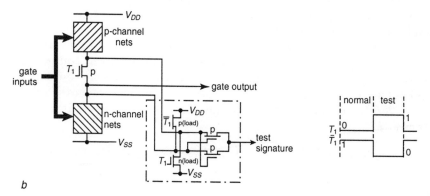

Figure 3.27 Possible CMOS test strategies based upon the duality of the p-channel and n-channel nets
  a separate p-channel and n-channel tests
  b comparison of the p-channel and n-channel responses, the test signature always being 0 for a fault-free circuit

However, the simultaneous switching of multiple I/Os with 100 Ω output impedance driving a 50 pF 100 Ω line may draw up to 5 A for 10 ns with an edge speed of 10 A/ns[1]. Further details of CMOS faults and their fault characteristics may be found in Reference 80.

Three types of I_{DDQ} tests have been proposed, namely:

(i)   every-vector I_{DDQ} test patterns;
(ii) selective I_{DDQ} test patterns;
(iii) supplemental test patterns.

In (i), a test set generated by ATPG based upon, say, the stuck-at model is used, the test sequence being interrupted after each test vector to perform an I_{DDQ} measurement. This is prohibitively time consuming except for small circuits; for example, if a VLSI test set was 200 000 vectors and the test time per vector including the I_{DDQ} measurement was 100 μs, it would take 20 seconds to complete the test of one circuit. Selective I_{DDQ} testing, however, uses a subset, possibly randomly selected, on which the I_{DDQ} measurements are made. It has been found that often less than one per cent of the full ATPG test can provide the same short-circuit cover as using the every-vector test sequence[80].

Finally, supplemental I_{DDQ} test patterns. These are test patterns specifically chosen and applied to the circuit separately from any other tests, I_{DDQ} measurements being made on every input test vector. If one randomly chosen test vector is applied to a circuit under test, then it is statistically possible for up to 25 % of the potential short-circuit faults in the circuit to be detected by this one test vector, since all gates must be in one or other of their two output states. A further randomly-chosen test vector may detect up to 25 % of the remaining faults, and hence a very short sequence of test vectors may provide a very high fault coverage for the circuit. However, this is not generally satisfactory as no exact quantification of the resultant fault coverage is available.*
Therefore, it is more satisfactory to determine a test set based upon short-circuit fault modelling which examines all possible paths from V_{DD} to ground.

Short-circuit fault modelling is undertaken at switch level, considering each p-channel and n-channel FET as a switch. For example, the circuit shown in Figure 3.28a is modelled by the arcs representing switches shown in Figure 3.28b. Using appropriate rules[1] the test set to detect any switch stuck-on is given by:

$$x_1 x_2 x_3 x_4 x_5 = 1\,1\,1\,-\,-,\ 1\,1\,-\,0\,0,\ 1\,1\,0\,1\,-,\ 0\,-\,1\,-\,-,\,-0\,1\,-\,-,\ 0\,-\,-0\,0$$
$$\text{and } -\,0\,-\,0\,0$$

which gives a final minimum test set of:

$$1\,1\,1\,1\,1,\ 1\,1\,1\,0\,0,\ 1\,1\,0\,1\,1,\ 0\,0\,1\,1\,1,\ 0\,0\,1\,0\,0.$$

Further details of this switch-level modelling will be found in References 1, 82–84.

An interesting feature of I_{DDQ} testing is that the test vectors for I_{DDQ} tests also detect certain open-circuit faults, since it has been found that a stuck-open fault in one transistor can increase I_{DDQ} due to certain physical secondary effects[80,85]. The tests will also detect device faults such as gate-oxide shorts and leaky pn junctions between source, drain and bulk substrate. It has further been found that I_{DDQ} measurements will detect circuits which are fully functional under normal logic criteria, but which have some imperfection in parametric performance. Although this imperfection may not give rise immediately to any system malfunction, from the long-term reliability point of view there is a problem present which may give rise to

---

* Notice, however, that many manufacturers do an initial quiescent current check on all CMOS circuits on power-up before commencing any further functioned or other tests. If excessive current is found, the circuit is immediately rejected before any time is wasted on further tests.

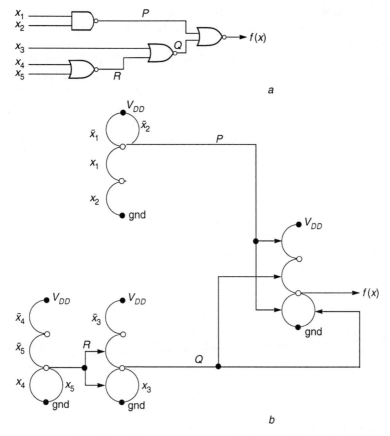

Figure 3.28 *A CMOS circuit assembly and its switch-level representation*
  a circuit
  b graphical representation

faulty operation at some later date. Overall, therefore, I_{DDQ} testing is being developed into possibly the most powerful single test procedure available for CMOS.

Looking, finally, at the means of measuring the I_{DDQ} current, two basic alternatives are possible, namely off-chip current measurement, monitoring the total current taken from the V_{DD} rail by the IC, or on-chip current measurement, using one or more circuits to monitor circuit consumption. In both cases it is necessary to ensure that all the I/Os are fully inoperative in order that the very large currents taken in output driving are completely dormant, which can be done if all the I/Os are three-state circuits. Alternatively, the I/Os may be given separate V_{DD} and ground supply rails, separate from the supply rails for the rest of the circuit.

Commercial a.c. and d.c. current probes are not generally suitable for off-chip I_{DDQ} measurements because of their inductance; a 5 nH probe inductance would cause a volt drop of about 5 V on a current pulse edge of

1 A/ns. Hence, some operational amplifier configuration is usually employed; typically, current measurements of microamps at a sampling frequency of 10 kHz or more are sought.

Figure 3.29a shows the basis of one proposal[86]. The series FET T1 is switched on during the application of the test vectors so as to present a low impedance in the supply rail, and switched to a high impedance when the switching transients have decayed. The I_{DDQ} current is then estimated by monitoring the volt-drop across T1. Capacitor C1 ensures that the supply voltage V_{DD} is held within acceptable limits during the test. An alternative circuit is shown in Figure 3.29b[86,87]; here the V_{DD} supply is open circuited following the application of the test vectors, and the time, Δt, for the voltage, V, to decay by a value ΔV is measured, this being a measure of the current taken by the circuit under test under quiescent input conditions.* However, calibration of both of these types of off-chip circuit to the required standard of accuracy is sometimes difficult.

On-chip current monitoring is usually referred to as built-in current (BIC) testing. In circuits of VLSI complexity, partitioning of the circuits and multiple current sensors is frequently proposed. The basic concept is shown in Figures 3.30a and b. A current sensor circuit is included in the supply connection of each partition, which will monitor the quiescent current through the partition. Each sensor is effectively a series component between the partition under test and the 0 V ground supply rail, the voltage across which is a measure of I_{DDQ}, but in order that its presence will not materially affect the normal performance of the circuit it is desirable that the connection between the partition and sensor shall be held at as near as possible to ground potential at all times.

Several BIC circuit configurations have been proposed, and many others are still under development. One possible configuration is shown in Figure 3.30c; here a substrate bipolar transistor Q1 is used as the voltage-drop device, the exponential V_{CE}/I_C characteristics of which make it particularly suitable for measuring I_{DD} values which may be several orders of magnitude greater under faulty conditions compared with the fault-free condition. Figure 3.30d illustrates the possible fault discrimination of this approach. Further circuit details of this configuration and others[88] may be found in the literature.

The disadvantage with BIC testing is clearly the additional circuit complexity which has to be built into the IC. This penalty is considered too high by some vendors, but it does have the advantage of enabling I_{DDQ} tests to be made faster than is generally possibly with off-chip current measurements. The discrimination settings of the sensor circuits may also be easier to determine when the circuit is partitioned, compared with the problem of dealing with the whole circuit at once.

---

* The initial rate of decay of V will be $C \delta V/\delta t = I_{DDQ}$, but the initial slope of an exponential is difficult to measure directly.

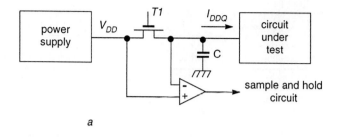

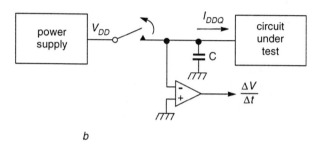

Figure 3.29 Off-chip I_{DDQ} measurements
   a sample and hold circuit
   b voltage decay circuit

Continuing developments based on current measurements are being pursued. One variation on the I_{DDQ} measurement considered above is to consider the current transients I_{DDQ} in supply rails[89]; clearly, if one could compare the total current waveform of a fault-free CMOS circuit with that of a faulty circuit (effectively the electrocardiogram of the circuit under test), many detailed differences would be present, but to measure them exactly when under test is a formidable problem. Another variation on I_{DDQ} test measurements has been to monitor the V_{SS} supply rail rather than the V_{DD} supply rail, giving I_{SSQ} current measurements instead of I_{DDQ} measurements. There may, in some circumstances, be practical reasons for measuring the current in the V_{SS} rail rather than the V_{DD} rail, but this is not the usual situation.

To summarise, I_{DDQ} testing has already been found to offer many advantages and to be able to detect more types of circuit fault than was initially envisaged. With continuing development current monitoring is likely to mature into the best testing strategy available for both digital-only and mixed analogue/digital CMOS circuits.

3.6 Delay fault testing

In all the preceding test methodologies no specific reference has been made to the possible need to confirm that the digital circuits will work at their

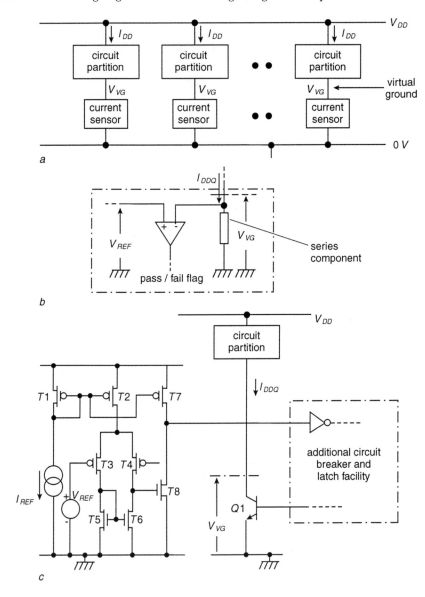

Figure 3.30 On-chip I_{DDQ} measurements
  a the built-in circuit (BIC) test
  b the principle of the current sensor
  c current sensor circuit of Kettler

intended maximum speed. Failures causing logic circuits to malfunction at the required clock speeds are referred to as delay faults or AC faults*, and are usually caused by some process parameter defect which may not affect the functionality at lower clock speeds.

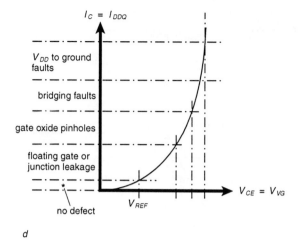

Figure 3.30 Continued
    d exponential V_{CE}/I_C characteristic of the series bipolar transistor, with its possible I_{DDQ} interpretation

We are not concerned here with delay through individual gates of the circuit under test, but instead with the cumulative delay of all gates in a path from primary inputs to primary outputs. Transitions from 0 to 1 and 1 to 0 are propagated, and a fault is recorded if the primary output(s) do not respond within a given time.

Although the propagation delay in only the shortest and longest path may be considered, a more satisfactory procedure is as shown in Figure 3.31 where every input/output path is involved. The input register is clocked with C1, and produces (possibly pseudorandom) test vectors to exercise the combinational network. The output responses are captured by the output register when triggered by clock C2, and checked by some further means such as a signature analysis technique (see following chapter). If the expected responses are not received by the time clock pulse C2 is applied, then some delay fault in the circuit is indicated.

The clock speeds C1 and C2 clearly provide a measure of the propagation delay; in general if this delay is unambiguously less than the period between the clock pulses of the circuit when in normal working mode the circuit is considered fault free. An important consideration, however, is the need to be able to generate the test vectors and record the output response at high speed, which may require special on-chip latch and flip-flop circuit configurations.

Further details of delay fault testing may be found published[11,47,90–93].

---

* The terminology AC which has been used in the literature in connection with delay testing[90,91] is unfortunate. It does not refer to the use of any alternating current (a.c.) test signals, but comes from the use by certain IC manufacturers in data sheets where DC data tables are the logic relationships and AC data tables are the timing and other parametric performance details in a comprehensive IC specification.

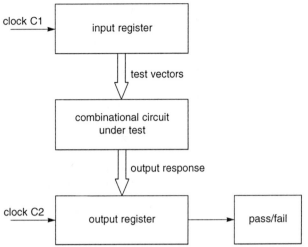

Figure 3.31 The test configuration for delay fault testing, C1 and C2 being interleaved clocks

However, it does not currently hold such an extensive place in digital testing as functional testing, although it may be crucial for VLSI circuits working at the leading edge of the performance of available technologies. It will, however, become more significant with deep submicron geometry VLSI circuits.

3.7 Fault diagnosis

In the foregoing sections, the principal objective was to detect faulty circuits but not to identify the precise location or cause of the failure. With ATPG programs based upon the stuck-at model, for example, no information as to the exact location of a stuck-at fault is generally available, since many s-a-0 or s-a-1 nodes can be covered by a single test vector. Most other functional tests have the same feature—that complete primary input to primary output paths are checked, with little information being available to indicate whereabouts along a path a failure has occurred. In a sense the generation of a minimum test set which detects all faults in a given fault list is at odds with the requirements of fault location.

The I_{DDQ} test method, however, can be considerably more diagnostic than conventional functional testing, since a high I_{DDQ} measurement following given input test vectors may point to a particular gate or macro in the circuit being faulty[80,85]. Figure 3.32 is a good example of a fault located in a CMOS circuit following an I_{DDQ} test procedure. However, in general, should fault location and thence defect diagnosis be required, it is necessary to undertake further specific tests once the circuit has been identified as faulty by some more general test procedure.

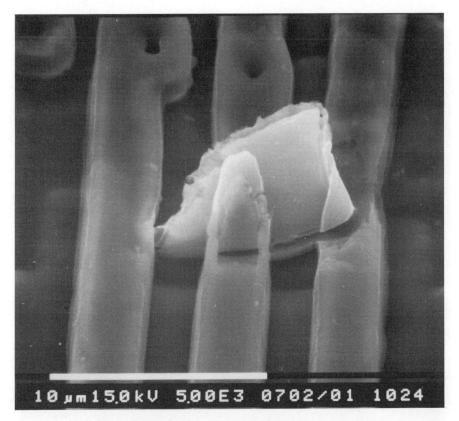

Figure 3.32 The identification of a bridging fault in the metalisation of a CMOS circuit (Photo courtesy Storage Technology Corporation, CA)

Procedures for distinguishing and locating the source of faults include:

(i) guided-probe testing;
(ii) simulation-based diagnostics;
(iii) diagnostic reasoning.

Not all may be appropriate for given situations.

Guided-probe testing monitors the internal logic signals in the circuit under test, working from a faulty output back towards the primary inputs while the input test vector which has detected the fault is held constant. The paths from the primary inputs to the faulty output are known from ATPG or simulation programs, together with the fault-free signals along the paths, and therefore the probing procedure works backwards until the interface between incorrect and correct signals is found.

The probe tester employed for this type of procedure is a precision instrument such as illustrated in Figure 3.33. Probe movements are usually controlled manually by the test engineer, but the tester may be an integral

Figure 3.33 A precision microprobe station featuring ultra-low capacitance probing and submicron resolution. GHz signal handling capability is possible (Photo courtesy Cascade Microtech, Inc., OR)

part of an automatic test equipment assembly containing the test generator, computer simulation facilities and VDU monitoring.

One factor, however, that complicates the microprobing of ICs is that completed circuits have a final silicon dioxide surface protection layer (passivation), and this (if present) has to be stripped off before probing can be effective.

Simulation-based diagnostic procedures take the faulty output which has been detected by a test vector and determine by simulation what circuit faults could give rise to this incorrect output. Using several further test vectors and their faulty or fault-free output response, it may then be possible to resolve the fault into one location or near neighbours. Considerable computational work may be necessary to achieve this resolution, although this effort is considerably eased if some form of partitioning is present in the original design which allows one faulty partition or circuit macro to be easily identified among the other fault-free partitions.

A variation on simulation-based diagnostics is to have already compiled a comprehensive fault dictionary for the circuit, which lists the circuit response under each possible fault condition. By matching the test set response to the fault dictionary data, identification of the source of the fault will be given. The problem in this case is the extremely large amount of computational effort that may be necessary in order to compile the fault dictionary data in the first place.

Finally, diagnostic reasoning using artificial intelligence (AI) has been proposed, based upon if-then rules such as:

if (ALU test passed) **then** (suspect register B)

This method is not easily applicable to fault location down to the gate level, even though this is precisely what a designer or test engineer may do to locate a gate fault, but is more relevant for macro diagnosis or system diagnosis where identifiable functional blocks are involved.

Further details of fault diagnosis may be found in References 12, 47, 94–96. Other techniques that we have not covered here, such as thermal mapping of ICs and laser techniques, have been researched[97-100] but as far as fault diagnosis on integrated circuits is concerned this must remain very much the province of the IC manufacturer rather than the OEM.

3.8 Chapter summary

This Chapter has considered several strategies for generating tests for digital circuits which, with the noticeable exception of I_{DDQ} testing, involve the monitoring of the 0 and 1 primary output response to appropriate input test vectors.

By far the greatest development effort has been concerned with combinational logic. Roth's D-algorithm forms the basis of many automatic test pattern generation programs, but the computational effort in producing ATPG programs for the increasing size and complexity of present-day VLSI circuits is becoming prohibitive. This has forced an increasing interest in other means of test pattern generation, which usually involves some partitioning of the circuit at the design stage so as to allow exhaustive, non-exhaustive or pseudorandom test patterns to be used. We will be concerned in much greater depth with these design for test or DFT concepts, in the following chapters.

The testing of sequential circuits is seen to be much less straightforward than the testing of combinational networks, and in general no useful testing algorithms have been developed. A guiding principle is to try to partition the sequential logic into smaller manageable blocks, and/or to separate the combinational logic gates from the sequential logic elements, but this immediately leads into a design for test strategy such as we will be considering later.

CMOS technology has been shown to have its own specific test requirements, brought about because of the dual nature of CMOS circuit configurations. The stuck-at fault model, which forms the basis of the majority of ATPG programs, has been found to be inadequate for catering with certain potential CMOS faults, and exhaustive testing may require more than 2^n test vectors to cater for memory faults in an n-input combinational circuit. On the other hand, current testing of CMOS rather than functional testing has been seen to be a most promising test strategy, one which will have much further development and acceptance.

The mathematics underpinning many of the developments considered in this Chapter has been given in such detail as is necessary for understanding the principles involved. A greater depth of the mathematics involved in simulation and test pattern generation may be found in the text of and further references contained in Abramovici et al.[11]; further mathematical understanding relating to pseudorandom test sequences may be found in Bardell et al.[47]. The one-dimensional cellular automata developments considered in this Chapter have not yet reached the maturity to figure in many textbooks; indeed research and developments are continuing such as that of Chowdhary et al.[101]. Further information on I_{DDQ} testing may be found in Gulati and Hawkins[80].

In summary, the simple testing of digital circuits by applying appropriate input test vectors and measuring the output response on each vector remains appropriate for circuits of LSI complexity, but not for complex VLSI circuits. We will, however, build upon the fundamentals covered in this Chapter in subsequent pages, particularly pseudorandom vector sequences which will be present in several later test strategies.

3.9 References

1　RAJSUMAN, R.: 'Digital hardware testing: transistor-level fault modelling and testing' (Artech House, 1992)
2　LALA, P.K.: 'Fault tolerant and fault testable hardware design' (Prentice Hall, 1985)
3　RUSSELL, G. (Ed.): 'Computer aided tools for VLSI system design' (IEE Peter Peregrinus, 1987)
4　BREUER, M.A., and FRIEDMAN, A.D.: 'Diagnosis and reliable design of digital systems' (Computer Science Press, 1976)

5   LEVENDEL, Y.H., and MENON, P.R.: 'Fault simulation methods—extension and comparison', *Bell Syst. Tech. J.*, 1981, pp. 2235-2258
6   GOEL, P.: 'Test generation cost analysis and projection'. Proceedings of IEEE conference on *Design automation*, 1980, pp. 77-84
7   LEVENDEL, Y.H., and MENON, P.R.: 'Comparison of fault simulation methods—treatment of unknown signal values', *IEEE J. of Digital Systems*, 1980, **4**, pp. 443-459
8   ARMSTRONG, D.B.: 'A deductive method of simulating faults in logic circuits', *IEEE Trans.*, 1972, **C-21**, pp. 464-471
9   ULRICH, E.G., and BAKER, T.G.: 'Concurrent simulation of nearly identical digital networks', *Computer*, 1974, **7**, April, pp. 39-46
10 WAICUKAUSKI, J.A., EICHELBURGER, E.B., FORLENZA, D.O., LINDBLOOM, E., and McCARTHY, T.: 'Fault simulation for structured VLSI', *VLSI Syst. Des.*, 1985, **6**, December, pp 20-32
11 ABRAMOVICI, M., BREUER, M.A., and FRIEDMAN, A.D.: 'Digital system testing and testable design' (Computer Science Press, 1990)
12 BREUER, M.A., and FRIEDMAN, A.D.: 'Diagnosis and reliable design of digital systems' (Computer Science Press, 1976)
13 WILLIAMS, T.W. (Ed.): 'VLSI Testing' (North-Holland, 1986)
14 HURST, S.L.: 'Custom VLSI microelectronics' (Prentice Hall, 1992)
15 SELLERS, F.F., HSIAO, M.Y., and BEARSON, C.L.: 'Analysing errors with the Boolean difference', *IEEE Trans.* 1968, **C-17**, pp. 676-683
16 CHIANG, A.C., REID, I.S., and BANES, A.V. 'Path sensitization, partial Boolean difference and automated fault diagnosis', *IEEE Trans.*, 1972, **C-21**, pp. 189-195
17 EDWARDS, C.R.: 'The generalised dyadic differentiator and its application to 2-valued functions defined on n-space', *IEE J. Comput. Digit. Tech.*, 1978, **1**, pp 137-142
17a EDWARDS, C.R.: 'The Gibbs differentiator and its relationship to the Boolean difference', *Comput Electr. Eng.*, 1978, **5**, pp. 335-344
18 CHING, D.K., and LIU, J.J.: 'Walsh-transform analysis of discrete dyadic-invarient systems', *IEEE Trans.*, 1974, **EMC-16**, pp. 136-139
19 KARPOVSKY, M.G. (Ed.): 'Spectral techniques and fault detection' (Academic Press, 1985)
20 ROTH, J.P.: 'Diagnosis of automata failure: a calculus and a method', *IBM J. Res. Dev.*, 1996, **10**, pp. 278-291
20a ROTH, J.P., BOURICIUS, W.G., and SCHNEIDER, P.R.: 'Programmed algorithms to compute tests to detect and distinguish between failures in logic circuits', *IEEE Trans.*, 1967, **EC-16**, pp. 1292-1294
20b ROTH, J.P.: 'Computer logic, testing and verification' (Computer Science Press, 1988)
20c BENNETTS, R.G.: 'Introduction to digital board testing' (Crane Russak, 1982)
20d MICZO, A.: 'Digital logic testing and simulation' (Harper and Row, 1986)
21 YARMOLIK, V.N.: 'Fault diagnosis of digital circuits' (Wiley, 1990)
22 THOMAS J.J.: 'Automated diagnostic test program for digital networks', *Comput. Des.*, 1971, **10**, August, pp. 63-67
23 BOWDEN, K.R.: 'A technique for automated test generation for digital circuits'. Proceedings of IEEE *Intercon*, 1975, pp. 15.1-15.5
24 BREUER, M.A., FRIEDMAN, A.D., and IOSUPAVICZ, A.: 'A survey of the art of design automation', *IEEE Trans.*, 1981, **C-30**, pp. 58-75
25 GOEL, P.: 'An implicit ennumeration algorithm to generate tests for combinational logic circuits', *IEEE Trans.*, 1981, **C-30**, pp. 215-222
26 GOEL, P., and ROSALES, B.C.: 'PODEM-X: an automatic test generation system for VLSI logic structures'. Proceedings of 18th IEEE conference on *Design automation*, 1981, pp. 260-268

27 BENNETTS, R.G.: 'Design of testable logic circuits' (Addison-Wesley, 1984)
28 BOTTORFF, P.S.: 'Test generation and fault simulation', in WILLIAMS, T.W. (Ed.): 'VLSI testing' (North-Holland, 1991)
29 FUJIWARA, H., and SHIMONO, T.: 'On the acceleration of test generation algorithms', *IEEE Trans.*, 1983, **C-23**, pp. 1137–1144
30 AGRAWAL, P., and AGRAWAL, V.D.: 'Probabilistic analysis for random test generation method for irredundant combinational networks', *IEEE Trans.*, 1975, **C-24**, pp. 691–695
31 SAVIR, J., DITLOW, G.S., and BARDELL, P.H.: 'Random pattern testability', *IEEE Trans.*, 1984, **C-33**, pp. 79–90
32 EICHELBURGER, E.B., and LINDBLOOM, E.: 'Random-pattern coverage enhancement and diagnostics for LSSD self-test', *IBM J. Res. and Dev.*, 1983, **27**, pp. 265–272
33 JAIN, S.K., and AGRAWAL, V.D.: 'Statistical fault analysis', *IEEE Des. Test Comput.*, 1985, **1** (1), pp. 38–44
34 TAMAMOTO, H., and NARITA, Y.: 'A method for determining optimal or quasi-optimal input probabilities in random test method', *Trans. Inst. Electron. Commun. Eng. Jpn. Part A*, 1982, **65**, pp. 1057–1064
35 SAVIR, J., and BARDELL, P.H.: 'On random pattern test length', *IEEE Trans.*, 1984, **C-33**, pp. 467–474
36 WILLIAMS, T.W.: 'Test length in a self-testing environment', *IEEE Des. Test Comput.*, 1985, **2** (2), pp. 59–63
37 WAGNER, K.D., CHIN, C.K., and McCLUSKEY, E.J.: 'Pseudorandom testing', *IEEE Trans.*, 1987, **C-36**, pp. 332–343
38 WUNDERLICH, H.J.: 'Multiple distributions for biased random test patterns'. Proceedings of IEEE international conference on *Test*, 1988, pp. 236-244
39 GOEL, P.: 'RAPS test pattern generator', *IBM Tech. Discl. Bull.*, 1978, **21**, pp. 2787–2791
40 KOHAVI, Z.: 'Switching and finite automata theory' (McGraw-Hill, 1970)
41 KOHAVI, Z., and LAVELEE, P.: 'Design of sequential machines with fault detection capabilities', *IEEE Trans.*, 1967, **EC-16**, pp. 473–484
42 HIBBARD, T.N.: 'Least upper bounds on minimal terminal state experiments for two classes of sequential machines', *J. Assoc. Comput. Mach.*, 1961, **8**, pp. 601–612
43 KOVIJANIC, P.G.: 'A new look at test generation and verification'. Proceedings of IEEE conference on *Design automation*, 1978, pp. 58–63
44 PETERSON, W.W., and WELDON, E.J.: 'Error correcting codes' (MIT Press, 1972)
45 MacWILLIAMS, F.J., and SLOANE, N.J.: 'The theory of error correcting codes' (North-Holland, 1978)
46 BRILLHART, J., LEHMER, D.H., SELFRIDGE, J.H., TUCKERMAN, B., and WAGSTAFF, S.S.: 'Factorization of $b^n \pm 1$, $b = 2, 3, 4, 5, 6, 7, 10, 11, 12$ up to higher powers'. Report American Mathematical Society, Providence, RI, 1983
47 BARDELL, P.H., McANNEY, W.H., and SAVIR, J.: 'Built-in test for VLSI: pseudorandom techniques' (Wiley, 1987)
48 GOLOMB, S.W.: 'Shift register sequences' (Aegean Park Press, 1982)
49 SARWATE, D.V., and PURSLEY, M.V.: 'Cross-correlation properties of pseudo-random and related sequences', *Proc. IEEE*, 1980, **68**, pp. 593–619
50 BARDELL, P.H.: 'Design considerations for parallel pseudorandom pattern generators', *J. Electron. Test., Theory Appl.*, 1990, **1** (1), pp. 73–87
51 WANG, L.T., and McCLUSKEY, E.J.: 'A hybrid design of maximum-length sequence generators'. Proceedings of IEEE international conference on *Test* conference, 1986, pp. 38–45
52 WOLFRAM, S.: 'Statistical mechanics of cellular automata', *Rev. Mod. Phys.*, 1983, **55**, pp. 601–644

53 HORTENSIUS, P.D., McLEOD, R.D., and CARD, H.C.: 'VLSI pseudorandom number generation for parallel computing'. Unpublished report of the Department of Electrical Engineering, University of Manitoba, 1985
54 PODAIMA, B., HORTENSIUS, P.D., SCHNEIDER, R.D., CARD, H.C., and McLEOD, D.: 'CALBO cellular automata logic block observation'. Proceedings of Canadian conference on *VLSI*, 1986, pp. 171–176
55 HORTENSIUS, P.D., CARD, H.C., and McLEOD, R.D.: 'Parallel random number generation for VLSI circuits using cellular automata', *IEEE Trans.*, 1989, **C-38**, pp. 1466–1473
56 HORTENSIUS, P.D., McLEOD, R.D., PRIES, W., MILLER, D.M., and CARD, H.C.: 'Cellular automata-based pseudorandom number generators for built-in test', *IEEE Trans.*, 1989, **CAD-8**, pp. 842–859
57 KHARE, M., and ALBICKI, A.: 'Cellular automata used for test pattern generation'. Proceedings of IEEE international conference on *Computer design*, (ICCD87), 1987, pp. 56–59
58 MARTIN, O., ODLYZKO, A.M., and WOLFRAM, S.S.: 'Algebraic properties of cellular automata', *Comm. Math. Phys.*, 1984, pp. 259–254
59 ZHANG, S., MILLER, D.M., and MUZIO, J.C.: 'Determination of minimum cost one-dimensional linear hybrid cellular automata', *Electron. Lett.*, 1991, **27**, pp. 1625–1627
60 SLATER, T., and SERRA, M.: 'Tables of linear hybrid 90/150 cellular automata'. VLSI Design and Test Group, Department of Computer Science, University of Victoria, Canada, report DCS-105-IR, January 1989
61 SERRA, M., SLATER, T., MUZIO, J.C., and MILLER, D.M.: 'The analysis of one-dimensional linear cellular automata and their aliaising properties', *Trans. IEEE.*, 1990, **CAD-9**, pp. 767–778
62 SERRA, M., and SLATER, T.: 'A Lanczos algorithm in a finite field and its application', *J. Comb. Math. Comb. Comput.*, 1990, **7**, April, pp. 11–32
63 SERRA, M.: 'Algebraic analysis and algorithms for linear cellular automata over GF(2), and their application to digital circuit testing'. Proceedings of *Congressus Numeratium*, Winnipeg, Canada, 1990, **75**, pp. 127–139
64 STEWART, G.W.: 'Introduction to Matrix Computations' (Academic Press, 1973)
65 BATE, J.A., and MILLER, D.M.: 'The exhaustive testing of CMOS stuck-open faults' *in* MILLER, D.M. (Ed.): Developments in Integrated Circuit Testing (Academic Press, 1987)
66 BATE J.A., and MILLER, D.M.: 'Fault detection in CMOS and an algorithm for generating Eulerian cylces in directed hypercubes'. Proceedings of *Congressus Numerantium*, Winnipeg, Canada, 1985, **47**, pp. 107–117
67 BATE, J.A., and MILLER, D.M.: 'Exhaustive test of stuck-open faults in CMOS', *IEE Proc. E*, 1988, **135**, pp. 10–16
68 REDDY, M.K., and REDDY, S.M.: 'Detecting FET stuck-open faults in CMOS latches and flip-flops', *IEEE Des. Test Comput.*, 1986, October, pp. 17–26
69 SODEN, J.M., TREECE, R.K., TAYLOR, M.R., and HAWKINS, C.F.: 'CMOS stuck-open fault electrical effects and design considerations'. Proceedings of IEEE international conference on *Test*, 1989, pp. 423–430
70 HENDERSON, C.L., SODEN, J.M., and HAWKINS, C.E.: 'The behaviour and testing implications of CMOS IC logic gate open circuits'. Proceedings of IEEE international conference on *Test*, 1991, pp. 302–310
71 RODRIGUEZ-MONTANES, R., SEGURA, J.A., CHAMPAC, V.H., FIGUERA, J., and RUBIO, J.A.: 'Current vs. logic testing of gate oxide short, floating gate and bridging failures in CMOS'. Proceedings of IEEE international conference on *Test*, 1991, pp. 510–519
72 HURST, S.L.: 'Open-circuit testing of CMOS circuits', *Int. J. Electron.*, 1987, **62**, pp. 161–165

73 CHEEMA, M.S., and LALA, P.K.: 'Totally self-checking CMOS circuit design for breaks and stuck-on faults', *IEEE J. Solid-State Circuits*, 1992, **27**, pp. 1203–1206
74 RAJSUMAN, R.: 'Design of CMOS circuits for stuck-open fault testability', *IEEE J. Solid-State Circuits*, 1991, **26**, pp. 10–21
75 JHA, N.K., and ABRAHAM, J.A.: 'Totally self-checking MOS circuits under realistic physical failures'. Proceedings of IEEE international conference on *Computer design*, 1984, pp. 665–670
76 LEVI, M.W.: 'CMOS is most testable'. Proceedings of IEEE international conference on *Test*, 1981, pp. 217–220
77 MALY, W., and NIGH, P.: 'Built-in current testing—a feasibility study'. Proceedings of IEEE *ICCAD*, 1988, pp. 340–343
78 FRITZEMEIER, R.R., SODEN, J.M., TREECE, R.K., and HAWKINS, C.F.: 'Increased CMOS IC stuck-at fault coverage with reduced IDDQ test sets'. Proceedings of IEEE international conference on *Test*, 1990, pp. 427–435
79 FERGUSON, F.T., TAYLOR, M., and LARRABEE, T.: 'Testing for parametric faults in static CMOS'. Proceedings of IEEE international conference on *Test*, 1990, pp. 436–443
80 GULATI, R.K., and HAWKINS, C.F. (Eds.): 'I_{DDQ} testing of VLSI circuits' (Kluwer, 1993)
81 SODEN, J.M., HAWKINS, C.F., GULATI, R.K., and MAO, W.: 'I_{DDQ} testing: a review', *J. Electron. Test., Theory Appl.*, 1992, **3**, pp. 291–303 (reprinted in [80])
82 BRZOZOWSKI, J.A.: 'Testability of combinational networks of CMOS cells', in MILLER, D.M. (Ed.): 'Developments in integrated circuit testing' (Academic Press, 1987)
83 RAJSUMAN, R., JAYASUMANA, A.P., and MALAIYA, Y.K.: 'Testing of complex gates', *IEE Electron. Lett.*, 1987, **23**, pp. 813–814
84 LEE, K.J., and BREUER, M.A.: 'On detecting single and multiple bridging faults in CMOS circuits using the current supply monitoring method'. Proceedings of IEEE international symposium on *Circuits and systems*, 1990, pp. 5–8
85 SODEN, J.M., and HAWKINS, C.F.: 'Electrical properties and detection methods for CMOS IC defects'. Proceedings of 1st European conference on *Test*, 1989, pp. 159–167
86 KEATING, M., and MEYER, D.: 'A new approach to dynamic IDD testing'. Proceedings of IEEE international conference on *Test*, 1987, pp. 316–321
87 WALLQUIST, K.M., RIGHTER, A.W., and HAWKINS, C.F.: 'Implementation of a voltage decay method for I_{DDQ} measurements on the HP82000'. Report of Hewlett Packard User Group meeting, CA, June 1992
88 MALY, W., and PATYRA, M.: 'Design of ICs applying built-in current testing', *J. Electron. Test., Theory and Applications*, 1992, **3**, pp. 397–406 (reprinted in [80])
89 SU, S-T., MAKKI, R.Z., and NAGLE, T.: 'Transient power supply current monitoring: a new test method for CMOS', *J. Electron. Test., Theory Appl.*, 1995, **6**, pp. 23–43
90 SAVIR, J., and BERRY, R.: 'AC strength of a pattern generator', *J. Electron. Test., Theory Appl.*, 1992, **3**, pp. 119–125
91 BARZILAI, Z., and ROSEN, B.K.: 'Comparison of AC self-testing procedures'. Proceedings of IEEE international conference on *Test*, 1990, pp. 387–391
92 BRAND, D., and IYENGAR, V.S.: 'Timing analysis using functional analysis', *IEEE Trans.*, 1988, **C-37**, pp. 1309–1314
93 IYENGAR, V.S., ROSEN, B.K., and WAICUKAWSKI, J.A.: 'On computing the sizes of detected delay faults', *IEEE Trans.*, 1980, **CAD-9**, pp. 299–312
94 ABRAMOVICI, M., and BREUER, M.A.: 'Multiple faulty diagnosis in combinational circuits based upon effect-cause analysis', *IEEE Trans.*, 1980, **C-29**, pp. 451–460

95 PURCELL, E.T.: 'Fault diagnosis assistant', *IEEE Circuits Devices Mag.*, 1988, **4**, January, pp. 47–59
96 WILKINSON, A.J.: 'A method for test system diagnosis based upon the principles of artificial intelligence'. Proceedings of IEEE international conference on *Test*, 1984, pp. 188–195
97 TAMAMA, T., and KUJI, N.: 'Integrating an electron-beam system into VLSI fault diagnosis', *IEEE Des. Test of Comput.*, 1986, **3**, pp. 23–29
98 SODEN, J.M., and ANDERSON, R.E.: 'IC failure analysis: techniques and tools for quality and reliability improvement', *Proc. IEEE*, 1993, **81**, pp. 793–715
99 GIRARD, P.: 'Voltage contrast', *J. Phys. IV, Colloq.*, 1991, **1**, December, pp. C6-259–C6-271
100 LEE, D.A.: 'Thermal analysis of integrated circuit chips using thermographic imaging techniques', *IEEE Trans. Instrum. and Meas.*, 1994, **43**, pp. 824–829
101 CHOWDHURY, D.R., SENGUPTA, I., and CHOUDHURI, P.P.: 'A class of two-dimensional cellular automata and their application in random pattern generation', *J. Electron. Test., Theory and Appl.*, 1994, **5**, pp. 67-82

Chapter 4
Signatures and self test

4.1 General introduction

As has been seen in previous chapters, the principal difficulties with digital network testing are (i) the volume of the simulation data, and (ii) the volume of data which has to be checked at network outputs when under test. In this Chapter we will consider means which have been proposed to ease the task of checking the output response when under test, so that the very large number of individual 0 or 1 output bits do not have to be compared step by step against the expected (fault-free) response, such as is present in the test arrangements shown in Figures 3.2b and c. Again we shall largely be considering combinational circuits rather than sequential, the latter still requiring their own special testing consideration as will be discussed in Chapter 5. Some of the concepts that we will review in this Chapter have yet to be reflected in common practice, but all are part of the whole research and development effort which has taken place and still continues on testing strategies.

An ideal test procedure would be as shown in Figure 4.1a when one additional input pin is added to the network under test so as to switch it from normal mode to test mode, with one additional output pin to show the final result of the test, this being a simple pass or fail (fault-free or faulty) indication. Also, the ideal situation would require no failure of the pass/fail check circuit, all this being done with minimum possible silicon area overhead!

A more realistic concept is to have some form of continuous (online) monitoring of the output response while the circuit is in its normal mode, as illustrated in Figure 4.1b. An obvious example of this is when dual or triple (or higher) system redundancy is present in the circuit under test, the response of the two or three (or more) separate circuit networks being monitored by the online monitoring circuit. Clearly, a very high silicon area

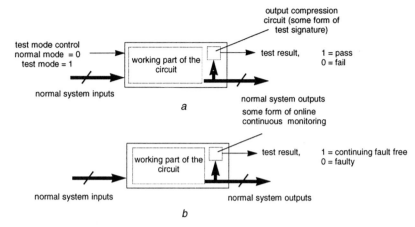

Figure 4.1 Ideal test configurations
    a one test mode input pin and one test result output pin
    b continuous online testing with one test result output pin, no separate test mode

overhead is present in such circumstances (≥200 %), but this may be necessary in safety-critical applications[1-3]. Somewhat simpler possibilities than complete system redundancy are however possible, and will be considered below.

Therefore, what we will be considering in the following sections are possible techniques which will compress the output data of the circuit under test into some simple test signature, rather than having to undertake the step by step comparison of the output response against a fault-free copy of the circuit response. If we can also find means of compressing the input set of test vectors into a smaller set, this would be advantageous. We will also find that if the input test vector sequence can be arranged in a specific order depending upon the output function under test, then a more satisfactory signature for the circuit may be possible than would otherwise be the case. Table 4.1 indicates the strategies which can be proposed, from which it will be seen that there may be a conflict between the simplest way of generating the input test vectors and the simplest signature when under test. These and other aspects will be considered.

4.2 Input compression

Although compression of the output response of a network under test is the most significant factor that we will be considering, there is also the possibility of some compression of the total number of input test vectors so that exhaustive testing may be facilitated. Recall that for an exhaustive global test of an n-input logic network, 2^n input vectors (ignoring any possible CMOS

Table 4.1 The testing of combinational logic networks, with test signatures as one possible means of test

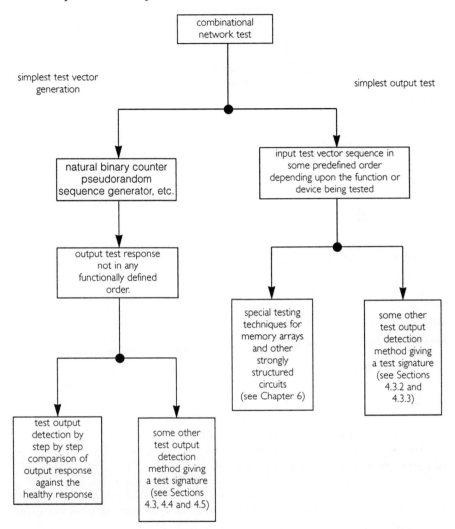

memory requirements) are conventionally required. Our ideal objective, therefore, is the realisation of the concepts shown in Figure 4.2.

Input compression, sometimes termed input compaction, relies upon the feature that in many logic networks with multiple outputs $z_1, z_2, ..., z_m$, individual outputs may only be dependent upon a subset of the n inputs. For example, if it was found that in a ten-input three-output logic network the three outputs were only dependent upon, say, 4, 5 and 6 of the inputs, then an exhaustive test of the network can be made with a maximum of

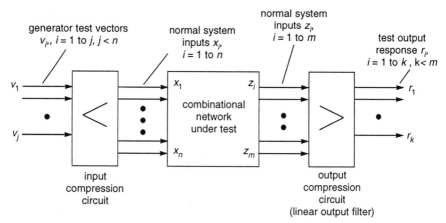

Figure 4.2 The concept of a linear output filter to combine primary outputs

$2^4 + 2^5 + 2^6 = 112$ test vectors instead of 2^{10}. This assumes that there is no incorrect interaction between outputs, and therefore is not quite as comprehensive as a fully exhaustive test of 2^n vectors.

The possibility of dividing the inputs into groups which control individual outputs is a form of circuit partitioning which the circuit designer may automatically recognise and capitalise upon in certain circumstances. However, it has been considered on a more theoretical basis, and the concept of input zones has been proposed as shown in Figure 4.3a. The difficulty here is in recognising the inputs that control the individual outputs, which may not be apparent from normal design procedures; indeed it has been demonstrated in several real-life designs that more inputs have been built into individual outputs than are functionally required, due to inefficient or incomplete minimisation of the output functions.

However, assuming that partitioning of the input variables into zones is undertaken, which may done by:

(i) a functional consideration and/or partitioning of the network;
(ii) minimisation of each output function to a minimised Boolean form from which the relevant x_i inputs can be seen;
(iii) some other technique such as determining the spectrum of each output (see Section 4.4.2 below);

then the determination of a reduced input test set which will exhaustively test each individual output can be undertaken. The possibility of compression of the output response of each output as well, see following sections, may also be considered.

One method that has been proposed for the generation of an appropriate reduced input test set is that of Akers[4], which extends the earlier

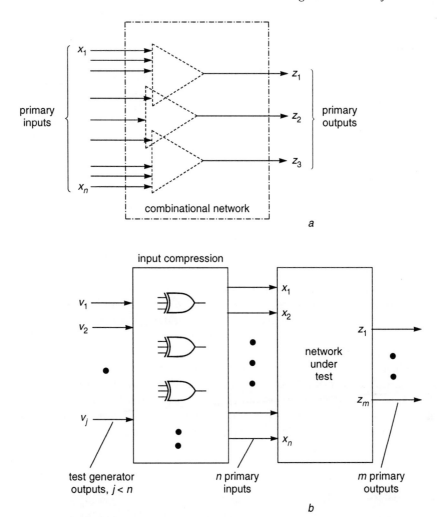

Figure 4.3 Input test vector compression
 a zones of influence of the x_i input variables
 b compression of the required n input variables into j test generator signals, $j < n$

considerations of Tang and Cheng[5], McCluskey[6] and others. The system total of n binary inputs is generated from j independent binary input test signals, $j < n$, by linear exclusive-OR relationships, as shown in Figure 4.3b. The value of j is determined by the maximum number of x_i input variables involved in any output function z_i, $i = 1$ to m, and the input independency between the m output functions. In the case of a simple three-bit test vector generator, a total of seven linearly related bit streams may be generated, see Table 4.2. The total number of bit streams possible from a j-bit test vector generator is $2^j - 1$.

Table 4.2 The seven possible test vector sequences available from three primary inputs v_1, v_2, v_3

Primary test inputs			Linearly related inputs			
v_1	v_2	v_3	$v_1 \oplus v_2$	$v_1 \oplus v_3$	$v_2 \oplus v_3$	$v_1 \oplus v_2 \oplus v_3$
0	0	0	0	0	0	0
0	0	1	0	1	1	1
0	1	0	1	0	1	1
0	1	1	1	1	0	0
1	0	0	1	1	0	1
1	0	1	1	0	1	0
1	1	0	0	1	1	0
1	1	1	0	0	0	1

The choice of any three of the seven vector sequences shown in Table 4.2 will give an exhaustive three-bit test sequence provided that all the subscripts 1, 2 and 3 appear in the three chosen vectors. If any subscript does not appear, then the sequence will be a function of the j-1 variables only. Hence, there is a considerable choice of test vector sequences for the exhaustive test of the individual outputs z_1 to z_m, the test of all outputs being completed within 2^j test vectors.

Published details cover the derivation of appropriate input test sets for given multiple-output networks, with emphasis upon minimising the width of the test vector generator v_1 to v_j and hence the test time[4-8]. However, this test strategy does not feature greatly in present-day VLSI test methodologies, except where it is obviously relevant from a partitioning of the circuit design and separate partition tests.

4.3 Output compression

Turning from the consideration of input compression to that of output compression, otherwise known as output compaction, here a number of avenues have been considered in order to reduce the data to be monitored when under test. However, the more the output response of a circuit is compressed, the extreme being one bit having the value, say, 0 when all responses are fault free and 1 when any fault is present, the greater will be the danger that this final output signature may not be valid. This problem that will be considered later involves what is termed aliasing, that is a fault-free output signature is given even though a fault is present in the network under test. The opposite signature failing, that of giving a fault-present signature from a fault-free circuit, is not of such great concern, since although it is wasteful to throw away a healthy circuit this is a safe-side failure and one which statistically can be tolerated.

The principal thrust of published means of output compression is usually to consider some exhaustive test input sequence, but not to have to do the step by step check of the response of every output function of the network. This form of compression has been termed space compression, since it does not modify the time to apply the input test vector sequence, as distinct from the input compression considered in the preceding section which may be termed time compression. One method that has been proposed is the use of a linear output filter to generate k test outputs from the m primary outputs of the network under test, where $k < m$. This is illustrated in Figure 4.4a. The difficulty here is to determine which primary outputs to combine, since it may readily be shown that two faulty output functions can produce a fault-free output response from the output filter. For example, should there be a faulty inverter gate somewhere within the circuit which causes wrong outputs to be given on the same input test vectors at two of the outputs exclusive-ORed together, then no fault will be present at this final output.

An extension of this output compression concept is shown in Figure 4.4b. Here an additional function, z_t, is added to the normal circuit, and all outputs including z_t are exclusive-ORed together to give a single test output r_1. If z_t is chosen appropriately, then the test output can be made a simple function when the network is fault free, for example always logic 0 or logic 1. Considering this in more detail, it will be appreciated that this test strategy will operate when the circuit is in normal system operation, and hence is a form of online self-checking test of the circuit network. However, it stills fails to detect the type of fault described in the previous paragraph, and therefore is certainly not a 100 % satisfactory self-test procedure.

We will make some further observations on the compression circuit of Figure 4.4a in the following section, and the concept of Figure 4.4b will be mentioned again in Section 4.6.

4.3.1 Syndrome (ones-count) testing

Conceptually, the most simple compacted signature of an output function is a count of the number of logic 1s (or 0s) which appears in the output bit stream when the full input test sequence is applied. Such a test is known as a ones-count test, the value of which clearly ranges from zero to a maximum of 2^n.

The syndrome of a Boolean function is a normalised ones-count, and therefore has the range of 0 to 1 obtained by dividing the ones-count by 2^n. This academic distinction between the ones-count and the syndrome value is of no immediate significance and will not be pursued here; instead here, as in other publications, we will consider syndrome testing to be the same as ones-count testing, the $1/2^n$ normalising factor being ignored for our present purposes. It may, however, be noted that certain authorities such as Savir[9,10] do maintain the normalising factor in some published works.

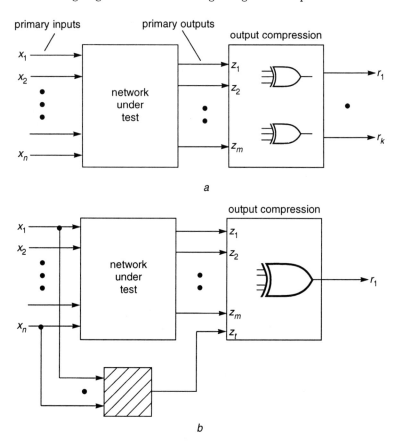

Figure 4.4 The use of a linear output filter to combine primary outputs
   a generation of k test outputs from m primary outputs
   b generation of one test output such that r_1 is, say, always 0 or 1 or some other simple function

A simple syndrome count of the response of a primary output is an insufficient test for general use. Certainly, if the syndrome count is not the same as expected, then the output function is faulty, but the converse is not true. A trivial example to illustrate the latter is that of an exclusive-OR (odd-parity) function; here the healthy syndrome count is $\frac{1}{2}2^n$, but this remains so even if a primary input variable is stuck-at 0 or stuck-at 1. Similarly, if the healthy syndrome count was 1, which indicates a single one-valued minterm in the output function truthtable, an inversion fault could change a literal in the minterm from to x_i to $\bar{x}_i$ or vice versa, but the syndrome count remains 1. In general, therefore, a simple syndrome count is an unsatisfactory signature for an output function.

A theoretical consideration of the probability of fault masking (that is giving the fault-free test output signature even when a fault is present) can be made. Consider a syndrome test with R input test vectors. Then the number of possible output sequences is 2^R with $2^R - 1$ being faulty sequences. Also the number of R-bit output sequences having s 1s is $\binom{R}{s}$, of which $\binom{R}{s} - 1$ will be faulty output sequences. Hence the probability of fault masking given any random set of input test vectors will be:

$$P_{fm} = \frac{\binom{R}{s} - 1}{2^R - 1} \times 100\% \qquad (4.1)$$

Since the coefficient $\binom{R}{s}$ has a bell-shaped distribution, eqn. 4.1 states that the probability of fault masking is low when s is near its extreme values, but increases rapidly as s nears the midpoint value of $\frac{1}{2}R$. However, this assumes that all errors in the output bit stream are equally likely, which is usually very far from real-life truth. As shown above, a syndrome count of 1 may easily be maintained under fault conditions, and therefore all the published analyses of fault masking (or aliasing) which are based upon equally probable faults must be treated with great care.

Many developments of syndrome testing which aim to give an acceptable signature test have been pursued. For example, in the compaction concept illustrated in Figure 4.4a syndrome counts have been suggested for the final output signatures. The choice of the functions to combine in the output linear filter is considered on an exhaustive basis, evaluating the syndrome values for all possible nontrivial combinations of the primary outputs z_i, $i = 1$ to m. From this exhaustive analysis a solution which embraces all z_i outputs and which gives near-maximum (or near-minimum) syndrome values is chosen[7,11,12]. An interesting technique for determining the syndromes for all exclusive-ORs of the z_i outputs is possible, using a $2^n \times 2^n$ Hadamard transform operating on a single column vector formed from the truthtable of the individual z_i outputs. As an example, consider a simple three-input function $f(x_1, x_2, x_3)$ with three outputs z_1, z_2 and z_3, as detailed below. Allocating a binary weighting to each output, a single decimal number may be given to represent each three-bit output, as follows:

| Outputs | | | Three-bit output in decimal notation |
z_3	z_2	z_1	
1	0	1	5
1	1	0	6
1	1	0	6
0	0	1	1
0	1	0	2
1	0	0	4
1	0	0	4
0	1	1	3

A column vector **T**] is now formed by recording the number of times that the decimal-notation output takes the value 0, 1, 2, ..., $2^m - 1$, which is then operated upon by the $2^n \times 2^n$ Hadamard transform matrix, using the −1 entries only in this transform matrix[8]. For the above example we have:

$$\begin{bmatrix} 1 & 1 & 1 & 1 & 1 & 1 & 1 & 1 \\ 1 & -1 & 1 & -1 & 1 & -1 & 1 & -1 \\ 1 & 1 & -1 & -1 & 1 & 1 & -1 & -1 \\ 1 & -1 & -1 & 1 & 1 & -1 & -1 & 1 \\ 1 & 1 & 1 & 1 & -1 & -1 & -1 & -1 \\ 1 & -1 & 1 & -1 & -1 & 1 & -1 & 1 \\ 1 & 1 & -1 & -1 & -1 & -1 & 1 & 1 \\ 1 & -1 & -1 & 1 & -1 & 1 & 1 & -1 \end{bmatrix} \begin{bmatrix} 0 \\ 1 \\ 1 \\ 1 \\ 2 \\ 1 \\ 2 \\ 0 \end{bmatrix} \begin{matrix} \text{(no zero)} \\ \text{(one 1)} \\ \text{(one 2)} \\ \text{(one 3)} \\ \text{(two 4s)} \\ \text{(one 5)} \\ \text{(two 6s)} \\ \text{(no 7)} \end{matrix} = \begin{bmatrix} |0| \\ |3| \\ |4| \\ |5| \\ |5| \\ |6| \\ |5| \\ |4| \end{bmatrix} \begin{matrix} \\ \text{syndrome for output } z_1 \\ \text{syndrome for output } z_2 \\ \text{syndrome for } z_1 \oplus z_2 \\ \text{syndrome for output } z_3 \\ \text{syndrome for } z_1 \oplus z_3 \\ \text{syndrome for } z_2 \oplus z_3 \\ \text{syndrome for } z_1 \oplus z_2 + z_3 \end{matrix}$$

$$[\text{Hd}] \qquad \text{T}]$$

Fast transform procedures are relevant for this computation. Finally, a selection involving maximum (minimum) syndrome values and all z_1 outputs at least once are chosen.

Other syndrome pursuits have included:

(i) checking to verify whether a syndrome count will detect all given faults on a fault list, such as all stuck-at faults;

(ii) modification to the network under test by the addition of test input(s) and further logic which effectively adds minterms when under test so as to give an acceptably secure syndrome count for all circuit failures;

(iii) constrained syndrome testing which involves more than one syndrome test, certain primary inputs being held constant during each individual syndrome count procedure;

(iv) weighted syndrome testing, where the individual z_i outputs are given weighting factors, the final syndrome test count being $\sum_{i=1}^{m} w_i S_i$ where w_i is the weight given to the syndrome count S_i from output z_i.

Further details of these concepts may be found published[12–16]. However it is always possible to postulate a faulty output sequence which will give the same syndrome count as a fault-free circuit; such postulated faults may not be physically possible in a given circuit design, but overall syndrome testing has not to date been viewed with favour as a VLSI test strategy since its security is often difficult to determine.

4.3.2 Accumulator-syndrome testing

A modification of the simple ones-count involved in syndrome testing has been proposed by Saxena and Robinson[17]. Here an integral of the syndrome count is considered, the final accumulator-syndrome count being effectively the area under the syndrome count over the exhaustive input sequence of 2^n test vectors. For example, if the output bit stream begins:

$$0 \quad 0 \quad 1 \quad 1 \quad 0 \quad 0 \quad 1 \quad 0 \quad 1 \quad \ldots$$

then the syndrome count S and the accumulator-syndrome count AS increase as follows:

S:	0	0	1	2	2	2	3	3	4	...
AS:	0	0	1	3	5	7	10	13	17	...

It will be appreciated that although the syndrome value S for any given circuit is independent of the order of application of the input test vectors, the accumulator syndrome value AS will depend upon the input test vector sequence. Therefore comparing the effectiveness of the signature AS with S, it is less likely that the former will give a fault-free count when the circuit is faulty than will the latter.

However, it is not obvious how to obtain the best (most secure) accumulator-syndrome signature for a given circuit, given that one is free to choose any sequence of the test input vectors. As before, it will be found that it is possible to formulate a faulty output sequence that will give the same count as the fault-free circuit, but again this may involve faults which are physically impossible in a given circuit. Further comments are given in the following section.

4.3.3 Transition count testing

An alternative to the preceding two count methods is to count the number of transitions from 0 to 1 (or vice versa) in the output bit stream when on test. Like the syndrome tests, this transition count test is usually made over a fully exhaustive input test set, but unlike the simple syndrome test this test is clearly dependent upon the order of application of the input test vectors.

However, as shown in Table 4.3, the transition count signature may not be secure against all possible faults unless specific care is taken (see later). Table 4.3 shows that the syndrome count S, the accumulator-syndrome count AS and the transition count T can each have their fault-free value even with certain faulty output bit streams. It is even possible for all three signatures to have their fault-free value under certain faulty sequences.

Table 4.3 *A simple example showing the failure of all three forms of output compression to provide, either individually or collectively, a unique fault-free signature*

	Output bit stream	Syndrome count S	Accumulator-syndrome count AS	Transition count T
(i) healthy	0 1 1 0 0 1 0 0	3	16	2
(ii) faulty	0 1 0 1 1 0 0 0	3	16	2
(iii) faulty	0 1 1 0 0 0 1 0	3	15	2
(iv) faulty	0 1 1 0 1 0 0 0	3	17	2
(v) faulty	0 0 1 1 0 1 1 0	4	16	2
(vi) faulty	1 0 0 1 0 1 0 0	3	16	3

There remains one final possibility which will yield a secure signature for a given combinational network, provided that we are allowed to arrange the test vector input sequence in a specific order dependent upon the output function. If the input sequence is arranged such that there is only one transition in the output bit stream, i.e. a block of all 0s followed by a block of all 1s (or vice versa), then this will detect any fault within either block of 0 or 1 output bits. However, this will fail if a block of logic 0s at the end of the healthy logic 0 bit stream becomes all 1s, or if a block of 1s at the beginning of the healthy logic 1 bit stream becomes all 0s, that is that the transition point between the 0 and 1 blocks shifts but remains just one transition. This shortcoming may very easily be overcome by repeating the initial test vector of the block of 0s and the initial test vector of the block of 1s at the end of their respective blocks, thus giving a total input sequence of $2^n + 2$ test vectors with a healthy transition count of one.

This secure and very elegant compression signature has two drawbacks, namely:

(i) the input test sequence has to be defined and generated in the required order;
(ii) if two or more primary outputs are present then separate test sequences are necessary for each output.

Hence, although the transition count signature has theoretical attractions, like syndrome counting it has not found industrial favour. Further details of these output compression techniques, including theoretical considerations of their fault-masking properties and combined use may be found References 9–27. We will return briefly to the syndrome count signature in Section 4.4.3, when it will be seen that the syndrome count is a particular case of a much wider range of numerical parameters that have been suggested for test purposes.

4.4 Arithmetic, Reed-Muller and spectral coefficients

Although the binary values of 0 and 1 and Boolean relationships are the normally accepted way of defining any two-valued digital logic function, there are alternative mathematical values and relationships which can be employed. These alternatives have been widely researched, and offer certain academic advantages for digital design and test in comparison with conventional {0,1} Boolean representations. We introduce them here for completeness, although they have not yet secured any widespread use in design or test practice.

Let us first define the several symbols which will be used in the following survey; some will be the same as previously used but the majority will be new to our discussions so far.

List of symbols used below

x_i, $i = 1$ to n	independent binary variables, $x_i \in \{0, 1\}$, x_1 least significant
$f(x_1, x_2, ..., x_n)$ or merely $f(X)$	binary function of the x_i input variables, $f(X) \in \{0, 1\}$
X]	truthtable column vector for function, $f(X)$, entries $\in \{0, 1\}$ in minterm decimal order
Y]	truthtable column vector for $f(X)$, entries $\in \{+1, -1\}$ in minterm decimal order
a_j, $j = 0$ to $2^n - 1$	coefficients in canonic minterm expansion for $f(X)$
A]	coefficient column vector of a_j in decimal order
b_j, $j = 0$ to $2^n - 1$	coefficients in arithmetic canonic expansion for $f(X)$
B]	coefficient column vector of b_j in decimal order
c_j, $j = 0$ to $2^n - 1$	coefficients in positive Reed-Muller canonic expansion for $f(X)$
C]	coefficient column vector of c_j in decimal order
r_j, $j = 0$ to $2^n - 1$	spectral coefficients defining $f(X)$, transformed from **X**]
R]	coefficient column vector of r_j in decimal order
s_j, $j = 0$ to $2^n - 1$	spectral coefficients defining $f(X)$, transformed from **Y**]
S]	coefficient column vector of s_j in decimal order
[**T**^{n}]	general $2^n \times 2^n$ transform matrix
[**Hd**]	$2^n \times 2^n$ Hadamard matrix
[**RW**]	$2^n \times 2^n$ Rademacher-Walsh matrix
[**I**]	$2^n \times 2^n$ identity matrix

4.4.1 Arithmetic and Reed-Muller coefficients

The canonic minterm expansion for any Boolean function $f(X)$ is the sum of products expression involving all 2^n minterms of $f(X)$, each of which may take the value 0 or 1. For any three-variable functions we have:

$$f(X) = a_0\bar{x}_1\bar{x}_2\bar{x}_3 + a_1x_1\bar{x}_2\bar{x}_3 + a_2\bar{x}_1x_2\bar{x}_3 + a_3x_1x_2\bar{x}_3 + a_4\bar{x}_1\bar{x}_2x_3$$
$$+ a_5x_1\bar{x}_2x_3 + a_6\bar{x}_1x_2x_3 + a_7x_1x_2x_3$$
$$a_j, j = 0 \text{ to } 2^n - 1, \in \{0,1\} \tag{4.2}$$

The vector **A]** of the a_j in correct canonic order therefore fully defines $f(X)$. This is a trivial case, since the a_j are merely defining the zero-valued and one-valued minterms of $f(X)$, with **A]** being identical to the normal output truth-table of $f(X)$ in minterm order. In matrix form we may write (see list of symbols):

$$[\mathbf{I}]\mathbf{X}] = \mathbf{A}], \text{ giving } \mathbf{A}] = \mathbf{X}] \tag{4.3}$$

However, a canonic arithmetic expansion for any function $f(X)$ is also available, but not widely known[24,25]. For any three-variable function we have:

$$f(X) = b_0 + b_1 x_1 + b_2 x_2 + b_3 x_1 x_2$$
$$+ b_4 x_3 + b_5 x_1 x_3 + b_6 x_2 x_3 + b_7 x_1 x_2 x_3$$
$$b_j, j = 0 \text{ to } 2^n - 1 \tag{4.4}$$

where the addition is arithmetic, not Boolean. The general expression for any $f(X)$ is given by:

$$f(X) = \sum_{j=0}^{2^n - 1} b_j X_j$$
$$X_j \in \{1, x_1, x_2, \ldots, x_n, x_1 x_2, \ldots, x_1 x_2 \ldots x_n\} \tag{4.5}$$

and where the binary subscript identifiers of the x_i product terms are read as decimal numbers j. As an example, the parity function $\bar{x}_1\bar{x}_2 x_3 + \bar{x}_1 x_2 \bar{x}_3 + x_1 \bar{x}_2 \bar{x}_3 + x_1 x_2 x_3$ may be written as:

$$f(X) = x_1 + x_2 - 2x_1 x_2 + x_3 - 2x_1 x_3 - 2x_2 x_3 + 4x_1 x_2 x_3$$

Note that b_j is not confined to positive values. The vector **B]** of the b_j coefficient is the arithmetic spectrum of $f(X)$. For any three-variable function ($n = 3$) it may determined by the transform:

$$\begin{bmatrix} 1 & 0 & 0 & 0 & 0 & 0 & 0 & 0 \\ -1 & 1 & 0 & 0 & 0 & 0 & 0 & 0 \\ -1 & 0 & 1 & 0 & 0 & 0 & 0 & 0 \\ 1 & -1 & -1 & 1 & 0 & 0 & 0 & 0 \\ -1 & 0 & 0 & 0 & 1 & 0 & 0 & 0 \\ 1 & -1 & 0 & 0 & -1 & 1 & 0 & 0 \\ 1 & 0 & -1 & 0 & -1 & 0 & 1 & 0 \\ -1 & 1 & 1 & -1 & 1 & -1 & -1 & 1 \end{bmatrix} \begin{bmatrix} b_0 \\ \vdots \\ b_7 \end{bmatrix} \mathbf{X}] = \mathbf{B} \tag{4.6}$$

The transform for any n is given by:

$$\mathbf{T}^n = \begin{bmatrix} \mathbf{T}^{n-1} & 0 \\ -\mathbf{T}^{n-1} & \mathbf{T}^{n-1} \end{bmatrix}$$

$$\mathbf{T}^0 = +1 \tag{4.7}$$

A further alternative is the Reed-Muller canonic expansion for $f(X)$ which involves exclusive-OR (addition modulo 2) relationships rather than the inclusive-OR. There are 2^n different possible Reed-Muller expansions for any given function $f(X)$, involving all possible permutations of each input variable, x_i, true or complemented; here we will confine ourselves to the positive polarity RM expansion for $f(X)$, which involves all x_is true and none complemented[28,30]. For any three-variable function we have:

$$f(X) = c_0 \oplus c_1 x_1 \oplus c_2 x_2 \oplus c_3 x_1 x_2 \oplus c_4 x_3$$
$$\oplus c_5 x_1 x_3 \oplus c_6 x_2 x_3 \oplus c_7 x_1 x_2 x_3 \tag{4.8}$$

$$c_j, j = 0 \text{ to } 2^n - 1, \in (0, 1)$$

where the addition is mod_2. The expression for any $f(X)$ is given by:

$$f(X) = \sum_{j=0}^{2^n-1} c_j X_j \tag{4.9}$$

X_j as previously defined.

The vector $\mathbf{C}]$ of the c_j coefficients is the Reed-Muller positive canonic spectrum of $f(X)$. For any three-variable function we have:

$$\begin{bmatrix} 1 & 0 & 0 & 0 & 0 & 0 & 0 & 0 \\ 1 & 1 & 0 & 0 & 0 & 0 & 0 & 0 \\ 1 & 0 & 1 & 0 & 0 & 0 & 0 & 0 \\ 1 & 1 & 1 & 1 & 0 & 0 & 0 & 0 \\ 1 & 0 & 0 & 0 & 1 & 0 & 0 & 0 \\ 1 & 1 & 0 & 0 & 1 & 1 & 0 & 0 \\ 1 & 0 & 1 & 0 & 1 & 0 & 1 & 0 \\ 1 & 1 & 1 & 1 & 1 & 1 & 1 & 1 \end{bmatrix} \mathbf{X} = \mathbf{C} \tag{4.10}$$
$$\text{mod}_2$$

where the matrix additions are modulo 2, not arithmetic.

The transform for any n is given by:

$$\mathbf{T}^n = \begin{bmatrix} \mathbf{T}^{n-1} & 0 \\ \mathbf{T}^{n-1} & \mathbf{T}^{n-1} \end{bmatrix}_{\text{mod}_2}$$

$$\mathbf{T}^0 = +1 \tag{4.11}$$

The relationships between the c_j coefficients for the positive canonic case (above) and any other polarity expansion may be given by a further matrix operation[30]; this will not concern us here.

4.4.2 The spectral coefficients

It is possible to consider the transformation of any binary-valued output vector of a function $f(X)$ into some alternative domain by multiplying this binary-valued vector by a $2^n \times 2^n$ complete orthogonal transform matrix. The transformation produced by such a mathematical operation is such that the information content of the original vector is fully retained in the resulting spectral domain vector, that is the spectral data is unique for the given function $f(X)$; the application of the inverse transformation enables the original binary data to be unambiguously recreated from the spectral domain data.

The mathematical theory of complete orthogonal transforms may be found in many mathematical texts[30–35]. However the simplest to apply for the transformation of binary data, and the most well-structured, is Hadamard transform, defined by the recursive structure:

$$\mathbf{Hd}^n = \begin{bmatrix} \mathbf{Hd}^{n-1} & \mathbf{Hd}^{n-1} \\ \mathbf{Hd}^{n-1} & -\mathbf{Hd}^{n-1} \end{bmatrix}$$

$$\mathbf{Hd}^0 = +1 \tag{4.12}$$

For example, given the function $f(X) = \bar{x}_1\bar{x}_2 x_3 + \bar{x}_1 x_2 \bar{x}_3 + x_1\bar{x}_2\bar{x}_3 + x_1 x_2 x_3$, we have the following transformation, where m_0 to m_7 are the minterms of $f(X)$ and r_0 to r_7 are the resulting spectral coefficients:

$$\begin{bmatrix} 1 & 1 & 1 & 1 & 1 & 1 & 1 & 1 \\ 1 & -1 & 1 & -1 & 1 & -1 & 1 & -1 \\ 1 & 1 & -1 & -1 & 1 & 1 & -1 & -1 \\ 1 & -1 & -1 & 1 & 1 & -1 & -1 & 1 \\ 1 & 1 & 1 & 1 & -1 & -1 & -1 & -1 \\ 1 & -1 & 1 & -1 & -1 & 1 & -1 & 1 \\ 1 & 1 & -1 & -1 & -1 & -1 & 1 & 1 \\ 1 & -1 & -1 & 1 & -1 & 1 & 1 & -1 \end{bmatrix} \begin{bmatrix} 0 \\ 1 \\ 1 \\ 0 \\ 1 \\ 0 \\ 0 \\ 1 \end{bmatrix} \begin{matrix} m_0 \\ m_1 \\ m_2 \\ m_3 \\ m_4 \\ m_5 \\ m_6 \\ m_7 \end{matrix} = \begin{bmatrix} 4 \\ 0 \\ 0 \\ 0 \\ 0 \\ 0 \\ 0 \\ -4 \end{bmatrix} \begin{matrix} r_0 \\ r_1 \\ r_2 \\ r_3 \\ r_4 \\ r_5 \\ r_6 \\ r_7 \end{matrix}$$

$$[\mathbf{Hd}] \qquad \mathbf{X}] \qquad \mathbf{R}] \tag{4.13}$$

However, for reasons which will be seen later, the spectral coefficients may be redesignated as follows:

above designation:	r_0	r_1	r_2	r_3	r_4	r_5	r_6	r_7
revised designation:	r_0	r_1	r_2	r_{12}	r_3	r_{13}	r_{23}	r_{123}
spectrum of $f(X)$:	4	0	0	0	0	0	0	-4

The two nonzero values $r_0 = 4$, $r_{123} = -4$, together fully define the given function $f(X)$.

It is equally relevant to transform the output truthtable vector $\mathbf{Y}]$ representing $f(X)$, where $\mathbf{Y}]$ is a recoding of $\mathbf{X}]$, recoded logic 0 → +1, logic

$1 \rightarrow -1$. **[Hd] Y]** then gives the spectral coefficient vector **S]** for $f(X)$, which for the above function would yield the spectrum:

$$\begin{bmatrix} 1 & 1 & 1 & 1 & 1 & 1 & 1 & 1 \\ 1 & -1 & 1 & -1 & 1 & -1 & 1 & -1 \\ 1 & 1 & -1 & -1 & 1 & 1 & -1 & -1 \\ 1 & -1 & -1 & 1 & 1 & -1 & -1 & 1 \\ 1 & 1 & 1 & 1 & -1 & -1 & -1 & -1 \\ 1 & -1 & 1 & -1 & -1 & 1 & -1 & 1 \\ 1 & 1 & -1 & -1 & -1 & -1 & 1 & 1 \\ 1 & -1 & -1 & 1 & -1 & 1 & 1 & -1 \end{bmatrix} \begin{bmatrix} +1 \\ -1 \\ -1 \\ +1 \\ -1 \\ +1 \\ +1 \\ -1 \end{bmatrix} \begin{bmatrix} m_0 \\ m_1 \\ m_2 \\ m_3 \\ m_4 \\ m_5 \\ m_6 \\ m_7 \end{bmatrix} = \begin{bmatrix} 0 \\ 0 \\ 0 \\ 0 \\ 0 \\ 0 \\ 0 \\ +8 \end{bmatrix} \begin{bmatrix} s_0 \\ s_1 \\ s_2 \\ s_{12} \\ s_3 \\ s_{13} \\ s_{23} \\ s_{123} \end{bmatrix}$$

$$\quad [\mathbf{Hd}] \qquad\qquad\qquad \mathbf{Y}] \qquad\qquad \mathbf{S}] \qquad\qquad (4.14)$$

Here, the single nonzero spectral coefficient value $s_{123} = +8$ fully defines $f(X)$. Note that if any coefficient has the maximum value of $\pm 2^n$, then all remaining coefficients must be zero-valued due to the orthogonality of the rows of the transform.

Both **[Hd] Y]** and **[Hd] X]** will be found extensively used in the literature[38]. The relationship between r_j and s_j spectral coefficients is linear, being:

$$r_0 = \frac{1}{2}(2^n - s_0) \; ; \; r_j, j \neq 0, = -\frac{1}{2}s_j \qquad (4.15)$$

There are a number of alternative complete orthogonal transforms to the Hadamard which are row reorderings[34,35,37]. The most prominent is the Rademacher-Walsh variant, **[RW]**, which directly generates the spectral coefficients in the logical order $r_0, r_1, r_2, r_3, r_{12}, r_{13}, \ldots$ rather than in the original Hadamard ordering. It will be appreciated that with such variants the recursive structure of the Hadamard transform is lost, but there is no difference in the information content of the resulting $r_j(s_j)$ spectral coefficients.

It should also be noted that the coefficients in the spectral domain may take both positive and negative values. For the majority of functions the coefficients will be nonzero; for example, the spectrum for $f(X) = x_1 \bar{x}_2 + \bar{x}_1 x_2 x_3$ is **R]** = 3, −1, 1, −3, −1, −1, 1, 1 or **S]** = 2, 2, −2, 6, 2, 2, −2, −2; the example shown in eqns. 4.13 and 4.14 is a special case of an odd-parity function.

Since all the above different sets of coefficients which can be proposed to define a given function $f(X)$ are obtainable by an appropriate matrix operation on the column vector representing $f(X)$, there exists appropriate matrix relationships between these alternative sets of coefficients[19]. For example, given the Hadamard spectral coefficients **R]** the other coefficient vectors are given by the following relationships.

(i) The r_j to a_j coefficient relationships:

$$\mathbf{A}] = \mathbf{X}] = [\mathbf{Hd}]^{-1} \mathbf{R}] \qquad (4.16)$$

where $\left[\mathbf{Hd}\right]^{-1} = \dfrac{1}{2^n}\left[\mathbf{Hd}\right]$

In particular:

$$a_0 = \dfrac{1}{2^n}\left[\sum_{j=0}^{2^n-1} r_j\right] \qquad (4.17)$$

Note that the syndrome count, S, without the normalising factor $1/2^n$ is merely:

$$S = \sum_{j=0}^{2^n-1} a_j, = r_0 \qquad (4.18)$$

(ii) The r_j to b_j coefficient relationships

$$\begin{aligned}\mathbf{B}] &= \left[\mathbf{T}^n\right]\mathbf{X}] \\ &= \left[\mathbf{T}^n\right]\left[\dfrac{1}{2^n}[\mathbf{Hd}]\,\mathbf{R}]\right] \\ &= \dfrac{1}{2^n}\left[\left[\mathbf{T}^n\right][\mathbf{Hd}]\,\mathbf{R}]\right] \end{aligned} \qquad (4.19)$$

where $[\mathbf{T}^n]$ is the transform given in eqn. 4.7. Evaluation of $[\mathbf{T}^n][\mathbf{Hd}]$, see eqns. 4.6 and 4.11, will give the result conversion matrix $[\mathbf{Trb}]$ from $\mathbf{R}]$ to $\mathbf{B}]$, giving:

$$\mathbf{Trb}^n = \begin{bmatrix} \mathbf{Trb}^{n-1} & \mathbf{Trb}^{n-1} \\ 0 & -2\mathbf{Trb}^{n-1} \end{bmatrix}$$

$$\mathbf{Trb}^0 = +1 \qquad (4.20)$$

Eqn. 4.19 may therefore be finally written as:

$$\mathbf{B}] = \dfrac{1}{2^n}[\mathbf{Trb}]\,\mathbf{R}] \qquad (4.21)$$

Again, the particular (simple) case of:

$$b_0 = \dfrac{1}{2^n}\left[\sum_{j=0}^{2^n-1} r_j\right] \qquad (4.22)$$

may be observed.

The inverse of $[\mathbf{Trb}]$ will provide the conversion matrix from $\mathbf{B}]$ to $\mathbf{R}]$, i.e.:

$$\begin{aligned}\mathbf{R}] &= \left[\mathbf{Trb}\right]^{-1}\mathbf{B}] \\ &= [\mathbf{Tbr}]\,\mathbf{B}]\end{aligned}$$

whence it readily follows from eqn. 4.20 that:

$$\mathbf{Tbr}^n = \begin{bmatrix} 2\mathbf{Tbr}^{n-1} & \mathbf{Tbr}^{n-1} \\ 0 & -\mathbf{Tbr}^{n-1} \end{bmatrix}$$

$$\mathbf{Tbr}^0 = +1 \qquad (4.23)$$

(iii) *The r_j to c_j coefficient relationships:*

From eqns. 4.10, 4.11 and 4.16, the Reed-Muller positive canonic c_j coefficients are related to the r_j spectral coefficients by:

$$\mathbf{C}] = \begin{bmatrix} \mathbf{T}^n \end{bmatrix}_{\mathrm{mod}\,2} \mathbf{X}]$$

$$= \begin{bmatrix} \mathbf{T}^n \end{bmatrix}_{\mathrm{mod}\,2} \left[\frac{1}{2^n} \begin{bmatrix} \mathbf{Hd} \end{bmatrix} \mathbf{R} \right] \qquad (4.24)$$

where $[\mathbf{T}^n]$ is the modulo 2 transform given in eqn. 4.11.

The modulo 2 (exclusive-OR) relationships imposed by $[\mathbf{T}^n]$ imply that all odd numbers are expressed as +1 and all even numbers expressed as 0. Hence, it is permissible to combine $[\mathbf{T}^n][\mathbf{Hd}]$ into a single conversion matrix $[\mathbf{Trc}]$ giving:

$$\mathbf{C}] = \frac{1}{2^n} \begin{bmatrix} \mathbf{Trc} \end{bmatrix} \mathbf{R}] \qquad (4.25)$$

where the final vector product summations are written as 1 or 0 depending upon odd or even summation values, respectively. From eqns. 4.11 and 4.12 the conversion matrix $[\mathbf{Trc}]$ is given by:

$$\mathbf{Trc} = \begin{bmatrix} \mathbf{Trc}^{n-1} & \mathbf{Trc}^{n-1} \\ 2\mathbf{Trc}^{n-1} & 0 \end{bmatrix}_{\mathrm{mod}\,2}$$

$$\mathbf{Trc}^0 = +1 \qquad (4.26)$$

The inverse of $[\mathbf{Trc}]$ will provide the conversion matrix $[\mathbf{Tcr}]$ from $\mathbf{C}]$ to $\mathbf{R}]$, i.e.:

$$\mathbf{R}] = [\mathbf{Trc}]^{-1}\mathbf{C}]$$

$$= [\mathbf{Tcr}]\mathbf{C}]$$

where it readily follows from eqn. 4.25 that:

$$\mathbf{Tcr}^n = \begin{bmatrix} 0 & \mathbf{Tcr}^{n-1} \\ 2\mathbf{Tcr}^{n-1} & -\mathbf{Tcr}^{n-1} \end{bmatrix}$$

$$\mathbf{Tcr}^0 = +1 \qquad (4.27)$$

Further information on the relationships between these coefficient vectors, and between the very many variants of complete orthogonal transforms which may be proposed instead of the Hadamard transform, may be found in References 19, 37–39.

4.4.3 Coefficient test signatures

To revert to digital testing considerations, looking back at the developments from eqn. 4.3, onwards it will be seen that the coefficients defining $f(X)$ in eqn. 4.2 give discrete information about the function, merely listing whether each individual minterm of $f(X)$ is logic 0 or logic 1. Looking at any one coefficient gives no information whatsoever about any of the other coefficients, and all 2^n have to be read to define $f(X)$. On the other hand, each of the spectral coefficients in eqn. 4.13 is a global parameter of $f(X)$, since all the entries in $f(X)$ influence the value of each coefficient. Between these extremes other coefficients contain an information content concerning $f(X)$ ranging from (i) discrete, (ii) some window on $f(X)$, through to (iii) some global parameter.

The interpretation of the meaning of the spectral coefficients as correlation coefficients has been widely published[37]. This interpretation is more readily appreciated when the {0,1} function vector **X**] is recoded into the alternative {+1, −1} vector **Y**]. This is illustrated in the following example, where the sixth row of the $n = 3$ Hadamard transform is shown operating on the function $f(X) = x_1 x_2 + x_1 \bar{x}_3 + \bar{x}_1 x_3$. The resultant coefficient value **S**] is clearly equal to the number of agreements minus the number of disagreements between the function vector **Y**] and the row in the transform, and therefore represents the correlation between the two. Further, if the rows in the transform are examined, and +1 is taken as logic 0 and −1 as logic 1, it will be appreciated that each transform row is a function of $f(X)$; the sixth row in the following example corresponds to the function $x_1 \oplus x_3$, the complete set of rows corresponding to 0, x_1, x_2, $x_1 \oplus x_2$, x_3, $x_1 \oplus x_3$, $x_2 \oplus x_3$ and $x_1 \oplus x_2 \oplus x_3$. Notice that the labelling of the resultant coefficients in **S**] shown in eqn. 4.14. reflects this functionality.

$$\begin{bmatrix} \bullet \\ \bullet \\ \bullet \\ \bullet \\ \bullet \\ 1 \quad -1 \quad 1 \quad -1 \quad -1 \quad 1 \quad -1 \quad 1 \\ \bullet \\ \bullet \end{bmatrix} \begin{bmatrix} 1 \\ -1 \\ 1 \\ -1 \\ -1 \\ 1 \\ -1 \\ -1 \end{bmatrix} = \begin{bmatrix} \bullet \\ \bullet \\ \bullet \\ \bullet \\ \bullet \\ 6 \\ \bullet \\ \bullet \end{bmatrix} \; s_{13}$$

$$[\text{Hd}] \qquad \qquad \mathbf{Y}] \qquad \mathbf{S}]$$

Therefore the spectral coefficients in **S**] of a given function represent the correlation between the function and its primary inputs and all possible exclusive-ORs of its primary inputs.

The same is true if the {0,1} function vector **X**] is used instead of **Y**], the actual coefficient values being scaled as indicated by eqn. 4.15. In the above example the given function is almost $x_1 \oplus x_3$, indicated by the value of s_{13} being close to maximum value, being one minterm different from exactly $x_1 \oplus x_3$. Any change (fault) in the given function will clearly change one or more of the spectral coefficient values.

The same considerations can be applied to all other coefficient vectors which represent a given function vector **X**]. All may therefore be regarded as some correlation value between the network output $f(X)$ and a further function of the primary inputs, the mathematical relationships between the actual values being as developed above. Hence, the commonality between many single-count signature proposals may be regarded as a search for the simplest test signature for $f(X)$ covering all fault conditions. At one extreme we have the situation where $f(X)$ is correlated with itself, see Figure 4.5a; clearly this is not a practical proposition except in exceptional circumstances, since the test generator is then a duplicate of, and therefore the same complexity as, the network under test. At the other extreme we have the situation where $f(X)$ is correlated with logic 1 (or 0), see Figure 4.5b, which is the syndrome count. The syndrome count is therefore the most simple (and the weakest) of all the correlation coefficients available in the spectral domain data. The two remaining examples shown in Figure 4.5 represent the use of higher-ordered spectral coefficients as signatures. In all cases a near-maximum count is usually considered to be an ideal objective.

The present status of the arithmetic, Reed-Muller and spectral coefficients as test signatures is generally as follows.

(i) *The b_i arithmetic coefficients*

The use of the b_i coefficients as test signatures has been proposed by Heidtmann[29]. This work has been shown to have a certain mathematical commonality with the so-called probabilistic treatment of combinational networks[40], but the results appear to offer little attraction in comparison with spectral coefficients, see below.

(ii) *The c_i Reed-Muller coefficients*

As far as is known the c_i coefficients do not appear to have been widely considered as test signatures. If the network heavily involves odd and even parity functions (exclusive-OR and exclusive-NOR relationships), then the c_i coefficients have obvious attractions, but in general little advantage has been found for their use as test signatures. Some further work and further references may be found in Upadhyaya and Saluja[25] and in Damarla and Karpovsky[41].

(iii) *The s_i (or r_i) spectral coefficients*

Because of the richness of information content in the spectral coefficients, a very great deal of research has been expended upon their possible use for

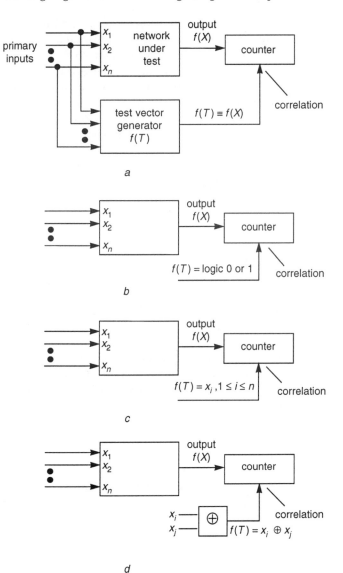

Figure 4.5 Test signatures based upon the correlation between the output f(X) and a test generator f(T) over the exhaustive input sequence of 2^n test vectors
  a the ultimate situation of f(X) compared against a copy of itself
  b f(X) compared against logic 0 or logic 1, which is effectively the syndrome count
  c f(X) compared against a single primary input x_i, = the spectral coefficient s_i (r_i)
  d f(X) compared against the exclusive-OR of two primary inputs $x_i \oplus x_j$, = the spectral coefficient s_{ij} (r_{ij})

analysis, synthesis and fault detection purposes. A disadvantage with the spectral domain which is often cited is the computation required to compute the fault-free spectrum of a given function $f(X)$ as n increases, which involves the $2^n \times 2^n$ orthogonal transform and the 2^n column vector defining $f(X)$. However, fast transform procedures are readily available which are similar to the fast Fourier transform procedure and which reduce the number of computations necessary from $N(N-1)$ additions and subtractions to $N\log_2 N$ where $N = 2^n$,[37,38], but in any case we are unlikely to require the online generation of all the spectral coefficients when a network is under test. If the spectral coefficients of each primary output are computed at the design stage, several advantages accrue, such as:

- confirming which primary inputs are necessary and which unnecessary in each primary output, an unnecessary x_i input being indicated by all the spectral coefficients containing i in their subscript identification being zero-valued;
- showing how like the primary outputs are to the primary inputs and all exclusive-OR combinations of the primary inputs, and also to each other, and hence perhaps suggesting an acceptable output signature using the methods shown in Figure 4.5;
- possibly even suggesting a simpler way to realise the primary output than the designer has considered.

Undoubtedly the spectral domain with its global information content parameters provides a much deeper insight into the logical structure of a combinatorial network than is available from discrete Boolean-domain data, but to date no generally acceptable test strategy has been developed for circuits of VLSI complexity. Recent research interests in this area have involved binary decision diagrams (BDDs), which may have significance in forwarding this area and proposing new test strategies. Further details of this extensive area of research, including the error masking (aliasing) properties and the use of alternative orthogonal transforms such as the Haar transform, may be found in the literature[8,19,29,36-39,42-53]. Reference 19, in particular, contains an extensive bibliography; BDDs are covered in References 48–54. An in-depth comparison of compressed signatures will also be found in Bardell et al.[10].

4.5 Signature analysis

In contrast to the coefficient signature test strategies of the previous section, which as far as is known have not been utilised in OEM applications, signature analysis has been developed and used as an industrial testing tool. Note that in the following we will still largely be dealing with combinational networks, sequential network testing still remaining outside our immediate considerations.

Signature analysis was developed by Hewlett-Packard as a method for both testing and fault finding in increasingly complex digital instruments, being suitable for use both within a factory environment and on site[55]. It consists basically of a 16-bit autonomous linear feedback shift register (LFSR), a probe connected to the input exclusive-OR of the LFSR, a hexadecimal numerical display and an appropriate test vector generator, see Figure 4.6. In use the following procedure is followed:

(i)   the network under test is first reset to a known initial state, and the LFSR also reset usually to 0 0 0 0;
(ii) the probe is manually applied to a chosen point in the network under test (a primary output or an internal node), and a start signal given which causes a given number of input test vectors to be generated and applied to the circuit, the same number of clock pulses also being applied to the LFSR;
(iii) after the given number of test vectors has been applied a stop signal is generated, and the resulting residual count in the 16-bit LFSR read out as four hexadecimal numbers.

The final hexadecimal output count of the LFSR is the test signature for the particular node under test. Note that this test gives a compressed signature, since only the residual bits in the LFSR are read at the end of a test and not the full LFSR output sequence. Fault finding is made possible by producing a comprehensive list of fault-free signatures for the nodes in the circuit, so that a manual step by step probing procedure can determine the interface between the fault-free part of the circuit and the faulty part. This was particularly relevant when there were accessible connections at nodes between discrete components and SSI, MSI and LSI circuits, as well as the final primary outputs.

The detailed action of the signature analysis is straightforward. Recall from our introductory theory in Section 3.4.3.1 of the previous chapter that when the exclusive-OR taps of a LFSR are appropriately chosen, a maximum length pseudorandom sequence of $2^n - 1$ states is generated. If, now, an additional data input is injected into the exclusive-OR feeding the first stage, the resulting sequence will be modified by every logic 1 received in the data bit stream. However:

(i)   if the LFSR is initialised to the all-zero state, then nothing changes in the LFSR until the first logic 1 is received; thereafter the (modified) pseudorandom sequence commences;
(ii) although the LFSR has an inherent maximum length property, it does not necessarily cycle through all the $2^n - 1$ possible states when data input bits are present. The sequences however remain pseudorandom, being subsets of the maximum length pseudorandom sequence;
(iii) although we have been considering combinational only circuit networks, this signature analysis test method can also be applied if there

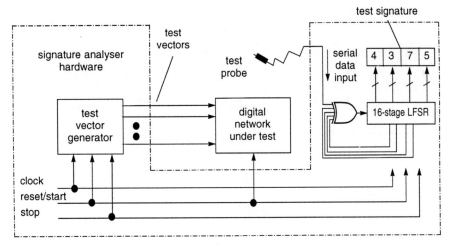

Figure 4.6 The signature analysis circuit, where the test probe may be manually applied to any accessible node of the network under test

are sequential elements in the network under test, provided that the network is full initialised at the beginning of each test, and (usually) that appropriate test nodes are available within the network[10,55].

The state diagram of a simple four-bit LFSR signature analyser is shown in Figure 4.7, with logic 0 and logic 1 as the data input at each state. Any cycle of movement around this state diagram is possible depending upon the data input. Only when the data input is 1, 0, 0, 0, 0, ... does the normal pseudo-random M-sequence result. Notice the return to the all-zero state from the 0 0 0 1 state if the data input is logic 1.

The theoretical fault masking (aliasing) property of this signature analyser is readily shown. Suppose the number of bits in the serial input stream to the LFSR is m (which in practice may be the result of some exhaustive input test sequence), and the number of stages in the LFSR is n, where $n < m$, then the total number of possible serial input sequences is 2^m, but the number of possible different residual signatures in the LFSR is only 2^n. Therefore, if all possible faults in the input bit stream are considered to be equally possible, we have one fault-free bit stream and the correct LFSR signature, with $2^m - 1$ faulty bit streams of which $2^{m-n} - 1$ give the same signature as the fault-free input. The probability of fault masking is therefore given by:

$$P_{fm} = \frac{2^{m-n} - 1}{2^m - 1} \times 100\%$$

which, if 2^m and 2^{m-n} are both $\gg 1$, gives:

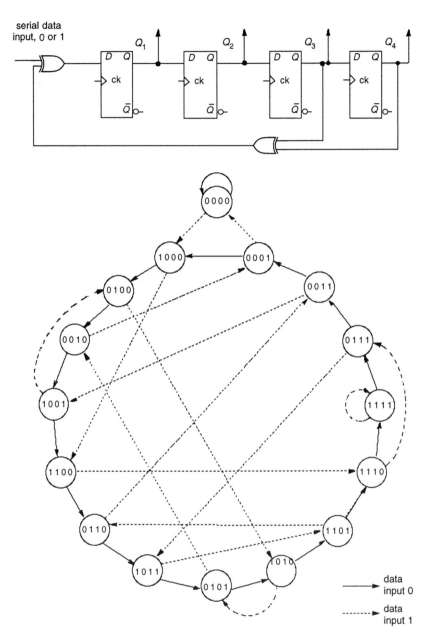

Figure 4.7 The state diagram for a simple four-stage signature analyser, with the characteristic polynomial $P(x) = 1 + x^3 + x^4$ in the LFSR. Practical circuits may contain, say, 16 stages rather than the four shown in this example. Notice also that it is always possible to move from one state to any other state with a maximum of n data input bits, where n is the number of stages in the LFSR (four in this simple example)

$$P_{fm} = \frac{2^{m-n}}{2^m}$$

$$= \frac{1}{2^n} \times 100\%$$

(4.28)

This result may also be developed by formal algebraic means, using polynomial relationships as introduced in Chapter 3 when considering maximum length autonomous LFSR generators. We are, however, now interested in the residual number remaining in the LFSR stages after completion of a test, rather than the output sequence produced during the test, which requires slightly different polynomial considerations from those in Chapter 3 as follows.

The polynomial relationship between an input bit stream, the LFSR output sequence from the final stage and the residue remaining in the LFSR comes from classic data encoding and error-correction theory[56], and is the polynomial division:

$$\frac{I(x)}{D(x)} = G(x) + \frac{R(x)}{D(x)}$$

(4.29)

where $I(x)$ is the data input bit stream expressed as a polynomial in x, $D(x)$ is the appropriate divisor polynomial, $G(x)$ is the final output sequence and $R(x)$ is the residue remaining in the LFSR.* For example, should $I(x)$ be $x^6 + x^4 + x^1 + 1$ and $D(x)$ be $x^4 + x^3 + 1$, then the polynomial division is as follows (note that we will drop the power of 1 in x^1 from here on):

$$\begin{array}{r}
x^2 + x + 0 \\
x^4+x^3+1 \overline{\smash{\big)}\, x^6 + x^4 + x + 1}\\
\underline{x^6 + x^5 + x^2 }\\
x^5 + x^4 + x^2 + x \\
\underline{x^5 + x^4 + x }\\
x^2 + 1
\end{array}$$

giving the quotient $G(x) = x^2 + x$ and the remainder $R(x) = x^2 + 1$. Multiplying $x^2 + x$ by $D(x)$ and adding $R(x)$ will check that $I(x)$ was $x^6 + x^4 + x + 1$. The different emphasis here compared with the polynomial developments in Chapter 3, and in particular the developments from eqn. 3.22 on, must be noted, namely:

(i) The polynomial division $G(x) = 1/P(x)$ of eqn. 3.22 gave directly the resulting sequence of logic 1s and 0s in the first stage of a type A

---

* An alternative way of expressing the residue $R(x)$, losing the information content of $G(x)$, is $R(x) = [I(x)] \bmod D(x)$, where $[I(x)] \bmod D(x)$ means the residue (remainder) which is present when the polynomial $I(x)$ is divided by the polynomial $D(x)$. This will be used later in Chapter 5, Section 5.5.1.

autonomous LFSR generator. This sequence is, of course, shifted along the n stages of the LFSR on subsequent clock pulses, disappearing n clock pulses later from the end of the shift register.

There was also no limit placed on the generated sequence, which merely repeated if the polynomial division of eqn. 3.22 was continued.

(ii) However, what we are now specifically concerned with is a serial input bit stream, the serial output bit stream which falls off the end of the n-stage LFSR, and the data finally remaining in the n stages of the shift register, which is what the polynomial division of eqn. 4.29 is generating.

There is also a finite limit placed on this division, so that if there are m bits in the serial input bit stream, $m - n$ bits will be in the quotient of eqn. 4.29 and n bits in the remainder.

Consider the two circuits that are shown in Figure 4.8. Figure 4.8a is what we have termed a type A LFSR (see Figure 3.20), and Figure 4.8b is the equivalent type B LFSR. The characteristic polynomial $P(x)$ of both circuits is $1 + x + x^5 + x^6 + x^8$, which happens to be a primitive polynomial, giving the same maximum length sequence in either circuit. However, as briefly mentioned but not pursued in Chapter 3, the data held in the n stages of the two circuit configurations at any particular time is not always the same except for the final bit. This is because the data bits in type A are shifted through the subsequent $n - 1$ stages unchanged, but in type B there are exclusive-OR changes in the shift of data, the final nth stage always being the same in both configurations.

Now the encoding and error-correction theory which gives us the polynomial relationships of eqn. 4.29 only gives the correct remainder $R(x)$ for the type B circuit configuration, and not the type A*. As a result, the published material in this area is invariably built upon the type B circuit configuration, which is sometimes termed a true polynomial divider. Looking at the circuit shown in Figure 4.8b, the divisor polynomial $D(x)$ for this configuration would be read off as $D(x) = x^8 + x^7 + x^3 + x^2 + 1$ where x^2, x^3 and x^7 identify the positions of the exclusive-OR gates. It will therefore be seen that with this polynomial division the divisor polynomial $D(x)$ is the reciprocal of the characteristic polynomial used in eqn. 3.22.

A more detailed discussion of characteristic polynomials and divisor polynomials may be found in Bardell et al.[10] Unfortunately, some other texts use the same designation for the two polynomials $D(x)$ and $P(x)$ without a clear distinction between them and therefore care must be taken when reading some published material.

To illustrate eqn. 4.29, consider the simple three-stage circuit shown in Figure 4.9a which has the divisor polynomial $D(x) = x^3 + x^2 + 1$. Assume that

---

* A correction may be applied to the value of $R(x)$ given by eqn. 4.29 to give the remainder in a type A circuit, see Yarmolik[18], but we will not pursue this here.

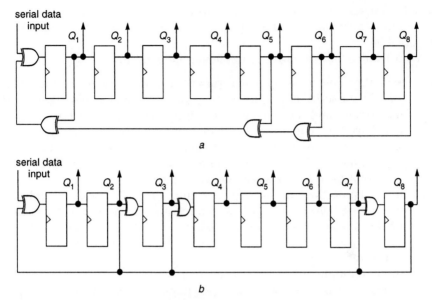

Figure 4.8 *The two principal LFSR circuit configurations (cf. Figure 3.20). The taps for maximum length pseudorandom sequence generation are shown*
a type A
b type B

the input bit stream is 1 1 1 1 0 1 0 1 0*, which may be expressed as:

$$I(x) = x^8 + x^7 + x^6 + x^5 + x^3 + x$$

The polynomial division is therefore:

$$x^3 + x^2 + 1 \overline{)x^8 + x^7 + x^6 + x^5 + x^3 + x}$$

or alternatively using the {0,1} notation:

```
              1 0 1 1 1 1
1 1 0 1)1 1 1 1 0 1 0 1 0   STOP!
        1 1 0 1
          1 0 0 1
          1 1 0 1
            1 0 0 0
            1 1 0 1
              1 0 1 1
              1 1 0 1
                1 1 0 0
                1 1 0 1
                  0 0 1
```

---

* Bit streams for polynomial division purposes are written with the first received bit on the left, with additional bits received later added on the right.

The final signature in the LFSR is therefore $Q_1 = 1$, $Q_2 = 0$, $Q_3 = 0$,. This may be confirmed by following the sequence through the state diagram shown in Figure 4.9b.

Four features of this simple example may be noticed:

(i) with the nine data input bits ($m = 9$) and three stages in the LSFR ($n = 3$), the serial output stream $G(x)$ which falls off the end of the LFSR has six bits, and the remainder has three bits. Hence, m bits of input data have been compressed into an n-bit signature;

(ii) if the above example input bit stream had a final logic 1 instead of a logic 0, there would have been an all-zero remainder at the end of the polynomial division; this is an entirely valid signature;

(iii) if we know the output sequence and the LFSR residue and the generating polynomial, then we can recreate the data input bit stream by the reverse of the above procedure—this is an essential feature of encoding and data transmission theory, but has no direct relevance to our present considerations;

(iv) there is nothing in eqn. 4.29 which dictates the choice of any particular divisor polynomial $D(x)$; any polynomial would yield an output sequence and a residue from a data input stream, provided that the number of bits in the data stream was greater than the number of stages in the shift register. The extreme case would be a shift register with no exclusive-OR taps, in which case $D(x)$ would be merely $x^n + 0$, i.e. 1000 for $n = 3$, and the input data would merely be shifted through the stages of the register unchanged, the final signature being the last n bits of the data input bit stream.

To revert for a moment back to the distinction between eqn. 3.22 with its characteristic polynomial $P(x)$ and eqn. 4.29 with its divisor polynomial $D(x)$, if the data input bit stream to the signature analyser was 1, 0, 0, 0, 0, 0, ..., then this should result in precisely the same bit sequence at the Q^n output of the signature analyser as at the Q^n output of the equivalent autonomous LFSR generator with the initial seed of 1. Looking again at Figure 4.8, for the autonomous LFSR signal generator of Figure 4.8a we have:

$$P(x) = 1 + x + x^5 + x^6 + x^8$$

giving the generated bit sequence of:

$$\left(\frac{1}{1 + x + x^5 + x^6 + x^8} \right)$$

$$= 1 1 0 0 0 1 1 0 1 \overline{)1 \cdots\cdots\cdots}$$

For the signature analyser circuit of Figure 4.8b we have the divisor polynomial:

$$D(x) = x^8 + x^7 + x^3 + x^2 + 1$$

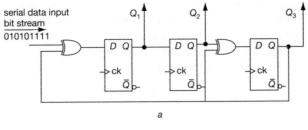

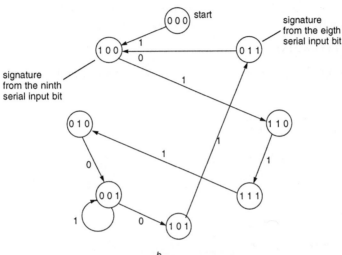

Figure 4.9 *A simple three-stage type B LFSR to illustrate the polynomial computation of the signature of a data input bit stream*
a the signature analyser
b state diagram and final signature. It is coincidental that all 2^n possible states of the shift register are present in this example

giving the polynomial division:

$$x^8 + x^7 + x^3 + x^2 + 1 \overline{\smash{)}1\,0\,0\,0\,0\cdots\cdots}$$
$$= 1\,1\,0\,0\,0\,1\,1\,0\,1 \overline{\smash{)}1\,0\,0\,0\,0\cdots\cdots}$$

Hence, because $D(x)$ is the reciprocal of $P(x)$ but is necessarily used in the reverse order in eqns. 3.22 and 4.29, the {0,1} numbers representing the polynomial divisors become identical. Remember, however, that the n-bit wide remainder involved in these polynomial divisions is only exact for the circuit configuration of Figure 4.8b.*

---

* It is left as an exercise for the reader to appreciate that the time events leading to the development of eqn. 3.22 were going into past time from left to right in the numerator of eqn. 3.22, but in eqn. 4.29 the numerator goes into future time from left to right. This is consistent with the polynomial division in each case.

Further algebraic developments have been extensively pursued. The linearity of the exclusive-OR operations, which allows the principle of superposition to be applied, enables many aspects to be investigated. For example, suppose that there are errors in the previous example data input bit stream such that we have the following data:

healthy input $I_h(x)$ =   1   1   1   1   0   1   0   1   0

faulty input $I_f(x)$ =   0   1   1   1   1   1   0   1   0

We may now specify an error bit stream, $E(x)$, formed by the bit by bit exclusive-OR of the above (mod$_2$ addition), giving:

$I_h(x) \oplus I_f(x) = E(x)$ =   1   0   0   0   1   0   0   0   0

Performing the polynomial division of $I_f(x)$ and $E(x)$ by the generating polynomial we have:

```
                       1 0 1 1 0
I_f(x):  1 1 0 1 ) 0 1 1 1 1 1 0 1 0
                   1 1 0 1
                   1 0 1 0
                   1 1 0 1
                     1 1 1 1
                     1 1 0 1
                         1 0 0
```

and

```
                     1 1 1 0 0 1
E(x):  1 1 0 1 ) 1 0 0 0 1 0 0 0 0
                 1 1 0 1
                 1 0 1 1
                 1 1 0 1
                   1 1 0 0
                   1 1 0 1
                       1 0 0 0
                       1 1 0 1
                         1 0 1
```

Therefore in total we now have:

							residual $R(x)$ (signature)		
$I_h(x)$:	1	0	1	1	1	1	0	0	1
$I_f(x)$:		1	0	1	1	0	1	0	0
$E(x)$	1	1	1	0	0	1	1	0	1

from which it will be seen that the three results are related by mod_2 addition. In particular, given the signature of the healthy input bit stream $I_h(x)$ and the signature of the error bit stream $E(x)$, the signature of the faulty bit stream $I_f(x)$ is the mod_2 addition of the two signatures.

The significance of the last statement is that the faulty bit stream signature will be identical to the healthy bit stream signature if and only if the signature of the error bit stream is all zero. In other words, the error bit stream must be exactly divisible by the divisor polynomial $D(x)$ for fault masking (aliasing) to occur. This result applies equally to both the type A and type B LFSR circuit configurations of Figure 4.8, and for any generating polynomial. When the LFSR degenerates into a simple shift register with no exclusive-OR taps, the above mathematics still holds, and aliasing will obviously occur when the final n bits of the input data bit stream are identical in the healthy and faulty cases, irrespective of the preceding $(m-n)$ input bits.

The rigorous proof of polynomial operations in GF(2) may be found in Peterson and Weldon[56]; detailed considerations of their application of signature analysis may be found in Bardell et al.[10] and elsewhere[18,57-63]. These considerations include analyses of the probability of fault masking, the security of the signature for one, two or more faults in the input bit stream and other aspects such as the choice of the polynomial which need not necessarily be that for maximum length pseudorandom sequence generation.

However, let us look at what can happen with the final n bits of a data input bit stream. As shown in Table 4.4 for a four-stage ($n = 4$) LFSR, the final n bits of an m-bit data input bit stream can always be chosen so that fault masking occurs, irrespective of the faults in the preceding $(m-n)$ bits. This agrees with the previous polynomial division of the error bit stream $E(x)$, since the final n bits of $E(x)$ can always be chosen such that the polynomial division $E(x)/D(x)$ leaves a zero remainder count.

However, if the input data bit stream is correct up to the $(m-n)$th bit, then it is not possible to have a fault-free signature if there are one or more errors in the final n bits. Also, if there is only one bit error anywhere in the m-bit input data, there is no way that aliasing can occur and a fault signature always results. Two faults, however, may cancel to give the fault-free signature. It is left as an interesting exercise for the reader to evaluate the polynomial division of $E(x)$ and $D(x)$ where $E(x)$ covers only one faulty bit, i.e. has only one 1 in its polynomial, and hence confirm that an all-zero residue $R(x)$ is impossible. Two bits faulty may be investigated by doing two single-bit fault divisions and adding mod_2 the two resulting $R(x)$ values, and so on.

The examples above have, for simplicity, been done with only three or four stages in the LFSR. The original Hewlett Packard analyser, as previously mentioned, had 16 stages, and therefore had a theoretical aliasing probability value of

$$\frac{1}{2^{16}} = 0.000015\ldots,$$

Table 4.4 *The data in a four-stage ($n = 4$) type A signature analyser with exclusive-OR taps on x^3 and x^4 back to the input, showing aliasing on the final signature*

Input bit	Fault free input	Fault free residue	Faulty input	Faulty residue	Faulty input	Faulty residue
•	•	•	•	•	•	•
•	•	•	•	•	•	•
$m-4$	•	0100	•	1111	•	1011
$m-3$	1	1010	1	1111	1	1101
$m-2$	1	0101	0	0111	1	0110
$m-1$	1	0010	0	0011	1	0011
m	0	1001	1	1001	1	1001

giving a fault detection efficiency of 99.9985...%. This is the often-quoted testing efficiency of a 16-bit signature analyser, but remember:

(i) it is based upon the assumption that all faults are equally likely in the data input bit stream, which is probably far from true;
(ii) the theoretical value of $1/2^n$ is independent of the nature of the feedback polynomial—it remains the same even when the LFSR becomes a simple shift register where the signature is merely the final n bits of the data stream with all the earlier $m - n$ bits irrelevant, which is clearly not valid;
(iii) it does not depend at all upon the fault-free data, but only on the error(s), i.e. $E(x)$.

Nevertheless, in spite of these serious doubts about the validity of the theoretical performance of the signature analyser, it has proved in practice to be an extremely efficient tool for catching faults in data bit streams and for diagnostic purposes. Clearly it is relevant for the generating polynomial, $D(x)$, of the LFSR to be a primitive polynomial, so that in the presence of a block of logic 0s in the input data stream the LFSR will cycle through a maximum length pseudorandom sequence rather than some shorter trivial sequence, but to select the 'best' polynomial for a given situation is an open problem. Clearly also, aliasing is less likely to occur if the number of stages, n, in the LFSR is increased, since then there is a greater number of possible states in the LFSR through which to sequence.

Further work on signature analysis including discussions on the validity of the theoretical aliasing property and the search for the 'best' polynomial, $D(x)$, to minimise aliasing may be found in the literature[10,18,59,61–66]. Yarmolik[18] also covers a comparison of the theoretical fault-masking probabilities of signature analysis with ones-counting (syndrome) testing and transition count testing, with the conclusion that signature analysis can offer a better aliasing property than the other two test compression methods. It is certainly the most widely known and used.

Signature analysis, however, only deals with one output bit stream at a time. For multiple-output networks signature analysis would need to be done on each output in turn for a complete system check. Therefore in the next chapter we will continue this general concept but with multiple data inputs to the LFSR; as will be detailed, this involves a multiple-input shift register (MISR) configuration, which will be seen to be an essential part of a built-in self test (BIST) strategy which has superseded the single-input signature analyser for VLSI test purposes.

4.6 Online self test

The concept of a digital circuit or network testing itself without the need to provide a separate external test vector generator and output monitoring circuit is clearly attractive, particularly when the size and complexity of the circuit under test grows and therefore requires increasingly complex and expensive test facilities.

The realisation of self test can be in one of two possible ways, namely:

(i) by the addition of some test means to the circuit, which is always functioning and continuously monitoring the output function(s);
(ii) by the incorporation of some test means within the circuit such that the circuit may be switched from its normal operating mode to a test mode for test purposes.

The first of the above possibilities we will call online self test, and it will be considered in the following pages; the second possibility is offline, but is usually referred to as built-in self test (BIST), and will be the major consideration in the following chapter. The full hierarchy of possible self-test techniques, including software check programs, is shown in Figure 4.10, with the left hand branch being our immediate concern.

Online self test, also referred to as concurrent or implicit testing, will be seen to divide into two main categories, namely information redundancy and hardware redundancy. An inherent advantage of any form of self test is that being continuously active it is possible to detect intermittent faults, something that is difficult with test strategies based upon conventional exhaustive or other offline functional test methods. Surveys that have been published[68,69] have also indicated that, due to the increasing overall reliability of microelectronic circuits, intermittent faults are becoming as important in VLSI circuits as permanent faults, and therefore testing techniques which can catch such faults are advantageous. This feature was briefly introduced in Chapter 2, Section 2.4.

4.6.1 Information redundancy

Considering first information redundancy, the principle here is to use more output lines than are required for normal system operation, with the digital

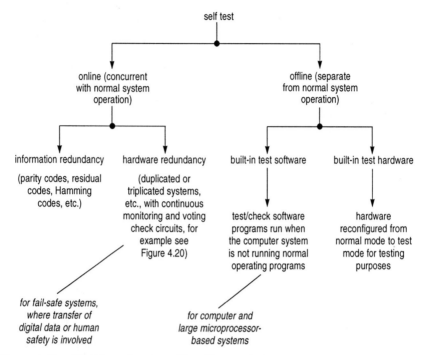

Figure 4.10 The hierarchy of possible self-test techniques

signals on the additional output line(s) being related to the healthy (fault-free) output signals by some chosen logical or mathematical relationship. Any disagreement in this relationship between the two sets of output signals will be detected as a circuit fault. This basic concept is illustrated in Figure 4.11, and clearly involves some additional hardware over and above the normal system complexity.

The simplest form of this self-checking concept is to generate one additional output from the check bit generation box shown in Figure 4.11, such that odd (or even) parity always exists when the system is fault free on the $m+1$ output lines (m lines from the normal operating system plus the one check bit line). For odd parity checks the number of logic 1 bits should always be an odd number; for even parity checks the number should always be even. If this is not present then there must be some fault in the operating system, or in the check bit generating circuit. (The circuit details of checker circuits themselves will be mentioned later.)

The single-bit parity check for a trivial example is shown below for both odd and even parity. Odd parity may be preferred in practical situations since it always results in at least one 1 appearing in the output word. Any error in one output bit will cause the $(m+1)$ bit output word to have the wrong parity, and hence an error can be detected.

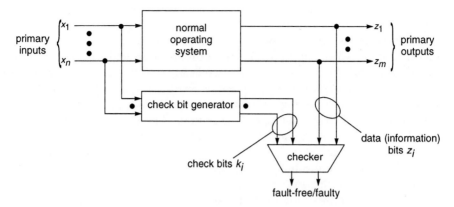

Figure 4.11 The organisation required for concurrent checking using information redundancy, the check bits being generated from the primary inputs and mathematically related to the fault-free system outputs

System output z_1 z_2	Odd-parity code z_1 z_2 z_k	Even-parity code z_1 z_2 z_k
0 0	0 0 1	0 0 0
0 1	0 1 0	0 1 1
1 0	1 0 0	1 0 1
1 1	1 1 1	1 1 0

This simple single-bit parity check is error detecting for any single bit fault, and for any odd number of faults, but is not error correcting. The addition of further check bits, however, enables two or more bit error detection and also error correction to be achieved.

Coding theory[56,70-75] considers the bits which contain the system output data as information bits, which together with the additional check bits make up an output code word. With m information bits and k check bits, the output word contains $(m+k)$ bits which the checker circuit finally interrogates. Notice that there is a maximum of 2^m valid outputs, which occur if every possible combination of the primary outputs is used, but 2^{m+k} theoretically possible output code words; hence the check bits are adding redundancy to the total output, as they are only required for error detection/correction purposes and not for any functional purpose in a fault-free system. The single-parity bit check considered above is effectively detecting unused combinations in the $(m+1)$ bit output word when a single fault is present.

Error detection and error correction thus depend upon having a greater number of output code words than is necessary to convey the system information; clearly if all the output code words are used to convey valid system information there is no possibility of detecting errors by examining the output patterns.

Consider the correct and incorrect code words shown in Figure 4.12a. Here the correct code words are 0 0 0 0 and 0 1 1 1, but are separated by two other code words which do not occur under fault-free conditions. The first incorrect code word, 0 0 0 1, differs from the first correct code word, 0 0 0 0, by one bit, and the second incorrect code word, 0 0 1 1, differs from the second correct code word, 0 1 1 1, also by one bit. If either 0 0 0 1 or 0 0 1 1 is present then either a single-bit fault or a two-bit fault is detected, but if we stipulate that only single-bit errors are possible then the output code 0 0 0 1 will also serve as an error correction, indicating that 0 0 0 0 is the proper fault-free output. Similarly, 0 0 1 1 will indicate that the correct code word should be 0 1 1 1. Thus the two intervening code words 0 0 0 1 and 0 0 1 1 act as an error detection for the one or two bits in error in the two code words, or error correction for the two code words for single-bit errors only. If the three least significant bits were in error, then no error detection or error correction would be possible. Figure 4.12b shows all the possible alternative pairs of nonvalid code words between 0 0 0 0 and 0 1 1 1 which have the same property as those in Figure 4.12a.

We may further illustrate and generalise these considerations as follows. Figure 4.13a shows the n-dimensional hypercube construction for a four-variable Boolean function, where each node on the hypercube represents one of the 2^n possible minterms of the function. Taking the example shown in Figure 4.12a, the two valid codes of 0 0 0 0 and 0 1 1 1 are shown by two heavy dots, and to traverse from one to the other along the shortest possible path involves traversing three edges in the hypercube construction. The two code words 0 0 0 0 and 0 1 1 1 are said to be a Hamming distance apart of three, which is identical, of course, to the number of bits which are of different value in the two code words. The two code words 1 1 1 1 0 0 and 0 0 0 0 0 1, for example, are a Hamming distance apart of five. The other alternative invalid code words shown in Figure 4.12b will be seen to be all the possible routes of distance three between 0 0 0 0 and 0 1 1 1. Further examples using the hypercube construction may be found in Wakerly[75].

Looking back at the simple single-parity bit example given above, the correct code words (including the single-parity bit) for odd parity are indicated on the $n = 3$ hypercube construction shown in Figure 4.13b. All these valid code words will be seen to be a Hamming distance apart of two, with the invalid codes a distance of one from adjacent valid codes. Hence, single-bit error detection is possible, but not error correction.

We can summarise the requirements for error detection and error correction as follows:

(i) the necessary and sufficient condition to detect E errors or fewer is that the Hamming distance between valid code words is $E + 1$;
(ii) the necessary and sufficient condition to be able to correct E errors or fewer is that the Hamming distance between valid code words is $2E + 1$; the latter condition is illustrated in Figure 4.14.

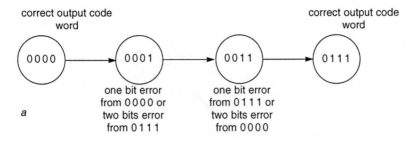

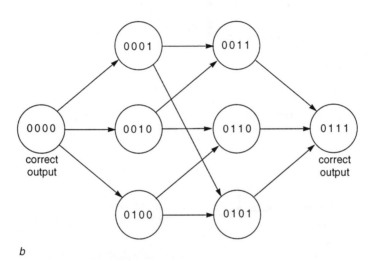

Figure 4.12 A graphical representation of error-free and error code words
  a one example
  b other possibilities between 0 0 0 0 and 0 1 1 1

4.6.1.1 Hamming codes

The particular system originally developed by Hamming and termed Hamming codes[70] involves the inclusion of k check bits for m-bit information data, where:

$$2^k \geq m + k + 1 \qquad (4.30)$$

Thus if $m = 4$, $k \geq 3$, resulting in an output code word of seven bits. There are therefore 16 possible fault-free output codes and $(128 - 16) = 112$ faulty output codes. In the Hamming code the check bits are interspersed among the data bits in power of two positions, i.e. $2^0, 2^1, 2^2, ...$; for example, with an output code word of seven bits the check bits would be in positions 1, 2 and 4, with the information bits in positions 3, 5, 6 and 7. Calling the check bits

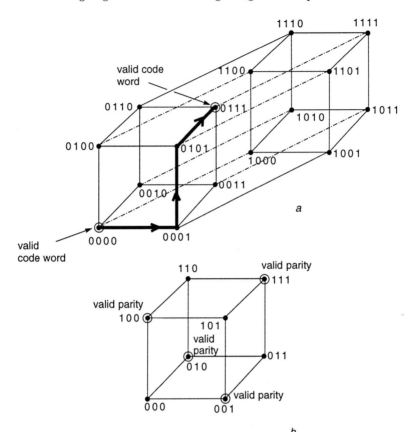

Figure 4.13 Hypercube constructions for n-variable Boolean functions
  a n = 4 with one of the six possible routes between 0 0 0 0 and 0 1 1 1 shown in bold
  b n = 3 construction, with odd parity coding

K_1, K_2 and K_3 and the information bits D_1, D_2, D_3 and D_4, the seven-bit output code would be:

bit positions:	1	2	3	4	5	6	7
check/data:	K_1	K_2	D_1	K_3	D_2	D_3	D_4

The check bits are calculated for each fault-free output such that:

$$K_1 = D_1 \oplus D_2 \oplus D_4$$
$$K_2 = D_1 \oplus D_3 \oplus D_4$$
$$K_3 = D_2 \oplus D_3 \oplus D_4 \qquad (4.31)$$

For example, if the information bits D_1, D_2, D_3, D_4 are 0 1 0 1, then the check bits K_1, K_2, K_3 will be 0 1 0 and the Hamming code word will be 0 1 0 0 1 0 1.

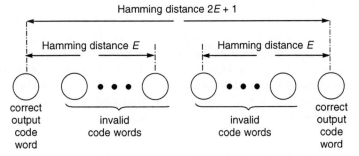

Figure 4.14   The necessary and sufficient conditions to correct E errors in an output code word. E + 1 to 2E errors will be detected but not corrected (Acknowledgements, Reference 72)

Notice that the three groups $K_1 D_1 D_2 D_4$, $K_2 D_1 D_3 D_4$ and $K_3 D_2 D_3 D_4$ each form an even-parity check under fault-free conditions, or in other words:

$$K_1 \oplus D_1 \oplus D_2 \oplus D_4 = 0$$
$$K_2 \oplus D_1 \oplus D_3 \oplus D_4 = 0$$
$$K_3 \oplus D_2 \oplus D_3 \oplus D_4 = 0 \qquad (4.32)$$

Suppose with the above example that the D_3 information bit was corrupted from 0 to 1, giving the codeword 0 1 0 0 1 1 1. If now the above parity eqns. 4.32 are computed we will find:

$$0 \oplus 0 \oplus 1 \oplus 1 = 0 \text{ (l.s.b.)}$$
$$1 \oplus 0 \oplus 1 \oplus 1 = 1$$
$$0 \oplus 1 \oplus 1 \oplus 1 = 1 \text{ (m.s.b.)}$$

Reading these resultant outputs as a three-bit binary number with the first equation giving the least significant bit, the number 1 1 0 = decimal 6 gives the location of the incorrect bit in the seven-bit code word. D_3, the sixth bit, may therefore be corrected. It is left as an exercise for the reader to confirm that any one-bit error, including errors in K_1, K_2 and K_3, is located by this procedure. The location number produced by the above is sometimes termed the error address.

This Hamming code has a Hamming distance of three between valid code words. It can therefore detect but will not correct two bits in error. If two bits are in error, the decimal number which results from computing the parity equations of eqn. 4.32 will not give sure information for the correction of one or other faulty bit. However, the addition of one further check bit, K_4, which checks for even parity over the whole eight-bit code word can be proposed which will detect when any odd number of bits are in error[75,76]. This may now be used with the previous fault detection capability given by K_1, K_2 and K_3 to distinguish between a single correctable fault and a two-bit incorrectable fault.

Hamming codes are widely used to detect and correct errors in memory systems, particularly where the word width is large. They are not so commonly used for general combinational logic networks, because the necessary overhead becomes unacceptable unless there are a very large number of output functions z_1 to z_m (see Figure 4.11). Table 4.5 indicates the overheads for an increasing number of information bits for single-error correction/double-error detection, with and without the parity bit; the overhead penalty obviously becomes less significant as m increases.

Table 4.5 *The increase in word length with single-error correction/double-error detection Hamming codes*

Information bits	Without the final overall even-parity check			With the final overall even-parity check		
	check bits	total bits	% increase	check bits	total bits	% increase
4	3	7	75	4	8	100
8	4	12	50	5	13	62.5
16	5	21	31.2	6	22	37.5
32	6	38	18.7	7	39	21.9
48	6	54	12.5	7	55	14.6
64	7	71	10.9	8	72	12.5

4.6.1.2 Two-dimensional codes

The principle of two-dimensional codes is illustrated in Figure 4.15a. Parity bits are added to each output word of m information bits, the rows, and also for a group of p output words, the columns. Additionally, further check bits may be added as a check on the check bits.

The simplest arrangement for a two-dimensional code is to have a single odd or even parity bit for each row and a single odd or even parity bit for each column, as shown in Figure 4.15a. Such an arrangement has a Hamming distance of four, thus providing error detection for any three-bit faults or error correction for any one-bit fault. If four-bit faults are present these may not be detected if they occur in a rectangular pattern such as shown in Figure 4.15b. With fewer than four-bit faults the row parity check and/or the column parity check will always detect such a situation.

To obtain a higher fault detection or error correction the row and column check bits can be increased from a single parity bit to form a Hamming code word, or by some other additional check information. However, the concept of two-dimensional coding is only relevant if the system being checked is itself two-dimensional in the sense that p separate words each of m bits are present. Again, it is in computer and processor memory that such a situation is encountered, rather than in general digital logic networks such as we have generally been considering in this chapter. Further details and discussion on two-dimensional codes may be found in Wakerly[74].

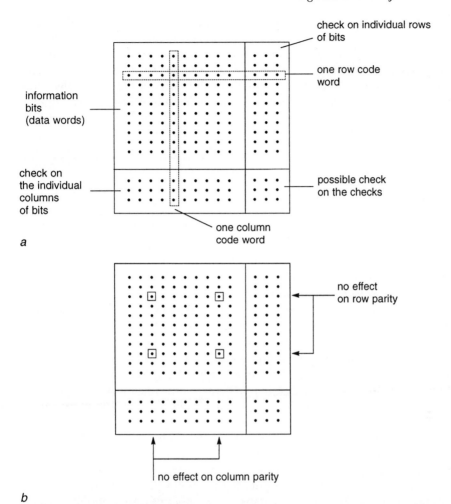

Figure 4.15 Two-dimensional coding checks
  a general arrangement
  b failure to detect four faults using single-bit odd or even parity in the row and column checks

4.6.1.3 Berger codes

Many other coding systems have been proposed for error detection and error correction, the original developments usually being for serial data transmission or data encoding rather than for digital testing applications. Among such possibilities have been m out of n codes, where m here is the number of ones in a word of n bits, $m < n$. The two out of five BCD code, for example, which has two 1s and three 0s in all its valid code words, is inherently error detecting if one, three, four or five 1s are detected. However,

m out of n codes are not convenient to consider for nonarithmetic system fault detection purposes, and cannot readily be adapted for fault correction.

Berger codes, on the other hand, have attractions for digital testing purposes. Standard Berger coding[77-79] involves the use of $\lceil \log_2(m+1) \rceil$ check bits for m information output bits, where the value of the check bits is the binary count of the number of logic 0s in the fault-free data. For example:

Fault-free output								Check bit value	Berger code			
0	0	0	0	0	0	0	0	8	1	0	0	0
0	1	0	1	0	1	0	1	4	0	1	0	0
1	0	1	1	1	1	0	1	2	0	0	1	0
0	1	1	1	1	1	1	1	1	0	0	0	1

For eight bits of information the total number of bits in the output checker is therefore 12; for 16 bits of information the total is 21, etc.

The Berger code generator is designed to generate the appropriate code from the primary inputs in the same way as we have previously considered, see Figure 4.11, thus making the output check bits continuously available alongside the normal primary outputs. The checker circuit also calculates this number by counting the number of zeros present in the m output bits, which it then compares with the generated check bits for agreement. Any disagreement between this output count and the number represented by the generated check bits indicates some internal circuit fault.

However, additional security has been proposed by including a parity check on the primary inputs, see Figure 4.16. The Berger code generator circuit now generates a binary output code which represents the expected number of zeros in the information output plus the input parity, which the checker circuit continuously compares against the number of zeros it monitors in the $m+1$ output data bits. The advantage of this expanded Berger code is that it enables detection of any single fault to be made whether internal or on a primary input line[80].

The particular strength of the Berger code as an online checker is for what have been termed undirectional faults in the normal operating circuit, that is additional 1-valued minterms are either added to or subtracted from blocks of 1s in the primary outputs under fault conditions. This will be recognised as classic failure characteristics of PLAs (growth faults, shrinkage faults, etc.), and hence Berger online checking has been advocated particularly for PLAs[88,89]. The Berger code generator and the input parity check circuit are made using additional product and sum lines in the PLA, with the checker circuit as a separate entity at the PLA output. We will come back to this again in Chapter 6, where we will be particularly concerned with the testing of strongly structured circuits such as ROMs and PLAs.

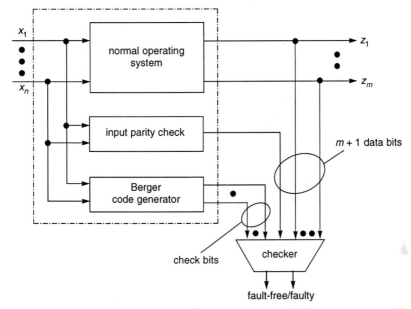

Figure 4.16 Online checking using Berger coding for the check bits with the addition of an input parity check, thus giving m + 1 *data bits*

4.6.1.4 Residue codes

In largely arithmetic or signal processing systems, specific online checking procedures have been advocated. These include:

- m out of n codes, as mentioned earlier;*
- Reed-Muller codes, which capitalise upon the linear properties of exclusive OR/NOR relationships;
- residue codes.

Details of the first two may be found in the literature[76,77,81,82], but here we will concentrate on residue coding as this has received considerable attention, involving lower overheads than, say, Hamming error detecting codes.

The properties of residue codes and their application to self test may be found in Sayers and Russell[83] and elsewhere[75,76,84–89]. In residue coding the output information bits and generated check bits maintain a mathematical relationship to each other when the network is fault free, this relationship failing in the event of an error in the output information bits or the check bits. The specific application of residue coding is for arithmetic circuits, where the primary inputs represent two binary numbers which have to be added or multiplied or have some other mathematical operation executed on

---

* These m out of n codes, and possible variants, may be referred to as nonseparable codes, since it is not possible to classify the bit order as either information bits or check bits as is the case in most of the other codes which we are considering.

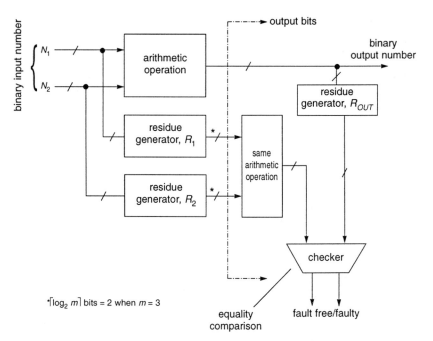

*⌈log₂ m⌉ bits = 2 when m = 3

Figure 4.17   *The application of residue coding for the automatic detection of errors in arithmetic systems*

them, rather than for general-purpose random logic circuits. This is illustrated in Figure 4.17.

Many residue codes are possible[84,89], but to keep the hardware overhead to a minimum a simple residue code such as follows is usually employed. The principle involved is to consider the information bits as a number N, and the check bits as a further number R which is mathematically related to N. The relationship between R and N is that R is the least significant digit of N when N is expressed in modulo m, $m \geq 2$. For example, with $m = 3$:

x_1	x_2	x_3	x_4	N	R
0	0	0	0	0	0
0	0	0	1	1	1
0	0	1	0	2	2
0	0	1	1	3	0
0	1	0	0	4	1
0	1	0	1	5	2
0	1	1	0	6	0
0	1	1	1	7	1
1	0	0	0	8	2
1	0	0	1	9	0
.	.	.	.	.	.
.	.	.	.	.	.
.	.	.	.	.	.

Expressed another way, R is the residue (remainder) when N is divided by m.

R is encoded in normal binary, thus requiring $\lceil \log_2 m \rceil$ bits; the above mod_3 example therefore has the six-bit binary coding:

Information bits				Check bits	
0	0	0	0	0	0
0	0	0	1	0	1
0	0	1	0	1	0
0	0	1	1	0	0
0	1	0	0	0	1
.	.	.	.	.	.
.	.	.	.	.	.

The generation of these binary check bits directly from the binary information bits may be done by a tree of simple circuits[88], but the hardware need not concern us at this stage.

Residue codes, like Berger codes, are separable codes in that the check bits may be treated separately from the information bits. Unlike Berger codes, residue codes have the property that the check bits of two input numbers N_1 and N_2 which are added or subtracted or multiplied may themselves be added, subtracted or multiplied to give the check bits of the output number, that is the check bits of the result can be determined directly from the check bits of the input operands. This property has been called independent checking[57]. For example:

Operation on inputs N_1 and N_2	Output N_{out}	Output residue code R_{out}
Add	$N_1 + N_2$	$\lvert R_1 + R_2 \rvert_{\mathrm{mod}_3}$
Subtract	$N_1 - N_2$	$\lvert R_1 - R_2 \rvert_{\mathrm{mod}_3}$
Multiply	$N_1 \times N_2$	$\lvert R_1 \times R_2 \rvert_{\mathrm{mod}_3}$

These relationships are illustrated for a simple numerical situation in Table 4.6, where the numbers in parentheses are the decimal equivalent of the numbers N and R. In practice some further binary coding of the decimal numbers N_1 and N_2 may be used to facilitate subtraction and division, such as two-complement encoding, but the principle illustrated here will be maintained[86].

Table 4.6 The use of residue coding in arithmetic operations, with $m = 3$

	Input		Output	
	information bits	check bits	information bits	check bits
Add:	1 0 1 0 (10)	0 1 (1)	1 1 1 0 (14)	1 0 (2)
	0 1 0 0 (4)	0 1 (1)		
Subtract:	1 0 1 0 (10)	0 1 (1)	0 1 1 0 (6)	0 0 (0)
	0 1 0 0 (4)	0 1 (1)		
Multiply:	1 0 1 0 (10)	0 1 (1)	1 0 1 0 0 0 (40)	0 1 (1)
	0 1 0 0 (4)	0 1 (1)		

The schematic for the self checking of addition and multiplication using residue coding is therefore as shown in Figure 4.18. This may be generalised for the self checking of a general-purpose arithmetic logic unit (ALU) by the addition of appropriate controls to cover addition, subtraction, multiplication and division of the two primary input numbers[86,88].

The application of residue coding to the Boolean operations of AND, OR, exclusive-OR is not so straightforward, and additional manipulation is necessary to convert the Boolean relationships into arithmetic ones. (This has been briefly encountered before in Section 4.4.1 in another context.) Considering the arithmetic operations of multiplication and addition and the Boolean operations of AND, OR and exclusive-OR (XOR), we have the relationships shown below for two bits n_1 and n_2. Note that we are using the symbol $\cap$ here to represent AND and $\cup$ to represent OR in order to differentiate between Boolean operations and arithmetic ones.

n_1	n_2	Arithmetic multiplication $n_1 \times n_2$	Arithmetic addition $n_1 + n_2$	Boolean AND $n_1 \cap n_2$	Boolean OR $n_1 \cup n_2$	Boolean XOR $n_1 \oplus n_2$
0	0	0	0	0	0	0
0	1	0	1	0	1	1
1	0	0	1	0	1	1
1	1	1	2	1	1	0

The Boolean operations are therefore defined by the following arithmetic relationships:

Boolean AND:

$$n_1 \cap n_2 = n_1 \times n_2$$

Boolean OR:

$$n_1 \cup n_2 = (n_1 + n_2) - (n_1 \times n_2)$$

Boolean XOR:

$$n_1 \oplus n_2 = (n_1 + n_2) - 2(n_1 \times n_2)$$

The inversion (complementation) of any bit b_i is clearly just $(1 - b_i)$ in the arithmetic field.

Therefore, given the numbers N_1 and N_2, the bit-wise AND, OR and XOR of these two is given by the above arithmetic operations, and hence the residue coding for these Boolean operations becomes as follows, still assuming modulo 3:

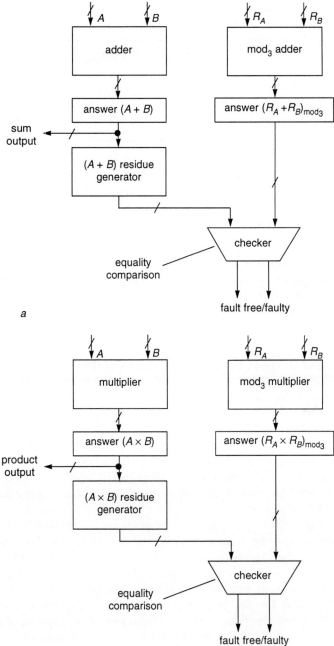

Figure 4.18 Self-checking addition and multiplication using mod_3 residue coding
 a addition
 b multiplication

Bit-wise operation on inputs N_1 and N_2	Output N_{out}	Output residue code R_{out}
$N_1 \times N_2$	$N_1 \cap N_2$	$\lvert N_1 \times N_2 \rvert_{\mod_3}$
$(N_1 + N_2) - (N_1 \times N_2)$	$N_1 \cup N_2$	$\lvert \lvert N_1 + N_2 \rvert_{\mod_3} - \lvert N_1 \times N_2 \rvert_{\mod_3} \rvert_{\mod_3}$
$(N_1 + N_2) - 2(N_1 \times N_2)$	$N_1 \oplus N_2$	$\lvert \lvert N_1 + N_2 \rvert_{\mod_3} - 2 \lvert N_1 \times N_2 \rvert_{\mod_3} \rvert_{\mod}$

The ability of residue coding to check arithmetic operations, and therefore also logical operations when the above arithmetic encoding of Boolean relationships is done, has been formally proved[89]. With the radix m odd all single-bit errors are detected, but cannot be identified.* With m even, some single-bit errors can escape detection, and therefore an odd-number radix is desirable. The actual choice of m odd involves the following considerations:

(i)   as radix m increases more bits are necessary to encode the residue;
(ii) the complexity of the checking hardware increases very considerably with increasing m;
(iii) the error detecting ability of the code only increases slowly with increasing m.

As a result $m = 3$ is usually proposed.

The number of check bits in residue coding is therefore very small, only two in the case of $m = 3$ irrespective of the number of bits in the input and output numbers, but only single-bit error detection and not error correction is possible. Single-bit error correction can be accomplished by generating two separate sets of check bits for each binary number involved, these two sets having a dissimilar basis m[76], but this at least doubles the number of check bits involved, and hence is not generally considered to be worthwhile or necessary. Further details of residue coding theory, applications and circuit schematics may be found in the published literature, particularly in References 83, 87 and 89.

Several other self-checking codes have been proposed, mostly for specific applications or specific hardware architectures. Many are for data transmission serial output bit streams rather than for logic network test purposes, and may involve synchronisation problems when identifying the beginning of words in the serial bit stream as well as bit errors[90–93]. One such recent proposal has been T-codes, being a nonseparable coding scheme which has the property of self synchronisation after the loss or corruption of received data[94–97]. However we will not consider such schemes or applications in this text, as they lie outside our principal interests in self test of logic networks. The parallels between the checking requirements of data transmission and online network testing, however, should always be borne in mind.

---
* The output code in fact has a Hamming distance of 2.

We will look into circuit details for the checker circuits involved in online checking later in Section 4.6.3.

4.6.2 Hardware redundancy

Turning from the information redundancy techniques of Figure 4.11 to hardware redundancy, the principle now is not to consider the bit structure of the input or output information data and thence determine some set of additional output check bits, but instead to incorporate duplication or other hardware addition(s) so that the output information itself becomes more secure. Hence the fundamental difference between information redundancy and hardware redundancy is:

- information redundancy is designed to give a signal that a fault in output is present, which may or may not then be corrected;
- but hardware redundancy is designed such that a faulty output is ideally never given.

Hardware redundancy is therefore a self-test mechanism in that it operates online and does not in service involve any separately generated test set vectors. Other terms may be encountered in this area, including:

- static redundancy;
- masking redundancy;
- online fault tolerance redundancy;
- online system availability;

and others, such terms arising when considering the reliability and availability of electronic and other systems. Details of many aspects of reliability engineering may be found in O'Connor[1] and elsewhere[3,98]. Here we will continue with simple digital hardware redundancy concepts.

Hardware redundancy can be considered at component level, but is increasingly irrelevant with the use of monolithic circuits of SSI, MSI, LSI and VLSI complexity. Component redundancy, however, is straightforward. If we consider the basic component, a resistor, and assume that it has two extreme failure modes, namely open circuit or short circuit, then it requires a net of resistors such as shown in Figure 4.19a to cater for any single resistor failure. With the first circuit configuration shown, the surrounding circuit has to be able to tolerate a resistance change from R to $\frac{2}{3}R$ or $2R$ under single short circuit or open circuit faults, respectively; with the second circuit the tolerance is from R to $\frac{1}{2}R$ or $\frac{3}{2}R$ under the same conditions. A larger number of series/parallel resistors will reduce this resistance change under single-fault conditions.

Series/parallel nets of other passive components are similar. Transistors used as switching devices may be arranged likewise, as shown in Figure 4.19b.

Such component redundancy may still be encountered where one small switching circuit has to have a fail-safe characteristic and be able to tolerate

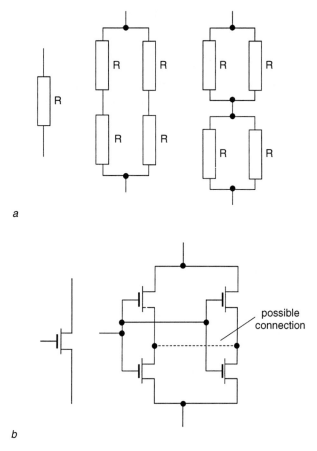

Figure 4.19 Component redundancy
 a passive resistor redundancy
 b active FET switches redundancy

individual component failure, but generally hardware redundancy will involve complete module or system redundancy.

Duplicated system redundancy was illustrated in Chapter 2, Figure 2.2 in connection with untestable nodes in circuits. However, from the information output point of view system duplication will only allow online error detection to be made when the system output information bits differ, but not any error correction. Hardware redundancy therefore almost invariably involves triple (or higher) level redundancy, such that majority voting on the output information can be made to distinguish between faulty and fault-free data. For triple redundancy it is assumed that never more than one system becomes faulty, and that this is repaired or replaced before a second system fault occurs.

Figure 4.20*a* illustrates the arrangement for triple redundancy, usually termed triple modular redundancy or TMR. The voting circuit, V, at the output continuously monitors the three module outputs, and gives a final

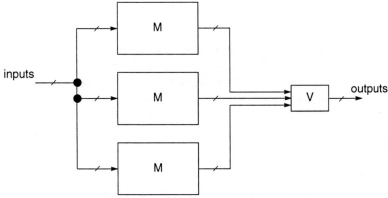

Figure 4.20   Triple modular redundancy (TMR), where the modules M may be very complex or less complex such as individual arithmetic logic units

output which is a majority verdict on its input data. Clearly the integrity of this voting circuit is crucial, but if its circuit complexity is considerably less than that of the modules it is monitoring, its reliability will be very much higher than the individual module reliability; redundancy circuit techniques may also be built into it without a high economic penalty.

If it is assumed that the failure rate of electronic circuits and systems in service is constant,* then the probability equations for equipment reliability and availability readily follow. The starting parameter is the failure rate (also known as the hazard rate) of a component or module, λ, where λ is defined as the number of failures per unit time as a fraction of the total population, which is normally expressed as a percentage failure rate per hour or per 1000 hours or per year. For example, a value of $\lambda = 0.05\ \%$ per 1000 hours (or $0.05 \times 10^{-3}\ \%$ per hour) means that in a population of 10 000 components we have an average failure of:

$$10000 \times \frac{0.05}{100} = 5 \text{ failures per 1000 hours}$$

$$= 0.005 \text{ failures/hour}$$

The mean time between failure, MTBF, is given by the reciprocal of the failure rate, which for the above example would be:

$$\text{MTBF} = \frac{1}{0.005} = 200 \text{ hours}$$

The reliability, $R(t)$, of a component or module is the probability of no failure over a given time, t, and is always calculated for λ constant over this period of time. With λ as the known failure rate of a component or module, the reliability is given by:

---

* This is generally true if early failures have been detected and corrected during assembly, test and installation, and that no final wear-out period has been reached. This gives the familiar bathtub characteristic of failure rate plotted against time[1].

$$R(t) = e^{-\lambda t}$$
$$= e^{-t/\text{MTBF}}$$

where λ is the failure rate per hour and t is in hours. This gives the familiar reliability curve shown in Figure 4.21a, with 100 % reliability (no failures) at time $t = 0$ but only a 36.8 % probability of no failures by the time $t =$ MTBF. A further parameter mean time to repair, MTTR, that arises in the servicing of equipment may also be encountered, which together with the MTBF gives the availability of a system[98,99], availability being defined as:

$$\frac{\text{MTBF}}{\text{MTBF} + \text{MTTR}} \tag{4.34}$$

We will, however, have no reason to consider MTTR here, but will continue with MTBF and $R(t)$ considerations only.

The theoretical reliability of any system consisting of identical components or modules in series or parallel may be readily determined. However, distinction has to be made between:

(i) systems in which every individual module must function if the complete system is to be functional;
(ii) systems in which one or more individual modules may fail without causing a complete system failure, i.e. some redundancy is present.

The simplest case of (i) above is a system with two or more modules in series. For two modules in series the system reliability will be the product of the individual reliabilities; if j identical modules each with a reliability, $R(t)_i$, are in series then the overall reliability will be the product:

$$R(t)_{overall} = \prod_{i=1}^{j} R(t)_i \tag{4.35}$$

Clearly, reliability falls with a series system, and therefore is detrimental.

Parallel systems constitute case (ii) above. Triple modular redundancy has the reliability factors following, again where $R(t)_i$ is the reliability of each individual module:

$$R(t)_{overall} = \left(R(t)_i\right)^3 + 3\left(R(t)_i\right)^2\left(1 - R(t)_i\right) + 3\left(R(t)_i\right)\left(1 - R(t)_i\right)^2 \tag{4.36}$$

The first term $(R(t)_i)^3$ is the probability of all three modules functioning until time t, the second term is the probability of two out of the three modules functioning until time t and the third term is the probability of only one system still being functional at time t. An alternative way of expressing this is to consider the unreliability $U(t)$ of a module, where $U(t)$ is the probability of the module being faulty at time t and where $R(t) + U(t) = 1$. Then, writing R for $R(t)_i$ and U for $U(t)_i$, we have the binomial expansion:

$$1 = R^3 + 3R^2U + 3RU^2 + U^3 \tag{4.37}$$

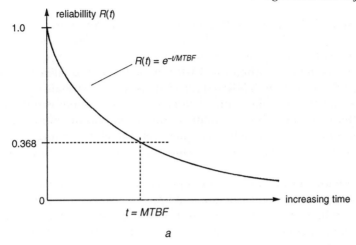

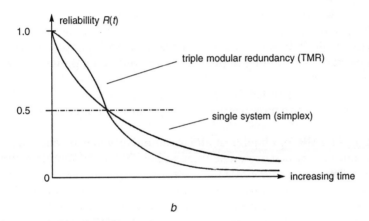

Figure 4.21 The reliability R(t) *against time characteristics*
 a single system
 b TMR system

where R^3 = all modules functional, $3R^2U$ = two modules functional, $3RU^2$ = one module functional, and U^3 = no module functional at time t.

For triple modular redundancy where the system provides a fault-free output provided two or three modules are fault free, the system reliability is therefore:

$$\begin{aligned}R(t)_{overall} &= R^3 + 3R^2U \\ &= R^3 + 3R^2(1-R) \\ &= 3R^2 - 2R^3\end{aligned} \quad (4.38)$$

This result assumes that the majority voting circuit, V, of Figure 4.20 has a very high reliability in comparison with the reliability value of the more complex modules, M.

If the TMR reliability eqn. 4.38 is plotted, it will be found to be as shown in Figure 4.21b. At low values of t a higher reliability is achieved in comparison with the single module situation, but where the module reliability has fallen to 0.5 then the TMR reliability has also fallen to 0.5. At increasing values of t the TMR system with its greater overall complexity proves less reliable than the simplex situation. Also, as soon as one of the three modules has failed, the reliability of the system with the two remaining modules becomes $(R(t))^2$ which is less than the reliability of a single module since now a fault in either of the two remaining modules can cause a system failure.

Hence, the use of triple modular redundancy is only appropriate when the operating time of the system (the mission time) is less than the MTBF of the individual modules, for example in flight computers or other systems which are required to operate for a specific period of time between routine offline test and maintenance.

In practice the reliability of TMR is usually much better than the above probability predictions indicate, since there may be faults in two modules but not the same faults[98]. Hence, on any particular primary input pattern two out of the three systems may still give the correct response even though there are faults in two modules.

The equations for the reliability, $R(t)$, of other forms of online redundancy, such as three out of five modular redundancy, may be found References 1–3, 76, 99.

Variations of TMR and higher modular redundancy may be encountered, including modules which are on standby rather than continuously active as considered above. In general, if a microelectronic module, M, is on standby its reliability value may not differ appreciably from that under active conditions, since there are no mechanical wear-out mechanisms present in microelectronics as there are in mechanical and electromechanical mechanisms. A mix of active and standby modules is sometimes termed hybrid modular redundancy, and may incorporate some switching matrix as shown in Figure 4.22 to switch in standby modules and thus maintain full redundancy in the active part of the system. Many other arrangements for monitoring and switching modules or complete computer systems in or out of service have been proposed[2,76,99–103], including:

- multiple complete computer plus memory arrangements;
- multiple processor and switching mechanisms for digital telephone exchanges;
- multiple processors for online financial transactions;
- multiple control computers and software for rail traffic control.

and others. These concepts are introduced in Lala[2]. However, the more components or modules or subsystems there are then the higher will be the incidence of failure somewhere in the complete system.

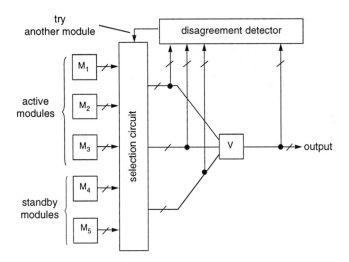

Figure 4.22 The concept of hybrid modular redundancy, where a faulty active unit may be switched out of service and replaced by a standby unit so as to maintain full active redundancy in the system

To summarise, online hardware redundancy is essential for systems which are required to operate as accurately as possible over a given period, and where economic considerations take second place. Aircraft blind landing systems, computer systems in space shuttle vehicles and other aerospace applications are well-known examples, but for less sophisticated applications information redundancy with its lower overheads is more appropriate. It should, however, be appreciated that to obtain the benefits of hardware redundancy, the individual components or modules must themselves be reliable; it is not possible to a make a reliable system with unreliable parts.

4.6.3 Circuit considerations

So far in the previous discussions of online self test using information redundancy or hardware redundancy, no details have been given on actual circuit designs. In particular, looking at the information redundancy schematics of Figures 4.11, 4.17 and 4.18, the final checker circuit shown in the trapezoidal box is clearly vital to the success of the overall system. Similarly, the voter circuit, V, of hardware redundancy schemes is the vital element.

The importance of this final output check circuit has led to the development of self-checking checkers, sometimes known as totally self checking (TSC) checkers, first considered in 1968 by Carter and Schneider[104] and later by Anderson and Metze[105]. The definition of totally self checking is that given a set of possible internal faults, F, in the checker, then any fault in F will sooner or later give a detectable erroneous output during normal operation with valid input data.

The checker circuits that we will be considering below will be totally self checking unless otherwise stated, and the possible internal faults will generally be single faults only and not multiple faults in the circuit.

Three principal variants of checkers are required, namely:

- parity checkers;
- equality checkers;
- m out of n checkers.

All are required to provide an output indication for fault-free valid information data, and another indication for faulty (invalid) information data. In a perfect world a checker could be a single-output circuit with an output, say, logic 1 for fault-free data and logic 0 for faulty data. However, this clearly is not secure, since a single stuck-at fault at the checker output will give incorrect results. Also, it is not possible to test for a stuck-at fault condition using the normal input information data, since if the checker output was stuck at 1, where 1 represents fault-free input information, it would require applying faulty data (a noncode word) to the checker inputs to test for this fault.

Hence two checker outputs are normally used as indicated in Figure 4.23, the two outputs giving a 1 out of 2 output code for all fault-free conditions. The checker outputs therefore are:

0 1 or 1 0:   valid system information (fault-free data)
0 0 or 1 1:   invalid system information (faulty data) or some checker circuit failure

Such an arrangement is often termed a two-rail checker. In fault-free operation the checker outputs will continuously change between 0 1 or 1 0 depending upon the particular information input word, and this dynamic action is itself a self test of the checker circuit.*

Self-checking parity check circuits are readily made using exclusive-OR (or exclusive-NOR) circuits, since any single failure will give a wrong answer (0 0 or 1 1) at the two-rail output under some fault-free input conditions. Figure 4.24 shows the circuit realisations for odd and even parity checks, the outputs always being 0 1 or 1 0 when the information data and the checker circuit are both fault free. In the general circuit for m information data bits shown in Figure 4.24c and d, the inputs to the two multiple-input exclusive-OR circuits must be disjoint, that is each data input must be connected to one or other but not both exclusive-ORs. Some additional circuit complexity within the exclusive-OR networks to cover certain open-circuit failings has been

---

* It may be noted that in many digital logic control systems where safety or fail-safe operation is involved, all critical circuits are dynamic, continuously switching under fault-free conditions. This ensures that any stuck-at 0 or stuck-at 1 failure will be a safe-side failure.

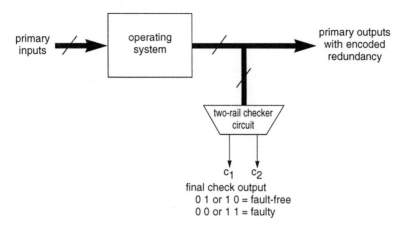

Figure 4.23   The two-rail checker requirements; see also Figures. 4.11, 4.17 and 4.18

proposed by Khakbaz[106], but the basic exclusive-OR relationships at the checker output are unchanged.

It is possible to combine the two-rail 1 out of 2 outputs from several separate checker circuits to give a final 1 out of 2 output code, as shown in Figure 4.25. From the truthtable given in Figure 4.25a it will be seen that the output code 0 1 or 1 0 only results when the two input pairs are themselves 0 1 or 1 0. The circuit is totally self checking, giving a noncode output of 0 0 or 1 1 for all single checker faults and certain other faults[106]. The tree topology of Figure 4.25b for more than two input pairs readily follows.

Equality between two data words may be checked by inverting all the data bits of one word and then using the checker circuits of Figure 4.25. This is illustrated in Figure 4.26, and can clearly be extended to any number of bits in the two words being compared.

Checker circuits for m out of n codes[105–110] are frequently based upon k out of $2k$ checkers, since the latter can readily be built from an assembly of simple cells. The extension to the more general case of m out of n, $n \neq 2m$, may then be done by appropriate changes or additions at the k out of $2k$ checker inputs. Recall that m out of n codes are nonseparable codes, that is there is no distinction between information bits and separate check bits, and hence all bits have identical importance. Some symmetry in the resulting check circuits is therefore to be expected.

The logic circuit for a 2 out of 4 checker is shown in Figure 4.27a. Since any single fault can only affect the output of one of the two subcircuits, the circuit is self checking for all single faults. It may readily be used as a 2 out of 3 checker by making one of the four inputs a steady logic 0, or a 1 out of 3 checker by making an input a steady logic 1.

An extension of this 2 out of 4 checker circuit to higher k out of $2k$ values readily follows[108]. Figure 4.27b shows a 3 out of 6 assembly, which is generalised to the k out of $2k$ case in Figure 4.27c. It will be seen that these

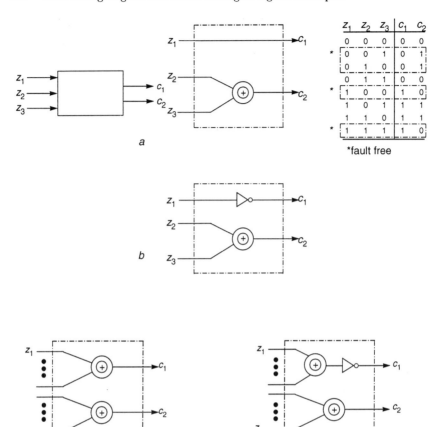

Figure 4.24   Self-checking parity circuits with 1 out of 2 output coding
a three-input odd parity
b three-input even parity
c m-input odd parity
d m-input even parity

cellular checker configurations use a total of $k(k-1)$ AND/OR cells in addition to the final k-input AND and OR gates. Other m out of n code checkers can be realised by connecting a steady logic 0 or logic 1 to appropriate inputs of these k out of $2k$ circuits.

Circuits such as shown in Figure 4.27 are one of a family of possible cellular arrays based upon AND/OR cells. The array shown in Figure 4.28 was developed for majority and threshold logic purposes[111], and has the characteristic that the outputs fill up from the bottom output according to how many inputs are at logic 1. Output $s_1 = 1$ when any one input is 1 (1 out of n or Boolean OR), outputs s_1 and s_2 are both 1 when any two inputs are 1 (2 out of n), outputs s_1 to s_m are all 1 when any m inputs are 1 (m out of n), through to all outputs being 1 when all inputs are 1 (n out of n or Boolean AND).

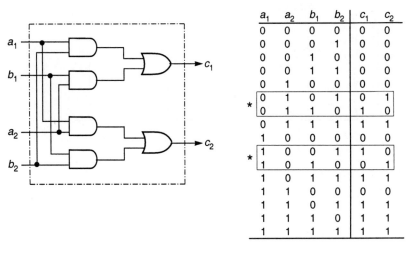

Figure 4.25 *Self-checking circuits to combine two or more 1 out of 2 codes and provide a single secure 1 out of 2 output*
a check for $a_1 a_2$ and $b_1 b_2$ both valid 1 out of 2 codes
b extension to more than two pairs of 1 out of 2 inputs

It is therefore possible to select any m out of n input condition by monitoring $s_m = 1$, $s_{m+1} = 0$ from such an array,* but like the circuits shown

---

* Note, unlike the previous checker circuits, this particular structure is not fault secure since it does not have independent subcircuits providing the two outputs, and therefore in practice cannot be considered for use as an m out of n self checking checker.

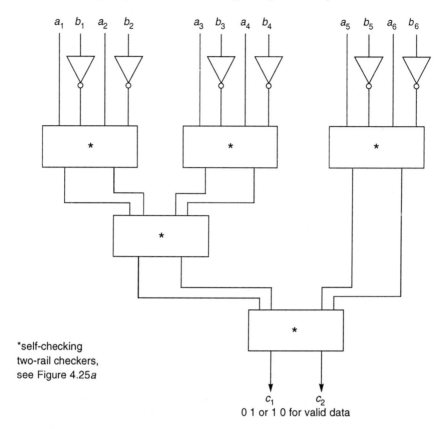

Figure 4.26 Self-checking checker circuit to check the equality of two information words **A** and **B**, where **A** = $a_1 a_2 a_3$ ... and **B** = $b_1 b_2 b_3$ ...

in Figure 4.27 there is a severe penalty in that the response time of all these cellular arrays is poor because of the number of gates in series between inputs and outputs. Because of this factor, alternative m out of n checker circuits have been proposed which minimise the number of levels of series gating.

The alternative nonarray circuits which have been disclosed for m out of n checker circuits[106–110,112] generally fall into two categories, namely:

- those which adopt appropriate AND/OR networks preceding a k out of $2k$ checker, these networks converting m out of n input words into k out of $2k$ code words, $k < m$;
- the design of specific AND/OR circuits to convert m out of n input words directly into 1 out of 2 output coding.

Comparisons between many of these proposals have been made by Lu and McCluskey[113] and others[89,114]. The first approach above may itself be divided into two possibilities, namely:

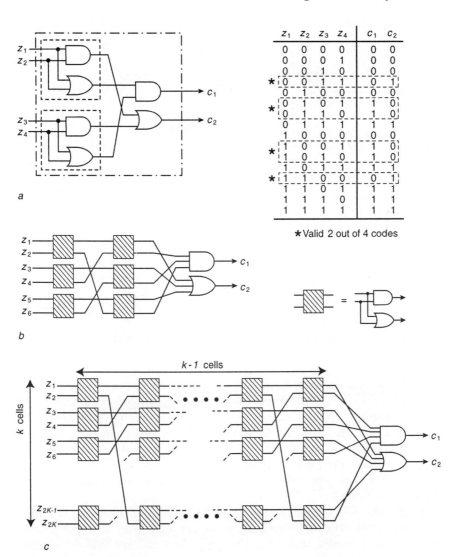

Figure 4.27 Self-checking k out of 2k code checker circuits
 a 2 out of 4 code checker
 b 3 out of 6 code checker
 c k out of 2k code checker

(i) the use of appropriate two-level AND/OR decoder circuits preceding a k out of 2k checker, the decoding circuits converting the m out of n inputs words into k out of 2k code words, $2 \le k < m$;

(ii) the use of more than two-level AND/OR decoder circuits preceding a 2 out of 4 checker, the decoding circuits converting the m out of n input words directly into 2 out of 4 code words.

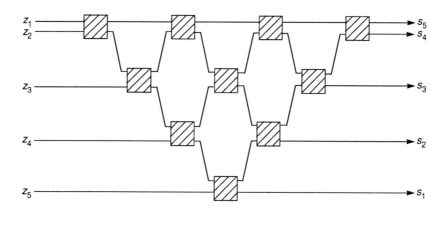

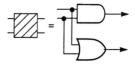

Figure 4.28 The AND/OR cellular array providing 1 out of n (OR) through to n out of n (AND) output information

These two alternatives are illustrated in Figure 4.29. In the first case if, say, a 2 out of 5 information code is present, then $\binom{5}{2} = 10$ AND gates are required to give a 1 out of 10 decode, each AND gate detecting one valid code word $z_1 z_2, z_1 z_3, z_1 z_4, \ldots, z_4 z_5$, with six five-input OR gates then translating this 1 out of 10 code into a 3 out of 6 code[2,76,106,109]. In the second case, the AND/OR decoder converts the given m out of n information code into a 1 out of p code where $p = 6$, but may be reduced to $p = 5$ for 2 out of n codes and $p = 4$ for m out of $(2m + 1)$ codes[2,76,109].

Figure 4.30 gives a circuit for a 2 out of 5 checker built upon the principles of Figure 4.29b. The AND network produces all the ten possible product terms of two variables, which the OR gates then translate into a 2 out of 4 code to the final checker. Notice that there are three levels of gating before the final 2 out of 4 checker circuit, but the total number of gates is less than would be the case for the equivalent circuit configuration of Figure 4.29a which in total would require ten AND gates and six OR gates to precede a 3 out of 6 checker circuit.

In contrast to Figure 4.30, the design of a minimal AND/OR checker circuit for a 2 out of 5 code is given in Figure 4.31[107]. With just two levels of gating this configuration clearly gives the potentially highest speed of the various circuit possibilities, and also uses fewer gates than the circuits of Figure 4.29. Notice that it requires the same number of AND gates as would Figure 4.29a, namely $\binom{5}{2}$, but the outputs from these gates have now been

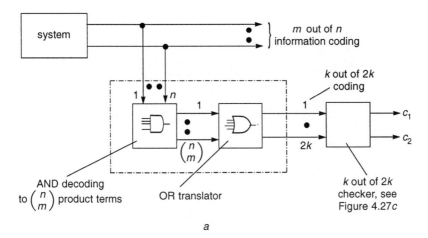

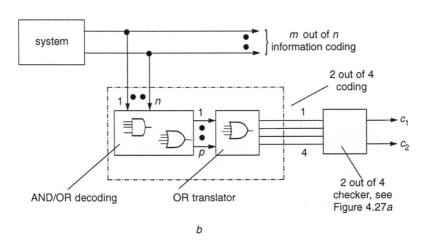

Figure 4.29 Circuits to preceed k out of 2k self-checking checkers to provide alternative m out of n checks
a two-level AND/OR with k out of $2k$ checker
b three (or more) level AND/OR with 2 out of 4 checker

partitioned such that a 1 out of 2 output code directly results for all valid input codes. Should there be less than two input bits at logic 1, then clearly the checker output will be 0 0; should there be more than two bits at logic 1, say z_1, z_2 and z_3, then it will be seen that there will be an input to the upper OR gate and an input to the lower OR gate, thus giving the checker output 1 1 to indicate the fault. Similar two-level circuit realisations are possible with AND and OR gates at the first level of logic, with one AND and one OR gate at the checker output[108].

In looking at these several examples of m out of n checker circuits, it will be apparent that there are strong structural and symmetrical relationships in

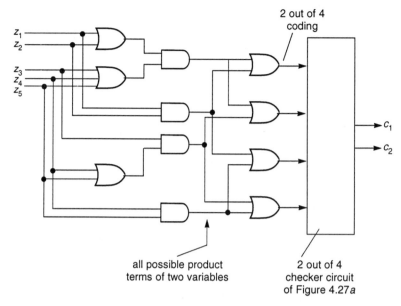

Figure 4.30 The circuit for a self-checking 2 out of 5 code checker using a 2 out of 4 output checker

the final circuits, particularly in the partitioning of the input information bits in input AND or OR gates. The two-level AND/OR circuit of Figure 4.31 is an obvious example, and also the AND/OR relationships flowing through the cells of Figure 4.27. Among the design algorithms which may be found for these self-checking checker designs are those of Anderson and Metze[105], Marouf and Friedman[109], Reddy[107], Smith[108] and others. We will not consider the details of these algorithms here, but the design results they give will be familiar from the above examples. If required, readers may find worked examples in Lala[2], Russell and Sayers[76], and Breuer and Friedman[89]. The later algorithms, which produce minimal checker circuits such as in Figure 4.31, are clearly more efficient than some of the earlier developments. Details of the minimum number of test vectors necessary to check completely m out of n self-checking checkers may also be found in the literature.

The checkers for Berger codes use a two-rail checker tree as shown in Figure 4.26 to check the equality of the Berger check bits against the contents of the output information. The schematic arrangement is shown in Figure 4.32a, where the check bit generator at the output calculates the number of logic 0s in the output data, which is then compared with the number represented by the Berger check bits[109]. If these two binary numbers agree then the output information is considered fault free.

The output check bit generator can readily be constructed from full-adder modules. However, if the number of bits in the output information is 2^{k-1},

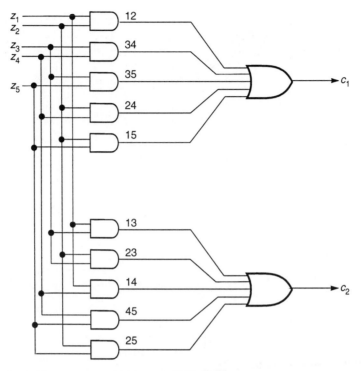

Figure 4.31 The circuit for a minimal AND/OR 2 out of 5 code self-checking checker

thus requiring k check bits, which is known as a maximal-length Berger code, it is particularly relevant to count not the number of 0s in the information bits but the number of 1s, this binary number then automatically being the bit-wise complement of the Berger code, i.e. if the information bits are, say, 0 1 1 1 1 0 0 then the 1s count is 100 which is the bit-wise complement of Berger check bit number 0 1 1. This means that the individual inverter gates shown in Figure 4.26 are no longer necessary since the bits are already complementary under fault-free conditions. The optimum topology of the full-adder modules for maximal-length Berger codes has been determined by Marouf and Friedman[109] such that it is self testing for all single faults and can be fully tested with a minimum number of test vectors. The circuit for seven data bits and therefore three check bits is shown in Figure 4.32b, and may readily be extended to $2^k - 1$, $k > 3$, data bits. For nonmaximal Berger codes, which is probably more common, then both full-adder and half-adder modules become necessary, and some complexity in the design of a self-checking circuit is present[2,109,115]. The two-rail output equality checker, however, remains unchanged.

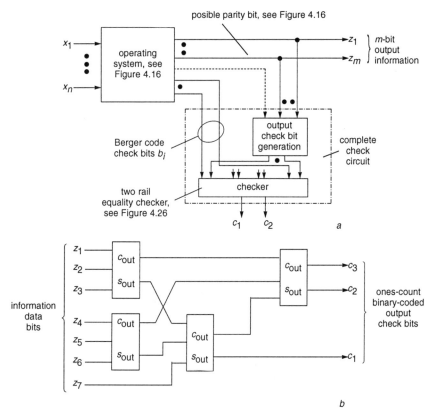

Figure 4.32 The self-checking checker for Berger codes
  a schematic arrangement
  b arrangement of full-adder modules for the output check bit generation with seven information bits

The design of voting circuits for hardware-redundancy systems is much more company specific and specialised than for self-checking two-rail checkers. For the triple modular redundancy (TMR) system of Figure 4.20, an assembly of majority gates is the basic requirement to give 2 out of 3 output information, as shown in Figure 4.33. For M out of N system redundancy N input majority gates are required, N normally odd.

The ideal requirement for a voting circuit is that it shall be 100 % reliable over the system operating time. This is not necessarily the same as requiring fail-safe operation, since it is desirable to maintain the correct output data rather than design for a specific output condition should a fault occur. A saving feature here is that the circuit complexity of majority gates is small compared with system or module complexity, and therefore the statistical probability of a failure is low compared with system or module failure. However, certain factors may be considered in the design of highly reliable circuits, namely:

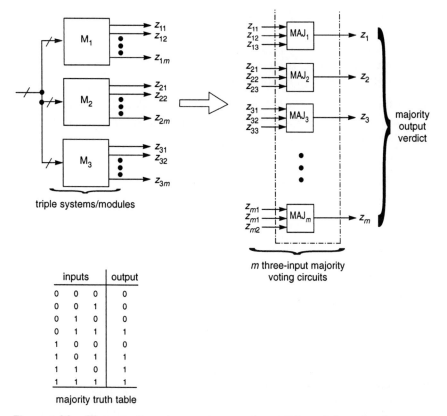

Figure 4.33   The majority voting requirements for a triple modular redundancy system

(i) the use of component-level redundancy, as illustrated in Figure 4.19;
(ii) extensive screening of standard production parts so as to ensure that any potential weakness is detected before being put into service.

The use of component-level redundancy has the inherent problem that (by definition) it is not possible to detect a faulty internal component from the accessible input and output terminals, and therefore it may remain in service with no redundancy left in the circuit. Component-level redundancy where the accessibility of every component is not available therefore is no longer considered good engineering practice, and reliable nonredundant voting circuits are to be preferred.

With microelectronic circuits it is probable that standard ICs in mass production will be more reliable than special circuits, since volume production allows process parameters to be matured and high quality manufacturing standards to be maintained. Any production line in continuous use is always better than a production line which has to be restarted from a shut-down condition. Hence enhanced screening of standard ICs which may include:

- acceleration (vibration) tests;
- temperature cycling;
- accelerated-stress life tests to catch potential failures and imperfections without overstressing perfect factors;
- detailed parameter measurements;

may be specified to ensure the use of as reliable a device as possible[1,99,116]. The American Military Standard Specification MIL-STD-883, for example, calls for target device failure rates as follows[117-119]:

- class A devices, $\lambda \simeq 0.001\%$ per 1000 hours;
- class B devices, $\lambda \simeq 0.005\%$ per 1000 hours;
- class C devices, $\lambda \simeq 0.05\%$ per 1000 hours;

which are achieved by increasing levels of screening of production circuits.

Therefore, the generally accepted design philosophy for highly reliable voting circuits is to use conventional logic design methods, but:

(i) use highly screened and therefore the potentially most reliable components possible;
(ii) perform routine offline maintenance to check that the voting circuits still correctly output a majority verdict with test input signals.

One minor modification may be found in some complex systems, as shown in Figure 4.34. This is used if the triplicated systems are partitioned into well-defined subsystems, but it does not eliminate the need for the voting circuits to be ideally completely reliable devices.

The basic circuit for a three-input majority voting circuit is shown in Figure 4.35a. Its simplicity gives it its high potential reliability. There is, however, one problem if a simple AND/OR circuit as shown here is used in practice, and that is that not all the signals from the triple modules or subsystems may arrive at the voter at exactly the same instant, and hence the probability of glitches in the voter output may be high. This is overcome in many systems by having a clocked voting circuit as shown in Figure 4.35b. The disadvantage now is that there will be a delay in the majority output decision given by the voter, but this is a known delay which can be built into the system operation.

Other simple circuits such as the TMR agreement/disagreement indicator shown in Figure 4.36 may be incorporated, but in general circuit sophistication in voting and monitoring circuits is to be avoided if at all possible. Further details and discussion of this highly specialised area of digital systems design may be found in Johnson[120] and elsewhere[121-125].

4.7 Self test of sequential circuits

The majority of the developments and procedures covered in the preceding sections of this chapter have involved combinational logic networks only; the

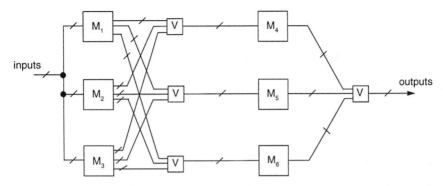

Figure 4.34 The inclusion of majority voting between TMR partitions of a complex system (cf. Figure 4.20)

signature analysis technique of Section 4.5 can, as mentioned, cover combined combinational and sequential networks but only if the network or system is initialised to a known state at the beginning of each signature test. The online information redundancy methods of Section 4.6.1 do not readily accommodate sequential elements in the network under test, but the hardware redundancy online methods of Section 4.6.2 do not have this constraint.

The use of coding and the concepts of self-checking design have, however, been specifically considered for sequential networks. The broad principal is that both the primary outputs and the state variables are codes, such that there are valid code words for each in the fault-free case, but invalid code words for one or the other or both if there is a system fault. This general concept is illustrated in Figure 4.37, when the outputs and the state variables are coded in, say, m out of n coding, and where code checkers such as we have previously considered monitor these code words.

Details of original developments by Ozguner[126], Diaz and Azema[127] and others may be found published[2,74]. However, this has not proved to be a profitable technique since it involves considerable redundancy in both the combinational and sequential logic, and therefore has been largely overtaken by the methods of testing which we will consider in detail in the following chapter. These design for test (DFT) methods involve partitioning the system into separate combinational and sequential blocks which are then tested separately, rather than attempting to do self testing on the combined network.

There is, however, one possible state assignment which lends itself to self test if required. This is the technique of one-hot state assignment, where only one of the state variables is at logic 1 at any time, the remainder all being at logic 0[128]. If N states in the sequential machine are required then there must be N storage elements (flip-flops) in the system instead of the minimum number of $\lceil \log_2 N \rceil$ if a binary-coded state assignment is used. This is clearly

190 *VLSI testing: digital and mixed analogue/digital techniques*

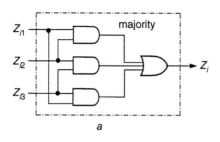

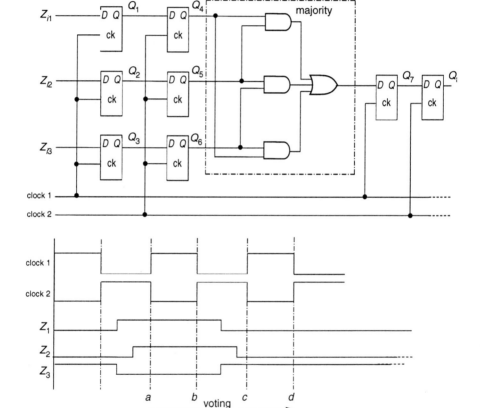

Figure 4.35 *Majority voting circuits for triple modular redundancy*
  a simple one-bit majority voter
  b synchronised majority voter using edge-triggered flip-flops on inputs and outputs to synchronise the input and output data

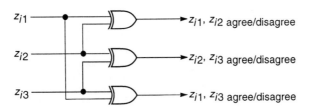

Figure 4.36 *Agreement/disagreement indicator circuit*

not economical if just the number of flip-flops is considered, but this to some extent may be compensated for by simplified logic in the combinational network. Self test of the one-hot sequential network is readily done by a 1 out of N self-checking checker.

4.8 Multiple-output overview

The early proposals on signatures and self test discussed in this chapter were largely concerned with just one primary output at a time. For example, the signature analysis covered in Section 4.5 requires each primary output of a circuit or system to be separately probed and its signature obtained. On the other hand, the later self-test concepts of Section 4.6 involving hardware redundancy normally consider all the primary outputs of a system under test simultaneously.

Multiple system outputs and data buses are clearly more usual than just one system output, and therefore testing techniques which are only able to handle one output at a time have restricted practical importance unless fault diagnosis is required. For completeness we will here briefly collate the signature and self-test methods and their ability to handle multiple system outputs, which will also point to the two following chapters where multiple outputs are invariably present.

Considering, therefore, the several techniques previously covered:

(i) *Syndrome (ones-count) testing*

This basically considers only one primary output and not multiple outputs. Proposals to combine two or more outputs to give a single syndrome count which were noted in Section 4.3.1[11-16] have not been found to be advantageous.

(ii) *Accumulator-syndrome testing*

The accumulator syndrome count could theoretically be applied to a combined syndrome count, see (i) above, but no known proposals for this have ever been published.

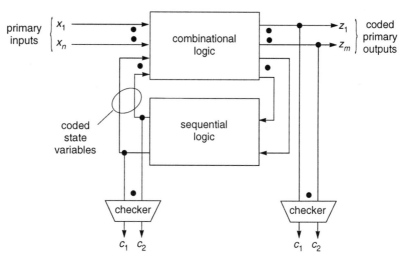

Figure 4.37 Self test for sequential networks with coded primary outputs and coded state variables (cf. Figure 3.14)

(iii) *Transition count testing*

There is no obviously advantageous way of combining multiple outputs and applying one transition count test, and hence this technique must be regarded as a single-output test method only.

(iv) *Arithmetic, Reed-Muller and spectral coefficients*

Arithmetic and Reed-Muller coefficients relate exclusively to single-output Boolean functions, and have no useful extension to more than one output. Spectral coefficients also define a single Boolean output, but some research work has been done to consider multiple outputs[47]. However, like the single-output spectrum, this work has not yet found any niche in the range of possible digital testing tools.

Further comments and additional references may be found in Serra and Muzio[129].

(v) *Signature analysis*

As noted above, signature analysis is strictly a single-output (or single-node) testing technique.

Certain proposals have been made for compressing the multiple outputs of a system or subsystem under test into a single-input signature analyser check circuit, as shown in Figure 4.38. The analysis of this off-line testing strategy may be found in Yarmolik[18]. However, as mentioned in Section 4.5, the concept of injecting multiple data inputs directly into the separate stages of a LFSR rather than just the one data input as used in this signature analysis configuration, is an obvious development. This gives us the MISR (multiple-input shift register) configuration, which will be considered in detail in the following chapter.

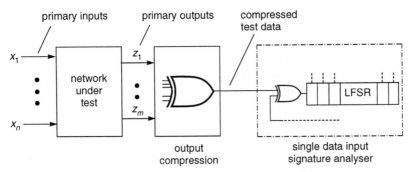

Figure 4.38 The proposal to compress multiple-output information data into one signature analyser input

(vi) *Information redundancy*

Odd or even parity checking of primary outputs is inherently multiple output, since the output word consisting of all outputs and check bit(s) in parallel is checked. Two-dimensional parity checks likewise involve multiple data bits. Hamming codes also involve all the output data in parallel, with the additional check bits interspersed with the information data bits. Parity checking in particular is extensively used in computer memory, and therefore we will return to this area in Chapter 6.

Berger codes, m out of n codes and residue codes are all also multiple-output testing strategies. Berger and residue codes are used particularly in applications where error correction is desirable, but m out of n codes generally have too high an overhead to be used except for specific applications such as BCD or similar coding.

(vii) *Hardware redundancy*

When the redundancy is applied at system or subsystem level, then multiple-output capability is clearly present. The disadvantage in comparison with information redundancy is the very high system overheads involved due to the several complete systems or subsystems now present.

Thus in spite of the obvious higher information content when considering all the outputs of a network rather than just one output, there is no simple secure way of providing a final fault-free/faulty output signal which acts as a test for all the outputs.

4.9 Chapter summary

This chapter has looked at several possible ways of testing a digital circuit network or system which avoids having to undertake a lengthy automatic test pattern generation procedure in order to determine some minimal test set, and then having to check the outputs on every input test vector for a fault-free/faulty response.

The concept of compressing the output response of a circuit under test into a single signature has obvious attractions, but as we have seen such output compression must involve the problem of aliasing, that is the possibility of giving a fault-free output signature when a fault is actually present. In practice the particular input test vector compression techniques of Section 4.2, time compression, and the output response compression techniques of Section 4.3, space compression, have not generally been adopted by industry except in very specific instances.

Arithmetic, Reed-Muller and spectral coefficients covered in Section 4.4 are prominent areas in the broad spread of digital logic theory, but their application to the specific role of circuit or system testing has not to date been found to be advantageous. Nevertheless, the spectral coefficients represent a powerful way of expressing digital functions (both binary and higher valued), and recent resurgence in interest in this area as a result of developments in binary decision diagrams (BDDs) may lead to a future reconsideration of their status.

The concepts of self test surveyed in this chapter have wide practical significance, particularly for complex systems where reliability and/or error detection/correction is necessary. Recall that in this chapter we have confined our consideration to self test which is carried out online, and which does not require the circuit or system to be switched from a normal mode to a test mode. This has the inherent advantage that circuits are operating under normal operating conditions and at normal speeds, and also that intermittent output faults will be detected.

An overview of self-testing techniques was given in Figure 4.10, from which it will be seen that we have now completed a consideration of the left hand hierarchy of possible self-testing methods. The right hand half of Figure 4.10 indicates the self-test techniques which involve switching the circuit or system under test from a normal mode to a test mode, and therefore are not tests carried out online simultaneously with normal operation. Some additional comments may be found in Reference 130.

Of the online self-test methods introduced in this chapter, parity checking has perhaps the greatest industrial use. The use of coding redundancy also has widespread applications, particularly where numerical calculations are involved. We will refer to these techniques again in Chapter 6. Hardware redundancy, particularly triple (or higher) modular redundancy, has essential use in safety-critical systems[120-125]; we will encounter this again in our considerations of microprocessor and microcontroller testing in Chapter 6.

Due to space constraints, the full mathematical developments underpinning error detection, error correction, coding theory and the probabilities of fault masking have not been given in this chapter. Further indepth treatment may be found in the literature particularly in References 10, 18, 56, 57, 76 and 89. However, many of the theoretical results, particularly in fault-masking (aliasing) probabilities, must be treated with caution since the mathematics takes no account of the physical construction of a circuit or system, and the likelihood or impossibility of a particular fault arising. Nevertheless, the mathematical background is an essential part of all this area of research, development and application.

The offline self-test methods indicated in Figure 4.10 will form a large part of the following chapter, being widely adopted for VLSI systems where sequential as well as combinational circuit elements are involved. As a summary we broadly find:

- the online techniques covered in this chapter are particularly relevant for specific applications in volume production, such as memory and processor ICs, or very special applications;
- the offline techniques to be considered later are much more useful tools for OEMs who are involved in the design of VLSI systems, and who need to simplify the resultant testing of their final circuits and completed systems.

4.10 References

1 O'CONNOR, P.D.T.: 'Practical reliability engineering' (Wiley, 1991)
2 LALA, P.K.: 'Fault tolerant and fault testable hardware design' (Prentice Hall, 1985)
3 ANDERSON, A., and LEE, P.: 'Fault-tolerance: principles and practice' (Prentice Hall, 1980)
4 AKERS, S.B.: 'In the use of linear sums in exhaustive testing'. Proceedings of IEEE conference on *Design automation*, 1985, pp. 148–153
5 TANG, D.T., and CHENG, C.L.: 'Logic test patterns using linear codes', *IEEE Trans.*, 1984, **C-33**, pp. 845–850
6 McCLUSKEY, E.J.: 'Verification testing: a pseudo-exhaustive test technique', *IEEE Trans.*, 1984, **C-33**, pp. 541–546
7 SALUJA, K.K., and KARPOVSKY, M.: 'Testing computer hardware through data compression in space and time'. Proceedings of IEEE international conference on *Test*, 1983, pp. 83–88
8 HURST, S.L.: 'Use of linearisation and spectral techniques in input and output compaction testing of digital networks', *IEE Proc.E*, 1989, **136**, pp. 48–56
9 SAVIR, J.: 'Syndrome-testable design of combinational circuits', *IEEE Trans.*, 1980, **C-29**, pp. 442–451, and correction pp. 1102–1103
10 BARDELL, P.H., McANNEY, W.H., and SAVIR, J.: 'Built-in test for VLSI: pseudorandom techniques' (Wiley, 1987)
11 AGARWAL, V.K.: 'Increasing effectiveness of built-in testing by output data modification'. Proceedings of 13th IEEE international symposium on *Fault tolerant computing*, 1983, pp. 227–233
12 AGARWAL, V.K., and ZORIAN, Y.: 'An introduction to an output data modification scheme', *in* [19]
13 BARZILAI, Z., SAVIR, J., MARKOVSKY, G., and SMITH, M.: 'Weighted syndrome sums approach to VLSI testing', *IEEE Trans.*, 1981, **C-30**, pp. 996–1000

14 SAVIR, J.: 'Syndrome testing of syndrome-untestable combinational circuits', *IEEE Trans.*, 1981, **C-30**, pp. 606–608
15 MARKOWSKY, G.: 'Syndrome-testability can be achieved with circuit modification', *IEEE Trans.*, 1981, **C-30**, pp. 604–606
16 SERRA, M., and MUZIO, J.C.: 'Testability by the sum of syndromes'. Proceedings of technical workshop on *New directions for IC testing*, University of Victoria, Canada, 1986, pp. 11.1–11.21
17 SAXENA, N.R., and ROBINSON, J.P.: 'Accummulator compression testing', *IEEE Trans.*, 1986, **C-35**, pp. 317–321
18 YARMOLIK, V.N.: 'Fault diagnosis of digital circuits' (Wiley, 1990)
19 MILLER, D.M. (Ed.): 'Developments in integrated circuit testing' (Academic Press, 1987)
20 HAYES, J.P.: 'Generation of optimal transition count tests', *IEEE Trans.*, 1978, **C-27**, pp. 36–41
21 REDDY, S.M.: 'A note on logic circuit testing by transition counting', *IEEE Trans.*, 1977, **C-26**, pp. 313–314
22 ROBINSON, J.P., and SAXENA, N.R.: 'A unified view of test compression methods', *IEEE Trans.*, 1987, **C-36**, pp. 94–99
23 SAVIR, J. and McANNEY, W.H.: 'On the masking probability with ones-count and transition count'. Proceedings of IEEE international conference on *Computer aided design*, 1985, pp. 111–113
24 FUJIWARA, H., and KIMOSHITA, K.: 'Testing logic circuits with compressed data'. Proceedings of 8th IEEE international symposium on *Fault tolerant computing*, 1978, pp. 108–113
25 UPADHYAYA, S.J., and SALUJA, K.K.: 'Signature techniques in fault detection and location', *in* [36]
26 MUZIO, J.C., RUSKEY, F., AITKEN, R.S., and SERRA, M.: 'Aliasing probabilities of some data compression techniques', *in* [19]
27 SAXENA, N.R.: 'Effective measures for data compression techniques under equally unlikely errors', *in* [19]
28 DAVIO, M., DESCHAMPS, J.P., and THAYSE, A.: 'Discrete and switching functions' (McGraw-Hill, 1978)
29 HEIDTMANN, K.D.: 'Arithmetic spectrum applied to stuck-at fault detection for combinational networks', *IEEE Trans.*, 1991, **C-30**, pp. 320–324
30 GREEN, D.H.: 'Modern logic design' (Addison-Wesley, 1986)
31 BRIKHOFF, G., and MACLANE, S.: 'A survey of modern algebra' (Macmillan, 1977)
32 AYRES, F.: 'Theory and problems of matrices' (Schaum's Outline Series, McGraw-Hill, 1962)
33 HOHN, F.E.: 'Elementary matrix algebra' (Macmillan, 1964)
34 BEAUCHAMP, K.G.: 'Transforms for engineers' (Oxford University Press, 1987)
35 AHMED, N., and RAO, K.R.: 'Orthogonal transforms for digital signal processing' (Springer-Verlag, 1975)
36 KARPOVSKY, M.G. (Ed.): 'Spectral techniques and fault detection' (Academic Press, 1985)
37 HURST, S.L., MILLER, D.M., and MUZIO, J.C.: 'Spectral techniques in digital logic' (Academic Press, 1985)
38 BEAUCHAMP, K.G.: 'Applications of Walsh and related functions' (Academic Press, 1984)
39 HURST, S.L.: 'The interrelationships between fault signatures based upon counting techniques', *in* [19]
40 KUMAR, S.K., and BREUER, M.A.: 'Probabilistic aspects of Boolean switching functions via a new transform', *J. Assoc. Comput. Mach.*, 1981, **28**, pp. 502–520

41 DAMARLA, T.R., and KARPOVSKY, M.G.: 'Fault detection in combinational networks by Reed-Muller transforms', *IEEE Trans.*, 1989, **C-38**, pp. 788-797
42 SUSSKIND, A.K.: 'Testing by verifying Walsh coefficients', *IEEE Trans.*, 1983, **C-32**, pp. 198-201
43 MILLER, D.M., and MUZIO, J.C.: 'Spectral fault signature for internally unate combinational networks', *IEEE Trans.*, 1983, **C-32**, pp. 1058-1062
44 MILLER, D.M., and MUZIO, J.C.: 'Spectral fault signatures for single stuck-at-faults in combinational networks', *IEEE Trans.*, 1984, **C-33**, pp. 745-769
45 MUZIO, J.C.: 'Stuck fault sensitivity of Reed-Muller and arithmetic coefficients'. Third international workshop on *Spectral techniques*, Dortmund, Germany, 1988, pp. 36-45
46 LUI, P.K., and MUZIO, J.C.: 'Spectral signature testing of multiple stuck-at faults in irredundant combinational networks', *IEEE Trans.*, 1986, **C-35**, pp. 1088-1092
47 MILLER, D.M., and MUZIO, J.C.: 'Spectral techniques for fault detection', *in* [36]
48 FUJITA, M., YANG, J.C-Y., CLARKE, E.M., ZHAO, X., and McGREER, P.: 'Fast spectrum computation for logic functions using binary decision diagrams'. Proceedings of IEEE conference *ISCAS*, 1994, pp. 275-279
49 MAIHOT, F., and DE MICHELI, G.: 'Algorithms for technology mapping based upon binary decision diagrams and Boolean operations', *IEEE Trans.*, 1993, **CAD-12**, pp. 599-620
50 MACII, E., and PONCINO, M.: 'Using symbolic Rademacher-Walsh spectral transforms to evaluate the agreement between Boolean functions', *IEE Proc., Comput. Digit. Tech.*, 1996, **143** (1), pp. 64-68
51 CLARKE, E.M., McMILLAN, K.L., ZHAO, X., FUJITA, M., McGREER, P., and YANG, J.: 'Spectral transforms for large Boolean functions with applications technology mapping'. Proceedings of international workshop on *Logic synthesis*, 1993, pp. 6b1-6b15
52 CHAKRAVARY, S.: 'A characterization of binary decision diagrams', *IEEE Trans.*, 1993, **C-42**, pp. 129-137
53 ISHIURA, N.: 'Synthesis of multi level logic circuits from binary decision diagrams', *IEEE Trans. Inf. and Systems*, 1993, **E76D**, pp. 1085-1092
54 SAUCIER, G., and MIGNOTTE, A. (Eds.): 'Logic and architecture synthesis: state of the art and novel approaches' (Chapman and Hall, 1995)
55 'A designer's guide to signature analysis'. Hewlett Packard application note 222, 1977
56 PETERSON, W.W., and WELDON, E.J.: 'Error correcting codes' (MIT Press, Cambridge, 1972)
57 ABRAMOVICI, M., BREUER, M.A., and FRIEDMAN, A.D.: 'Digital systems testing and testable design' (Computer Science Press, 1990)
58 RAJSUMAN, R.: 'Transistor-level fault modelling and testing' (Artech House, 1992)
59 SKILLING, J.K.: 'Signatures take a circuit's pulse by transition counting or PRBS', *Electron. Des.*, 1980, **28** (2), pp. 65-68
60 BHAVSAR, D.K., and HECKELMAN, R.W.: 'Self-testing by polynomial division'. Proceedings of IEEE international conference on *Test*, 1984, pp. 208-216
61 FUJIWARA, H.: 'Logic testing and design for testability' (MIT Press, 1985)
62 WILLIAMS, T.W., DAEHN, W., GRUETZNER, M., and STARKE, C.W.: 'Aliasing errors in signature analysis registers', *IEEE Des. Test Comput.*, 1987, **4** (2), pp. 39-45
63 DAVID, R.: 'Testing by feedback shift registers', *IEEE Trans.*, 1980, **C-29**, pp. 668-673
64 DAMAINI, M., OLIVO, P., FAVALLI, M., and RICCO, B.: 'An analytical model for the aliasing problem in signature analysis testing', *IEEE Trans.*, 1989, **CAD-8**, pp. 1133-1144

65 PRADHAN, D.K., and GUPTA, S.K.: 'A new framework for designing and analysing BIST techniques and zero aliasing compression', *IEEE Trans.*, 1991, **C-40**, pp. 743–763
66 SMITH, I.F.: 'Measures of the effectiveness of fault signature analyses', *IEEE Trans.*, 1980, **C-29**, pp. 510–514
67 YARMOLIK, V.N., and DEMIDENKO, S.N.: 'Generation and application of pseudorandom sequences in test and check-out systems' (Wiley, 1988)
68 TENDOLAKER, N., and SWANN, R.: 'Automated diagnostic methodology for IBM 3081 processor complex'. *J. Res. Dev.*, 1982, **26**, pp. 78–88
69 TASAR, O., and TASAR, V.: 'A study of intermittent faults in computers'. Proceedings of conference *AFIPS*, 1977, **46**, pp. 807–811
70 HAMMING, R.W.: 'Error-detecting and error-correcting codes', *Bell Syst. Tech. J.*, 1950, **29**, pp. 147–160
71 SELLER, F.F., HSIAO, M.Y., and BEARNSON, L.W.: 'Error detecting logic for digital computers' (McGraw-Hill, 1968)
72 WILKINSON, B.: 'Digital system design' (Prentice Hall, 1992)
73 LIU, S., and COSTELLO, D.J.: 'Error control coding: fundamentals and applications' (Prentice Hall, 1983)
74 WAKERLY, J.F.: 'Error-detecting codes, self-checking circuits, and applications' (Elsevier/North Holland, 1978)
75 WAKERLY, J.F.: 'Digital design: principles and practice' (Prentice Hall, 1994)
76 RUSSELL, G., and SAYERS, I.L.: 'Advanced simulation and test methods for VLSI' (Van Nostrand Reinhold, 1989)
77 JOHNSON, B.W.: 'Design and analysis of fault tolerant systems,' (Addison-Wesley, 1989)
78 BERGER, J.M.: 'A note on error detection codes for asymmetric channels', *Inf. Control*, 1961, **4**, pp. 68–73
79 SERRA, M.: 'Some experiments on the overhead for concurrent checking'. Record of 3rd technical workshop on *New directions in IC testing*, Halifax, Canada, 1988, pp. 207–212
80 WESSELS, D., and MUZIO, J.C.: 'Adding primary input error coverage to concurrent checking for PLAs'. Record of 4th technical workshop on *New directions in IC testing*, Victoria, Canada, 1989, pp. 139–153
81 ANDERSON, A., and LEE, P.: 'Fault-tolerance: principles and practice' (Prentice Hall, 1980)
82 PRADHAN, D.K. (Ed): 'Fault tolerant computing: theory and techniques' (Prentice Hall, 1986)
83 SAYERS, I., and RUSSELL, G.: 'A unified error detection scheme for ASIC design' *in* MASSARA, R.E. (Ed.): 'Design and test techniques for VLSI and WLSI circuits' (Peter Peregrinus, 1989)
84 GARNER, H.L.: 'The residue number system', *IRE Trans. Electronic Computers*, 1959, **EC-8**, pp. 140–147
85 AVIZIENIS, A.: 'Arithmetic error codes: cost and effectiveness studies for applications in digital system design', *IEEE Trans.*, 1971, **C-20**, pp. 1322–1331
86 MONTERRO, P., and RAO, T.R.N.: 'A residue checker for arithmetic and logical operations'. Proceedings of 2nd IEEE symposium on *Fault tolerant computing*, 1972, pp. 8–13
87 RAO, T.R.N.: 'Error coding for arithmetic processors' (Academic Press, 1974)
88 SAYERS, I.L., RUSSELL, G., and KINNIMENT, D.J.: 'Concurrent checking techniques: a DFT alternative', *in* [19]
89 BREUER, M.A., and FRIEDMAN, A.D.: 'Diagnosis and reliable design of digital systems' (Pitman, 1977)
90 FERGUSON, T.J., and RABINOWITZ, J.H.: 'Self-synchronizing Huffman codes', *IEEE Trans.*, 1984, **IT-30**, pp. 687–693

91 HATCHER, T.R.: 'On a family of error correcting and synchronizable codes', *IEEE Trans.*, 1969, **IT-15**, pp. 366–371
92 MAXTED, J.C., and ROBINSON, J.P.: 'Error recovery for variable length codes', *IEEE Trans.*, 1985, **IT-31**, pp. 794–801
93 TANAKA, H.: 'Data structure of Huffman codes and its application to efficient encoding and decoding', *IEEE Trans.*, 1987, **IT-33**, pp. 154–156
94 TICHENER, M.R.: 'Digital encoding by means of a new T-code to provide improved data synchronization and message integrity', *IEE Proc.E*, 1984, **131**, pp. 151–153
95 TICHENER, M.R.: 'Construction and properties of augmented and binary depeletion codes', *IEE Proc. E*, 1985, **132**, pp. 163–169
96 TICHENER, M.R.: 'Synchronization process for the variable length T-codes', *IEE Proc. E*, 1986, **133**, pp. 54–64
97 HIGGIE, G.R., and WILLIAMSON, A.G.: 'Properties of low augmentation level T-codes', *IEE Proc. E*, 1990, **137**, pp. 129–132
98 YORK, G., SIEWIOREK, D.P., and ZHU, Y.Z.: 'Compensating faults in triple modular redundancy'. Proceedings of 15th IEEE symposium on *Fault tolerant computing*, 1985, pp. 226–231
99 FUQUA, N.B.: 'Reliability engineering for electronic design' (Marcel Dekker, 1987)
100 SIEWIOREK, D.P., and McCLUSKEY, E.J.: 'An iterative cell switch design for hybrid redundancy', *IEEE Trans.*, 1973, **C-22**, pp. 290–297
101 TOY, W.: 'Fault-tolerant design of local ESS processors', *Proc. IEEE*, 1978, **66**, pp. 1126–1145
102 SU, S.Y.H., and DuCASSE, E.: 'A hardware redundancy reconfiguration scheme for tolerating multiple module failures', *IEEE Trans.*, 1980, **C-29**, pp. 254–257
103 DeSOUSA, P.T., and MATHUR, F.R.: 'Sift-out modular redundancy', *IEEE Trans.*, 1978, **C-27**, pp. 634–627
104 CARTER, W.C., and SCHNEIDER, P.R.: 'Design of dynamically checked computers', *Proc. IFIP*, 1968, **2**, pp. 873–883
105 ANDERSON, D.A., and METZE, G.: 'Design of totally self-checking check circuits for *m*-out-of-*n* codes', *IEEE Trans.*, 1973, **C-22**, pp. 263–269
106 KHAKBAZ, J.: 'Self-checking embedded parity trees'. Proceedings of 12th IEEE symposium on *Fault tolerant computing*, 1982, pp. 109–116
107 REDDY, S.M.: 'A note on self-checking checkers', *IEEE Trans.*, 1974, **C-23**, pp. 1100–1102
108 SMITH, J.E.: 'The design of totally self-checking check circuits for a class of unordered codes', *Design Automation and Fault Tolerant Computing*, 1977, pp. 321–343
109 MAROUF, M.A., and FRIEDMAN, A.D.: 'Efficient design of self-checking checker for any *m*-out-of-*n* code', *IEEE Trans.*, 1978, **C-27**, pp. 482–490
110 GASTINIS, N., and HALATIS, C.: 'A new design method for *m*-out-of-*n* TSC checkers', *IEEE Trans.*, 1983, **C-32**, pp. 273–283
111 HURST, S.L.: 'Digital-summation threshold-logic gates: a new circuit element', *Proc. IEE*, **120**, pp. 1301–1307
112 REDDY, S.M., and WILSON, J.R.: 'Easily testable cellular realizations for (exactly *p*)-out-of-*n* and (*p* or more)-out-of-*n* logic functions', *IEEE Trans.*, 1974, **C-23**, pp. 98–100
113 LU, D.J., and McCLUSKEY, E.J.: 'Quantative evaluation of self-checking circuits', *IEEE Trans.*, 1986, **CAD-3**, pp. 150–155
114 CROUZET, Y., and LANDRAULT, C.: 'Design of self-checking MOS:LSI circuits', *IEEE Trans.*, 1980, **C-29**, pp. 532–537
115 ASHJAEE, M.J., and REDDY, S.M.: 'On totally self-checking checkers for separable codes', *IEEE Trans.*, 1977, **C-26**, pp. 737–744

116 GLASER, A.B., and SUBAK-SHARPE, G.E.: 'Integrated circuit engineering: design, fabrication and applications' (Addison-Wesley, 1979)
117 'Test methods and procedures for microelectronic devices'. National Technical Information Service, Springfield, VA, US MIL-STD-883
118 'Electronic system reliability: design thermal applications'. National Technical Information Service, Springfield, VA,US MIL-HDBK-251
119 'Reliability predictions for electronic systems'. National Technical Information Service, Springfield, VA, US-MIL-HDBK-217
120 JOHNSON, B.W.: 'Design and analysis of fault-tolerant digital systems' (Addison-Wesley, 1989)
121 VILLEMEUR, A.: 'Reliability, availability, maintainability and safety assessment' (Wiley, 1992)
122 HOYLAND, A., and RAUSAND, M.: 'Reliability theory models and statistical methods' (Wiley, 1994)
123 NELSON, V.P., and CARROLL, B.D.: 'Tutorial: fault-tolerant computing' (IEEE Computer Society Press, 1986)
124 JOHNSON, B.W.: 'A course on the design of reliable digital systems', *IEEE Trans.*, 1987, **E-30**, pp. 27–36
125 RENNELS, D.A.: 'Architectures for fault-tolerant spacecraft computers', *Proc. IEEE*, 1978, **66**, pp. 1255–1268
126 OZGUNER, F.: 'Design of totally self-checking asynchronous and synchronous sequential machines'. Proceedings of IEEE international symposium on *Fault tolerant computing*, 1988, pp. 124–129
127 DIAZ, M., and AZEMA, P.: 'Unified design of self-checking and fail-safe combinational and sequential machines', *IEEE Trans.*, 1979, **C-28**, pp. 276–281
128 HAYES, J.P.: 'Introduction to digital logic design' (Addison-Wesley, 1994)
129 SERRA, M., and MUZIO, J.C.: 'Space compaction for multiple-output circuits', *IEEE Trans.*, 1988, **CAD-7**, pp. 1105–1113
130 SERRA, M.: 'Fault detection using concurrent checking'. Proceedings of Canadian conference on *VLSI*, Halifax, NS, 1988, pp. 74–79

Chapter 5
Structured design for testability (DFT) techniques

5.1 General introduction

Because of the difficulties or complexities encountered when formulating acceptable tests for integrated circuits as they become larger and more complex, it is now essential to consider testing at the design stage of a VLSI circuit or system using VLSI parts, and not as an afterthought once the design has been completed. The old-fashioned separation between a design engineer who designs a circuit or system and a test engineer in a separate office who takes the design and then attempts to formulate an acceptable test strategy for it is no longer viable.

Design for testability, sometimes called design for test and almost always abbreviated to DFT, is therefore the philosophy of considering at the design stage how the circuit or system shall be tested, rather than leaving it as a tack-on exercise at the end of the design phase.

DFT techniques normally fall into three general categories, namely:

(i) *ad hoc* design methods;
(ii) structured design methods;
(iii) self test.

The first two of these methods usually require the use of some external comprehensive test facility, such as the tester illustrated in Figure 1.4 for VLSI circuits, but the third method usually minimises to a considerable extent the use of external test resources.

We will look at these three DFT categories in the following sections.

5.2 Partitioning and *ad hoc* methods

Partitioning of a circuit or system and other *ad hoc* design methods consist largely of a number of recommended or desirable practices which a designer should consider at the design phase. There are, therefore, no underpinning mathematical developments or equations to support these practices, but much more a list of recommended do and don'ts which the experienced designer follows. The general objective is simple, being to provide increased controllability and observability of the circuit or system being designed.

One of the most obvious and intuitive techniques for easing the testing problem is to partition the overall circuit or system into functional blocks, each of which may be independently tested. It is generally accepted that test costs are proportional to the square of the number of logic gates in a circuit and hence partitioning a circuit into, say, four equal parts reduces the testing problems of each part by a factor of sixteen compared with the overall circuit. How this may be done and how many partitions are practical obviously depends upon the specific circuit being designed.

The penalty for the partitioning and separate test of partitions is that additional I/O pins may be required to give access to the partition boundaries. This in turn may require the incorporation of multiplexers within the circuit to switch lines from their normal mode to a test mode, for example as shown in Figure 5.1. This may be particularly necessary in the design of complex standard cell custom VLSI circuits incorporating large macros such as PLAs, memory, microprocessors, etc., where the vendor will have known techniques for testing each macro. The OEM designer will therefore have no need to worry about how to test these macros, but the overall circuit design must provide the appropriate test access to them.

Individual multiplexers may also be necessary in order to give controllability or observability of individual nodes in a circuit, where the designer requires access for particular tests. This is shown in Figure 5.2, and clearly is a technique which may be added at the design stage when

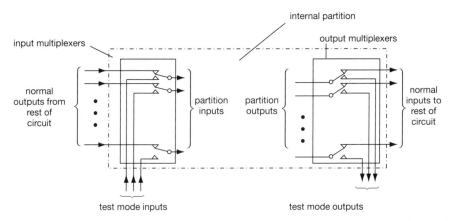

Figure 5.1 *The use of multiplexers to provide controllability and observability of internal macros or partitions of a complex circuit. The multiplexers are switched from normal mode to test mode under the control of an appropriate normal mode/test mode control signal (not shown)*

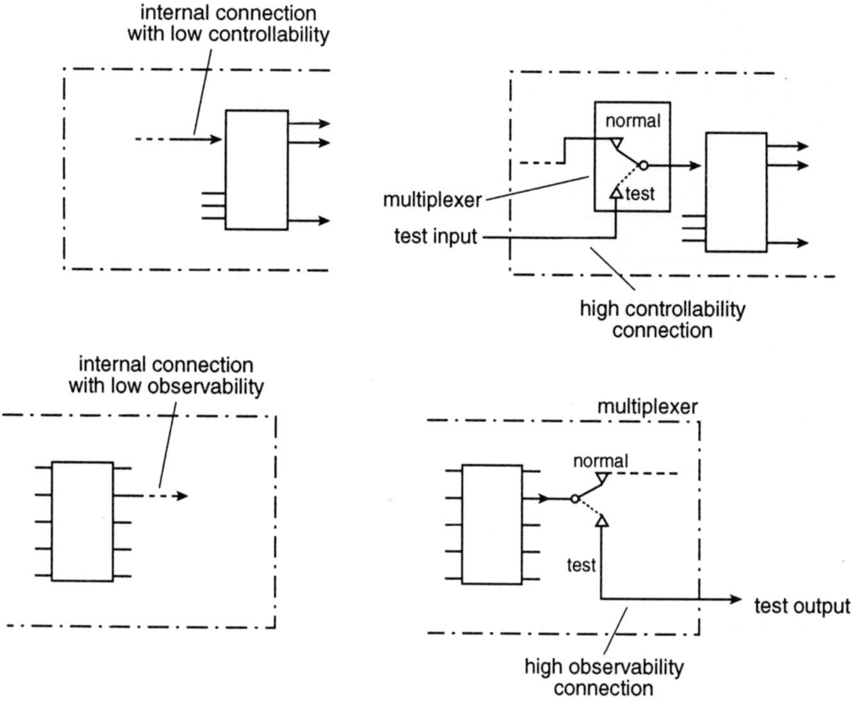

Figure 5.2 The use of individual multiplexers to provide controllability or observability of difficult internal nodes

considered necessary. Again the penalty to be paid, as in all these DFT techniques, is additional I/O requirements.

As well as the *ad hoc* partitioning of the complete circuit to facilitate testing, should any long counter assemblies be present then these should be divided into smaller groups of stages in order to reduce the number of clock pulses required for a test of each stage. This is illustrated in Figures 5.3a and b. It may also be necessary to clock a circuit or part of a circuit when under test at a different clock speed than normal, possibly one clock pulse at a time or to stop the clock to facilitate some surrounding test procedure, as illustrated in Figure 5.3c.

Other partitioning which may be considered is the segregation of analogue and digital parts of a circuit, so that one does not have to be accessed via the other. The testing of analogue circuits will form the subject of Chapter 7, and we will therefore have particular reason to come back to this partitioning requirement later.

Apart from these general considerations, which relate to controllability and observability of all the nodes of the circuit when finally under test, the following factors are some of the points which a designer should bear in mind in order to facilitate the final test procedures:

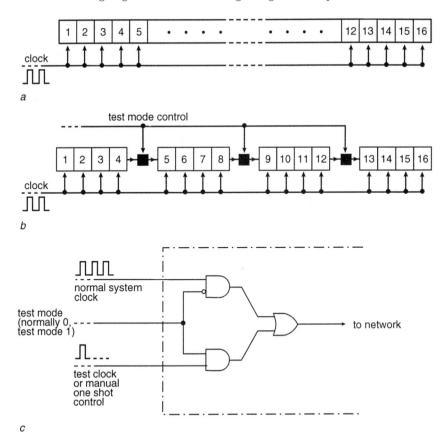

Figure 5.3 *The need to partition very long counters in order to reduce testing time, or to incorporate a separate test clock*
     a example 16-stage counter, outputs not shown, requiring 65 536 clock pulses for an exhaustive test
     b partitioned into four four-stage counters, test mode control details not shown, each partition requiring only 16 clock pulses for an exhaustive test
     c the provision of a separate clock which may be required to step through or stop part of the sequential circuit network separately from other parts

- never build in logical redundancy unless it is specifically required; by definition such redundancy can never be directly checked from the normal I/O pins;
- make a list of key nodes during the design phase, and check that they can easily be accessed;
- if possible use level-sensitive flip-flops in preference to edge-triggered types, since the behaviour of circuits using the former is more secure than circuits using the latter, not being dV/dt sensitive;

Structured design for testability (DFT) techniques 205

- at all cost avoid the use of on-chip monostable circuits; these must be off-chip if really necessary. On-chip use some form of clocked counter-timer circuit;
- if at all possible avoid the use of asynchronously operating macros; make all sequential circuits operate under rigid clock control unless absolutely impossible;
- ensure that all storage circuit elements, not only in counters but in PLAs and random-access memory can be initialised before test; this also applies to any internal cross-coupled NANDs and NORs which act as local latches;
- limit the fan-out from individual gates as far as possible so as not to degrade performance and to ease the task of manual or automatic test pattern generation;
- provide the means to break all feedback connections from one partition of a circuit to another partition (global feedback) so that each partition may be independently tested;
- consider the advantages of using separate test clocks (see Figure 5.3c) to ease certain sequential checks;
- avoid designing clever combinational modules which perform different duties under different circumstances; in general design a module as simply as possible to do one job;
- keep analogue and digital circuits on a chip physically as far apart as possible, with completely separate or very decoupled d.c. supply connections;
- when designing a complex VLSI custom IC ensure that the vendor can supply details of the appropriate vectors necessary for the test of large macros such as PLAs, memory, etc., and that it is possible to apply these vectors and monitor the test outputs;
- consider adding some online parity or other check bits if fault detection or fault correction would be advantageous;
- remember that a vendor's 100 % toggle test on internal nodes of a custom IC does not guarantee a fully fault-free circuit;
- controllability and observability numbers produced by analysis programs such as SCOAP, HI-TAP, etc. do not give any direct indication on how to design an alternative more easily tested circuit configuration;
- some commercial IC testers may not cater for don't care conditions and may require to know whether the actual design should give a logic 0 or a logic 1 under given test conditions;
- homing or synchronising sequences before the commencement of a sequential network test are not always possible with some testers, or acceptable; a specific test signal to initialise (reset) all sequential elements at the beginning of a test should always be used;
- statistical programs which estimate the fault coverage of a particular test set should be treated with caution; they may work well with random

combinational logic networks, but not so accurately for bus-structured networks. Their value is to give an indication of the progress being made during the determination of a set of test vectors for a network rather than a final absolute indication;
- the problem of open-circuit (memory) faults in CMOS is rarely covered with presently available ATPG software programs;
- finally, never waste any spare I/Os on a custom chip and its package—use them up if possible as additional I/O test points.

Further comments and discussion on these basic design considerations which remain valid in spite of increasing design and CAD sophistication may be found in Bennetts[1] and elsewhere[2-6].

The major DFT guidelines may therefore be summarised as follows:

(i) maximise the controllability and observability of all parts of the circuit or system by partitioning or other means;
(ii) provide means of initialising all internal latches and flip-flops at the beginning of any test;
(iii) keep analogue and digital circuits physically and electrically as separate as possible;
(iv) avoid asynchronous working and redundancy if at all possible.

As will be appreciated, these guidelines apply equally to the more formal DFT techniques which we will consider below and to the informal *ad hoc* DFT techniques just considered.

5.3 Scan-path testing

The techniques discussed in the preceding section, which employed multiplexers as the means of improving the controllability and/or observability of internal nodes of a circuit when under test, have the severe disadvantage of requiring many more I/Os to be provided on the circuit to give this improved testability. Scan-path testing, sometimes referred to as scan test is a means of overcoming this disadvantage and reducing the additional I/Os required to a minimum irrespective of the size of the circuit being tested. As will be seen, the penalty will be an increase in the time to test the circuit, plus additional housekeeping in order to keep track of the serial input and output logic signals now involved.

Scan-path testing fundamentally covers sequential logic networks. Recall from Figure 3.14 that all such networks can be modelled by a combinational logic network and a storage (memory) network, with secondary inputs and outputs linking the two halves. The primary outputs may be a function of the storage circuit states only (a Moore model) or a function of both the storage circuit states and the primary inputs (a Mealy model), but this distinction will not concern us here.

Scan-path testing of such a network involves switching all the storage elements of the circuit from their normal mode to a test mode shift register configuration. A scan-in I/O and a scan-out I/O allow data to be read into and read out from this reconfiguration for test purposes, thus providing controllability and observability of internal nodes which would not otherwise be readily accessible. Figure 5.4 shows the general schematic arrangement, the secondary inputs to the memory elements and the secondary outputs back to the combinational elements being present in both normal mode and test mode.

The testing procedure for a circuit proceeds as follows:

(i) the circuit is switched from its normal mode to scan mode, which converts the storage elements into a scan-path shift register;
(ii) the switching and storage action of this shift register is first checked by clocking through a pattern of 0s and 1s under the control of the test clock* ;
(iii) if this initial test is all correct, an input test vector is applied to the primary inputs, and a chosen pattern of 0s and 1s is serially loaded into the shift register under the control of the test clock, the latter becoming the secondary inputs to the combinational logic;
(iv) the circuit is switched back to its normal mode, and the clock operated once so as to latch the resultant secondary outputs from the combinational logic back into the shift register;
(v) the circuit is switched back again to its test mode, and the test clock operated so as to scan out the latched data from the shift register to the scan-out I/O for checking.

Steps (iii), (iv) and (v) are repeated as many times as necessary in order to test all the combinational logic circuits.

It will be appreciated that step (iii) is applying a test input to the combinational logic which with the primary (accessible) inputs can provide any necessary test vector for the combinational logic. Step (iv) with the primary (accessible) outputs is the response of the combinational logic to each such test vector. The storage elements are all assumed to be fault free if the scan-in/scan-out test of step (ii) is correct, the scan-in serial data usually being something like:

...0 1 0 1 0 0 0 1 0 0 0 1 1 1 0 1 1 1...

which will test that every stage is capable of switching from 0 to 1 and vice versa, and able to maintain a 0 or 1 state when its near neighbours are in the opposite state. (The precise serial test data may be influenced by the physical layout of the circuit, or by vendors' experience.) However, this does assume that the means to switch the sequential elements from their normal mode to

---

* Different vendors have slightly different circuits, including possibly only one clock which is switched to different duties when in normal or test mode.

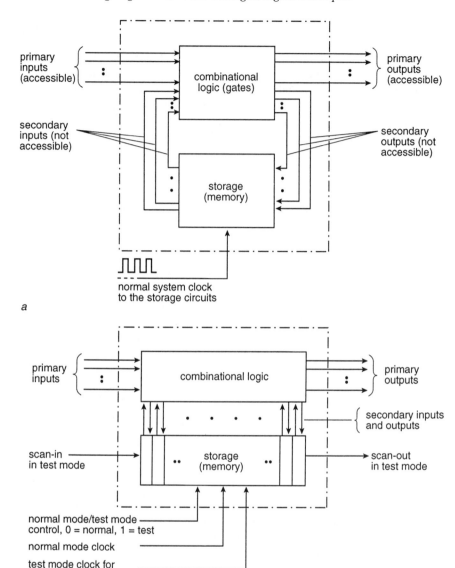

Figure 5.4 The principle of scan-path testing of sequential networks
    a repeat of Figure 3.14, emphasising the nonaccessibility of the secondary input and output signals
    b reconfiguration of the storage elements into a shift register scan-in/scan-out configuration when in test mode, the combinational logic and the secondary inputs and outputs remaining unchanged

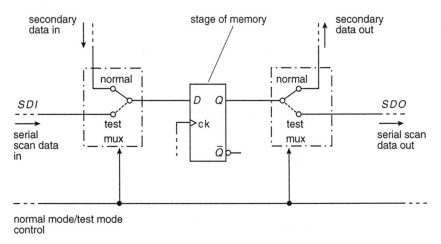

Figure 5.5 *The multiplexing requirements on each stage of the storage network*

scan-test mode, which involves some multiplexer switching such as shown in Figure 5.5, is satisfactory, but any fault in these normal-mode/test-mode switches will almost invariably result in either the scan test or the combinational logic test failing. Note that no fault location or diagnostic information is available from these tests; if step (ii) fails then steps (iii) and (iv) will not be executed.

The principle behind scan-path testing is therefore simple: it provides controllability and observability of r internal nodes of the circuit when in test mode, where r is the number of stages in the shift register reconfigurement, with only four additional I/Os being basically required to provide this facility. However, it requires an ATPG program such as PODEM or FAN to determine a minimum set of test vectors to test the combinational logic, and (usually) a further software program to assemble the required test data into the serial form required for each input/output test. No ATPG for the sequential logic, which is the most difficult and sometime insolvable area for ATPG programs to handle, is now required. The ATPG for the combinational logic may itself be considerably simplified by some partitioning of the complete circuit under test, as shown in Figure 5.6, which divides the test vector generation requirements into more readily handled blocks of logic. Indeed, with partitioning of the combinational logic and with a little more complexity in the control signals and housekeeping of the serial bit streams, it is possible to inject test vectors into one partition simultaneously with reading out data from another partition. This is sometimes termed simultaneous scan-test controllability and observability[6], but clearly requires meticulous attention to the ordering and interpretation of the input and output bit streams.

210 VLSI testing: digital and mixed analogue/digital techniques

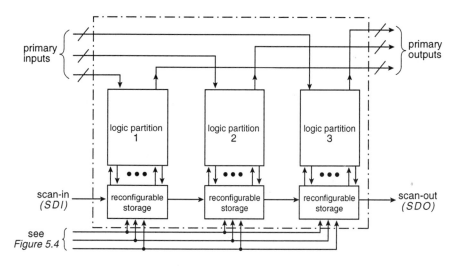

Figure 5.6 *Partitioning of a very large circuit into smaller blocks for test, using scan test to give controllability and observability of each partition*

The penalties that accompany the use of scan-path testing, however, should be appreciated. In summary they are:

- each of the storage elements now becomes more complex due to the need to switch from normal mode to test mode;
- chip size will increase, not only due to these larger storage elements but also due to the routing of the additional interconnect;
- the maximum available speed may be lower due to the extra series gates in the storage element circuits, possibly a loss of 1 or 2 ns in the maximum toggle speed of the flip-flops;
- its adoption imposes constraints on the chip designer, with careful consideration of the timing of clock pulses and also placement and routing to optimise the chip layout for both normal-mode and test-mode interconnections;
- a considerable amount of data has to be sequenced by the external test resource in order to perform the large number of individual load and test operations, and hence the total time to test will be much longer than where parallel vectors can be directly applied to the circuit under test.

In general, scan-path design and test is not necessary for ICs of less than, say, one or two thousand gates; on the other hand some vendors recommend its adoption for custom ICs of from five thousand gates upwards[7-11].

Many company variants on scan-path testing have been developed. Details of the most well known one, the IBM LSSD (level-sensitive scan design) circuit and certain others will be given below, which will show the complexity and hence silicon area overheads which can build up with scan-path IC testing. Further general details may be found in References 6, 12–15.

5.3.1 Individual I/O testing

It is particularly important that the input signals provided by the I/Os at the perimeter of any IC are correct to drive the subsequent on-chip logic. In total a vendor with test resources such as illustrated in Figure 1.4 may measure:

(i) the output voltage of each output pin under some worst-case loading or supply conditions;
(ii) the satisfactory performance of each input pin under low voltage input conditions, V_{IL}, and high voltage input conditions, V_{IH}, which equals the threshold design limits of the I/O circuits.

Testing over a range of temperatures may also be performed, particularly for military or avionics applications.

However, it may still be advantageous to provide a quick means of checking that all input I/Os remain functional after an IC has been received by an original equipment manufacturer (OEM), and to provide this facility some vendors have added a simple scan path through all input cells as shown in Figure 5.7. To check that each input cell responds correctly to low and high input voltages $(N+1)$ tests at V_{IL} (low) followed by the same $(N+1)$ tests at V_{IH} (high) are applied, where N is the total number of input cells, the resulting signal at the single output pin confirming the action of each of the input cells being monitored[9].

If the I/Os are bidirectional and have to operate as either inputs or outputs, then an additional test-enable pin has to be provided to disable the output portion of the bidirectional cells when in test mode. The same V_{IL} and V_{IH} patterns may then be applied and the response checked through the NAND cascade.

However, this simple scan-path testing of just the input I/Os does not address the greater problem of testing the internal logic of the IC. Its arguable advantage is to save the OEM from starting any more detailed tests if the circuit fails this basic test, but in general the following scan testing methods which address the functional logic have found greater favour. The I/Os are only a small (but vital) part of a VLSI circuit, and therefore statistically are less likely to be faulty than one of the many internal gates or macros of the complete circuit.

5.3.2 Level-sensitive scan design (LSSD)

Looking in more detail at the various circuit configurations that have been proposed and used in scan-path designs to ease the test of complex digital circuits, a commonality is that the normal storage elements must be clocked and must easily be reconfigured into a shift register. Hence almost without exception D-type circuits have been used, with various detailed circuit configurations and the means to implement the normal mode/test mode changes. Although all the storage circuits are clocked, the term latch rather than flip-flop is generally used in this context, since data is latched into the

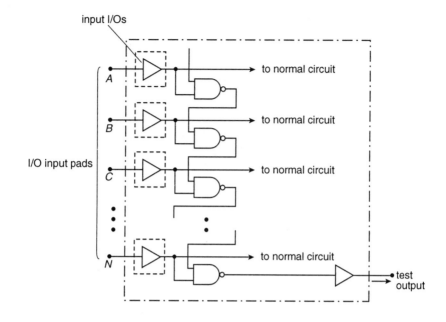

Figure 5.7 Simple scan test of input I/Os, with the input vectors being applied at the threshold limits of V_{IL} and V_{IH}

various stages during the testing phase. We will, therefore, for conformity use the term latch in the following rather than perhaps the more customary term flip-flop.

However, there still remains the choice between edge-triggered D-type circuits, which switch on either the positive-going or negative-going clock edge but are unresponsive to changing data on the data input, D, when the clock is at steady logic 0 or 1, or level-sensitive circuits which respond to changing input data whilst the clock is at logic 1 (or logic 0). The latter, sometimes referred to as d.c. coupled or level sensitive, are independent of the rise and fall times, t_r and t_f, of the clock edges, and hence can give more reliable operation if t_r and t_f and other dynamic characteristics of the circuit when under test cannot be guaranteed.

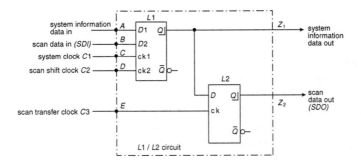

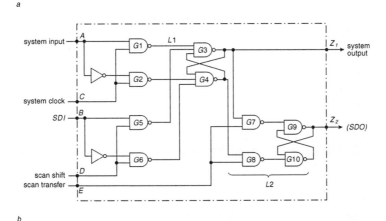

Figure 5.8 An early single-latch IBM L1/L2 scan-path storage circuit
    a schematic; the information data-in signals are the inaccessible secondary outputs from the combinational logic, and the information data-out signals are the inaccessible secondary inputs to the combinational logic, see Figure 5.4
    b NAND realisation

The most widely documented and known scan design is the level-sensitive scan design (LSSD) of IBM first developed in the late 1970s. Several variants of the first LSSD design were developed, differing mainly in the details of the special storage circuit elements[15–18]. One of the early disclosed LSSD circuits is shown in Figure 5.8; this will be seen to use two latches $L1$ and $L2$ per stage controlled by three separate clocks $C1$, $C2$ and $C3$, with separate system data out and scan data out output lines.* Each stage is therefore similar to a master-slave flip-flop, but with separate clocks and separate outputs.

In normal mode both the scan shift clock, $C2$, and the scan transfer clock, $C3$, are inoperative, with the system clock, $C1$, operating the $L1$ latches only.

---

* Care should be taken in considering the LSSD circuit diagrams given in some publications which show a triangular sign on the clock input of the latches, and which normally indicates edge-triggered flip-flop action. Edge-triggered action is not present in these circuits.

However, since there is only this one level-sensitive latch between the information data input and the information data output, race conditions may be set up around $L1$ and the combinational logic, as indicated by Figure 5.9a. Hence, uncertain final logic signals may result unless design precautions are taken.

The usual precaution taken is to segregate the combinational logic into two blocks as shown in Figure 5.9b and to use two interleaved system clocks $C1a$ and $C1b$. By this means it is guaranteed that no feedback loop can exist, as no changing input signal to any $L1$ can take place while its particular clock is active. The full LSSD scan test circuit therefore becomes as shown in Figure 5.9c, with clocks $C1a$ and $C1b$ being interleaved, and clocks $C2$ and $C3$ also being interleaved. Further details and discussion of the importance of the clocks and their timing may be found in Williams[15].

The action of the complete circuit in test mode, therefore, is as follows:

(i) with system clocks $C1a$ and $C1b$ off, scan in a test vector using clocks $C2$ and $C3$, which is directly fed from $L1$ into the combinational logic;
(ii) switch off clocks $C2$ and $C3$ when scan-in is complete, and with the appropriate primary input test vector record the primary output response from the combinational logic;
(iii) at the same time, apply a single clock pulse $C1a$ followed by $C1b$ to load the secondary output test response back into the $L1$ circuits;
(iv) scan out this latched secondary output data by means of clocks $C2$ and $C3$;
(v) check (ii) and (iv) for fault-free/faulty response of the combinational network.

Notice that during the scanning out of the response data in (iv) the secondary input data to the combinational logic from $L1$ will be changing, but this will not influence the data in the $L1$, $L2$ latches because the system clocks $C1a$ and $C1b$ are inoperative. Notice also that the two nonoverlapping clocks, $C2$ and $C3$, are critical for correct scan path action; if only one clock was used or if $C2$ and $C3$ clock pulses overlapped, then both latch $L1$ and latch $L2$ would be simultaneously open to data, and a completely transparent scan path through all stages would result. This will be recognised as the same situation that has to be guarded against in a normal master-slave flip-flop, and which is conventionally guarded against by using clock Ck to clock the input master circuit and $\overline{Ck}$ to clock the output slave circuit.

A major disadvantage of this early $L1$, $L2$ LSSD circuit is that the $L2$ latch plays no part in the normal system operation, being idle except when in test mode. A more economical LSSD circuit termed the $L1/L2*$ ($L1/L2$ star) circuit which uses both $L1$ and $L2$ in the normal system operation is shown in Figure 5.10. This is also a single latch normal mode configuration since only one latch, $L1$ or $L2*$, is present in each system data path when operating in normal mode.

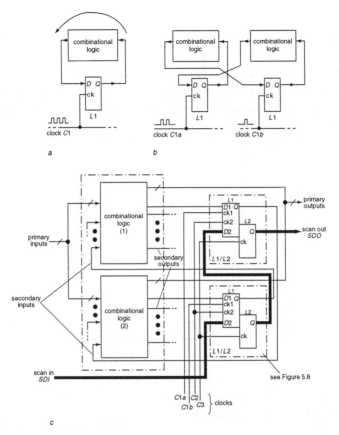

Figure 5.9 Single-latch race condition prevention
  a the problem of a race condition using one clock $C1$ and a single latch $L1$
  b race-free conditions using interleaved clocks $C1a$ and $C1b$ and two blocks of combinational logic
  c the complete single-latch $L1/L2$ architecture with normal system clocks $C1a$ and $C1b$ and scan test clocks $C2$ and $C3$. For simplicity only two $L1/L2$ circuits are shown rather than many

The action of this $L1/L2^*$ circuit should be clear from the previous $L1/L2$ circuit. Notice that four clock lines are still required, but the number of $L1/L2^*$ circuits for a given network is halved in comparison with the previous $L1/L2$ circuit where only one of the two latches was used in normal mode. Strict segregation of the signals from $L1$ and $L2^*$ back to the combination logic partitions must be maintained as in the previous $L1/L2$ circuit in order to avoid the possibility of any race conditions around combinational logic/storage loops. However, there is now a problem with the scan out of the response data from the combinational logic when in test mode, namely that it is not possible to scan out response data held in $L1$ and $L2^*$ at the same time. Consider the circuit shown in Figure 5.10a: if test data from the

216 VLSI testing: digital and mixed analogue/digital techniques

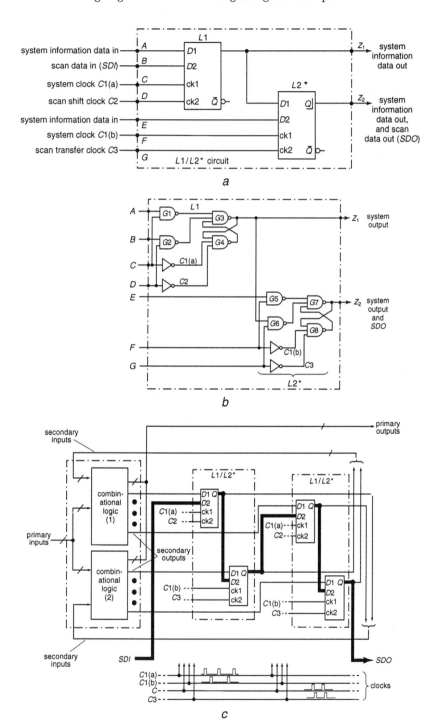

Structured design for testability (DFT) techniques 217

combinational logic is clocked into $L1$ and into $L2^*$ by the operation of clocks $C1a$ and $C1b$, then in scan-out mode one or other of these bits of information will be destroyed. For example, consider clock $C2$ occurring before $C3$, then the data in the previous $L2^*$ circuit will be shifted into $L1$, destroying the information in $L1$; if clock $C3$ occurs before $C2$ then the data in $L1$ will be shifted into the following $L2^*$ circuit destroying the information in this latch. Hence, whichever way we look at it, one half of the test response information from the combination logic will be lost. The way this problem is overcome is to test each half of the combinational logic separately; the test data from, say, combinational logic block 1 are first clocked into the $L1$ circuits, see Figure 5.10c, and scanned out using clocks $C2$ and $C3$, followed by the test data from combinational logic block 2 being clocked into the $L2^*$ circuits and then scanned out by using clocks $C2$ and $C3$ again. Clearly, this is yet a further complexity in the overall testing procedure.

As well as this detailed sequencing, there is also the high silicon overhead penalty which has to be paid for in the adoption of these LSSD design and test techniques. A somewhat artificial way in which this overhead has been calculated in the literature is to consider:

(i) the number of logic gates in the combinational logic network (see Figure 5.4), divided by the number of stages in the sequential network;
(ii) the number of gates in each stage of the sequential network with and without LSSD.

The first ratio, usually designated K, is therefore the number of logic gates in the combinational logic per storage element in the sequential network, and in most sequential circuits has a value between about five and 15.

Considering the later circuit shown in Figure 5.10, we could use this storage element in two possible ways, first to use only $L1$ for the system data as was done in Figure 5.9, or use both $L1$ and $L2^*$ for system data as was done in Figure 5.10. Counting the number of NAND gates then necessary per stage we have:

(a) *Single normal-mode latch per stage*

three gates $G1$, $G3$, $G4$ used in normal mode
seven gates $G1$, $G2$, $G3$, $G4$, $G6$, $G7$, $G8$ used in test mode.

We may ignore gate $G5$ since this gate would not be incorporated if only a single normal-mode latch was being designed. The literature also does not count the inverter gates, but only the NAND gates.

Figure 5.10   *The later single-latch $L1/L2^*$ LSSD development*
    a schematic, cf. Figure 5.8a
    b minimum NAND realisation
    c race-free topology with two blocks of combinational logic, cf. Figure 5.9c. For simplicity only two $L1/L2^*$ circuits are shown

(b) *Two normal-mode latches per stage*

six gates G1, G3, G4, G5, G7, G8 used in normal mode for two latches eight gates G1 to G8 used in test mode.

Therefore, with case (a) above we have per normal mode latch:

- K gates in the combinational logic plus three in the latch, $=(K+3)$ gates total;
- four additional gates required for LSSD purposes;
- therefore overhead $= 4/(K+3) \times 100$ %.

For case (b) above, we have per normal mode latch:

- K gates in the combinational logic plus three in the latch, $=(K+3)$ gates total;
- one additional gate required per latch for LSSD purposes;
- therefore overhead $= 1/(K+3) \times 100$ %.

These results are shown in Figure 5.11, but should be treated with caution since they are based upon a single NAND gate count, and do not consider the silicon area required for the place and route of the additional clocks and scan paths through the circuit. A figure of 25–30 % additional silicon area may possibly be more realistic in most circumstances. Some concluding comments on scan path design will be given at the end of the following section.

5.3.3 Further commercial variants

Alternative scan testing methods for VLSI circuits have been proposed, the three most widely publicised ones being:

(i)   the scan-path design method of NEC;
(ii) the scan-set design method of Sperry-Univac;
(iii) the random-access scan (RAS) design method of Fujitsa.

We will briefly look at these three techniques as they provide further illustrations of the way in which scan design and test may be built in at the initial design stage of a complex chip.

The scan-path method of NEC has a number of similarities to the IBM LSSD method. It employs what NEC term a raceless master-slave D-type flip-flop with scan path, the circuit of which is shown in Figure 5.12a. As will be seen, the circuit closely resembles a conventional level-sensitive master-slave D-type circuit (see Figure 2.10), but with the addition of scan-in and scan-out terminals and a second clock, $C2$, to control the circuit when in test mode. Notice also that the master-slave action is controlled by clock and not-clock in both the normal mode and scan mode, and not two pairs of nonoverlapping clocks as in the IBM LSSD circuits.

The system schematic is shown in Figure 5.12b. Both clocks are normally at logic 1, switching to logic 0 to clock the circuit. In normal mode the scan

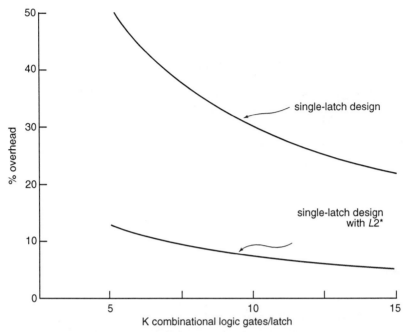

Figure 5.11   *Theoretical overheads for single-latch LSSD design based upon a simple NAND gate count only*

clock is inoperative, and in scan mode the system clock is inoperative. However, although the slave circuit does not respond to the master circuit until the system or scan clock returns to logic 1, there is the reported possibility that all master circuit inputs may not be fully cut off as the slave circuits are opened under some faulty tolerancing conditions, and hence a race condition around the combinational logic/storage circuits may result. This potential race condition could be guarded against by additional circuit complexity, but will not have the safety of the previous LSSD system with its two nonoverlapping clocks for all data transfer.

Further details of this scan-path method may be found published[8,11,19–21].

The scan-set design technique of Sperry-Univac is dissimilar from the LSSD and scan-path test methods in that it does not employ reconfigurable shift register latches (flip-flops) in the normal circuits; instead it basically leaves the normal-mode sequential elements unchanged and adds an entirely separate shift register scan chain to give the necessary controllability and observability of internal (inaccessible) nodes. The schematic circuit configuration is shown in Figure 5.13.

The test data which is scanned into the shift register can be loaded into any node of the normal system, and not just the secondary inputs to the sequential elements. Similarly, the response loaded back into the shift register can be from any node and not just the secondary outputs. Hence,

220 VLSI testing: digital and mixed analogue/digital techniques

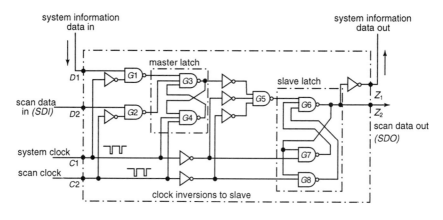

Figure 5.12 The scan-path design method
  a NAND realisation of each master-slave circuit
  b system schematic

there is a flexibility available to the system designer as to which working nodes should be accessed, but in practice it is sensible to provide test access to and from secondary inputs and outputs so that the sequential elements and the combination logic may be considered separately.

The test access to the chosen nodes from the shift register is usually provided by multiplexer switches, as indicated in Figure 5.13*b*. The test procedure for one test of the sequential elements to which access has been made is as follows:

(i)   scan appropriate test data into the shift register;
(ii) with the scan check off, parallel load this secondary input data into the system sequential elements with one system clock;

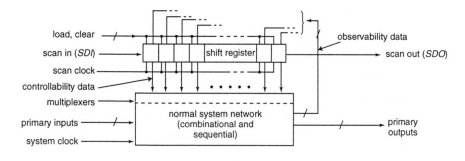

Figure 5.13 The scan-set design method
    a the principle
    b multiplexing signals into the normal system

(iii) parallel load the resulting secondary input data back into the shift register;
(iv) scan out and check this test response.

The combinational logic may be checked in a similar manner to that done in the LSSD circuit, using the accessible primary data from the sequential elements, but additionally the designer may choose to access certain internal nodes in the combinational logic as well as the accessible primary nodes.

In comparing this scan-set technique with the previous scan test methods there are certain advantages and disadvantages.

The advantages include:

(i) it does not require complex reconfigurable storage circuits in the normal working network or additional clock lines in the network;
(ii) being separate from the normal working network, the scan-in/scan-out shift register will introduce no additional race hazards or partitioning in the normal-mode circuits;

(iii) the scan-in and scan-out speed will be higher than systems using reconfigurable storage circuits;
(iv) the working system may be monitored under normal running conditions by taking a snapshot of the parallel data input to the scan shift register.

The latter feature is particularly significant, since it allows the system performance in normal mode and at normal operating speed to be monitored, so that some measure of the dynamic performance of the network may be possible.

The disadvantages, however, include:

(i) the silicon area overheads to build in a completely separate scan shift register and route all the individual connections to and from the working network may be higher than with LSSD or scan-path testing;
(ii) it may be difficult to formulate a good test strategy and decide which nodes of the working system should be controlled and observed when in test mode, bearing in mind that the normal system latches are not disengaged when in test mode.

However, on balance this scan-set technique does offer acceptable test advantages for certain applications of medium complexity, but not generally for circuits of VLSI complexity. The serial shadow register, which we will cover in Chapter 6, will be seen to use a separate scan-in/scan-out shift register for test purposes, and is a good example of the use of a separate scan shift register as introduced here.

Further details and discussion of scan-set testing may be found published[1,19,20,22].

Finally, random-access scan: unlike all three preceding methods, random-access scan (RAS) does not employ a shift register configuration to give internal controllability and observability. Instead it employs a separate addressing mechanism which allows each internal storage circuit to be selected individually so that it can be either controlled or observed. This mechanism is very similar to random-access memory (RAM), and consists of a matrix of addressable latches which are used as the system flip-flops when in normal mode, or which may be individually set or reset to provide controllability when in test mode. The state of each latch (flip-flop) may also be individually read out when required. The general schematic arrangement is shown in Figure 5.14a.

The individual latches in the RAS array may be polarity-hold addressable latches, as shown in Figure 5.14b, or set-reset addressable latches as shown in Figure 5.14c. Both perform identical duties, being individually selected by the X,Y address lines. The action of the second of these alternative circuits is briefly as follows.

Under normal-mode conditions, the preset and clear lines, which are common to all RAS latches are ineffective (preset = logic 0, clear = logic 1),

Structured design for testability (DFT) techniques 223

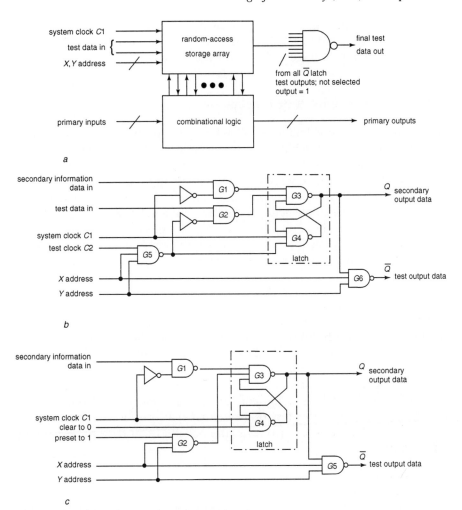

Figure 5.14 Random-access scan (RAS) testing
   a overall schematic
   b polarity-hold addressable latch
   c alternative set-reset addressable latch

and secondary input information from the combinational logic is free to enter the latches under the control of the system clock. The Q output is then the normal system secondary output back into the combinational logic. This output data may be individually read out, if required, by addressing each individual latch with its X, Y address, the test data out being $\bar{Q}$ rather than Q. In test mode all latches are first reset to $Q = 0$ by means of the clear line and then individually set as required by the preset line and individual X, Y addresses. The full RAS array may therefore be loaded with appropriate 0,1 data by sequentially addressing all the latches which are required to be set to $Q=1$, leaving the nonselected latches as $Q=0$. This data therefore provides any required secondary input test vector for the combinational logic. The circuit of Figure 5.13b provides identical controllability and observability, but will be seen to use a test clock, $C2$, to control the selected test data input.

The advantages of the RAS technique are:

(i) it provides individual controllability and observability of all system latches, with observability in normal-mode also being available;
(ii) it may also be used to provide observability of any additional critical node(s) within the combinational logic if required;
(iii) there is the flexibility to access some but not all of the RAS latches when in test mode; also the action of reading out data from the latches is not destructive as it is in the case of scan-in/scan-out shift register configurations, and thus switching repeatedly between normal mode and test mode may be more easily or quickly carried out.

The disadvantages are that this pseudoscan action is not as simple as a shift register configuration, requiring the X, Y address data to be sequenced to give scan-in/scan-out of test data, and marginally more I/O pins are required than in the previously considered scan-test methods. Further details may be found in References 19, 20, 22, 23 and 23a. A broad ranging discussion on scan-test principles may also be found in Abramovici et al.[6].

In summarising the scan-test methods which have now been considered, all have been designed to give the additional controllability and observability necessary to facilitate the test of sequential networks, with the minimum number of additional I/Os. However, the penalties that have to be paid for this include:

- increased silicon area;
- increased time to test;
- possibly some small loss in system performance due to additional logic gating;
- additional complexity at the circuit design stage, particularly in the increased place and route requirements.

Also, although this has segregated the difficulty of testing a combined combinational/sequential network into separate parts, there still remains the requirement for formulating appropriate test vectors for what may still be a very large combinational testing problem.

Current industrial reactions suggest that although the principles of full scan test remain valid, the above techniques have not proved to be always justified in practice, particularly in the silicon area overheads and potential loss of maximum speed performance. Partial scan, see Section 5.3.4, may be more acceptable for certain circuits.

5.3.4 Partial scan

The requirement to scan in a very lengthy serial bit stream to act as a test vector, and to serially scan out a test response, is a serious constraint in the previous scan-test methods. To some degree this time factor may be reduced by partitioning the circuit or system under test as shown in Figure 5.6 and providing separate scan-in and scan-out terminals for each partition, or multiplexed inputs and outputs, but this is adding yet further complication in the design layout.

However, when the interaction between the combinational logic and the sequential (storage) elements of many sequential systems is examined, it is found that the test vectors for the combinational logic often require very few secondary inputs, with many states being don't care for a particular combinational test. Similarly, the state of many stages can remain unaltered for a number of combinational test vectors. Equally, any fault in the combinational network is likely to propagate via the secondary inputs to a very small number of storage elements, and therefore all these factors point to some detailed consideration of the actual circuit under test so as to minimise the test efforts involved in using long scan-in/scan-out procedures.

The technique of partial scan has therefore been introduced to address these considerations, and hopefully to provide a faster test procedure without adding any additional I/Os. As will be seen, there may still be small sequential machines left in the individual test procedures, since not all of the sequential elements may be reconfigured into a scan-test configuration, and hence purely combinational logic test-pattern generation is no longer necessarily present.

To introduce and illustrate the advantages that may be gained by some partitioning of the circuit or network under test into smaller scan-test blocks, consider the circuit shown in Figure 5.15. The combinational test requirements are:

block $L1$: 50 test patterns required, eight bits from scan register $R1$ (secondary inputs), four output response bits (secondary outputs) in addition to the accessible primary outputs;

block $L2$: 25 test patterns required, six bits from scan register $R2$, nine secondary output response bits in addition to accessible primary outputs.

Consider blocks $L1$ and $L2$ tested separately. Then, ignoring $R2$, $L2$ and $R3$, we may test $L1$ by scanning in 50 individual test vectors into $R1$, each eight bits long, thus requiring 400 scan-in clock cycles. The four outputs may be loaded into $R4$ and scanned out simultaneously with the next loading of $R1$ if the system and scan clocks to the separate registers or other means are appropriately arranged. A total of 404 clock cycles are therefore required, with only the appropriate blocks of four response bits from the scan-out terminal being checked.

To test $L2$ separately from $L1$ we need to scan in 25 individual test vectors into $R2$, each six bits long. However, there are nine secondary outputs, and therefore if scan-out of the test response from $R3$ is to occur simultaneously with the loading of the next test vector into $R2$ there must be nine clock cycles per scan-in/scan-out test = $(25 \times 9) = 225$ clock cycles. If we allow an additional 12 clock cycles to begin the scan-in loading of $R2$ through $R1$ and complete the scan-out of $R3$ through $R4$, this gives a total of 237 clock cycles to test $L2$, = 641 clock cycles to test $L1$ and $L2$. If a single scan-in/scan-out register 14 bits long was used, then $14 \times 50 = 700$ clock cycles would be necessary to scan in the test vectors, and 700 clock cycles to scan out the resulting test responses.

The above separate test of $L1$ and $L2$ is sometimes termed a separate-mode scan-test. As an alternative possibility there is an overlapped-mode scan-test in which scan testing of the combinational partitions $L1$ and $L2$ takes place as far as possible simultaneously. Clearly, the block with the largest required number of test patterns still dominates the number of clock cycles required for scan-in, but with some care (and complexity) it is possible to interleave scan-in/scan-out tests for the separate blocks so as to achieve a reduction in the total number of clock cycles required for test. More details of separate-mode and overlapped-mode scan tests may be found in Breuer et al.[24] and elsewhere[6,25–27].

However, the above example still involves all the sequential elements being reconfigurable from normal mode to scan mode, with the attendant increase in complexity per stage, and does not address the possibility of leaving some storage elements as normal-mode circuits, changing only a certain proportion into scan-in/scan-out elements as partial scan usually implies.

A number of examples have been published for selecting a subset of storage elements in a circuit and making them the reconfigurable scan-test elements. If this subset is substantially smaller than the total number of storage elements, then clearly the circuit when in test mode may still have some cyclic sequential action, although not of the same complexity and difficulty to test as the normal circuit. However, the problem of choosing which storage elements shall be made reconfigurable and which left unaltered is not readily defined, and may in practice be made on an *ad hoc* basis using the expertise of the circuit designer.

One publicised example of partial scan design is the balanced structure scan test (BALLAST) methodology of Gupta *et al.*[27], which sets out certain guide lines for the selection of the storage elements which shall be reconfigurable, and attempts to leave the storage circuits which are not reconfigured in an acyclic configuration. The ideal grouping of the storage elements proposed in BALLAST therefore is such that:

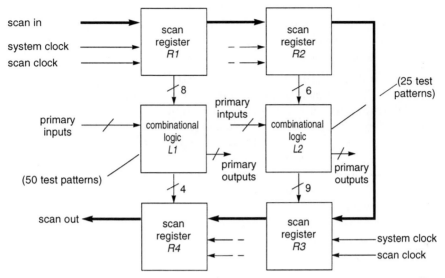

Figure 5.15 A simple example to illustrate separate-mode scan-path design, where the separate blocks of combinational logic may be individually scan tested. Possible additional multiplexers or other means (not shown) may be present to achieve separate partition testing

(i) each group of storage elements must have the same clock and control lines to all storage elements (this is usually the case and is nothing extraordinary);
(ii) each group must have secondary input data from only one partition of the combinational logic;
(iii) each group must feed secondary output data to only one partition of the combinational logic;
(iv) there shall be no cyclic sequential actions within the groups which are not reconfigured into scan-test configurations, that is the non-reconfigurable groups shall be acyclic;
(v) all paths between two combinational logic partitions separated by non-reconfigurable groups of storage elements shall have the same number of clocked storage elements in series such that the signal delay in all paths is the same (termed balanced).

Notice that if the above conditions are met, there is no cyclic sequential action to be handled by the test pattern generation, the testing of the logic being purely combinational but with a time factor for clocking test data between the combinational partitions.

The example circuit shown in Figure 5.16a shows the combinational logic divided into four partitions $L1$, $L2$, $L3$ and $L4$; the sequential elements have been divided into six groups $R1$ to $R6$, of which $R1$, $R2$, $R3$ and $R4$ have been made acyclic, with $R5$ and $R6$ reconfigurable scan-in/scan-out elements. In scan-test mode the secondary inputs to $L1$ and $L4$ become available, and the secondary (response) outputs from $L4$ may be scanned out at the end of each test. This is shown in Figure 5.16b.

The procedure for scan test is therefore as follows:

(i) scan in appropriate test bits into $R5$ and $R6$;
(ii) load this test data from $R5$ and $R6$ into $L4$ and $L1$, and apply the remaining test bits to all the primary inputs;
(iii) keeping all these test bits unchanged, clock the nonreconfigurable storage elements $R1$ and $R4$ an appropriate number of times so as to transfer the response data from $L1$ to $L2$ and $L3$ and on to $L4$. Notice that because $R1$ to $R4$ are acyclic and the primary and secondary inputs are held constant, steady-state conditioning throughout the circuit $L1$ to $L4$ will result;
(iv) check the test response on all primary outputs;
(v) scan out and check the secondary test response from $L4$;
(vi) repeat using the next test vector.

The precise control signals necessary for this action will depend upon the actual circuit details of the reconfigurable scan-in/scan-out storage elements.

It will be seen that this scan-test technique is dissimilar from the techniques previously considered in that the scan-in test data has to be held while several system clock pulses are applied to propagate response data through the partitions before scan-out is done. The combinational logic partitions that are separated by acyclic groups of storage elements are effectively being partitioned in time. However, it may not be possible or economical to partition the circuit exactly as above, and alternative constructs have been suggested. Further details of this particular partial scan test procedure may be found published[6,27,29].

A number of other partial scan proposals will be found[30–40]. The difficulty of choosing which storage elements to include in the scan chain and which to leave unaltered is the key problem in all methods, and has been approached in several general ways, namely:

- a consideration of the duty and criticality of each flip-flop in the circuit;
- selection so as to break all cycles and make the resulting partition acyclic;
- starting with all flip-flops in the scan chain, remove them one at a time;
- starting with no flip-flops in the scan chain add flip-flops one at a time.

In the latter two methods, various heuristics and graph structures may be used to decide which flip-flops shall finally be included in the scan chain, but so far there is no general consensus of which is the best method, or of the relative efficiencies of the resulting test procedures. An overview of the present status and further references may be found in Reference 41.

Finally, it will be appreciated that although the adoption of partial scan may ease the difficulties of test pattern determination, it may involve more complexity in organising the test data input and monitoring the accessible

Structured design for testability (DFT) techniques 229

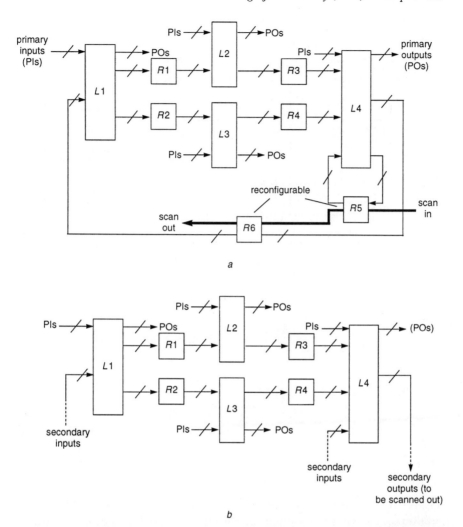

Figure 5.16 An example BALLAST partial scan test methodology, with balanced acyclic partitions of the sequential storage elements (based upon Reference 6)
  a the partition, omitting all clock and control lines for clarity; R5 and R6 constitute the partial scan elements
  b the all-combinational test having loaded in the secondary test bits from the scan chain

and scan data output. The external test equipment, therefore, requires additional resources to manage the sequencing of events and providing the scan-in data, resources which may not be available on all commercial VLSI testers. Figure 5.17 shows typical scan memory requirements, with additional timing complexity being present within the timing generator box to control the overall testing sequence.

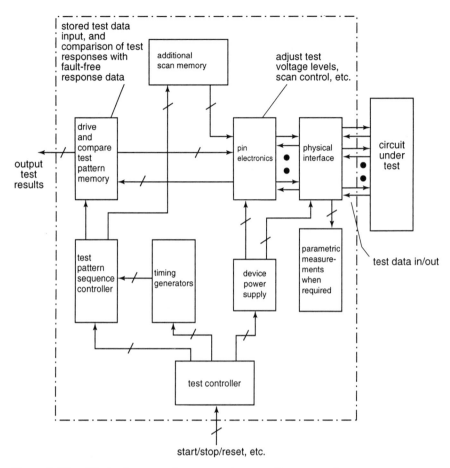

Figure 5.17   The basic external tester requirements for scan testing

In summary, scan testing provides a means of testing VLSI circuits without the addition of many extra test points on the circuit. Partial scan provides a range of possibilities ranging from full scan to nonscanning, with the advantage of reducing time to test if some acceptable partial scan rather than full scan can be formulated, but at the expense of somewhat greater conceptual difficulties and test sequencing. Although partial scan has been adopted by some vendors for specific applications it remains a minority test methodology, having been widely superseded by built-in logic block observation (BILBO) for VLSI circuits as will be covered in Section 5.5.

5.4 Boundary scan and IEEE standard 1149.1

Boundary scan is primarily concerned with delivering test signals to and collecting response signals from the I/Os of several integrated circuits which have been assembled into a complete working system, rather than the testing of an individual unconnected IC. It is therefore a board-level testing tool for the use of the OEM, enabling faulty items and interconnections to be detected and hopefully replaced or repaired.

The development of boundary scan originated with OEM designers in Europe who were involved with the design of very complex printed circuit boards (PCBs) and their test after assembly, and where decreasing spacing between conductors and components was making test probing more difficult (see Chapter 1, Section 1.5). As a result the Joint European Test Action Group (JETAG) was established in 1985, which subsequently grew into the Joint Test Action Group (JTAG) when North American companies later joined. Their work resulted in the IEEE Standard 1149.1-1990, entitled 'standard test access port and boundary scan architecture'[42-44]. The IEEE 1149 working group now maintains and updates this standard, which has been accepted internationally by all IC manufacturers. In all that follows the circuits, test procedures and terminology will be based upon standard 1149.

The essence of boundary scan is to include a boundary scan cell in every I/O circuit, as shown in Figure 5.18. During normal-mode working, system data passes freely through all the boundary cells from *NDI* to *NDO*, but in test mode this transparency is interrupted and replaced by a number of alternatives, providing in total the following seven possible actions:

(a) transparent mode, normal system operation;
(b) test mode: scan in appropriate test data to the input boundary cells in cascade to act as an input test vector for the core logic;
(c) test mode: use the test data (b) to exercise the core logic, and capture the resulting test response in the output boundary cells;
(d) test mode: scan out the test results (c) captured in the output boundary cells;
(e) test mode: scan in appropriate test data to the output boundary cells to act as input test data for the surrounding interconnect and/or components*;
(f) test mode: use the test data (e) to exercise the surrounding interconnect or components, and capture the test response in further input boundary cells (or monitor directly if accessible);
(g) test mode: scan out the test results (f) captured in the input boundary cells.

It will be seen that tests (b) to (d) constitute the usual serial bit stream load and test procedure associated with the scan testing of an IC, but in tests (e) to (g) the boundary cells on the IC outputs enable the same load and test technique to be applied to the system interconnect and circuitry surrounding

---

* Testing of the interconnections between assembled ICs may be particularly significant if surface-mount IC packages are used where soldered connections may not be visible.

an IC, the normal IC output data being blocked in the output boundary cells during these tests. Notice also that modes (b), (d), (e) and (g) above are scan modes with all boundary cells on the PCB in cascade, and that only two control signals, TMS and TCK, are used for all control purposes as will be detailed below. With the test data in and test data out signals, TDI and TDO, this makes a total of only four test terminals necessary on a PCB for the incorporation of boundary scan.

Multiple ICs on an assembled PCB are all under the control of these four input signals, TMS, TCK, TDI and TDO, forming what is called the JTAG four-wire test bus. Figure 5.19a shows that each IC requires a test access port (TAP) controller which receives this four-wire test data, and controls the normal or

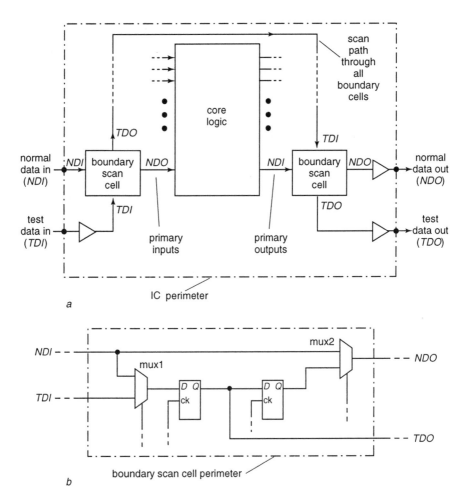

Figure 5.18 Boundary scan controllability and observability principles
    a provision of a boundary scan cell on every I/O of an IC
    b simplified details of one boundary cell circuit, omitting clock and control lines

test mode action required in the IC. For PCBs with a very large number of ICs it may be advantageous to group the ICs into more than one serial scan path as illustrated in Figure 5.19*b* and *c*, maintaining common control lines as far as possible.

The TAP controller can be on-chip as indicated in Figure 5.19*a* if the IC is designed to be used with boundary scan, or it can be part of a separate commercially-available IC containing one or more bytes of boundary cells plus the TAP controller and which is assembled adjacent to the LSI or VLSI IC to provide the boundary scan facility. The circuit of one such commercial IC carrying two nine-bit data words (two bytes plus two parity bits) is shown in Figure 5.20. There are effectively three register paths between the single test data input, *TDI*, and the single test data output *TDO*, namely:

(i) the boundary scan cells, each of which contain memory latches (see Figure 5.20*b*), and which collectively are referred to as the data register, *DR*;
(ii) the instruction register, *IR*, which can be loaded with data from *TDI* to control the boundary cell states of a following *DR* register;
(iii) a single-stage bypass register, that allows data to be transferred from *TDI* to *TDO* on a single clock pulse rather than the many clock pulses which would be necessary via the previous two registers.

The latter facility is essential to speed up the test procedures where a number of VLSI ICs are involved, allowing test access to individual circuits to be made very much more quickly than would otherwise be the case. In some TAP controllers there may be additional memory for miscellaneous purposes, but we will not consider this here.

The precise circuit details of the TAP controller and boundary scan cells vary between IC vendors, but all have to obey the protocols laid down in specification 1149.1. The TAP controller with its two inputs, *TDI* and *TCK*, is a 16-state Moore finite-state machine, each state controlling one of the conditions necessary to provide the normal mode and test mode actions. All transitions between states take place on the rising edge of the clock signal *TCK*, and all actions in the registers take place on a falling clock edge. The state diagram is shown in Figure 5.21, from where it will be seen that the 16 states broadly divide into:

- a normal mode, where no test scan-in or scan-out action takes place, the boundary scan cells all being transparent;
- states which control the data register, *DR*;
- states which control the instruction register *IR*.

Action can therefore be undertaken in the instruction register without affecting the data register, and vice versa.

Considering these 16 states $S0$ to $S15$, see Figure 5.21, in more detail, the action is as follows:

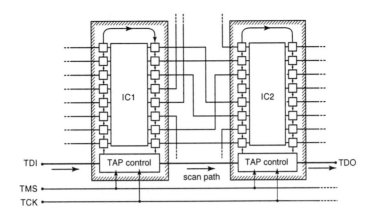

Figure 5.19 Boundary scan interconnections
  a the scan path through a number of ICs in series
  b possible parallel-series chains
  c multiple independent chains with common *TMS* and *TCK* controls
  d a complete PCB system test (see also Figure 5.22)

S0, test-logic-reset:

Normal-mode action, with all scan action disabled.

S1, run-test/idle:

A control state which exists between scan operations, where some internal test action (see later) may be executed.

S2, select-DR-scan:

A temporary controller state, leading into *S3* as soon as *TMS* is made 0.

S3, capture-DR:

Data from the core logic, see Figure 5.18*a*, is loaded in parallel into the boundary scan data registers using shift-DR and clock-DR, see Figure 5.20*b*.

S4, shift-DR:

The data in the boundary scan cells is shifted one position on each clock pulse, data being read out at *TDO*, new data coming in at *TDI*.

S5, exit 1-DR:

In this state, all data registers hold their state; the scan process can now be terminated if desired by making *TMS* logic 1.

S6, pause-DR:

All registers still hold their state, but further action elsewhere such as checking or waiting or assembling new test data may be necessary.

S7, exit 2-DR:

A return to shift-DR may be made from *S7* if desired, or exit to *S8*.

S8, update-DR:

The final state of scan latch *L1* in the output boundary scan cells, see Figure 5.20*b*, is transferred into latch *L2* for future reference.

The return from *S8* to normal mode can be made by *TMS* = 1,1,1, or 0,1,1,1. It is always possible to return to normal mode from any state in the test sequence by making *TMS* = 1 for a maximum of five consecutive clock periods.

The action of the seven states in the information register control is similar to that above for the data register control, briefly being:

S9, select-IR-scan:

Temporary state leading into *S10*.

236 VLSI testing: digital and mixed analogue/digital techniques

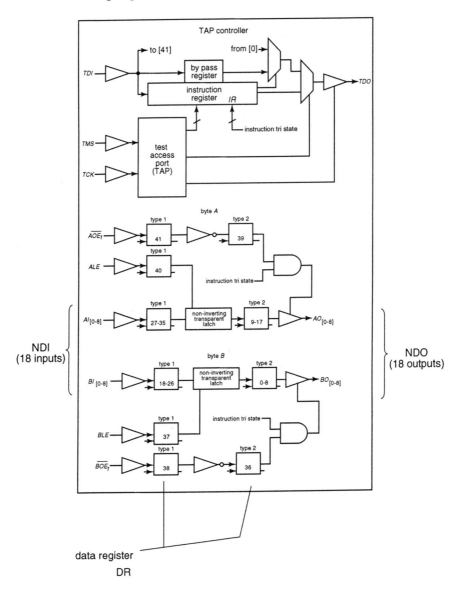

a

Figure 5.20 An example commercial circuit taken from the SCAN1800 family of National Semiconductor, providing 18 input and 18 output boundary scan cells and the TAP control
a general schematic; the scan chain is through cells 41, 40, 39, . . . 2, 1, 0 to TDO

Structured design for testability (DFT) techniques 237

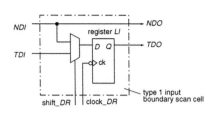
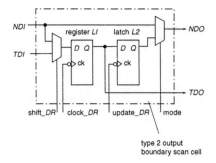

b

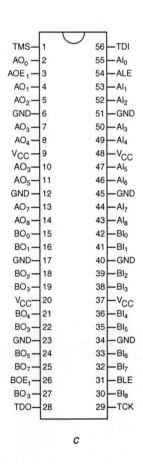

c

Figure 5.20   Continued
      b input and output boundary cells (data registers)
      c 56-pin DIL packaging (Acknowledgements, based upon National Semiconductor data sheet SCAN18373T)

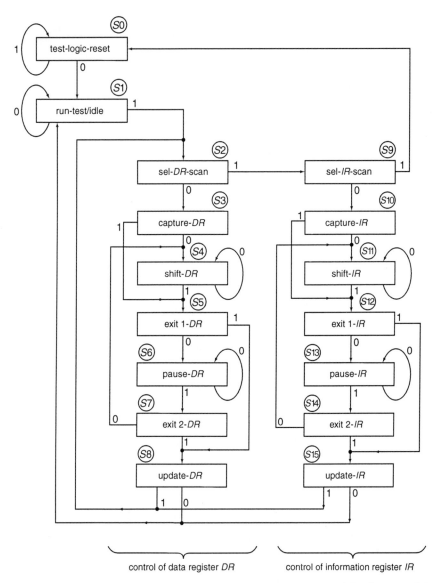

Figure 5.21 The state diagram for a TAP controller. The value of TMS required for each state transition is shown on the link between states, all transitions being on the rising edge of clock TCK

S10, capture-IR:

Load any fixed data into the information register; possibly scan out to check veracity of the information scan chain.

S11, shift-IR:

Shift in information data from *TDI* on each clock pulse.

S12, exit 1-IR:

Hold last state, exit if desired by making *TMS* logic 1.

S13, pause-IR:

Similar to *S6*.

S14, exit 2-IR:

Return to shift-IR if desired; otherwise exit to *S15*.

S15, update-IR:

Load final bits which are in the shift register *IR* latches into the parallel *IR* latches, see further comments below.

These states together with the condition of the tristate output terminal *TDO* are summarised in Table 5.1.

Table 5.1 The logic operations controlled by the 16-state TAP controller

State	Action	TDO output	Instruction register IR	Scan data register DR
S0	test-logic-reset	high impedance	initialise to 1000 0001	set to normal mode
S1	run-test/idle	high impedance	retain last state	execute test or retain last state
S2	select-DR-scan	high impedance	retain last state	retain last state
S3	capture-DR	LSB of register *DR*	retain last state	load *L1* registers from parallel system data
S4	shift-DR	LSB of register *DR*	retain last state	shift towards *TDO*
S5	exit 1-DR	LSB of register *DR*	retain last state	retain last state
S6	pause-DR	LSB of register *DR*	retain last state	retain last state
S7	exit 2-DR	LSB of register *DR*	retain last state	retain last state
S8	update-DR	high impedance	retain last state	load shadow output latch *L2*
S9	select-IR-scan	high impedance	retain last state	retain last state
S10	capture-IR	LSB of register *IR*	load in fixed instruction data	retain last state
S11	shift-IR	LSB of register *IR*	shift towards *TDO*	retain last state
S12	exit 1-IR	LSB of register *IR*	retain last state	retain last state
S13	pause-IR	LSB of register *IR*	retain last state	retain last state
S14	exit 2-IR	LSB of register *IR*	retain last state	retain last state
S15	update-IR	high impedance	load parallel *IR* latches	retain last state

The run-test/idle state *S1* in the above is a state where some built-in self test (if present) within the active logic can be run. This may be, for example, a built-in logic block observation (BILBO) test program such as will be covered in Section 5.5, and is separate from any test vectors that may be serially loaded into the boundary scan cells for scan-path tests. There needs to be boundary scan I/O cells on the active logic perimeter to activate any built-in self test, the signals to initiate this being scanned in to the appropriate input boundary cells on state *S1*. As many *TCK* clock pulses as necessary may be applied during state *S1* as long as *TMS* is held at 0, as is the case during the shift-out and pause states *S4*, *S6*, *S11* and *S13*.

The information register *IR* is also somewhat more complex than we have so far detailed. It consists of an eight-stage shift register through which instructions or other data may be clocked through all such instruction registers in cascade in the system, plus a nonshifting eight-stage latch in parallel. When the appropriate data have been clocked into the shift register stages, the parallel eight-stage latch is loaded on state *S15*, releasing the shift register stages for further use. The data in the eight-stage latch are decoded as required to control the required scan path or other activity, and remains unchanged until *S9* onwards is next selected. With the data 1 1 1 1 1 1 1 1 stored in this register the bypass register is activated, this being one of the specification 1149.1 protocols.

There remains further complexity in some of the TAP circuits and their timings, but the broad concept of being able to scan in test data and scan out response data from the perimeter of ICs in an assembled system should be clear from the above discussions. The reader is referred to IC vendors' data brochures for specific commercial information[45–48], and to Bleecker *et al.*[44] and elsewhere for wider information and discussions[49–57].

Boundary scan has become an accepted test methodology for digital assemblies which would otherwise be difficult to test, but a very great deal of housekeeping is necessary to define the required sequence of test signals and their fault-free response. However, because of the protocols imposed by IEEE standard 1149.1, commercial hardware and software aids are available for OEM use, usually with links to the CAD used for system design. Figure 5.22 illustrates a resource which provides automatic boundary-scan test pattern generation, with diagnostic software to provide faulty mode location and other facilities. In the end, however, there still remains the penalty of time to test, which must inherently be much longer with boundary scan than if test access to internal modes was freely available.

Finally, two further comments on standards:

(i) A working party is currently engaged in considering a further standard, IEEE Standard P1149.2 'Shared input/output scan test architectures'. This is intended to be a standard for boundary scan which incorporates certain features not present in standard 1149.1, namely shared input/output cells, an optional parallel update stage to supplement or replace serial scan-in data, and a high impedance input pin condition. However, the considerations so far have not produced a standard which is compatible with the existing standard 1149.1, and hence further deliberations are necessary before P1149.2 becomes an acceptable boundary-scan specification. The basic concepts and practice of standard 1149.1 will, however, remain unchanged.

(ii) Secondly, a new standard high-speed on-chip bus to provide very fast on-chip communication has recently been announced by a consortium of IC vendors. This is the 'peripheral interconnect bus', (PI-Bus), and is intended to be a standard for use on submicron VLSI circuits and modular architectures. Possible uses include interconnecting processor cores, on-chip memories, I/O controllers and other functional blocks. However, to date this has not been specifically linked with the use of boundary scan, although it would seem that there may be advantages if the two are considered together in the design and test of complex VLSI assemblies.

5.5 Offline built-in self test

As has been seen, scan-path test methodologies do not eliminate the need to prepare some set of acceptable test vectors for the circuit under test, or the need for appropriate external hardware to apply the test stimuli and monitor the resulting response. They do, however, ease the problem of automatic test pattern generation by making the resulting test mode as nonsequential as possible, but at the expense of assembling the serial test data and the time required to execute the resulting tests. Boundary scan itself is not a direct testing procedure, but merely a recognised means of serially accessing internal modes and interconnections on a PCB or other system assembly.

Turning therefore to offline self test methodologies, the concept here is to build in appropriate means whereby the circuit under test can be switched from a normal mode of operation to a test mode, with the required test stimuli and the resulting response both being done on-chip without the need to generate any ATPG data based upon stuck-at fault modelling or other means, or apply it in series or parallel using some external hardware resource. The circuit under test becomes its own test vector generator and test response detector. A further advantage of this concept is that on-chip tests can be done using the normal system clock(s) and at the full speed of the normal system.

This built-in self test (BIST) methodology becomes increasingly significant or necessary as VLSI chip size and complexity increases. We will look principally at two possible BIST architectures in the following pages, the first, built-in logic block observation (BILBO), being a well established technique, the second, cellular automata logic block observation (CALBO), being still largely in the research and development stage. As will be seen, both refer

242 *VLSI testing: digital and mixed analogue/digital techniques*

Figure 5.22   A PC-based boundary-scan test resource with boundary-scan test pattern generation, test vector editing, and diagnostic capabilities
  a the hardware resource

back to the fundamentals of pseudorandom sequence generation and signature analysis introduced in Chapter 3, Section 3.4, and Chapter 4, Section 4.5, respectively. They may both be contrasted with the online methods of self test using parity and other checks which were covered in Chapter 4, Section 4.6; the latter tended to be applied to signal processing, data processing and arithmetic applications rather than more general-purpose digital logic networks.

5.5.1 Built-in logic block observation (BILBO)

The basic concept of the built-in logic block observation test method is to partition the circuit under test into blocks of combinational logic which are each separated by blocks of sequential logic elements. The sequential blocks are referred to as BILBO registers, the output test signatures being observed in these blocks. Figure 5.23 indicates the general schematic arrangement.

Structured design for testability (DFT) techniques 243

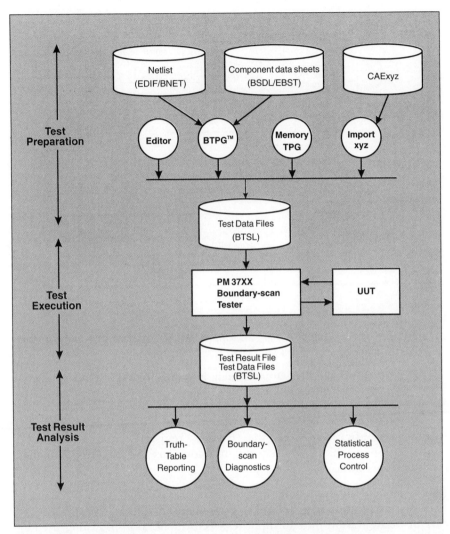

Figure 5.22 *Continued*
   b functional schematic; BSTL (boundary scan test specification language) is a hardware-independent high-level language (Acknowledgements, JTAG Technologies B., Eindhoven)

In normal mode each BILBO register forms part of the normal sequential logic for the working circuit, with the usual secondary inputs and outputs linking the sequential and the combinational logic. In test mode, a pair of BILBO registers, one preceding and one following each block of combinational logic, is used to test the block, the first reconfigured to provide the input test vectors and the second reconfigured to provide a test output signature from the combinational logic.

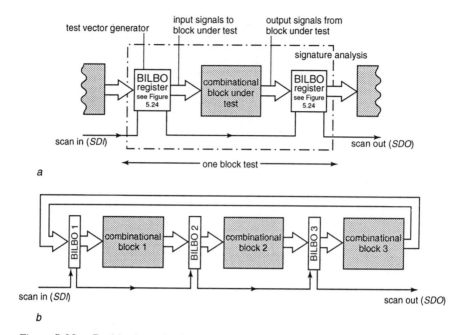

Figure 5.23 Paritioning of a logic network by BILBO registers; the control signals have been omitted for clarity
  a basic BILBO concept
  b example BILBO partitioning into three blocks in test mode, each block tested individually

This test signature can then be read out in serial form for checking by a further reconfiguration of the BILBO registers into a scan-path configuration. Each BILBO register therefore has the following principal modes of operation:

(i)   normal-mode operation;
(ii) reconfiguration into a scan-in/scan-out scan path;
(iii) reconfiguration into an input test-vector generator, providing pseudo-random test vectors;
(iv) reconfiguration into an output test signature register.

It will also be seen that the first BILBO register may act as the output signature register for the last block of logic, as indicated in Figure 5.23*b*.

The BILBO registers are therefore fairly complex circuits to provide this resource. Figure 5.24 shows the usual circuit arrangement, four stages only being shown. Two control signals, $C1$ and $C2$ control the action of a register as follows:

$C1 = 0$, $C2 = 1$: reset mode, to set all the D-type flip-flops to the reset state of $Q = 0$ on receipt of a clock pulse;

$C1 = 1$, $C2 = 1$: normal operating mode with each flip-flop operating independently in accordance with the secondary input data from the combinational logic and providing secondary output data back to the combinational logic;

$C1 = 0$, $C2 = 0$: scan path mode with the flip-flops all in a shift register configuration and with the input data from combinational logic blocked;

$C1 = 1$, $C2 = 0$: linear-feedback shift register mode, to provide either pseudorandom test input sequence to the following block under test or to provide multiple-input test signature analysis.

The clock signal to all flip-flops remains the normal system clock in all reconfigurations. The scan-in and scan-out terminals on each BILBO register may be connected in series to provide one long scan chain, or kept as several scan chains if the total size of the circuit under test is large.

To test the flip-flops in the BILBO registers, the registers are reconfigured into the shift register scan path mode, see Figure 5.24c, and an appropriate bit stream fed through and checked at the output. This is a first test to ensure that all flip-flops are in working order, and may be referred to as the flush test. It is also possible to scan in serial test vectors to the registers which can then be switched to become secondary input test data to the combinational network; the output response may then be captured by the registers and scanned out for checking. This will be appreciated as being the normal scan-path test method discussed previously in Section 5.3, see Figure 5.4, but this is not the principal testing mechanism in BILBO.

Following the satisfactory completion of a flush test, consider the BILBO register preceding a block of conventional logic switched to the condition shown in Figure 5.24d, and the BILBO register following this block of combinational logic switched to the condition shown in Figure 5.24e. The first BILBO register will be a maximum-length pseudorandom test vector generator, working as detailed in Chapter 3, Section 3.4.3.1, and illustrated in Figures 3.16, 3.17 and 3.18, providing (ideally) an exhaustive input test sequence to the combination logic block. The following BILBO register will be a signature analyser for the resulting output responses from the combinational logic block similar in principal to that discussed in Chapter 4, Section 4.5, and illustrated in Figures 4.6 to 4.9, but with one very important distinction.

Looking back at, say, Figure 4.7, there is in this signature analyser only one data input to the LFSR to modify its pseudorandom sequence and provide the final compressed signature. In BILBO, however, there are as many test response inputs as there are stages in the BILBO register, giving as we briefly mentioned in Chapter 4 a multiple-input signature register (MISR)

246 VLSI testing: digital and mixed analogue/digital techniques

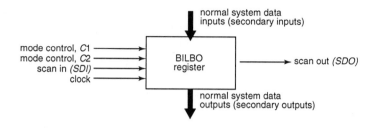

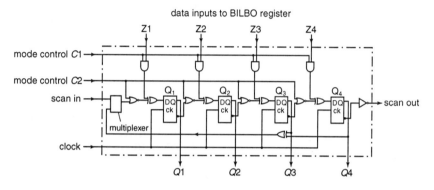

a

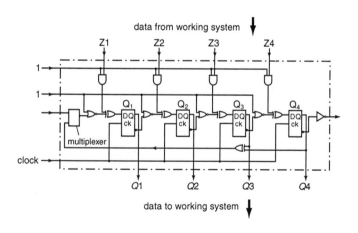

b

Figure 5.24   *The functional attributes of a BILBO register, showing four stages only for clarity*
  a block schematic and overall circuit arrangement
  b normal mode with $C1 = 1$, $C2 = 1$

Structured design for testability (DFT) techniques   247

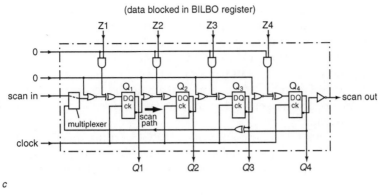

c

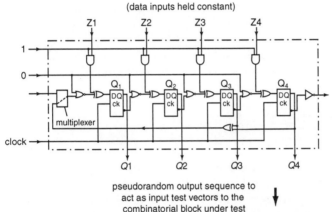

d

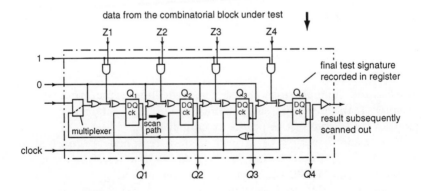

e

Figure 5.24   Continued
    c scan-path mode with $C1 = 0$, $C2 = 0$
    d autonomous pseudo-random test vector generation mode with $C1 = 1$, $C2 = 0$ (cf. Figure 3.17)
    e output test signature capture mode with $C1 = 1$, $C2 = 0$ (cf. Figure 4.7). Note, $C1 = 1$, $C2 = 0$ (not shown) resets all stages to zero. Detailed variations on this circuit arrangement may be found

configuration. Hence, as the BILBO input register applies its sequence of input test vectors, the output BILBO register simultaneously clocks through a random sequence depending upon the parallel-input test response data, giving a final test signature at the end of the sequence.

This signature is then scanned out for verification again using the reconfiguration of Figure 5.24c. Notice that all this action is under the control of the normal system clock, and hence is executed at the full speed of the normal system.

Details of this development of BILBO testing may be found published[6,15,19,58–60]. However, the question of fault masking (aliasing) of the MISR test signature has to be addressed to show the viability of this test procedure.

The aliasing property of the single-input signature analyser was discussed in Section 4.5 of the previous chapter, and shown by eqn. 4.28 to have a theoretical aliasing probability of $P_{fm} = 1/2^n \times 100\,\%$, where n is the number of latches (flip-flops) in the signature register. A similar analysis for the multiple-input shift register signature may be found[19,61–63], which gives a theoretical fault-masking value of:

$$P_{fm} = \frac{2^{m-1} - 1}{2^{m+n-1} - 1} \times 100\,\% \tag{5.1}$$

where n is as above and m is the number of applied test vectors to give the resulting output signature. If the number of stages in the signature register and the number of applied test vectors are reasonably large, we have:

$$\begin{aligned} P_{fm} &= \frac{2^{m-1}}{2^{m+n-1}} \times 100\,\% \\ &= \frac{1}{2^n} \times 100\,\% \end{aligned} \tag{5.2}$$

which is the same aliasing value as the single-input signature analyser. As in Chapter 4, the developments leading to eqn. 5.2 assume that all possible faults in the output test response are equally likely.

This result may also be arrived at by a reasoned approach. Consider any fault or faults occurring in the n-bit output response such that the signature in the MISR is incorrect at any step excluding the final one, then it is always possible to formulate a further incorrect test response which will return the MISR to a fault-free signature condition. For example, considering the simple four-stage circuit of Figure 5.24e with linear feedback from the third and fourth stages to the first stage and with test response data being fed into all four stages, we could have the situation as follows:

	Fault free		Faulty	
test data in	signature		test data in	signature
•	•		•	•
•	•		•	•
•	•		•	•
•	1 0 0 0		•	0 1 0 0
				(wrong signature)
0 1 1 0	0 0 1 0		0 1 1 1	0 1 0 1
				(wrong signature)
0 1 1 1	1 1 1 0		0 1 0 0	1 1 1 0
				(aliasing)
1 0 0 0	0 1 1 1		1 0 0 0	0 1 1 1
•	•		•	•
•	•		•	•
•	•		•	•

The state diagram of these MISR signatures is shown in Figure 5.25.

In this simple example the faulty input response 0 1 0 0 is seen to correct the MISR data, and hence produce aliasing in the final MISR signature. Therefore we may argue that given a fault has been registered, then if a subsequent faulty test response occurs, of the $2^n - 1$ possible faulty words there will be one which produces aliasing, and hence the probability of aliasing is

$$\frac{1}{2^n - 1} \cong \frac{1}{2^n} \times 100\%$$

This argument is not completely satisfying, but it does indicate the general probability of aliasing in the MISR register. Notice that one n-bit input word is always sufficient to correct the MISR signature; this may be contrasted with the single data input signature analyser where (a maximum of) n serial input bits will always correct the signature in this analyser, see caption of Figure 4.7.

Considering the above fault masking situation a little further, it is clear that if the signature held in the MISR had been scanned out after the first fault (or faults) had been trapped, but before the second fault had corrected it, then the faulty circuit response would have been detected. Hence, aliasing may be reduced by performing an intermediate scan-out operation before all the input test vectors have been applied as well as at the completion of the test. If all errors are still considered to be equally probable, then the theoretical probability of aliasing is now given[19] by:

$$P_{fm} = \frac{2^{m-2n} - 1}{2^n - 1}$$
$$\cong \frac{1}{2^{2n}} = \frac{1}{4^n} \times 100\%$$
(5.3)

Again, this result should be treated with some caution.

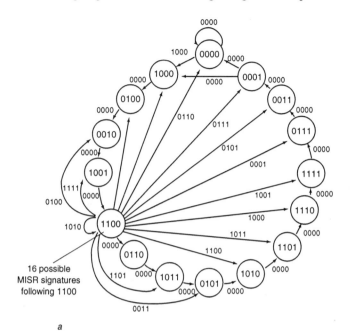

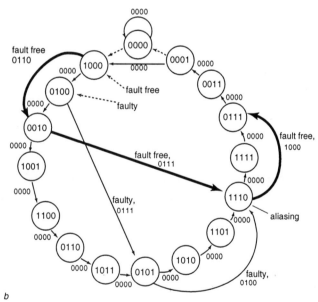

Figure 5.25   The state diagram for the four-stage MISR test signature register of Figure 5.24e with its characteristic polynomial $P(x) = 1 + x^3 + x^4$
  a state diagram showing all 16 states, but the routes from only one of the 16 states shown for clarity (cf. Figure 4.7 for the signature analyser)
  b the fault masking (aliasing) example in the text, which corrects the MISR signature on receipt of the faulty input 0 1 0 0

Structured design for testability (DFT) techniques

A more mathematical treatment of the BILBO registers may be pursued, similar to that covered in Chapter 4 and in particular eqn. 4.29. It will be recalled from earlier developments that there are two autonomous LFSR generator configurations possible, type A where the exclusive-OR feedback taps all feed the initial shift register stage, and type B where the exclusive-OR logic is included between stages, see Figures 3.20a and b respectively. The latter is the true polynomial divider discussed in the developments in Chapter 4, Section 4.5.

Therefore, although the BILBO register shown in Figure 5.2.4 is reconfigured into a type A autonomous LFSR generator when used to provide test vectors and also when reconfigured to act as the signature register, in the following we use the true polynomial divider configuration, the type B rather than the type A. This is illustrated in Figure 5.26a.

Theoretical analyses using the true polynomial divider configuration have been approached in several different ways, see Bardell et al.[19], Rajsuman[29] and Yarmolik[64] and further references cited therein. One approach shows that the test signature held in a MISR with a given divisor polynomial $D(x)$ is the same as the test signature held in an identical single-input signature analyser when:

(i) the latter is fed with a serial input bit stream which is the modulo-2 summation of time-shifted copies of the MISR inputs;
(ii) both are reset to an initial all-zero state before the commencement of the test.

For example, consider the following inputs applied to the circuit of Figure 5.26a:

	Data input					MISR signature after a clock pulse				
	i_0	i_1	i_2	i_3	i_4	Q_1	Q_2	Q_3	Q_4	Q_5
Reset						0	0	0	0	0
First vector	1	0	0	1	1	1	0	0	1	1
Next vector	0	1	1	1	1	1	1	0	1	0
Next vector	0	0	1	0	0	0	1	0	0	1
Next vector	1	0	1	0	1	0	1	1	0	1

The time-shifted mod_2 summation of these four data vectors is:

```
      1 0 0 1 1
        0 1 1 1 1
          0 0 1 0 0
            1 0 1 0 1
      1 0 1 1 0 1 0 1
```

Feeding this in as a serial bit stream, with the first input bit on the right in the above mod_2 summation, we have the following response in the circuit of Figure 5.26b:

	Q_1	Q_2	Q_3	Q_4	Q_5
Reset:	0	0	0	0	0
Input bit: 1	1	0	0	0	0
0	0	1	0	0	0
1	1	0	1	0	0
0	0	1	0	1	1
1	1	0	1	0	1
1	0	0	1	1	0
0	0	0	0	1	1
1	0	1	1	0	1

which is the same signature as in Figure 5.26a.

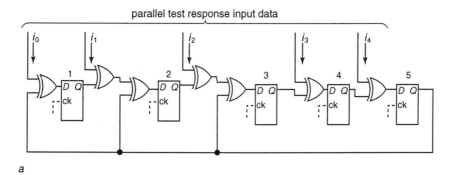

a

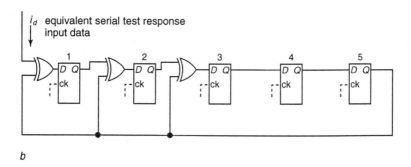

b

Figure 5.26 Equivalent type B MISR and single-input signature registers
  a example five-stage MISR with the divisor polynomial
  $$D(x) = x^5 + x^2 + x + 1$$
  b an equivalent register with a serial input bit stream, see text

Structured design for testability (DFT) techniques 253

It is left as an exercise for the reader to add further input vectors to the above tabulation, and confirm that the same signature results in Figure 5.26a and b when the above procedure is undertaken. Note, however, that strictly speaking using normal polynomial mathematics we should write the above vectors in reverse order to that shown here, shift the following vectors to the right in doing the mod$_2$ addition, and then read off from the left instead of the right. This is then in accordance with the rules used in Section 4.5 of Chapter 4. Published literature is not always consistent in this respect, but the final least significant data input i_0 in the MISR sequence must obviously be the final serial bit input in the equivalent single-input signature analyser. The above equivalence between a MISR and a single-input analyser allows further properties of the MISR to be confirmed which are closely analogous to the single-input signature analyser properties discussed in the previous chapter. However, we may pursue a mathematical development as follows, which does not (directly) involve any equivalence between the two types.

Using the same designation as employed in Chapter 4, in particular see eqn. 4.29, consider the type B LFSR circuit shown in Figure 5.27a with test response data inputs $i_0, i_1, \ldots, i_{n-1}$. Let the test data input polynomial after clock pulse t be:

$$I_t(x) = i_{t,n-1}x^{n-1} + i_{t,n-2}x^{n-2} + \ldots + i_{t,1}x + i_{t,0}$$

and the state of the register also after clock pulse t be:

$$R_t(x) = r_{t,n}x^n + r_{t,n-1}x^{n-1} + \ldots + r_{t,2}x^2 + r_{t,1}x$$

(Note, this state of the register is what we have termed the residue in Chapter 4, Section 4.5.)

Now the resulting D inputs to the flip-flops $Q_1, Q_2, \ldots, Q_n$, just prior to the next clock pulse $t+1$ are given by:

$$[I_t(x) + R_t(x)] \bmod D(x) \tag{5.4}$$

where $[X] \bmod D(x)$ is the residue when any polynomial $[X]$ is divided by the divisor polynomial $D(x)$. The next state of the register $R_{t+1}(x)$ is when these inputs to the flip-flops have been clocked by the next clock pulse, transferring them from the D inputs to the Q outputs of the flip-flops, thus giving:

$$R_{t+1}(x) = x\{[I_t(x) + R_t(x)] \bmod D(x)\} \tag{5.5}$$

For example, consider the five-stage MISR shown in Figure 5.27b with the data inputs and states shown, giving the present-state polynomials:

$$I_t(x) = x^2 + x + 1$$
$$R_t(x) = x^5 + x^3 + x^2 + x$$

Therefore:

$$[I_t(x) + R_t(x)] = x^5 + x^3 + 1$$

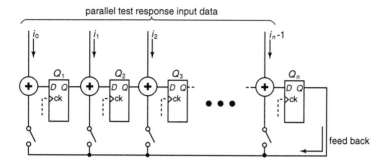

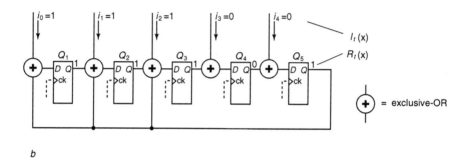

Figure 5.27 The BILBO register model used for polynomial developments
a general model with latches (flip-flops) Q_1 to Q_n and test data inputs i_0 to i_{n-1}, cf. Figures 3.20*b* and 4.8*b*
b particular case with five latches and divisor polynomial as in Figure 5.26*a*

and the resulting inputs to the flip-flops is given by the remainder (residue) in the polynomial division:

$$\begin{array}{r} 1 \\ x^5+x^2+x+1 \overline{\smash{\big)}\, x^5 +x^3 +1} \\ \underline{x^5 +x^2+x+1} \\ x^3+x^2+x \end{array}$$

The resulting next state of the MISR is therefore:

$$R_{t+1}(x) = x(x^3 + x^2 + x)$$
$$= x^4 + x^3 + x^2$$

i.e. $Q_5 = 0$, $Q_4 = 1$, $Q_3 = 1$, $Q_2 = 1$, $Q_1 = 0$. This result may easily be confirmed by considering the signals in Figure 5.27*b*.

Should the present-state output of Q_5 have been 0 instead of 1, with all other logic values unchanged, we would have had:

$$I_t(x) + R_t(x) = x^3 + 1$$

giving us:

$$
\begin{array}{r}
0 \\
x^5+x^2+x+1 \overline{) x^3 +1} \\
0 0 \\
\hline
x^3 +1
\end{array}
$$

with the resulting next-state of

$$R_{l+1}(x) = x^4 + x$$

i.e. $Q_5 = 0$, $Q_4 = 1$, $Q_3 = 0$, $Q_2 = 0$, $Q_1 = 1$, the difference between these two results being of course the absence of any feedback signal in the latter compared with the former.

Consider now a sequence of test response data inputs. Assuming the initial state of the MISR is all-zeros, then following the first clock pulse the MISR states will be:

$$R_1(x) = x\{[I_0(x) + 0] \bmod D(x)\}$$
$$= x\{I_0(x)\}$$

Continuing, we have:

$$R_2(x) = x\{[I_1(x) + R_1(x)] \bmod D(x)\}$$
$$= x\{[I_1(x) + x\{I_0(x)\}] \bmod D(x)\}$$
$$R_3(x) = x\{[I_2(x) + R_2(x)] \bmod D(x)\}$$
$$= x\{[I_2(x) + xI_1(x) + x^2 I_0(x)] \bmod D(x)\}$$
$$R_4(x) = x\{[I_3(x) + R_3(x)] \bmod D(x)\}$$
$$= x\{[I_3(x) + xI_2(x) + x^2 I_1(x) + x^3 I_0(x)] \bmod D(x)\}$$

$$\vdots$$

$$R_m(x) = x\{[I_{m-1}(x) + xI_{m-2}(x) + \ldots\ldots + x^{n-2} I_1(x) + x^{n-1} I_0(x)] \bmod D(x)\} \quad (5.6)$$

The last expression is the signature in the MISR after m input tests have been clocked into the circuit. The term within the square brackets may be referred to as the cumulative input polynomial. The validity of the above development lies in the linear (exclusive-OR) relationships which hold in every part of the MISR circuit.

As an illustration of the above consider the same five-stage MISR shown in Figure 5.27b with an initial state of Q_1 to $Q_{5=0}$, and with the following sequence of data inputs:

	i_4	i_3	i_2	i_1	i_0
$I_0(x)$:	1	0	0	1	0
$I_1(x)$:	0	1	1	1	1
$I_2(x)$:	1	0	0	1	0
$I_3(x)$:	0	0	1	0	1
$I_4(x)$:	1	1	0	0	0

The final signature in the MISR following the fifth clock pulse therefore is:

$$R_5(x) = x\{[(x^4 + x^3) + x(x^2 + 1) + x^2(x^4 + x) + x^3(x^3 + x^2 + x + 1)$$
$$+ x^4(x^4 + x)] \bmod D(x)\}$$
$$= x\{[x^8 + x] \bmod D(x)\}$$

Evaluating the remainder (residue):

$$\begin{array}{r} x^3 + 1 \\ x^5 + x^2 + x + 1 \overline{\smash{\big)} x^8 + x } \\ \underline{x^8 + x^5 + x^4 + x^3 } \\ x^5 + x^4 + x^3 + x \\ \underline{x^5 + x^2 + x + 1} \\ x^4 + x^3 + x^2 + 1 \end{array}$$

giving the final state:

$$R_5(x) = x^5 + x^4 + x^3 + x$$

i.e. $Q_5 = 1$, $Q_4 = 1$, $Q_3 = 1$, $Q_2 = 0$, $Q_1 = 1$

Now from the fundamentals covered in Chapter 3 and 4 we may consider each faulty polynomial as the linear addition of the fault-free polynomial plus an error polynomial. Therefore, recalling the technique employed in Chapter 4 to deal with the error bit stream of the single-input register, we may consider MISR error polynomials in the same way as we have considered the full information polynomials $I(x)$ in eqn. 5.6. This gives us the error signature in the MISR of:

$$R_m^e(x) = x\{[E_{m-1}(x) + xE_{m-2}(x) + \ldots x^{n-2}E_1(x) + x^{n-1}E_0(x)] \bmod D(x)\} \quad (5.7)$$

Structured design for testability (DFT) techniques 257

where $R_m^e(x)$ is the difference between the fault-free signature in the MISR and the actual signature after the sequence of m data inputs has been clocked in.

Clearly for fault masking (aliasing) to occur, the value of $R_m^e(x)$ must be zero, that is the resulting cumulative polynomial

$$[E_{m-1}(x) + xE_{m-2}(x) + \ldots + x^{n-2}E_1(x) + x^{n-1}E_0(x)]$$

must be exactly divisible by the divisor polynomial $D(x)$. This will be seen to be closely analogous to the situation with the single-input signature analyser, where the error bit stream $E(x)$ has to be exactly divisible by $D(x)$ for aliasing to occur.

As an example, consider the previous worked example of five data inputs $I_0(x)$ to $I_4(x)$, and assume that under fault conditions we have the following:

	i_4	i_3	i_2	i_1	i_0
$I_0^h(x)$ (fault free):	1	0	0	1	0
$I_0^f(x)$ (faulty):	1	0	0	0	0
$E_0(x)$:	0	0	0	1	0
$I_1^h(x)$ (fault free):	0	1	1	1	1
$I_1^f(x)$ (faulty):	1	0	0	1	0
$E_1(x)$:	1	1	1	0	1
$I_2^h(x)$ (fault free):	1	0	0	1	0
$I_2^f(x)$ (faulty):	1	0	0	1	1
$E_2(x)$:	0	0	0	0	1
$I_3^h(x)$ (fault free):	0	0	1	0	1
$I_3^f(x)$ (faulty):	1	1	1	0	0
$E_3(x)$:	1	1	0	0	1
$I_4^h(x)$ (fault free):	1	1	0	0	0
$I_4^f(x)$ (faulty):	1	0	0	1	1
$E_4(x)$:	0	1	0	1	1

Taking the five faulty polynomials and the five error polynomials, we can demonstrate that fault masking occurs as follows:

(i) Taking the faulty data polynomials $I_0^f(x), \ldots, I_4^f(x)$, the MISR signature following the fifth clock pulse is given by:

$$R_5^f(x) = x\{[(x^4 + x + 1) + x(x^4 + x^3 + x^2) + x^2(x^4 + x + 1) + x^3(x^4 + x) + x^4(x^4)] \bmod D(x)\}$$
$$= x\{[x^8 + x^7 + x^6 + x^5 + x^4 + x^2 + x + 1] \bmod D(x)\}$$

Evaluating the residual:

$$
\begin{array}{r}
x^3+x^2+x \\
x^5+x^2+x+1 \overline{\smash{\big)}\, x^8+x^7+x^6+x^5+x^4+x^2+x+1} \\
\underline{x^8+x^5+x^4+x^3} \\
x^7+x^6+x^3+x^2+x+1 \\
\underline{x^7+x^4+x^3+x^2} \\
x^6+x^4+x+1 \\
\underline{x^6+x^3+x^2+x} \\
x^4+x^3+x^2+1
\end{array}
$$

giving the final state:

$$R_5^f(x) = x^5 + x^4 + x^3 + x$$

which has the same signature as for the fault-free condition.

(ii) Taking the error polynomials $E_0(x), ..., E_4(x)$, the MISR signature is now given by:

$$R_5^e(x) = x\{[(x^3+x+1)+x(x^4+x^3+1)+x^2(1)+x^3(x^4+x^3+x^2+1)+x^4(x)] \bmod D(x)\},$$

$$= 3\{[x^7+x^6+x^5+x^4+x^2+1] \bmod D(x)\}$$

Evaluating the residual:

$$
\begin{array}{r}
x^2+x+1 \\
x^5+x^2+x+1 \overline{\smash{\big)}\, x^7+x^6+x^5+x^4+x^2+1} \\
\underline{x^7+x^4+x^3+x^2} \\
x^6+x^5+x^3+1 \\
\underline{x^6+x^3+x^2+x} \\
x^5+x^2+x+1 \\
\underline{x^5+x^2+x+1} \\
0
\end{array}
$$

Hence, the final state of the MISR is the all-zeros state, illustrating that for fault masking to occur the cumulative error polynomial $R_m^e(x)$ must be exactly divisible by the divisor polynomial of the LFSR.

Looking at the cumulative error polynomial in eqn. 5.7 of

$$[E_{m-1}(x) + xE_{m-2}(x) + ... + x^{n-1}E_0(x)]$$

it is a polynomial ranging in powers of x from $x^0 = 1$ through to $(x^{n-1})(x^{m-1}) = x^{m+n-2}$. Calling this polynomial $E(x)$, it therefore has a maximum number of possible values including the all-zeros case of 2^{m+n-1}. The divisor polynomial

$D(x)$ ranges in powers of x from x^0 through to x^n, and hence to divide exactly into $E(x)$ the highest possible power in the quotient is $2^{(m+n-2)-n} = 2^{m-2}$. There are therefore 2^{m-1} possible exact quotients, including the all-zeros case. If we now consider only the nonzero multiples, the theoretical probability of fault masking, which is the same as the probability of $D(x)$ dividing exactly into $E(x)$, is given by:

$$P_{fm} = \frac{2^{m-1}-1}{2^{m+n-1}-1} \times 100\,\%$$
$$\cong \frac{2^{m-1}}{2^{m+n-1}} = \frac{1}{2^n} \times 100\,\% \tag{5.8}$$

which is the result originally quoted in eqn. 5.2. As previously, the above development has as its basis the probability that all faults are equally probable, which may not be true in real life.

Further algebraic considerations may be found in the literature, including the effects of choosing different divisor polynomials $D(x)$ on the fault-masking properties[19,64-67]. However, let us here conclude with some additional considerations of the circuit action and use of BILBO registers in their normal and test mode configurations.

First in normal mode, it will be seen from Figure 5.24 that there are additional logic gates in the data flow from the combinational logic through the flip-flops than would otherwise be present, and therefore some maximum speed penalty is present. However, there is no additional loading on the Q outputs, since in test mode the $\bar{Q}$ outputs are used. The use of $\bar{Q}$ for the scan path also means that there is no logic inversion within the BILBO register between stages; some other published circuits for BILBO registers may use Q rather than $\bar{Q}$, in which case there will be scan path inversions unless some additional signal inversion(s) are also present.

Looking at the BILBO register in autonomous LFSR pseudorandom generation mode, Figure 5.24d, and in MISR signature capture mode, Figure 5.24e, the same control signals C1 = 1, C2 = 0 are used for both duties. However, to give the required circuit actions, additional conditions must also be imposed, namely:

(i) seed the LFSR pseudorandom generator with a known initial state at the beginning of a self test, such as 0 0 ... 0 1;
(ii) simultaneously reset the LFSR MISR to 0 0 ... 0 0;
(iii) block off data from the combination logic from entering the pseudorandom generator but allow it to enter freely into the MISR.

Clearly, additional control is necessary over and above the C1, C2 inputs. Several different detailed circuit arrangements to provide these conditions have been published, and they are to a large extent the choice of the system designer. Conditions (i) and (ii) above can be achieved with a scan-in after the scan path has been initially checked and before switching to test mode, or alternatively just the single seed to the generator can be scanned in following the reset condition of C1 = 0, C2 = 1 (not shown in Figure 5.24). However, if there are logic inversions between stages in the scan path then, when such a configuration is switched to pseudorandom generation, we have the interesting fact that the forbidden state of the autonomous LFSR is no longer the all-zeros state (see previous comments in Chapter 3, Section 3.4.3.1), and therefore reset to 0 0 ... 0 0 is now a possible seed condition.

Condition (iii) above, which is the principal distinction between the BILBO registers in generator mode and MISR mode, requires some additional control facility to deactivate the data inputs from the combination logic. This may be done in several ways, including adding a test latch to each BILBO register, the latch in one state causing the register to operate in generator mode by blocking the data inputs, and in the other state allowing data in thus realising MISR conditions. This test latch can be part of the complete scan chain, so that it may be set to reset by normal scan-in data[68]. Alternatively, should the system under test be a bus-oriented system, where the information data is carried on the bus highways, then tristate bus drivers and/or tristate I/Os may be present which can be switched to their high impedance state to deactivate the data inputs to the BILBO registers. Finally, of course, an additional logic gate or additional fan-in can be inserted in each data input path as shown in Figure 5.28, which involves a third control signal C3.

Several further variants on the BILBO register configuration shown in Figure 5.24 have been published. These include:

- the concurrent built-in logic block observation (CBILBO) registers proposed by Wang and McCluskey[69,70], which effectively has two D-type flip-flops per stage and can act simultaneously as an autonomous LFSR generator and a MISR;
- the built-in digital circuit observer (BIDCO) proposed by Fasang[71], which is a BILBO structure adapted specifically for printed-circuit board testing;

and others. Some further built-in self-test proposals which do not relate so closely to BILBO have also been proposed, and will be briefly noted in Section 5.5.3.

A further circuit feature which has received attention is that, when the LFSR is reconfigured to act as an autonomous test vector generator, there is the usual maximum-length sequence of $2^n - 1$ test vectors, but not the one forbidden test vector which is normally 0 0 ... 0 0. It may be advantageous to be able to apply this test vector when the generator is running so as to provide the fully exhaustive sequence of 2^n test vectors, and to this end a circuit addition to the maximum-length pseudorandom circuit may be added. This will provide the all-zeros state in the counter, making the counter what may be now termed a complete autonomous LFSR generator[70,72], or a de Bruijn counter[19,73]. A circuit that will increase the sequence length from $2^n - 1$ to 2^n is shown in Figure 5.29; as may be seen, when the counter is in

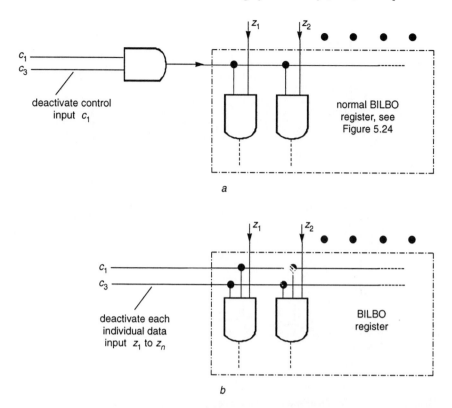

Figure 5.28 Additional controls necessary to deactivate the data inputs when in autonomous test vector generation mode, otherwise known as input disable, see Figure 5.24d
a interruption of input C1 when C3 is made 0
b individual deactivation of each data input when C3 = 0

the $0\,0\,0\ldots 0\,0\,1$ state the resulting linear feedback to the first stage is modified from logic 1 to logic 0 by the presence of signal $M = 1$, which results in the all-zeros state on the next clock pulse. This destroys the feedback from the final nth stage, but does not destroy the $M = 1$ signal which therefore provides a seed to enable the sequence to continue rather than stall in the all-zeros state. A reversible de Bruijn counter is also possible with the addition of a little further gating[19].

Whether this additional $0\,0\ldots 0\,0$ vector is necessary or not depends upon the circuit under test. A fully exhaustive test of the combinational circuit driven by the BILBO register will equally well be given if there are more stages in the BILBO register than secondary output lines; for example, if there are n stages in the BILBO register but only $n-1$ output lines, then the $n-1$ output lines will be automatically driven almost twice through a fully exhaustive sequence.

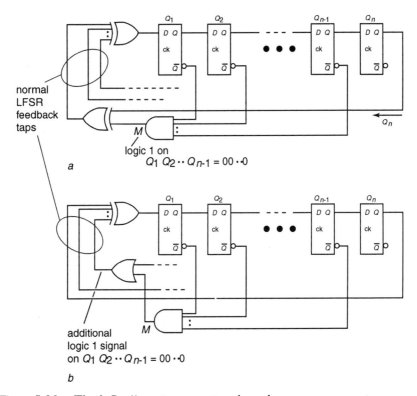

Figure 5.29   *The de Bruijn autonomous pseudorandom sequence generator*
a signal M cancelling the nth stage feedback
b an alternative circuit which does not require any increase in the XOR fan-in to the first stage

Two final problems must also be noted, namely:

(a) how to partition a complex circuit at the design stage into an appropriate number of BILBO registers and combinational logic blocks;
(b) how to determine the fault-free signatures in the MISRs for all the combination blocks.

To a large degree circuit partitioning will depend upon the actual circuit being designed, and whether it automatically breaks down into functional blocks. With a pipelined structure the choice of partitions may be obvious, but for nonstructured architectures it is a design choice influenced largely by the number of storage elements and their secondary interconnections. Figure 5.30 illustrates some of the configurations which may have to be considered.

Structured design for testability (DFT) techniques 263

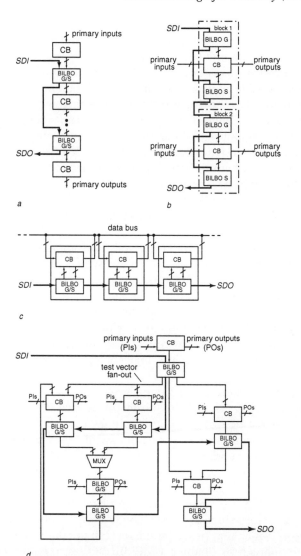

Figure 5.30 *Some concluding outline BILBO configurations*
 a pipelined configuration
 b the duplex method where it is not possible to cascade BILBO registers as in Figure 5.23*b*
 c bus-structured configurations
 d a duplex situation where it is not possible to segregate the interconnections between blocks completely, necessitating multiplexing of fan-in and fan-out of signals when in self test[70]
 Key: CB = combinational block
 BILBO G/S = normal BILBO register
 BILBO G = BILBO register used as a test vector generator only when in test mode
 BILBO S = BILBO register use as a signature register only when in test mode

There is also the question of whether all BILBO registers shall be one scan path, or whether two or more scan paths may be advantageous. Again, this depends heavily upon the circuit being designed.

Turning to the problem of determining the fault-free signature for each MISR there are two possible approaches, namely simulation or using a known good circuit. Simulation has the disadvantage of requiring computer time to execute; the alternative has the difficulty of ensuring a fault-free circuit from which to generate the signature. In practice possibly both methods may be used: simulation at the design stage and physical determination at the prototype stage to see whether both agree. Knowing what the correct response of a MISR should be, there is also the possibility of presetting the MISR to some starting condition such that with a fault-free response the final signature in the MISR is all-zeros or all-ones, which may possibly be easier to check after the test.

In conclusion, the BILBO concept has proved to be an acceptable means of building self test into a complex circuit or system. The silicon area overhead may be high, possibly 20 % additional chip area or thereabouts, but the tests are executed at normal system speed and are usually exhaustive or near exhaustive. There is, however, the inability to apply test vectors in other than the fixed pseudorandom test sequence, which may be unfortunate if CMOS memory faults are present. Notice also that although here, as in most published literature, we have taken the logic blocks between the BILBO registers to be purely combinational logic, there is nothing in the BILBO test strategy which precludes there being some simple sequential circuit action within these blocks. The essential requirement is that if there is any sequential action it must be unambiguously initialised (reset) at the beginning of each block test, and also that the number of input test vectors are sufficient to give a comprehensive test of this sequential machine action. This will be recognised as being an identical situation to that discussed in Section 4.5 when dealing with the single-input signature analyser test methodology.

The theoretical calculations of the aliasing probabilities of MISR circuits are not entirely satisfactory, since they are all built upon the equal probability of all possible errors, which may be far from a real-life situation. Nevertheless, like the signature analyser in Chapter 4, the BILBO test method has been found to be fully acceptable in practice.

Further discussion on BILBO techniques may be found in References 6, 19, 64, 74 and 75. Some additional developments using multiple autonomous LFSR test vector generators will also be mentioned later in Section 5.5.3.

5.5.2 Cellular automata logic block observation (CALBO)

The autonomous cellular automaton pseudorandom generator, which is an alternative to the LFSR generator, is introduced in Chapter 3, Section 3.4.3.2. Figures 3.21 to 3.23 illustrate the underlying concept of type 90 and type 150 cells, which when appropriately chosen produce a maximum-length pseudorandom sequence, an M-sequence of $2^n - 1$ states, where n is the total number of cells in the string of cells. Like the autonomous LFSR generator, the forbidden state of $0\,0\,0\ldots0\,0$ is present with the normal circuit interconnections, but unlike the LFSR generator the pseudorandom output sequence does not exhibit any shift register sequence of states in its output vectors.

It readily follows that reconfigurable cellular automata (CA) generators can be used in a BILBO-like configuration to provide on-chip self test, as illustrated in Figure 5.31. As in the BILBO configurations, one reconfigurable CALBO register can provide the input test vectors to the following combinational logic block, with a second CALBO register acting as the test response signature register.

The circuit for the CALBO register is shown in Figure 5.32, and follows closely that of the BILBO register illustrated in Figure 5.24. The two controls $C1$ and $C2$ provide the four reconfigurable conditions:

(i) $C1 = 1$, $C2 = 1$, normal mode operation;
(ii) $C1 = 0$, $C2 = 0$, reconfiguration into a scan-in/scan-out scan path;
(iii) $C1 = 1$, $C2 = 0$, reconfiguration into an input test vector generator, providing pseudorandom test vectors;
(iv) $C1 = 1$, $C2 = 0$, reconfiguration into an output MISR test response register;

plus a reset condition of all flip-flop outputs $Q = 0$ when $C1 = 0$, $C2 = 1$. As with BILBO self test, additional means (not shown) have to be provided to deactivate the combinational logic inputs from entering the register when in autonomous test vector generation mode, but allow this data in when in signature register mode. Additionally, not present in the BILBO self test, the multiplexers between every stage in the CALBO register have to be switched as follows to provide the above reconfigurations:

(i)   in normal mode, don't care, overridden by $C2 = 1$;
(ii) in scan-path mode, through connection from outputs Q to inputs D;
(iii) in test vector generation mode, local feedback to inputs D;
(iv) in signature register mode, local feedback to inputs D;
(v)   in reset mode, don't care, overridden by $C2 = 1$.

The control signal $C1$ will be seen to be able to provide this multiplexer control, being $C1 = 0$ for scan-path mode and $C1 = 1$ for all other conditions.

A CALBO register is somewhat more complex than a corresponding BILBO register (Figures 5.24a and 5.32), and therefore the question to be asked is whether the testing properties of CALBO are sufficiently better than BILBO to justify this additional overhead. Certainly the pseudorandom output sequence of a CALBO generator does not possess the high cross correlation between individual output bits of the BILBO LFSR generator, this high cross correlation being inherent in any normal shift register action, but

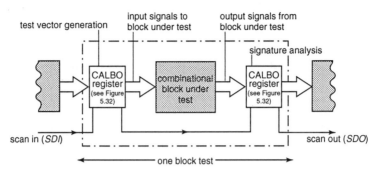

Figure 5.31 *Partitioning of a logic network by CALBO registers; the control signals have been omitted for clarity (cf. Figure 5.23 for the corresponding BILBO configuration)*

whether this is significant or not is difficult to quantify. Intuitively the CALBO sequence is more random than a BILBO sequence, and hence (again intuitively) its test vector generation must be at least as good if not better than that of the BILBO test vectors.

However, when we consider the CALBO signature capture properties the question is how does its fault masking (aliasing) compare with that of the BILBO MISR? We may again approach this question in a reasoned manner or an analytical manner, as was done for the BILBO registers. Consider first a reasoned argument. The simple CALBO register shown in Figure 5.33a, has a 90/150/90/150 cell configuration. Its state diagram with zero data input (equivalent to the autonomous test vector generation) is shown in Figure 5.33b. The sequence from one of these states with all possible input test responses is shown in Figure 5.33c, which illustrates the fact that, like the BILBO register, it is possible to shift from any state to any other state on receipt of the appropriate n-bit data input.

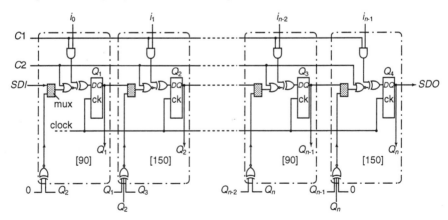

Figure 5.32 *The circuit arrangement of a CALBO register, showing four stages only for clarity (cf. Figure 5.24a for the corresponding BILBO configuration)*

Structured design for testability (DFT) techniques 267

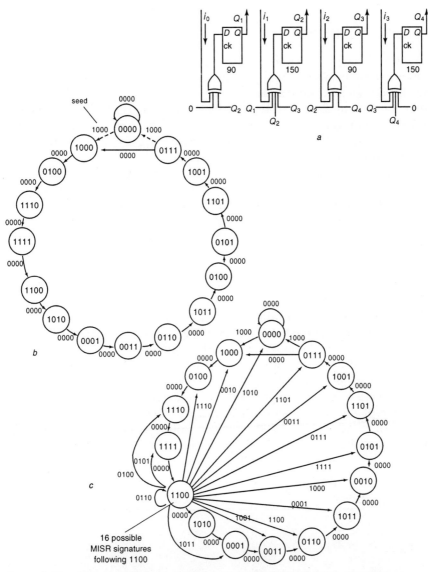

Figure 5.33 The state diagram for a four-stage 90/150/90/150 CALBO register
  a circuit diagram, omitting unnecessary detail
  b state diagram showing all sixteen possible states, with data inputs 0000 except to escape from and to return to the all-zeros state
  c routes from one of the sixteen states to all other possible states, cf. Figure 5.25a for the corresponding BILBO register

Hence, there is always one of the 2^n possible test responses that will correct a faulty CALBO signature in a manner similar to that illustrated in Figure 5.25b, and hence the probability of fault masking is exactly the same as for the BILBO MISR, namely:

$$P_{fm} = \frac{1}{2^n - 1} \cong \frac{1}{2^n} \times 100\,\% \tag{5.9}$$

Therefore it seems that the increased randomness of the CALBO register does not result in any theoretical improvement in fault masking; indeed using the above argument we may say that any n-stage counter, pseudorandom or otherwise, has the same aliasing probability when used as a signature register. Reflection on this result shows that it must be so if we assume all wrong data inputs are equally probable, and it therefore highlights the question of whether this assumption is truly valid in real life.

The mathematical analysis of the probability of CALBO fault masking has been considered by several authorities[76-83]. The statistical properties of the CALBO test vector sequences, their randomness, have been extensively considered by Podaima and McLeod[77] and elsewhere[83], but this does not directly address the aliasing properties of the CALBO register when in MISR test-capture mode. This is, however, specifically addressed by Serra et al.[78], using in the mathematical developments the isomorphism (similarity) which exists between BILBO and CALBO registers, as introduced in our discussions in Chapter 3, Section 3.4.3.2. Unfortunately, there are no polynomial relationships that may be directly applied to the CALBO registers, and therefore the polynomial division over GF(2) and the error polynomials considered in the previous section are not applicable.

The theoretical considerations of CALBO aliasing therefore take as their basis the formal proof that LFSR and CA generators are mutually isomorphic, that is given a maximum-length LFSR generator there is a corresponding CA generator of the same length which cycles through a sequence which is a fixed permutation of the LFSR sequence, and vice versa. For example, the isomorphism between the LFSR generator of Figure 5.25 and the CA generator of Figure 5.33 is the following one to one mapping:

LFSR generator		CA generator	
decimal	binary	decimal	binary
0	0 0 0 0	0	0 0 0 0
8	1 0 0 0	8	1 0 0 0
4	0 1 0 0	4	0 1 0 0
2	0 0 1 0	14	1 1 1 0
9	1 0 0 1	15	1 1 1 1
12	1 1 0 0	12	1 1 0 0
6	0 1 1 0	10	1 0 1 0
11	1 0 1 1	1	0 0 0 1
5	0 1 0 1	3	0 0 1 1
10	1 0 1 0	6	0 1 1 0
13	1 1 0 1	11	1 0 1 1
14	1 1 1 0	2	0 0 1 0
15	1 1 1 1	5	0 1 0 1
7	0 1 1 1	13	1 1 0 1
3	0 0 1 1	9	1 0 0 1
1	0 0 0 1	7	0 1 1 1

Structured design for testability (DFT) techniques

For any single-input bit stream, that is test data i_0 applied to the first stage of the BILBO and CALBO registers, with zero data to the remaining stages, it formally follows[78] that if the BILBO register aliases (that is finishes up on the same signature with a healthy bit stream and a faulty input bit stream), then the CALBO register will also alias, the one to one mapping between the two circuits always holding. As a trivial example, consider the input bit streams 1,0,1,0,1,1 and 0,1,1,1,1,0 fed into i_0 of the isomorphic BILBO and CALBO registers of Figures 5.25 and 5.33 respectively, the remaining MISR inputs i_1, i_2 and i_3 being held at zero. The BILBO and CALBO sequences with MISR inputs 1 0 0 0, 0 0 0 0, 1 0 0 0, 0 0 0 0, 1 0 0 0 and 1 0 0 0 representing the healthy test response and 0 0 0 0, 1 0 0 0, 1 0 0 0, 1 0 0 0, 1 0 0 0 and 0 0 0 0 representing the faulty test response may be traced through Figures 5.25a and 5.33b, and will be found to give the following sequences:

Healthy data input	BILBO signature	CALBO signature
1 0 0 0	1 0 0 0 (8)	1 0 0 0 (8)
0 0 0 0	0 1 0 0 (4)	0 1 0 0 (4)
1 0 0 0	1 0 1 0 (10)	0 1 1 0 (6)
0 0 0 0	1 1 0 1 (13)	1 0 1 1 (11)
1 0 0 0	0 1 1 0 (6)	1 0 1 0 (10)
1 0 0 0	0 0 1 1 (3)	1 0 0 1 (9)

Faulty data input	BILBO signature	CALBO signature
0 0 0 0	0 0 0 0 (0)	0 0 0 0 (0)
1 0 0 0	1 0 0 0 (8)	1 0 0 0 (8)
1 0 0 0	1 1 0 0 (12)	1 1 0 0 (12)
1 0 0 0	1 1 1 0 (14)	0 0 1 0 (2)
1 0 0 0	0 1 1 1 (7)	1 1 0 1 (13)
0 0 0 0	0 0 1 1 (3)	1 0 0 1 (9)

The one to one mapping between the two signatures is seen to hold in all cases, both aliasing on the final faulty input. The probability of aliasing of the CALBO register is therefore identical to that of the BILBO register under these conditions.

Aliasing for the true MISR situation with parallel data inputs $i_0, i_1, i_2, \ldots$, does not follow convincingly from further mathematical considerations. Unfortunately, it does not follow that the parallel data inputs into a CALBO register can be replaced by a serial input bit stream to the first stage to give the same signature, where this bit stream is a time-shifted bit by bit addition of the parallel data inputs as was possible for the BILBO MISR (see previous section and Figure 5.26a. This is demonstrated in Figure 5.34 using the same CALBO register as previously; taking just a single parallel input test vector 1 0 1 1 for simplicity, this produces the signature of 1 0 1 1 when operating in true MISR mode, but not this signature when a single data bit stream of 1,1,0,1 is fed into i_0, data inputs i_1, i_2 and i_3, being held at logic 0.

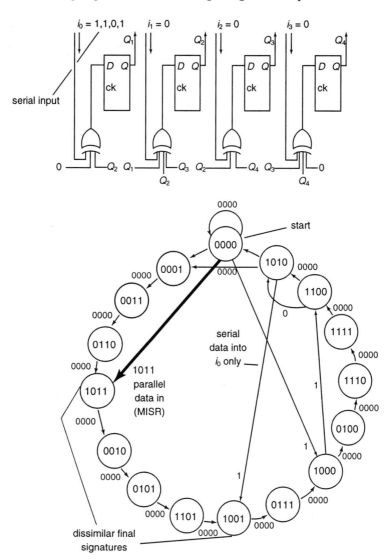

Figure 5.34 The failure of the CALBO register of Figure 5.33 to have the same signature with a parallel data input word of 1 0 1 1 and a serial input bit stream of i_0 = 1, 1, 0, 1, i_1 to i_3 being held constant at logic 0 for the latter case

The reason why this equivalence between parallel data inputs and single-bit serial data inputs does not hold is that there is no shift register action through all the CALBO stages as there is in the BILBO register, and hence to reformat and sequence the parallel data inputs i_1, i_2, i_3, ..., i_n into the first data input, i_0, does not give the same final signature. Therefore it remains to be formally proven that the CALBO MISR has the same probability of aliasing as the BILBO MISR; further academic consideration of this may be expected[78,83,84], but the outcome has to be that the probability of aliasing of the two registers must be the same if all faulty data inputs are considered to be equally likely.

Given, therefore, that the CALBO signature register is no better as far as its theoretical aliasing is concerned than the BILBO (or any other) signature register of the same length, its justification as a built-in self-test method must primarily be based upon its pseudorandom test vector generating performance rather than its signature capture ability. We have previously shown that LFSR test vectors are not successful in testing for CMOS memory faults, see Figure 3.25, and the fact that the same values appear in the LFSR test vector bits time shifted with respect to each other also proves to be disadvantageous in certain other structural logic networks[85]. Hence, it is possible that the greater randomness of the CALBO test vector generator, with its lower cross correlation between individual output bits, may prove to be more satisfactory for the testing of certain combinational networks than BILBO. No general quantification of this is possible, since it is dependent upon the logic and structure of the particular network under test, and hence the circuit designer must consider each case on its merits should a BILBO versus CALBO discussion arise. Alternatively, or in addition, the network synthesis may be undertaken such that pseudorandom pseudoexhaustive testing is facilitated[86].

We can, however, continue a little further here, and consider some circuit details and silicon area penalties involved in CALBO.

As in the BILBO case, the CALBO can provide the near exhaustive test sequence for a combinational network, lacking only the all-zeros test vector. This can, if required, be provided by resetting the CALBO register by the condition $C1 = 0$, $C2 = 1$, or alternatively we may incorporate a circuit addition which diverts the CALBO sequence from 0 ... 0 1 to 0 ... 0 0 and then back into its normal sequence of 0 ... 1 1. One possible circuit addition for providing this is shown in Figure 5.35, which may be contrasted with the BILBO circuit additions shown in Figure 5.29.

Considering the silicon area overheads for CALBO in comparison with BILBO, we may revise slightly the circuit per CALBO cell shown in Figure 5.32 to economise in the number of logic gates necessary. These modifications are shown in Figure 5.36, which also highlights the additions required over and above a BILBO cell.

A CMOS realisation of the three cells shown in Figure 5.36 is given in Figure 5.37, from which the number of transistors between stages of both a

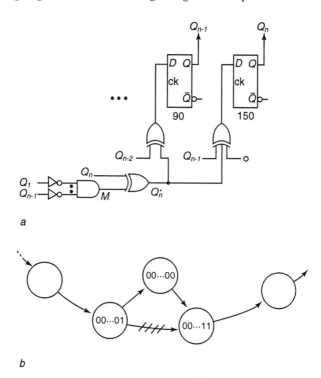

Figure 5.35 An addition to the autonomous CALBO generator to provide the all-zeros state and hence a maximum-length sequence of 2^n instead of $2^n - 1$. Data inputs have been omitted for clarity
  a the circuit addition; the inverter gates on the Q signals may be omitted if signals are taken from $\bar{Q}_1$ to $\bar{Q}_{n-1}$ as in Figure 5.29
  b portion of the state diagram showing the diversion now present between 0 ... 0 1 and 0 ... 1 1

BILBO and a CALBO string of flip-flops may be found. Notice that it is possible to eliminate one of the inverters and hence save one transistor pair in circuits b and c if Q_{j-1} and $\bar{Q}_{j-1}$ are routed into b and if Q_{j+1} and $\bar{Q}_{j+1}$ are routed into c, but this would probably require a greater silicon area than required by an inverter.

The total number of transistor pairs involved for BILBO and CALBO registers is therefore detailed in Table 5.2. This includes the additional transistor pairs required in the feedback of a BILBO LFSR circuit, namely an m-input exclusive-OR and a multiplexer, see Figure 5.24a, where the former is assumed to be made from $(m-1)$ two-input exclusive-ORs thus requiring $5(m-1)$ transistor pairs, the multiplexer also requiring five transistor pairs. These results are summarised in Figure 5.38, where, the exclusive-OR fan-in, is averaged as three (two minimum, four maximum m, see Appendix A).

Structured design for testability (DFT) techniques 273

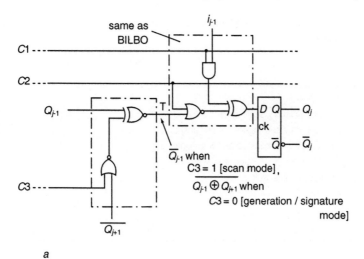

a

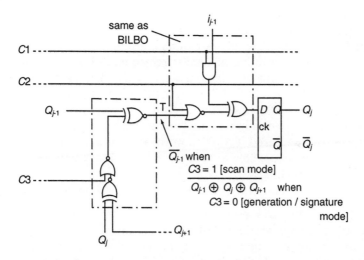

b

Figure 5.36 Possible circuit details for the interstage logic of CALBO registers
a type 90
b type 150 See also Figure 5.37

274 VLSI testing: digital and mixed analogue/digital techniques

Table 5.2 The number of transistor pairs required for n-stage BILBO and CALBO registers

Circuit	Number of transistor pairs		
	BILBO	CALBO 90	CALBO 150
Typical D-type flip-flop	10	10	10
Additional common circuitry, see Figure 5.37a	8	8	8
Further CALBO circuitry, see Figures 5.37b and c	–	6	10
Average further CALBO circuitry, assuming equal number of type 90 and type 150 cells	–	8	
Further BILBO circuitry, see text,			
multiplexer	5	–	
exclusive-OR	$5(m-1)$	–	
Total for an n-stage assembly	$18n + 5m$	$26n$	

CALBO is therefore seen to require some 40 % more transistor pairs than BILBO. However, this does not take into account the silicon area required for routing purposes, which should be simpler in the CALBO case compared with BILBO since all CA feedback connections are local compared with the end to end feedback routing of LFSR circuits; in CALBO never more than two local cell interconnections and hence routing channels are necessary.

Further consideration of CALBO versus BILBO silicon area overheads may be found published[87,88]. However, the recent disclosure that never more than two type 150 cells are required in any length CALBO register (see Appendix B) reduces the required CALBO overheads from previously published figures, but even so possibly a 20 % additional silicon area penalty above BILBO may be anticipated.

Hence any justification for CALBO must come back to the two factors:

(i) is the autonomous pseudorandom test vector generation sequence more effective than the BILBO pseudorandom sequence?
(ii) is the ability to be able to extend or reduce the length of the CALBO register more easily than is possible with the BILBO register useful?

5.5.3 Other BIST techniques

From the previous detailed discussions it will be evident that there are many possible variants on the circuits specifically discussed, and individual vendors may propose their own strategies. As far as the MISR signature register is concerned, there is no advantage, based upon theoretical aliasing, of any one type of n-stage register compared with another, and therefore ease of realisation is a more relevant consideration. In this respect the type A LFSR circuit with its simple exclusive-OR feedback to the first stage is the simplest synchronously-operating counter circuit possible when an exhaustive or near exhaustive sequence of 2^n or $2^n - 1$ states is required, being simpler than a

Structured design for testability (DFT) techniques 275

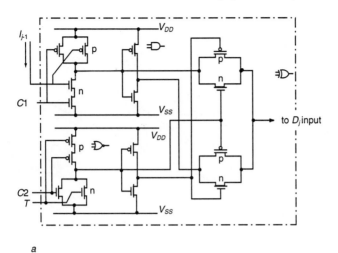

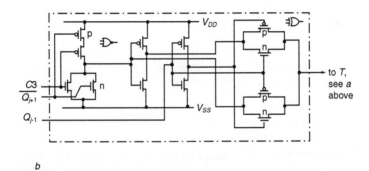

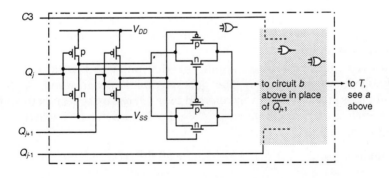

Figure 5.37 Possible CMOS realisations for the additional circuits shown in Figure 5.36
  a the common intercircuit details for both CALBO 90 and 150 circuits, eight transistor pairs
  b the addition for each CALBO 90 circuit, six transistor pairs
  c the addition for each CALBO 150 circuit, ten transistor pairs

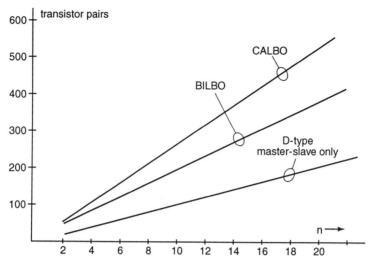

Figure 5.38 *The number of transistor pairs required for BILBO and CALBO with increasing n compared with simple D-type flip-flop circuits*

synchronous binary counter particularly as n increases. The LFSR therefore remains predominant in all built-in self-test proposals.

Reviewing other published BIST strategies, these may be divided into two possible categories namely:

(i) some form of hybrid built-in self test, which combines both scan-path testing and BILBO self test;
(ii) some partitioning or other modification of the BILBO pseudorandom generators to simplify the silicon layout or improve the overall testing performance.

It is likely that this trend into company-specific or other variants of (particularly) the BILBO concept will continue, since there is no single best solution to VLSI testing.

Let us first briefly look at some of the published hybrid BIST methodologies, which will show the general philosophy involved. These hybrid methods may be collectively referred to as self test using scan path and signature analysis, or S^3, techniques[15,74,89,90]; they in turn may be divided into two categories, namely:

(i) S^3 internal (or *in situ*) structures, where all the required test circuitry is on-chip;
(ii) S^3 external (or *ex situ*) structures, where part of the test circuitry is off-chip.

S^3 *in situ* proposals include the following four configurations:

- HILDO (highly integrated logic device observer), which uses one register that reconfigures to provide test pattern inputs, a scan mode and output signature analysis[91];

- LOCST (LSSD on-chip self test), which uses a pseudorandom test pattern generator, LSSD scan paths and a signature analysis register[92];
- SASP (signature analysis and scan path), which conceptually is similar to LOCST[93];
- STR (structured test register), which is a form of BILBO but with certain additional operating modes including scan test[94].

Other variants may also be found, including those for use in gate array structures[95–97].

Details of HILDO, LOCST and SASP may be found surveyed in Russell and Sayers[74], with additional information elsewhere[6,19,97]. However, let us here consider only one of these, namely LOCST, which will illustrate the concepts that may be present in many of these BIST derivatives.

The LOCST architecture, developed by IBM, is based upon level-sensitive scan design (LSSD) scan-path testing, but has provision for on-chip test vector generation and signature analysis so as to reduce the complexity which accrues with simple scan-in/scan-out of all the required test data. Its application is to embrace system-level testing as well as IC testing, and therefore boundary scan is available as well as self test for each individual IC.

The block schematic is shown in Figure 5.39, and incorporates the following facilities:

- LSSD scan-path;
- boundary scan;
- BILBO-type self test;
- on-chip monitoring.

In self-test mode twenty stages, which in one particular example constitute the PRBS (pseudorandom bit sequence) circuit, are reconfigured into a maximum-length pseudorandom generator with the characteristic polynomial $x^{20} + x^{17} + 1$, and the 16 stages which constitute the SAR (signature analysis register) are configured into a MISR with the characteristic polynomial $x^{16} + x^9 + x^7 + x^4 + 1$. Test bits from PRBS are scanned in to fill the input boundary scan register, the output test response being captured by the output boundary scan register and then scanned into SAR to give an initial signature. This procedure is repeated until the required number of test patterns have been applied to the circuit under test, after which the final signature in SAR may be scanned out for checking or confirmed by the on-chip monitoring circuit.

It will be seen that in this LOCST strategy it is not possible to parallel load the circuit under test with pseudorandom test vectors at normal system speed as is possible in BILBO, but instead we are back to one test at a time as in scan test. However, we do not have to scan test data in and test data out through one lengthy scan chain around a complete PCB or other assembly, but instead now we have relatively short local scan-in paths present. It has been reported[92,98], that fewer than 5000 random test patterns have provided

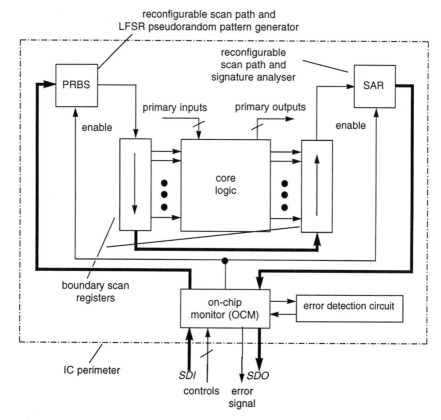

Figure 5.39 The LOCST architecture of IBM, which provides individual on-chip test vector generation and signature capture with system scan-path resources

acceptable test coverage; also it is possible to seed the PRBS circuit with alternative starting seed via the scan path, which can provide improved fault coverage and lower aliasing for given conditions.

One feature which is not possible with LOCST is the means to check interconnections between ICs on a completed assembly, which normal boundary scan provides. Its principle advantage, therefore, is to provide on-chip test vectors and hence eliminate the need for separate hardware resources for the test vector generation and the test response capture. Its disadvantages in speed of test may be minimised by simultaneously loading several of the PRBS circuits in a complete system assembly, and doing several circuit tests at once—this is specifically done in the SASP method[93], which uses several scan-path elements that are simultaneously active in test mode.

Further details of LOCST may be found in the literature[92,98]. Some comparative studies of silicon area and time to test of (i) scan-path, (ii) BILBO and (iii) S^3 *in situ* test strategies may be found in Reference 74, which quantify the reduced time to test of the latter two methods in comparison with the first, but with scan path having the lowest overall silicon area penalty of the three methods.

Turning from *in situ* self-test proposals to *ex situ* methods, the most widely publicised structure is the STUMPS (self testing using MISR and parallel shift register sequence generator) architecture, which was original developed for board level testing and then extended to the IC testing level[99]. The principle of *ex situ* self test is shown in Figure 5.40a, and the STUMPS realisation of this principle in Figure 5.40b.

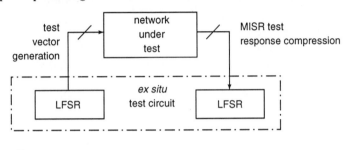

a

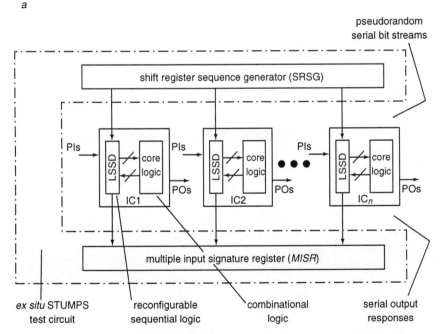

b

Figure 5.40   Ex situ self test methodology
          a basic principle, with off-chip test vector generation and test response analysis
          b the STUMPS architecture catering for several ICs per system. Each IC has a LSSD or equivalent test mode, see Figures 5.8 to 5.10. The clocks and normal mode/test mode control signals have been omitted for clarity

The STUMPS architecture has multiple LSSD paths, generally one scan path per IC, which are all driven in parallel from the one common pseudorandom test generator (SRSG). The individual IC test responses are captured by their own output scan path, and scanned into the common output MISR. The test procedure is therefore as follows:

(i) reconfigure each IC into a level-sensitive scan design (LSSD) mode using the normal on-chip latches (flip-flops), see Section 5.3.1;
(ii) scan in a pseudorandom test vector from the SRSG simultaneously to each IC to load the input latches;
(iii) exercise each circuit under test with the loaded test vector;
(iv) capture the test response in the output latches of each IC;
(v) scan out the resulting data simultaneously from every IC into the one common MISR circuit;
(vi) repeat (ii) to (v) as many times as necessary to complete the required test coverage.

Notice that in order to scan in a completely new test vector on each procedure (ii) above, L clock pulses are required, where L is the length of the longest LSSD scan path. This means that some excess bits will be fed into shorter scan paths, which may overflow into the MISR circuit. A potentially more troublesome problem, however, is that the test bits serially fed into a LSSD scan path are determined by the shift register action of the pseudo-random generator, and hence cannot have certain patterns; for example, the simple four-stage LFSR of Figure 3.18 can only provide the bit sequence of:

1 1 1 1 0 1 0 1 1 0 0 1 0 0 0

from any one of its Q outputs, and hence may not be able to test for, say, certain stuck-at faults when loaded into a LSSD circuit. Also, the test bits shifted into one LSSD scan path are a time-shifted copy of those in another scan path, rather than being completely uncorrelated. These factors may cause testing difficulties if not fully understood[85]; it may be necessary to include some additional exclusive-OR gates in the SRSG outputs to break up these relationships in the serial input but streams[100]. (Notice that a CALBO generator would not have these problems.)

Further details of STUMPS may be found published[15,19,85,99–100]. Details of and references to several other proposed built-in self variants, including CEBS (centralised and embedded BIST architecture with boundary scan), RTD (random test data), CATS (cyclic analysis testing systems) and CSTP (circular self test path) may be found in Reference 6.

Finally, returning to variants on the basic BILBO concept without involving additional scan-test structures, there have been several practical variations which have considered partitioning or otherwise modifying the autonomous pseudorandom LFSR generator in some manner. One published method is to modify the usual pseudorandom output of the LFSR generator so as to produce weighted random patterns, which may result in the possibility of using fewer than $2^n - 1$ test vectors.

It will be recalled from Chapter 3 that in the LFSR pseudorandom sequence there is an equal probability of a 0 or a 1 in each bit, all vectors (ignoring the all-zeros case) being equally likely. The disadvantage of this distribution for testing can be illustrated by considering, say, a five-input NAND gate; if all inputs are equally likely then the probability of detecting a stuck-at 0 fault on one of the gate inputs by one random test vector is 1 in $2^5 \cong 0.031$. However, if the test vectors are weighted so as to increase the probability of a bit being 1 and decrease the probability of being 0, then the probability of detecting this s-a-0 condition is greatly increased.

Weighted pseudorandom patterns may be generated from a normal pseudorandom sequence by combining output bits through an AND or NAND gate, as illustrated in Figure 5.41a. Since the probability of a 1 on any NAND input is 0.5, the probability of three 1s simultaneously present is $(0.5)^3 = 0.125$, and hence the probability of the output of the NAND gate remaining at logic 1 is 0.875.*

The weighting of this final output bit may be made programmable by using an AND/OR/exclusive-OR network as shown in Figure 5.41b. Here one input, two inputs or three inputs from the LFSR are selected as required, the final exclusive-OR being included so as to be able to provide an increased probability of 0s or 1s in the final output as follows:

AND gate(s) selected	Output probability of 0s	Output probability of 1s
G1	0.5	0.5
G2	0.75 or 0.25	0.25 or 0.75
G3	0.875 or 0.125	0.125 or 0.875
G1 and G2	0.375 or 0.625	0.625 or 0.375
G1 and G3	0.437 or 0.563	0.563 or 0.437
G2 and G3	0.756 or 0.344	0.344 or 0.756
G1, G2 and G3	0.328 or 0.672	0.672 or 0.328

Other circuit arrangements may be found to weight the output test vectors[101–103]. One practical application may be found published[104], which introduces the term WBILBO (weighted built-in logic block observation) for the register circuit configuration. However, the question of what optimum weightings should be placed on the test vector bits, and how great a reduction in the number of test vectors is now possible to give an acceptable fault coverage, is still an ongoing research and development subject. Existing work addressing this problem may be found published[101–102,105–106]; additional comments and references may be found in References 6, 19 and 85.

A different variant of the BILBO strategy than above is to use two or more autonomous LFSR generators instead of one to provide the test vectors. A

---

* Strictly speaking, with the four-stage ($n = 4$) LFSR shown in Figure 5.41a, since it only has 15 states in its cyclic sequence the probability of three ones is 2 out of 15 and not 2 out of 16, see Figure 3.18b, giving a probability of 0.133 instead of 0.125. This difference between 2^n and $2^n - 1$ becomes negligible with higher n.

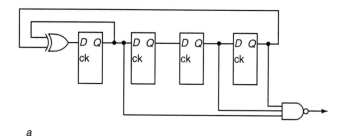

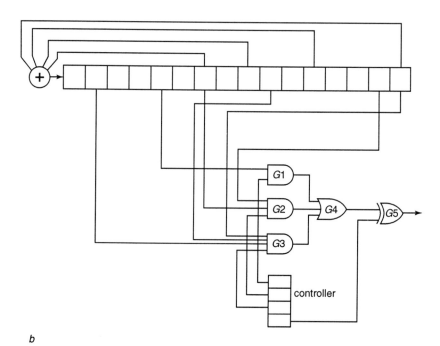

Figure 5.41   *The weighted BILBO concept*
   a a simple example giving an increased probabilty of output 1 (see Figure 3.18*b*)
   b programmable example, giving a choice of output weighting

good practical application of this technique has been described by Illman and Clarke for the testing of a 35 000-gate CMOS circuit, with further applications towards 250 000-gate size ICs[107].

The architecture of the CMOS circuit described in Reference 107 is given in Figure 5.42*a*, from which it will be seen that the conventional BILBO topology of test registers preceding and following blocks of combinational logic is maintained. The BILBO registers, however, are not each single autonomous LFSR generators with a maximum length sequence of $2^n - 1$ states, instead being arranged as follows.

Structured design for testability (DFT) techniques

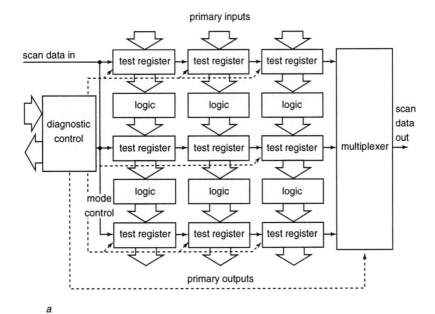

Figure 5.42 The BIST architecture of Reference 107 which uses partitioned LFSR test generation
  a the overall architecture
  b two pseudorandom sequence generators providing test vectors

Consider the two maximum-length pseudorandom sequence generators shown in Figure 5.42*b*, with a total of $2^{n_A} - 1$ test vectors available from A and $2^{n_B} - 1$ available from B. The combined test vector sequence from the two generators will repeat when both simultaneously return to their initial starting stage, assuming both are synchronously initialised and clocked.

The combined sequence will therefore repeat when the number of patterns generated is an integer multiple of S_A and S_B where $S_A = 2^{n_A} - 1$ and $S_B = 2^{n_B} - 1$. To produce the longest possible combined sequence, that is $(S_A \times S_B)$, the highest common factor of S_A and S_B must be unity.

Now the highest common factor of two terms $2^F - 1$ and $2^G - 1$ is $2^H - 1$, where H is the highest common factor of F and G. Therefore, if for example $n_A = 6$, $2^{n_A} - 1 = 63$, and $n_B = 9$, $2^{n_B} - 1 = 511$, the highest common factor is $2^3 - 1 = 7$, and the combined length of the sequence is therefore $63/7 \times 511/7 = 4599$. Thus, for the maximum possible combined sequence from S_A and S_B it follows that n_A and n_B must be coprime, that is their highest common factor is unity.

This is known as the coprime rule for the lengths of pseudorandom generators which are working separately in order to provide one maximum-length combined test vector sequence. Considering n_A and n_B at least ten stages long, we have the following combinations which yield the maximum sequence of $(2^{n_A} - 1) \times (2^{n_B} - 1)$ test vectors.

n_A	n_B
10	11, 13, 17, 19, ...
11	12,13,14,15,16,17,18,19,...
12	13,17,19,...
13	14,15,16,17,18,19,...
14	15,17,19,....
15	16,17,19,...
16	17,19,...
17	18,19,...
.	.
.	.
.	.

Two LFSRs of length ten and 11 would therefore provide a total of 2 094 081. This may be contrasted with the maximum length of a 21-stage LFSR, which is 2 097 151, but the test vectors across the coprime pair of LFSRs have better testing properties than the single 21-stage LFSR, and are easier to route in an IC layout than one very long LFSR. We are also not restricted to two partitions as shown above; three or more maximum-length LFSRs may be combined, provided that the number of stages obeys the coprime rule. (Failure to obey this rule will, of course, result in a reduced-length combined test vector sequence.)

Although the coprime rule gives the maximum combined-length sequence, noncoprime pairs may also be acceptable in given circumstances, provided that they are not of equal length. Of the possible 45 different pairs of LFSRs with lengths between $n = 10$ and 19, 32 pairs are coprime and 13 are noncoprime. The lowest combined length noncoprime sequence is with $n = 10$ and 15, giving 1.08×10^6 test vectors, and the highest is with $n = 16$ and 18, giving 5726×10^6 test vectors.

The use of these pseudorandom test vectors disclosed in Reference 107 is also slightly different from that which we have considered before. Previously we have used the BILBO registers when in test vector generation mode to apply the near-exhaustive sequence of $2^n - 1$ test vectors (or exhaustive if the all-zeros state is also used) to n combinational logic inputs. However, this normally requires partitioning of the logic into matching sizes of LFSR generators and combinational logic blocks, which may not always be convenient to implement. In the testing strategy discussed in Reference 107, this is done for a few logic blocks, but other blocks are fed from unmatched coprime assemblies of two or more LFSRs, more than one block possibly being fed from one coprime assembly.

The theoretical fault coverage given by coprime-prime assemblies is based upon the probability that random and not pseudorandom test patterns are applied to the combinational network. (Random test patterns may contain the same test pattern more than once; a maximum-length pseudorandom sequence will not.) Given a combinational network with n primary inputs the probability that a single random test vector will detect a nonredundant fault is given by:

$$P_1 = \frac{1}{2^n} \times 100\,\% \tag{5.10}$$

The probability of detecting the fault with M random test vectors is:

$$P_M = \left\{1 - \left(1 - 2^{-n}\right)^M\right\} \times 100\,\% \tag{5.11}$$

If M is made greater than 2^n then it follows that the fault coverage closely approaches one hundred per cent.

Therefore, to test many of the combinational logic blocks, test vectors are selected from coprime LFSR assemblies which have in total more stages than the maximum number of primary inputs per logic block. A cited example is a 20-stage coprime LFSR assembly, 17 connections from which are taken to test the largest combinational block. Hence $n = 17$ and $M = 20$ in eqn 5.11, giving a theoretical fault coverage of 99.9 % with 2^{20} test vectors. The actual fault coverage may be better than these theoretical values, since other theoretical studies have indicated that these values represent the lower bound for nonredundant faults, with true fault coverage being higher[107,108].

The thrust of this built-in test strategy, therefore, is to permit flexibility in partitioning and in routing of test vectors, with a very high probability of fault detection at the expense of more test vectors than used in an exact exhaustive test set. Standard MISR techniques are used to capture the test results, which are scanned off-chip for confirmation of fault-free performance. More than one block is tested at a time; no scan-test ATPG programs are used; a total of 3 800 000 clock cycles are used to check the chip fully, with 40 000 clock cycles to scan in and scan out test data. A factor not fully discussed in Reference 107 is that although the method of test vector generation guarantees near 100 % fault coverage, see eqn. 5.11, the on-chip test results are still compressed in MISR configurations, and therefore the probability of aliasing, see eqns. 5.2 and 5.8 above, is still present. Hence, although high fault coverage is achieved by the autonomous test vector generation, we are still dependent upon the performance of the output compression for final fault-free/faulty decisions. More than one scan-out of the MISR signature during a test may be a way of minimising the possibility of aliasing.

Other segmented LFSR generators have been discussed in published literature[109,110], but as far as is known have not been used to date in practical applications. An analysis of the properties of these generators and further variants may be found in Reference 19. Sharing of the individual outputs of a single LFSR between several circuit inputs is also considered in Reference 111.

To summarise these various BIST techniques, all will be seen to involve (i) some form of autonomous pseudorandom test vector generator and (ii) some form of MISR test response capture. An objective in all cases is to avoid having to use expensive off-chip test hardware as far as possible, and also not to have to undertake any comprehensive ATPG programming. Further discussions on principles and theory may be found in References 16, 19.

We will return to some additional considerations of built-in self test in the following chapter when we consider the testing of specific devices such as PLAs, memory and other strongly-structured architectures.

5.6 Hardware description languages and test

Increasing complexity in the design of VLSI circuits has led to the introduction of high-level hardware description languages (HDLs) for the design phase. The evolving international standard is VHDL, the very high speed integrated circuit hardware description language, originally developed in 1983 as a language for documenting the circuit action of VLSI circuits and subsequently enhanced to become a design tool as well as a documentation record. Details of VHDL may be found in several publications[112-114]. It is, therefore, appropriate that VHDL should be considered as a means of incorporating some form of self test for the circuits being designed during the design phase.

A number of other HDLs and expert systems have been used for VLSI design and test specification activities, although most have been applied to improve the efficiency of ATPG programs rather than to build in any self-test mechanism. Details of many of these, including SUPERCAT[115], HITEST[116], CATA[117], TDES[118], IDT[119] and others may be found reviewed and referenced in Russell and Sayers[74]. However, here we will comment principally on VHDL and its potential use in self test, since this generally represents the most active basis outside certain commercial enterprises.

The VHDL language provides a hierarchical top-down design environment from behavioural level through architectural level and register transfer level down to macro, gate or other primitive level, using a common database throughout. The VHDL descriptions of the design at the various levels may be computer simulated with selected timing data to confirm the correct functionality (system validation), with printout in textual, waveform or other format as appropriate. Provided that the original high-level behavioural descriptive statements are correct and complete, so correct by construction (conformity) of the lower hierarchical levels of design should follow. This hierarchical process is illustrated in Figure 5.43.

The data within the VHDL design environment can be offloaded into separate commercial test-vector generating software which uses fault

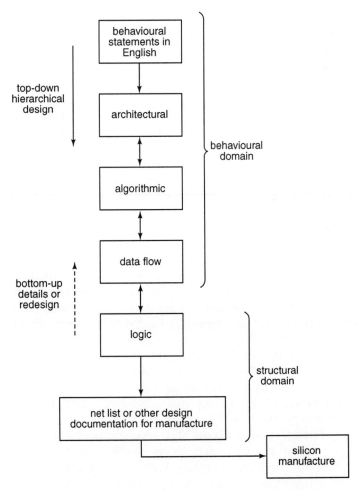

Figure 5.43   The outline top-down design hierarchy in VHDL

modelling to generate appropriate test vectors (see Chapter 3, Section 3.2), but offloaded ATPG procedures are not what we are primarily concerned with in this chapter—there are also reported interface problems between VHDL and separate ATPG programs which again need not concern us here[120,121]. Rather, here we are wishing to consider the use of VHDL to realise some built-in test methodology such as discussed in previous sections of this chapter.

In the VHDL design environment any logic circuit or macro is represented as a design entity. A design entity may be as simple as an AND or OR or inverter gate, or as complex as a microprocessor; it may be expressed in Boolean logic terms (if simple) or in algorithmic form if it has to perform a given (more complex) function. Another feature of VHDL is that it can also be used in a bottom-up design mode as well as a top-down mode, and therefore design entities can be built up from gate level when required. Hence, any design entity may be specifically simulated and verified correct, and then held in the VHDL library for use as many times as required in fleshing out the complete circuit or system being designed.

It is, therefore, conceptually straightforward to include in this VHDL library design entities such as boundary scan elements, level-sensitive scan design (LSSD) elements, linear-feedback shift register (LFSR) macros and other self-test circuit requirements. Notice that VHDL is not an intelligent CAD tool, and hence it still relies upon the designer to supply the design knowhow, the VHDL merely providing the platform to keep and manipulate all the design data and system simulation at any level of abstraction within one standard database. The latter feature has the advantage that more than one designer or design team (or even company) may simultaneously work on different partitions of a complete design, knowing that the separate design data will be fully compliant with each other within the CAD environment.

Boundary scan for completed PCBs and other assemblies was introduced in Section 5.4. A boundary scan description language (BSDL) has been developed to support the inclusion of boundary scan during a system design phase, which specifies the required order of the scan components in the system and the test data requirements. BSDL has been specifically written to be a subset of VHDL, and therefore is designed to be included within the VHDL design environment. Full conformity with IEEE standard 1149.1 is maintained[42], and hence the facilities detailed in Section 5.4 become part of the VHDL design activity. Further details of this subset of VHDL may be found in Reference 44; increasing availability from CAD vendors of this resource may also be noted.

Built-in self-test techniques have also received attention, but tend to remain a part of the normal system design activity rather than automatic insertion into an evolving design. However, on-going research and development activities are considering the possibility of automatic insertion into VHDL design data, and subsets of VHDL are awaited[122-125]. Details of the present status of VHDL may be found[120,121], but certain conflicts still remain concerning the OEM adoption of the full VHDL IEEE standard 1076, or reduced subsets of 1076, or competitive commercial alternatives particularly Verilog™ of Cadence Design Systems, Inc., CA.

High-level hardware description languages, therefore, will not introduce any new testing concepts beyond those which we have previously covered; what they may do is to provide a unified platform for the incorporation of chosen test strategies into a circuit or system concurrent with the normal functional design activity. Difficulties of conforming with standards, such as IEEE 1076 and 1149 and others*, and also commercial interests, will undoubtedly continue to cause problems for OEM designers seeking to use the latest CAD tools for DFT and BIST purposes.

5.7 Chapter summary

Referring back to Figure 4.10, it will be appreciated that we have now covered the extreme right-hand branch of this range of self-test strategies; in Chapter 4 we were concerned with online methods, which were particularly appropriate for critical systems that could not afford to have a faulty output or where a fault had to be immediately flagged, whereas in this present chapter offline techniques with a normal mode of system operation and a separate test mode of operation have been our principal interest. In general, fault detection but not fault location (except to within a separate circuit or subsystem) has been involved.

Design for testability (DFT) procedures have been seen to become necessary as circuit complexity has increased to the VLSI level. Automatic test pattern generation (ATPG) programs have been found to be increasingly unable to formulate an acceptable set of test vectors within a given CPU time for large circuits, and testers to apply test to complex circuits and monitor the results are themselves of increasing complexity and cost.

Partitioning of a large circuit into smaller parts which may be separately tested is therefore a natural solution to reduce the testing problem. However, as we have seen, the precise partitioning may be largely dictated by the actual circuit being designed, but separation of the combinational logic and the sequential (memory) elements is a basic consideration in most DFT strategies. Four aims and objectives are therefore present, namely:

---

* Other standards include IEEE standard P1149.5, standard module test and maintenance bus, which is a dedicated test and maintenance interface independent of the module's functional interface; IEEE standard 1212, standard command and status register architecture, which covers a range of system management functions and microprocessor backplanes; and ISO10164, open standards, covering a range of open systems standards for computers and telecommunication systems. The latter will also address test management and define standard classes of system test.

(i) to attempt to use the on-chip sequential elements to provide test stimuli and record the test results, so as to minimise the need for costly external test hardware;
(ii) to attempt to do the on-chip tests at normal system speed if possible;
(iii) to eliminate the need to determine any minimum set of test vectors based upon fault modelling or other basis;
(iv) to minimise the number of additional I/Os necessary to undertake the built-in test procedure(s).

Level-sensitive scan design (LSSD) has been shown to be a powerful means of feeding test data into a circuit and extracting the test results. It minimises the number of additional I/Os necessary but at the expense of the time necessary to feed in and feed out serial test data. It does not, therefore, satisfy clause (ii) above. Boundary scan, however, has become the internationally recognised means of providing controllability and observability of the I/Os and interconnections of separate ICs or other components on completed assemblies such as PCBs, and as such is a standard DFT procedure for OEM use.

Built-in self-test methods are also becoming increasingly accepted for complex circuits, in spite of the silicon overheads which are inherently involved. BILBO has been the most widely publicised BIST method to date, but as we have seen incurs a silicon area penalty of possibly 20 % additional chip area. The probability of aliasing of the compressed test signature in MISR registers has been shown to be theoretically equal to $1/2^n$ where n is the number of stages in the register, but experience indicates that provided that $n \geq 16$, acceptable fault detection is achieved.

The autonomous linear feedback shift register (LFSR) circuit is currently the most widely used form of on-chip test vector generator, but does not provide an entirely satisfactory test sequence because of the time-shifted correlation between its output bits. The autonomous cellular automata (CA) generators used in the CALBO proposals have better test properties, but at the expense of further additional silicon area, and we may therefore see some swing away from LFSRs to CA or other on-chip test generators in the future. Continuing research and development work such as that of Chen and Gupta may be noted[126].

We have given in this chapter a sufficient mathematical coverage to substantiate the use of the built-in self-test proposals, which may be read in association with the coverage given in preceding chapters. There is, however, a considerably greater width and depth of mathematics available in the subject area, particularly dealing with the choice of characteristic polynomials, test length considerations and aliasing. Readers who are particularly interested in a wider appreciation may find this coverage in Bardell et al.[19] and Yarmolik[64], and the many further references cited therein.

5.8 References

1. BENNETTS, R.G.: 'Design of testable logic circuits' (Addison-Wesley, 1984)
2. GIACOMO, J.D.: 'Designing with high performance ASICs' (Prentice Hall, 1992)
3. NEEDHAM, W.M.: 'Designer's guide to testable ASIC devices' (Van Nostrand Reinhold, 1996)
4. Parker, K.P.: 'Integrating design and test: using CAE tools for ATE programming' (IEEE Computer Society Press, 1987)
5. 'Design for testability'. Microelectronics for industry publication PT.505DFT, The Open University, UK, 1988
6. ABRAMOVICI, M., BREUER, M.A., and FRIEDMAN, A.D.: 'Digital systems testing and testable design' (Computer Science Press, 1990)
7. 'Design for testability' in 'Design handbook' (Intel Corporation, 1986)
8. 'Testing and testability' (NEC Electronics Corporation, 1986)
9. 'Data book and design manual' (LSI Logic Corporation, 1986)
10. BERGLUND, N.C.: 'Level sensitive scan design test chip/board systems', *Electronics*, 1979, **52**, 15 March, pp. 118–120
11. FUNATSU, S., WAKATSUKI, N., and YAMADA, A.: 'Easily testable design of large digital circuits', *NEC Res. Dev.*, 1979, (54), pp. 49–55
12. FUJIWARA, H.: 'Logic testing and design for testability' (MIT Press, 1985)
13. MICZO, A.: 'Digital logic testing and simulation' (Harper and Row, 1986)
14. PYNN, C.: 'Strategies for electronic test' (McGraw-Hill, 1986)
15. WILLIAMS, T.W. (Ed.): 'VLSI testing' (North-Holland, 1986)
16. WILLIAMS, T.W., and PARKER, K.P.: 'Design for testability—a survey', *Proc. IEEE*, 1983, **71**, pp. 98–112
17. EICHELBERGER, E.B.: 'Latch design using level sensitive scan design'. Proceedings of *COMCON*, 1983, pp. 380–383
18. DASGUPTA, S., GOEL, P., WALTHER, R.G., and WILLIAMS, T.W.: 'A variation of LSSD and its implications on design and test pattern generation in VLSI'. Proceedings of IEEE international conference on *Test*, 1982, pp. 63–66
19. BARDELL, P.H., McANNEY, W.H., and SAVIR, J.: 'Built-in test for VLSI: pseudorandom techniques' (Wiley, 1987)
20. WILKINS, B.R.: 'Testing digital circuits: an introduction' (Van Nostrand Reinhold, 1986)
21. GUTFREUND, K.: 'Integrating the approaches to structure design for testability', *VLSI Des.*, 1983, **4** (6), pp. 34–37
22. STEWART, J.H.: 'Future testing of large LSI circuit cards'. Digest of IEEE *Semiconductor test* symposium, 1977, pp. 6–15
23. ANDO, H.: 'Testing VLSI with random-access scan'. Digest of IEEE *COMPCON*, 1980, pp. 50–52
23a. LALA, P.K.: 'Fault tolerant and fault testable hardware design' (Prentice Hall, 1986)
24. BREUER, M.A., GUPTA, R., and LIEN, J.C.: 'Concurrent control of multiple BIT structures'. Proceedings of IEEE international conference on *Test*, 1988, pp. 431–442
25. ABADIR, M.S., and BREUER, M.A.: 'Scan path with look-ahead shifting'. Proceedings of IEEE international conference on *Test*, 1986, pp. 699–704
26. ABADIR, M.S.: 'Efficient scan-path testing using sliding parity response compaction'. Proceedings of IEEE international conference on *Computer aided design*, 1987, pp. 332–335
27. GUPTA, R., GUPTA, R., and BREUER, M.A.: 'The BALLAST methodology for structural partial scan design', *IEEE Trans.*, 1990, **C-39**, pp. 538–544

28 CHEN, K.T., and AGARWAL, V.D.: 'A partial scan method for sequential circuits with feedback', *IEEE Trans.*, 1990, pp. 544–548
29 RAJSUMAN, R.: 'Digital hardware testing: transistor-level fault modelling and testing' (Artech House, 1992)
30 MA, H.-K.T., DEVADAS, S., NEWTON, A.R., and SANGIOVANNI-VINCENTELLI, A.: 'An incomplete scan design approach to test generation for sequential machines'. Proceedings of IEEE international conference on *Test*, 1988, pp. 730–734
31 TRISCHLER, E.: 'Incomplete scan path with an automatic test generation methodology'. Proceedings of IEEE international conference on *Test*, 1980, pp. 153–162
32 CHICKERMANE, V., and PATEL, J.H.: 'A fault-oriented partial scan design approach'. Proceedings of IEEE international conference on *Computer aided design*, 1991, pp. 400–403
33 PARK, S., and AKERS, S.B.: 'A graph theoretic approach to partial scan design by K-cycle elimination'. Proceedings of IEEE international conference on *Test*, 1992, pp. 303–311
34 ASHER, P., and MALIK, S.: 'Implicit computation of minimum-cost feedback-vertex sets for partial scan and other'. Proceedings of IEEE conference on *Design automation*, 1994, pp. 77–80
35 MIN, H.B., and ROGERS, W.A.: 'A test methodology for finite state machines using partial scan design', *J. Electron. Test., Theory Appl.*, 1992, **3**, pp. 127–137
36 ANDERSON, T.L., and ALLSUP, C.K.: 'Partial scan and sequential ATPG', *Test*, 1993, **16**, July/August, pp. 13–15
37 KUNZMANN, A., and WUNDERLICH, H.-J.: 'An analytical approach to the partial scan problem', *J. Electron. Test., Theory Appl.*, 1990, **1**, pp. 163–174
38 TAI, S.-E., and BHATTACHARYA, D.: 'A three-stage partial scan design method using the sequential circuit flow graph'. Proceedings of IEEE international conference on *VLSI testing design*, 1994, pp. 101–106
39 KIM, K.S., and KIME, C.R.: 'Partial scan using reverse direction empirical testability'. Proceedings of IEEE international conference on *Test*, 1993, pp. 498–506
40 MARTLETT, R.A., and DEAR, J.: 'RISP (reduced intrusion scan path) methodology', *J. Semicust. ICs*, 1988, **6** (2), pp. 15–18
41 POMERANZ, I., and REDDY, S.M.: 'Synthesis of testable sequential circuits: an overview'. Digest of papers, test synthesis seminar, IEEE international conference on *Test*, 1994, paper 1.3
42 IEEE standard 1149.1–1990 'Standard test access port and boundary scan architecture'
43 Supplement to IEEE standard 1149.1, P.1149.1a/D9, ibid., 1993
44 BLEEKER, H., VAN DEN EIJNDEN, P., and DE JONG, F.: 'Boundary scan: a practical approach' (Kluwer, 1993)
45 'Testability primer'. Texas Instruments, Inc., SSYA.002A, CA, 1991
46 'The ABCs of boundary-scan test' (Philips Test and Measurement, Eindhoven, 1994)
47 'SCOPE: an extended JTAG architecture'. Texas Instruments SCL162,CA, 1989
48 'SCAN: IEEE 1149.1 (JTAG) compliant logic family'. National Semiconductor Databooklet, UK, 1992
49 MAUNDER, C., and BEENKER, F.: 'Boundary scan: a framework for structured design-for-test'. Proceedings of IEEE international conference on *Test*, 1987, pp. 714–723
50 PARKER, K.P., and ORESJO, S.: 'A language for describing boundary scan devices', *J. Electron. Test., Theory Appl.*, 1991, **2** (1), pp. 43–54

51 BRAVSAR, D.: 'An architecture for extending IEEE standard 1149.1 test access port to system backplanes'. Proceedings of IEEE international conference on *Test*, 1991, pp. 768–776
52 DE JONG, F.: 'Testing the integrity of the boundary scan test infrastructure'. Proceedings of IEEE international conference on *Test*, 1991, pp. 106–112
53 ROBINSON, G.D., and DESHAYES, J.G.: 'Interconnect testing of boards with partial boundary scan'. Proceedings of IEEE international conference on *Test*, 1990, pp. 572–581
54 BASSETT, R.W., TURNER, M.E., PANNER, J.H., GILLIS, P.S., OAKLAND, S.F., and STOUT, D.W.: 'Boundary scan design principles for efficient LSSD ASIC testing', IBM *J. Res. Dev.*, 1990, **34**, pp. 339–353
55 OAKLAND, S.F., MONZEL, J.A., BASSETT, R.W., and GILLIS, P.S.: 'An ASIC foundary view of design and test'. Digest of papers, test synthesis seminar, IEEE international conference on *Test*, 1994, paper 4.2
56 ESCHERMANN, B.: 'An implicity testable boundary scan TAP controller', *J. Electron. Test., Theory Appl.*, 1992, **3**, pp. 159–169
57 LINDER, R.: 'Getting to grips with boundary scan', *Test*, 1994, **16**, July/August, pp. 17–19
58 KONEMANN, B., MUCHA, J., and ZWIEHOFF, G.: 'Built-in logic block observation technique'. IEEE international conference on *Test*, 1979, digest of papers, pp. 37–41
59 KRASNIEWSKI, A., and ALBICKI, A.: 'Automatic design of exhaustively self-testing chips with BILBO modules'. IEEE international conference on *Test*, 1985, pp. 362–371
60 KONEMANN, B., MUCHA, J., and ZWIEHOFF, G.: 'Built-in test for complex integrated circuits', *IEEE J. Solid-State Circuits*, 1980, **SC-15**, pp. 315–318
61 SMITH, J.E.: 'Measures of effectiveness of fault signature analyses', *IEEE Trans.*, 1980, **C-29**, pp. 510–514
62 CARTER, J.: 'The theory of signature testing for VLSI'. Proceedings of 14th ACM symposium on *Theory of computing*, 1982, pp. 111–113
63 MUZIO, J.C., RUSKEY, F., AITKEN, R.C., and SERRA, M.: 'Aliasing probabilities of some data compression techniques', *in* MILLER, D.M. (Ed.): 'Developments in integrated circuit testing' (Academic Press, 1987)
64 YARMOLIK, V.N.: 'Fault diagnosis of digital circuits' (Wiley, 1990)
65 DAVID, R.: 'Signature analysis of multi-output circuits'. IEEE symposium on *Fault tolerant computing*, 1984, digest of papers, pp. 366–371
66 SRIDHAR, T., HO, D.S., POWELL, T.J., and THATTE, S.M.: 'Analysis and simulation of parallel signature analysers'. Proceedings of IEEE international conference on *Test*, 1982, pp. 656–661
67 DAMIANI, M., OLIVO, P., FAVALLI, M., ERCOLANI, S., and RICCO, B.: 'Aliasing in signature analysis testing with multiple input shift registers' *IEEE Trans.*, 1990, **CAD-9**, pp. 1344–1535
68 HUDSON, C.L., and PETERSON, G.D.: 'Parallel self-test with pseudorandom test patterns'. Proceedings of IEEE international conference on *Test*, 1987, pp. 954–963
69 WANG, L.T., and McCLUSKEY, E.J.: 'Concurrent built-in logic block observation (CILBO)'. IEEE international symposium on *Circuits and systems*, 1986, pp. 1054–1057
70 WANG, L.T., and McCLUSKEY, E.J.: 'Circuits for pseudoexhaustive test pattern generation', *IEEE Trans.*, 1988, **CAD-7**, pp. 1068–1080
71 FASANG, P.P.: 'BIDCO: built-in digital circuit observer'. Proceedings of IEEE international conference on *Test*, 1980, digest of papers, pp. 261–266
72 McCLUSKEY, E.L., and BOZORGUI-NESBAT, S.: 'Design for autonomous test', *IEEE Trans.*, 1981, **C-30**, pp. 860–875

73 DE BRUIJN, N.G.: 'A combinatorial problem', *Nederlands Akad. Wetensch. Proc. Ser. A*, 1946, **49** (2), pp. 758–764
74 RUSSELL, G., and SAYERS, I.L.: 'Advanced simulation and test methodologies for VLSI Design' (Van Nostrand Reinhold, 1989)
75 BHAVSAR, D.K., and HECKELMAN, R.W.: 'Self-testing by polynomial division'. IEEE international conference on *Test*, 1981, digest of papers, pp. 208–216
76 HORTENSIUS, P.D., and McLEOD, R.D.: 'Cellular automata-based signature analysis for built-in self-test'. Proceedings of 3rd Canadian technical workshop on *New directions for IC testing*, Halifax, NS, 1988, pp. 117–128
77 PODAIMA, B.W., and McLEOD, R.D.: 'Weighted test pattern generation for built-in self-test using cellular automata'. Proceedings of 3rd Canadian technical workshop on *New drections for IC testing*, 1988, pp. 195–205
78 SERRA, M., SLATER, T., MUZIO, J.C., and MILLER, D.M.: 'The analysis of one-dimensional linear cellular automata and their aliasing properties', *IEEE Trans.*, 1990, **CAD-9**, pp. 767–778
79 ZHANG, S., and MILLER, D.M.: 'A comparison of LFSR and cellular automata BIST'. Proceedings of Canadian conference on *VLSI*, 1990, pp. 8.4.1–8.4.9
80 MILLER, D.M., ZHANG, S., PRIES, W., and McLEOD, R.D.: 'Estimating aliasing in CA and LFSR based signature registers'. Proceedings of IEEE *ICCD*, 1990, pp. 157–160
81 WILLIAMS, T.W., DAEHN, W., GRUETZNER, M., and STARKE, C.W.: 'Bounds and analysis of aliasing errors in linear feedback shift registers', *IEEE Trans.*, 1988, **CAD-7**, pp. 75–83
82 SERRA, M.: 'Algebraic analysis and algorithms for linear cellular automata over GF(2) and the application to digital testing'. Congressus Numerantium, Winnipeg, Canada, 1990, **75**, pp. 127–139
83 BARDELL, P.H.: 'Analysis of cellular automata used as pseudorandom pattern generators'. Proceedings of IEEE international conference on *Test*, 1990, pp. 762–768
84 DAMIANI, M., OLIVO, P., FAVALLI, M., EROLANI, S., and RICCO, B.: 'Aliasing in signature analysis with multiple-input shift-registers'. Proceedings of European conference on *Test*, 1989, pp. 346–353
85 SAVIR, J., and BARDELL, P.H.: 'Built-in self-test: milestones and challenges', *VLSI design*, 1993, **1**, pp. 23–44
86 KRASNIEWSKI, A.: 'Logic synthesis for efficient pseudo-exhaustive testability'. Proceedings of IEEE conference on *Design automation*, 1991, pp. 66–72
87 HURST, S.L.: 'VLSI custom microelectronics: analog, digital and mixed-signal' (Marcel-Dekker, 1998)
88 HURST, S.L.: 'A hardware consideration of CALBO testing'/ Proceedings of 3rd technical workshop on *New directions for IC Testing*, Halifax, Canada, 1988, pp. 129–146
89 EL-ZIQ, Y.M.: 'S^3: VLSI testing using signature analysis and scan path techniques'. Proceedings of IEEE *ICCAD*, 1983, pp. 73–76
90 KOMONYSTKY, D.: 'LSI self-test using level-sensitive scan design and signature analysis'. IEEE international conference on *Test*, 1982, digest of papers, pp. 414–424
91 BEUCLER, F.P., and MANNER, M.J.: 'HILDO: the Highly Integrated Logic Device Observer', *VLSI Des.*, 1984, **5** (6), pp. 88–96
92 LeBLANC, J.J.: 'LOCUST: a built-in self-test technique', *IEEE Des. Test Comput.*, 1984, **1** (4), pp. 42–52
93 SAYERS, I.L., KINNIMENT, D.J., and RUSSELL, G.: 'New directions in the design for testability of VLSI circuit'. Proceedings of IEEE *ISCAS*, 1985, pp. 1547–1550
94 PARASKEVA, M., KNIGHT, W.L., and BURROWS, D.F.: 'A new test structure for VLSI self-test: the structured register', *Electron. Lett.*, 1985, **21**, pp. 856–857

95 COSGROVE, B.: 'The UK5000 gate array', *IEE Proc. G*, 1985, **132**, pp. 90–92
96 RESNICK, D.R.: 'Testability and maintainability with a new 6K gate array', *VLSI Des.*, 1983, **4** (2), pp. 34–38
97 TOTTON, K.A.E.: 'Review of built-in test methodologies for gate arrays', *IEE Proc. E*, 1985, **132**, pp. 121–129
98 EICHELBERGER, E.B., and LINDBLOOM, E.: 'Random pattern coverage enhancement and diagnosis for LSSD logic self-test', *IBM J. Res. Dev.*, 1983, **27** (3), pp. 265–272
99 BARDELL, P.H., and McANNEY, W.H.: 'Self-testing of a multi-chip module'. IEEE international conference on *Test*, 1982, digest of papers, pp. 302–308
100 BARDELL, P.H., and McANNEY, W.H.: 'Pseudorandom arrays for built-in tests', *IEEE Trans.*, 1986, **C-37**, pp. 653–658
101 GLOSTER, C.: 'Synthesis for BIST with PREWARP'. Proceedings of 3rd annual *OASIS* research review, Research Triangle Park, NC, May 1990
102 McANNEY, W.H., and SAVIR, J.: 'Distributed generation of non-uniform patterns for circuit testing', *IBM Tech. Discl. Bull.*, 1988, **35** (5), pp. 113–116
103 CHIN, C.K., and McCLUSKEY, E.J.: 'Weighted pattern generation for pseudorandom testing', *IEEE Trans.*, 1987, **C-36**, pp. 252–256
104 MARTINEZ, M., and BRACHO, S.: 'Concatenated LFSR makes a weighted built-in logic block observation', *Microelectron. J.*, 1994, **25**, pp. 219–228
105 WUNDERLICH, H.J.: 'Self-test using unequiprobable random patterns'. Proceedings of IEEE international conference on *Test*, 1988, pp. 236–244
106 WAICUKAUSKI, J.A., and LINDBLOOM, E.: 'Fault detection effectiveness of weighted random patterns'. Proceedings of IEEE international conference on *Test*, 1988, pp. 245–255
107 ILLMAN, R., and CLARKE, S.: 'Built-in self-test of the MACROLAN chip', *IEEE Des. Test Comput.*, 1980, **6** (2), pp. 29–40
108 WAGNER, K.D., CHIN, C.K., and McCLUSKEY, E.J.: 'Pseudorandom testing', *IEEE Trans.*, 1987, **C-36**, pp. 332–343
109 HURD, W.J.: 'Efficient generation of statistically good pseudonoise by linearly-connected shift registers', *IEEE Trans.*, 1974, **C-23**, pp. 146–15
110 BARDELL, P.H., and SPENCER, T.H.: 'A class of shift-register Hurd generators applied to built-in test'. IBM technical report TR00.3000, NY, September 1984
111 KANZMANN, A., and KRINBEL, S.: 'Self-test with determinstic test pattern generators', *in* SAUCIER, G., and MIGNOTTE, A. (Eds.): 'Logic and architecture synthesis: state-of-the-art and novel approaches' (Chapman and Hall, 1995)
112 ARMSTRONG, J.A., and GRAY, F.G.: 'Structured logic design with VHDL' (Prentice Hall, 1993)
113 RUSTON, A.: 'VHDL for logic synthesis' (McGraw-Hill, 1995)
114 IEEE standard 1076-1987, 'Standard VHDL language reference'
115 BELLON, C., ROBACH, C., and SAUCIER, G.: 'An intelligent assistant for test program generation: the SUPERCAT system'. IEEE international conference on *Computer aided design*, 1984, digest of papers, pp. 32, 33
116 ROBINSON, G.D.: 'HITEST: intelligent test generation'. IEEE international conference on *Computer aided design*, 1983, digest of papers, pp. 311–323
117 ROBACH, C., MALECHA, P., and MICHEL, G.: 'CATA: a computer-aided test analysis system', *IEEE Des. Test Comput.*, 1984, **1** (2), pp. 68–79
118 ABADIR, M.S., and BREUER, M.A.: 'A knowledge based system for designing testable VLSI chips', *IEEE Des. Test Comput.*, **2** (4), pp. 56–68
119 SHUBIN, H., and ULRICH, J.W.: 'IDT: an intelligent diagnostic tool'. Proceedings of *AAAI-82*, 1982, pp. 90–295
120 LEUNG, E., QURESHY, N., RHODES, T., and TSAI, T.-S.: 'VHDL integrated test development', *Electron. Prod. Des.*, 1995, **16** (10), pp. 53–56

121 HARDING, W. (Ed.): 'HDLs: a high-powered way to look at complex designs', *Comput. Des.*, 1990, **29** (5), pp. 74–84
122 JONES, N.A., and BAKER, K.: 'Knowledge-based system tool for high-level BIST design', *Microprocess. Microsyst.*, 1987, **11** (1), pp. 35–48
123 AVRA, L.J., and McCLUSKEY, E.J.: 'High-level synthesis of testable designs: an overview of university systems'. Digest of papers, test synthesis seminar, IEEE international conference on *Test*, 1994, pp. 1.1–1.8
124 ROY, R.K., DEY, S., and POTKONJAK, M.: 'Test synthesis using high-level design information'. Digest of papers, test synthesis, seminar, IEEE international conference on *Test*, 1994, pp. 3.1–3.6
125 VISHAKANTAIAH, J.A., ABRAHAM, J.A., and ABADIR, M.: 'Automatic test knowledge extraction from VHDL'. Proceedings of IEEE conference on *Design automation*, 1992, pp. 273–278
126 CHEN, C.-A., and GUPTA, S.K.: 'BIST test pattern generators for two pattern testing—theory and design algorithms', *IEEE Trans.*, 1996, **C–45**, pp. 257–269

Chapter 6
Testing of structured digital circuits and microprocessors

6.1 Introduction

The majority of the material covered in previous chapters has been general in its concepts, and has not specifically considered certain families of circuits or circuit architectures. The self-test considerations of the previous chapter, for example, did not involve any specific circuit structure or layout in the core logic or sequential (storage) networks, apart from the requirement in the majority of cases to separate the combinational logic elements from the sequential elements.

However, in this chapter we will consider particular circuits and circuit architectures, and how they may be tested. The fundamental principles of digtal logic testing will still be relevant, but specific fault mechanisms, circuit failures and/or testing procedures may now be involved. Additionally, many of the required test procedures will have been considered in detail by the IC manufacture (vendor), and therefore the OEM will not have the problem of formulating tests for such circuits from scratch as may be necessary with a system assembly of simpler ICs or other components—the problem of accessibility of the I/Os of such circuits for test purposes in a completed system assembly will, of course, still be present.

The main areas which will be considered in the following pages are:

- microprocessors and memory;
- programmable logic devices;
- cellular arrays.

Memory is an essential part of all processor-based systems; both RAM and ROM have very strongly structured architectures, the testing of which is crucial to the performance of the processor system. Similarly, programmable logic devices have distinctive silicon layouts which, like cellular array structures, lead to specific testing considerations. We will, therefore. consider each of these in turn, beginning with programmable logic devices which are of increasing significance to original equipment manufacturers.

6.2 Programmable logic devices

The importance of programmable logic devices (PLDs) is due to their increasing logic capabilities and reduced cost, this being reflected in the number of publications now available which address this area of digital logic[1-9]. The programmable logic array is the architecture which has received the greatest attention as far as testing is concerned[9-17], but we will also consider other programmable structures in the following sections.

6.2.1 Programmable logic arrays

Programmable logic arrays (PLAs) provide a ready means of realising random combinational logic, and are increasingly used as the glue logic in processors, DSP circuits and other applications. PLA structures may also be found as a logic partition within more complex VLSI ICs, to provide local logic requirements in a more structured way than that provided by separate logic gates. Sequential logic (storage or memory) elements are not provided on PLAs, but are found on other programmable logic devices which will be considered later.

Off the shelf PLAs of small to medium complexity pose no problems for test, since exhaustive testing can be performed. However, for large PLAs with tens of primary inputs there is a need to find some minimum test set and to have available ATPG programs for this purpose; a 30-input PLA, for example, would require over 10^9 input test vectors for an exhaustive test at minterm level, but internally there may be only, say, 50 to 100 product terms generated by the AND matrix, see Figure 6.1. Also the marching and galloping test vectors used for random-access memory testing (see Section 6.4) are not appropriate for testing PLAs since the internal architecture of the two is completely dissimilar.

The particular architecture of a PLA does, however, lead to certain fundamental properties which can simplify the testing requirements. Recall that the PLA is a two-level AND–OR structure as shown in Figure 6.1a, with the individual outputs given in some minimised sum of products form rather than always in minterm form as in read-only memory (ROM or PROM) devices. In practice the PLA circuit may be a NAND/NAND or a NOR/NOR/INVERT structure depending upon the technology used, rather than the logically-equivalent AND/OR structure shown in Figure 6.1a, but this does not affect the following general discussions and developments.

The PLA structure does not have any reconvergent fan-out topology, see Figure 6.1b, which second-generation ATPG programs such as PODEM and FAN can accomodate, but it is possible for the product terms in the AND array to have a very high fan-in from the primary inputs which conventional ATPG programs find very time consuming to handle in their backward tracing procedures. This high fan-in also militates against the use of random

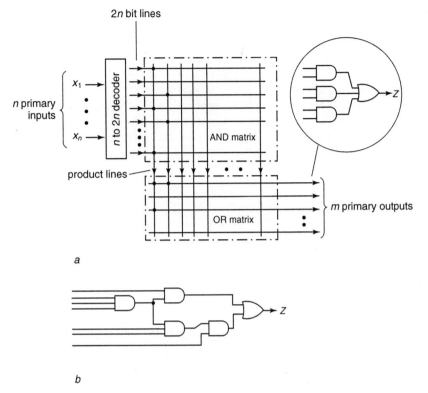

Figure 6.1 The PLA architecture
  a the two-level AND/OR realisation produced by the programmable AND matrix and the programmable OR matrix
  b a reconvergent fan-out topology with more than two levels of logic which is never present in any PLA architecture

or pseudorandom test patterns, since to sensitise a particular product term in the AND array requires the application of all 1s to the product term inputs, (all 0s if the PLA is a NOR realisation), and the probability of random input vectors achieving this is correspondingly very low.

Hence from many aspects it becomes much more appropriate to consider how to formulate test patterns to test a PLA from the specific architecture of the device itself.

The programmed (dedicated) output functions of a PLA are dependent upon the presence or absence of the connections at the crosspoints in the AND and OR arrays. The fault model that has therefore been accepted for PLA structures is not the stuck-at fault model, but instead is the crosspoint fault model, since this more accurately reflects the possible failures which may occur. The crosspoint faults have been classified by Smith[18] as follows:

- growth fault—a missing connection in the AND array, causing a product term to loose a literal and hence to have twice as many minterms as it should have, e.g. loss of x_3 in $x_1 x_2 x_3 x_4$ would give $x_1 x_2 x_4$;
- shrinkage fault—an erroneous extra connection in the AND array, causing a product term to have an extra literal and hence loose half of its required minterms, e.g. gain of x_2 in $x_1 x_4 x_5$ would give $x_1 x_2 x_4 x_5$;
- disappearance fault—a missing connection in the OR array, causing one complete product term to be lost in one or more OR outputs;
- appearance fault—an erroneous extra connection in the OR array, causing an additional literal or product term to appear in one or more OR outputs.

If an erroneous extra connection in the AND array happens to connect the complement of an existing literal to an AND term, eg. $\bar{x}_2$ and x_2, this may sometimes be referred to as a vanishing fault19, since this variable disappears from the resulting product term. It is, in fact, a special case of a growth fault, but we will have no need to refer specifically to this particular situation.

These various individual crosspoint faults are illustrated in Figure 6.2. Notice that all of them either add minterms to or subtract minterms from the fault-free output function, but do not cause a complete shift of minterms to another part of the complete n-space of $f(X)$. Note also that missing crosspoint connections are equivalent to some stuck-at faults, but extra connections cannot generally be modelled by a stuck-at fault condition. Also there are $2nm + km = (2n + k)m$ possible single crosspoint faults, where n is the number of individual input variables, k is the number of product terms, and m is the number of output functions.

A crosspoint fault is also termed a functional fault if it alters at least one output function of the dedicated PLA. This qualification is necessary since an appearance fault, for example, may merely introduce an additional but redundant product term in one of the sum of products outputs. Only functional faults are of significance in our following discussions.

The full hierarchy of possible PLA testing strategies has been compiled by Breuer et. al., and is shown in Figure 6.3[17,20]. Some of these possibilities remain academic, and therefore we will here only consider selected ones. Further reading and references may however be found in Reference 20.

Test pattern generation based upon the possible crosspoint faults and leading to ATPG programs has received considerable attention. Essentially such processes look at:

(i) individual possible crosspoint faults and the determination of appropriate input test vectors to detect their existence if present;
(ii) the relationships between the types of crosspoint fault and the effect on the total test vector requirements.

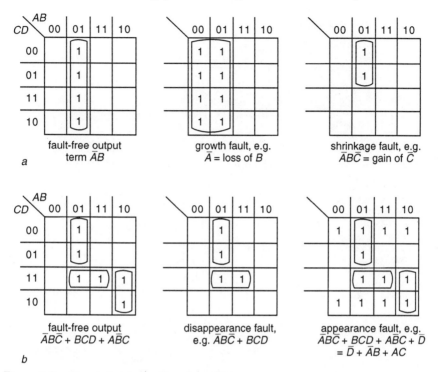

Figure 6.2 Crosspoint fault effects in a PLA
  a AND matrix growth and shrinkage faults
  b OR matrix disappearance and appearance faults

The earliest considerations were those of Smith[18], Cha[20] and Ostapka and Hong[19], the latter developing the PLA test pattern program TPLA. Subsequent variations have been published, including the PLA/TG algorithm of Eichelberger and Lindbloom[21] and others. It has been shown[22,23] that a complete test coverage for all single crosspoint faults also covers most stuck-at faults on input and output lines, many shorts and a high percentage of multiple faults. Other research has also related this single crosspoint fault coverage to AND bridging faults in the architecture as well as to multiple crosspoint faults[24,25].

Details of these ATPG proposals may be found discussed and further referenced in Russell and Sayers[13]. In general most programs involve the selection of each individual product term in the AND array, and consider the crosspoint faults which affect this term. For example, consider the simple case illustrated in Figure 6.4, and suppose we loose the literal x_2 in the product term $x_1 x_2$ due to a crosspoint fault. This will result in a growth fault in the output $f_1(X)$, giving us $f_1(X) = \bar{x}_2 x_4 + x_1 + x_3$ instead of $\bar{x}_2 x_4 + x_1 x_2 + x_3$.

The difference (growth) in the product term due to this crosspoint fault is given by the Boolean equation:

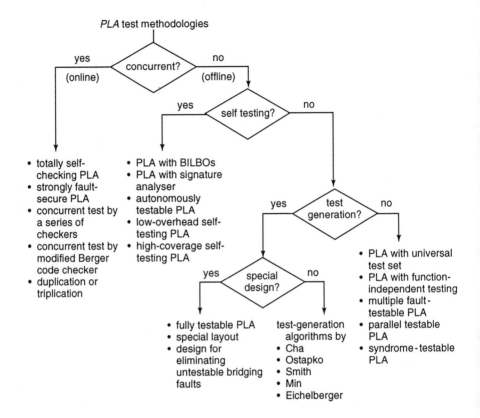

Figure 6.3 The hierarchy of possible test methods for programmable logic arrays (Acknowledgements, based on Reference 26)

[(product term with the fault present) . $\overline{\text{(product term without the fault)}}$] (6.1)

which in this particular case is

$$[(x_1) . (\overline{x_1 x_2})]$$
$$= [x_1(\bar{x}_1 + \bar{x}_2)]$$
$$= x_1 \bar{x}_2$$

This is shown dotted in Figure 6.4b*. To detect the effect of this fault on the final OR ouput $f_1(X)$ we need to perform a similar operation:

[(growth term with the fault present) . $\overline{\text{(sum of the remaining fault-free product terms)}}$]

(6.2)

---

* The operation $A.\bar{B}$ between two Boolean terms A and B is sometimes referred to as the Sharp operation, and expressed as $(A \# B)$. However, this is not a term often used in engineering circles and we will not use it here.

Testing of structured digital circuits and microprocessors 303

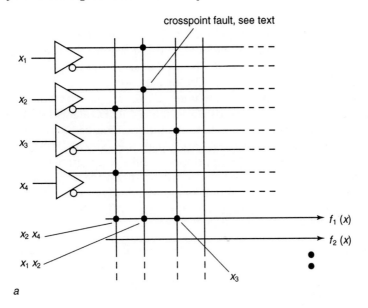

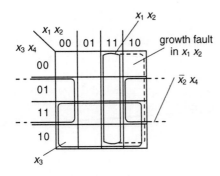

Figure 6.4 A simple PLA example with a crosspoint growth fault in the product term $x_1 x_2$
  a circuit
  b Karnaugh map for $f_1(X)$

which in this case is

$$\left[x_1 \bar{x}_2 \left(\overline{\bar{x}_2 x_4 + x_3} \right) \right]$$

$$= \left[x_1 \bar{x}_2 \left(x_2 \bar{x}_3 + \bar{x}_3 \bar{x}_4 \right) \right]$$

$$= x_1 \bar{x}_2 \bar{x}_3 \bar{x}_4$$

This is confirmed in Figure 6.4b. If the crosspoint fault in $x_1 x_2$ had been the loss of x_1 rather than x_2, it will be seen that the effect of this growth fault on $f_1(X)$ would have been the additional output $\bar{x}_1 x_2 \bar{x}_3$.

The test vector to check for the loss of x_2 in $f_1(X)$ would therefore be $x_1 x_2 x_3 x_4 = 1\ 0\ 0\ 0$, which would give $f_1(X) = 0$ when fault free and 1 when faulty. Similarly, to test for the loss of x_1 would require the test vector 0 1 0 –. Should the product term $x_1 x_2$ also be used in any other output $f_2(X)$, ..., then some alternative test vectors may also be appropriate.

Hence it is possible to determine a set of test vectors by Boolean consideration of all relevant crosspoint failures. ATPG algorithms based upon these considerations with a search for commonality to reduce the final test set have therefore been developed. Shrinkage and growth faults are usually considered first, followed by appearance and disappearance faults. A high proportion of the latter are detected by the shrinkage and growth fault test vectors, which successfully minimises the length of the final test set.

There are, however, certain disadvantages with this procedure and the resulting ATPG programs, namely:

- multiple crosspoint faults are not specifically considered, although they may be with certain extensions to the algorithms;
- the test patterns are entirely dependent upon the dedicated functions in the PLA, and are not generalised for any possible dedication of the device;
- the test patterns do not take into account the possibility of other than crosspoint faults, for example a loss of inversion in an input decoder or an output buffer;
- if the PLA is part of a complex VLSI circuit, controllability and observability of the PLA inputs and outputs may not be fully available for such tests, see Figure 6.5;
- as PLA size grows towards 50 or more inputs and hundreds of product terms, so the number of crosspoint faults increases such as to make ATPG programs based upon this fault model prohibitively expensive in CPU and execution time.

Hence, although useful ATPG programs have been developed based upon the crosspoint fault model[13,17], other approaches have been actively pursued. These include:

(i) syndrome and other offline output compression test methods;
(ii) online concurrent error detection by the addition of checker circuits using Berger codes or other methods;
(iii) additions to the basic PLA architecture to provide some built-in self test facility to ease the testing problem.

Looking at the first of the three above broad categories, and considering a ones-count (syndrome) test as covered in Chapter 4, Section 4.3.1. If we could guarantee that all faults in the PLA would result in growth or shrinkage faults at the outputs, then syndrome testing would be a powerful tool. For example, in the simple function illustrated in Figure 6.4, the syndrome count for $f_1(X)$ would be as shown in Table 6.1. However, a fundamental

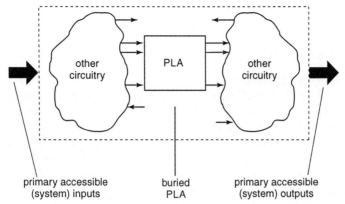

Figure 6.5 The increasing use of large buried PLAs within VLSI ICs, with the consequent difficulties of controllability and observability

disadvantage is that to determine the full syndrome count of each output, a fully exhaustive input set of test vectors is necessary, which may not be acceptable particularly with a large number of primary inputs on the PLA.

Table 6.1 The effect of growth and shrinkage faults in the circuit of Figure 6.4a

Growth or shrinkage fault in Figure 6.4	$f_1(X)$ Syndrome count
Fault-free	12
$\bar{x}_2$ missing in $\bar{x}_2 x_4$	13
x_4 missing in $\bar{x}_2 x_4$	14
x_1 missing in $x_1 x_2$	14
x_2 missing in $x_1 x_2$	13
x_3 missing in x_3	16
Additional x_1 in $\bar{x}_2 x_4$	11
Additional x_3 in $\bar{x}_2 x_4$	10
.	.
.	.
.	.

It has been formally shown that all single faults in the AND and the OR arrays of PLAs are detected by a syndrome count, including bridging faults between product lines and between output lines[27-29]. However some multiple faults and I/O faults can result in the fault-free syndrome count still being produced, to overcome which the possiblity of weighted syndrome summation (WSS) testing, see Section 4.3.1, has been proposed. Nevertheless, the basic need for a fully exhaustive input test set remains, and hence circuit additions to the basic PLA architecture to give a design which may be tested with some nonexhaustive test set are potentially more attractive.

6.2.1.1 Offline testable PLA designs

We are now considering the special design category defined in Figure 6.3, where the basic AND/OR structure of the PLA is augmented in some way so as to provide a more easily testable circuit. In the following discussions we will assume that the PLA is an AND/OR structure, and all logic signals of 0 and 1 will be based on this premise. In practice some of the polarities of these logic signals may need to be opposite from that detailed here, depending upon the specific NAND or NOR circuit configuration, but this will not affect the general test strategy being described.

The principle needs when considering an easily-testable PLA design include the following:

- controllability of the $2n$ x_i and $\bar{x}_i$ input literals to the AND array;
- separate controllability and perhaps observability of each of the k product lines (the OR array inputs);
- some observability and/or check of the connections in the OR array which provide the final PLA outputs.

Typical ways in which these requirements may be approached by additional circuitry are shown in Figure 6.6. As will be seen, there are three principal ideas, namely:

(i) alter the normal x_i input decoders so as to be able to modify the x_i and $\bar{x}_i$ input literals to the AND array;
(ii) add scan path shift registers so as to be able to shift specific 0 or 1 signals into the AND or OR arrays, or possibly to scan out data;
(iii) add additional product column(s) to the AND array and/or additional row(s) to the OR array so as to provide parity or other check resources.

However there is no one accepted best method of augmenting the basic PLA architecture, and all involve an additional silicon area overhead which may not be acceptable to some vendors or OEMs.

Published examples of offline testable PLA designs include those of Fujiwara and Kinoshita[30], Hong and Ostapko[31], Khakbaz[32], and others[32-38]; these may be found reviewed in Zhu and Breuer[26] and elsewhere[10,13,17]. As an example, Figure 6.7 shows the early proposed circuit of Fujiwara and Kinoshita which incorporates many of the principles that will be found in later disclosures, and which is designed to enable the correct dedication of all crosspoints in the AND and OR arrays to be verified.

The input decoders in Figure 6.7 allow the input literals to be controlled as follows, where x_i may be 0 or 1:

Decoder inputs			AND array input literals		
x_i	Y_1	Y_2	x_i	$\bar{x}_i$	
x_i	0	0	x_i	$\bar{x}_i$	(normal mode)
x_i	0	1	x_i	1	
x_i	1	0	1	$\bar{x}_i$	
x_i	1	1	1	1	(not used)

The test procedure is as follows:

(i) Set all stages in the shift register to logic 0 so as to disable all the product lines, and check that both odd parity outputs Z_1 and Z_2 are zero. This checks for any stuck-at 1 fault in the parity check circuits.

(ii) Shift a 1 into the first stage of the shift register so as to select the first product column only leading into the OR array; set all x_i inputs to 0 and $Y_1 Y_2$ to 1 0 which will give a 1 on all $2n$ AND input literals, and check that the odd parity check on column 1 is 1 by observing the output Z_2. Note that if there are no programmed crosspoints to any output $f(X)$, i.e. the column is spare, there will only be the parity bit in the OR array to give the output $Z_2 = 1$. Check also that the odd-parity output Z_1 is also 1 which indicates that the single column is energised and not any additional column.

(iii) Repeat (ii) with all x_i inputs set to 1 and $Y_1 Y_2$ set to 0 1, which again will give a 1 on all AND array inputs.

(iv) Repeat (ii) and (iii) for each of the remaining product columns in turn by shifting a 1 along the shift register. At the completion of these tests all OR matrix crosspoints will have been checked.

(v) Set all stages in the shift register to 1 so as to energise all the product columns. With $Y_1 Y_2 = 0\ 1$ set input x_1 to 0 and all remaining x_i inputs to 1 so as to deactivate the x_1 row in the AND array but activate all remaining $2n-1$ rows. With this loss of all p (p odd) programmed crosspoints in the first row of the AND array including the parity check, and hence loss of p product terms in the $k+1$ product columns, check that the odd parity output Z_1 remains 1. Failure will indicate a loss of odd parity on this first row in the AND array.

(vi) Repeat with $Y_1 Y_2 = 1\ 0$, $x_1 = 1$ and all remaining x_i inputs at 0 so as now to deactivate the $\bar{x}_i$ row in the AND array, checking $Z_1 = 1$ as before.

(vii) Repeat (v) and (vi), deactivating each of the remaining x_i and $\bar{x}_i$ rows in the AND array in turn, checking that $Z_1 = 1$ on each test. Tests (ii) to (vi) will also have completely checked the decoder circuits.

With the completion of these $2n + k + 1$ tests it is claimed that as well as all single crosspoint faults all single stuck-at faults will have been covered. An additional initial check of the shift register may also be included, which increases the total number of tests to $2n + 3k^{10}$. Notice also that this test

308   VLSI testing: digital and mixed analogue/digital techniques

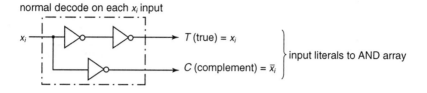

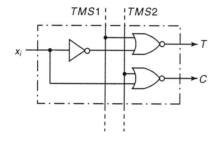

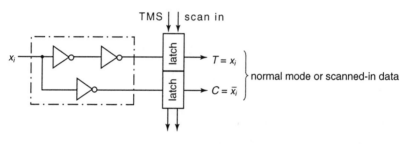

Figure 6.6   Possible circuit additions for easily-testable PLAs
    a control of the AND array input literals, TMS being the test-mode signal(s)
    b shift register to allow removal of product terms from the AND array

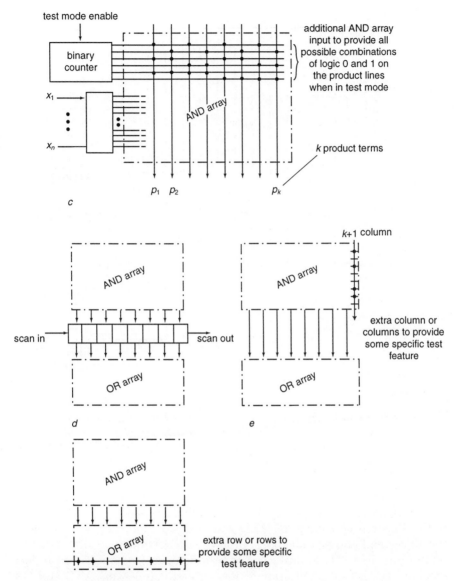

Figure 6.6 Continued
    c an alternative to *b*
    d shift register to provide direct controllability of the inputs to the OR array, and/or observability of the product term outputs from the AND array
    e the addition of extra column(s) and/or row(s) to the AND and OR arrays so as to provide parity or other checks

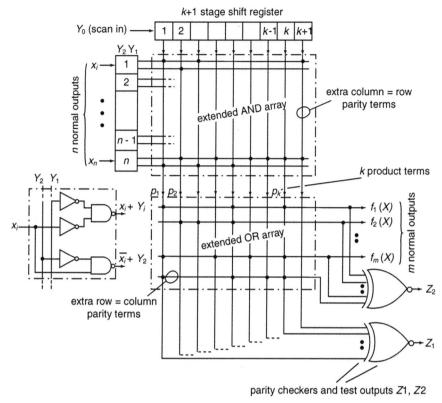

Figure 6.7 The offline test additions of Fujiwara and Kinoshita. The $(k + 1)$ input exclusive-OR gate shown here for output Z_1 would be a cascade of k two-input XOR gates, and similarly for output Z_2 there would be a cascade of m two-input XOR gates

strategy does not rely upon the output functions of the dedicated PLA, but relies instead on the two parity checkers for fault detection.

The PLA testing strategy of Hong and Ostapko[31] has many similarities to the above, and also provides a test procedure independent of the dedicated output functions. The shift register to provide controllability of the product column lines into the OR array is now located between the AND and OR arrays, rather than as shown in Figure 6.7, and somewhat more complex input decoders are present. However, these and further similar proposals all require cascades or trees of exclusive-OR gates for the parity checking, which impose a timing constraint on the offline tests and require a considerable silicon area overhead for the additional circuit details.

A much reduced silicon area penalty can be achieved by not using parity checking, and making the test patterns function dependent rather than independent. The K-test scheme of Khakbaz[32] uses a shift register above the AND array to select individual product lines as in Figure 6.7, but instead of any parity checks uses an additional row in the OR array to provide a test output Z^* which is the OR of all the product lines. Knowing the dedication of the PLA, a test set can be formulated which tests (i) the correct operation of the shift register, (ii) the correct presence of each product term by applying the appropriate input test vectors x_1 x_2 ... x_n, and (iii) the correct outputs from the OR matrix. In test (ii) the appropriate x_i inputs are applied to activate the product term on the selected product line, and its presence checked by observing the Z^* output. All inputs are then individually inverted in turn; if the particular x_i is a don't care term in the selected product column then Z^* should remain 1; if x_i is not a don't care input then Z^* should change from 1 to 0 when each x_i is inverted from its correct value. The OR array crosspoints are checked by observing that the programmed outputs only respond when the correct AND terms are present on the product lines from the array. It is claimed that all faults, crosspoint, stuck-at and bridging, which give rise to incorrect output functions are detected, using a total of $2.5n + 2k + 5$ tests[10,13,32].

A number of other function-dependent and function-independent test strategies may be found[33-38], the former including the proposals of Rajski and Tyszer[33] and Reddy and Ha[34]. Further details may be found in the cited references, and comparisons of the fault coverage and circuit overheads may be found in References 10, 13 and 26. In general, these offline test strategies have not found very great favour with vendors and OEMs, being unnecessary for relatively small PLAs, with online self tests currently receiving greater attention and acceptance for larger PLAs as will be covered in the next section.

6.2.1.2 Online testable PLA designs

Concurrent or online test represents the left-hand extreme of Figure 6.3. Clearly it is possible to duplicate or triplicate a complete PLA structure under certain circumstances, say triple modular redundancy (TMR) with output voting as was covered in Section 4.6.2 of Chapter 4; such replication provides an extremely powerful online method of PLA checking, particularly if:

- the primary x_i inputs are allocated in a different order in the replicated PLAs;
- the outputs are likewise dissimilarly allocated;
- different product terms are used if possible to generate the output functions in the replicated PLAs, or one PLA always generates the complements of another PLA's outputs for check purposes.

By such means any functional fault in a PLA may be detected, and the effects of possible manufacturing faults or weaknesses which may affect all the PLAs can be minimised. However, there still remains the situation that the primary x_i inputs may be faulty and also that fail-safe voting or other check circuits at the duplicated outputs are necessary, plus the obvious greatly increased silicon overhead involved in such replication of complete PLA structures.

For the majority of applications duplication or triplication is not justified, and therefore other means of online checking (but not correction) of PLA outputs are more appropriate. Figure 6.8 shows the online test strategy proposed by Khakbaz and McCluskey[39], which relies upon the following features:

(i) each pair of literal lines in the AND array forms a two-rail code, i.e. x_i and $\bar{x}_i$;
(ii) the product signals on the k product lines are arranged such that only one product line is energised at a time in normal operation, so as to provide a 1 out of k code;
(iii) additional output lines are added to the OR array so that some output error detection capability is present.

From Figure 6.8 it will be seen that three checker circuits C1, C2 and C3 are present. Checker C1 is a totally self checking (TSC) two-rail checker that will test for all single faults on the bit lines and the input decoders during normal operation; checker C2 is a TSC 1 out of k checker which checks that only one of the product lines is energised at a time, and checker C3 is a specific output

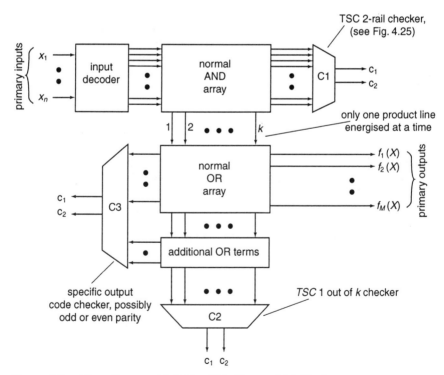

Figure 6.8 The online testable PLA of Khakbaz and McCluskey

code checker depending upon some chosen coding scheme across all the OR array outputs. This may be a parity check or some p out of q code word check.

All crosspoint and most other PLA faults are caught by one or more of these checkers, but the overall circuit is not totally self checking and some faults may be masked. Additional function-dependent tests have therefore been proposed[39] to check the PLA after the initial dedication, following which the online checks in service are claimed to be satisfactory. However, the requirement that only one product line leading into the OR array shall be energised at a time means that a minimum sum of products for the several output functions may not be possible. For example, consider the trivial case shown in Figure 6.9. The minimum sum of products realisations for the two functions are:

$$f_1(X) = \bar{x}_1 x_3 + \bar{x}_1 x_4 + x_2 x_4$$
$$f_2(X) = x_1 \bar{x}_3 + x_1 \bar{x}_2$$

but this would mean that product lines $x_2 x_4$ and $x_1 \bar{x}_3$, for example, are both energised when the input vector is $x_1 x_2 \bar{x}_3 x_4$. To ensure that only one product line is energised at a time requires the functions to be realised as:

$$f_1(X) = \bar{x}_1 x_3 + \bar{x}_1 \bar{x}_3 x_4 + x_1 x_2 x_3 x_4 + x_1 x_2 \bar{x}_3 x_4$$

and

$$f_2(X) = x_1 x_2 \bar{x}_3 \bar{x}_4 + x_1 \bar{x}_2 + x_1 x_2 \bar{x}_3 x_4$$

or some other disjoint factorisation. If static hazard-free realisations of the output functions are required, which involve a deliberate overlap between product terms, then it will not be possible to achieve only one product line energised at a time as this self-test method requires.

An alternative online test strategy proposed by Chen et al.[40] does not have this particular constraint. It maintains the same two-rail checker circuit, C1, as in Figure 6.8, but not the same C2 and C3 checkers. Instead, additional rows in the OR array and possibly additional products from the AND array are incorporated such that odd (or even) parity checks on the total energised product terms can be made. A maximum of three additional rows in the OR array are required in order to give these parity checks.

A different approach to the same objective may be found in the proposal of Mak et al.[41], as shown in Figure 6.10. Neither of the previously described categories of test circuits specifically considers the situation where all the PLA outputs are simultaneously zero, there being no programmed output of $f(X) = 1$ on certain primary input vectors. In Figure 6.10 this situation is specifically covered by adding false terms to the AND array, which give energised product lines on all otherwise unused input vectors. The output check is then based upon generating additional OR functions from all the true and false product terms such that p out of q coding or Berger coding may be used in the checker circuit. A modified Berger coding was also introduced by Mak et al.[41,42] which uses fewer bits than the normal Berger code (see

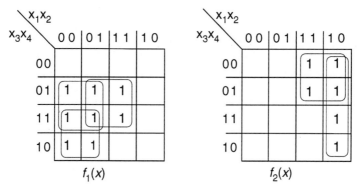

Figure 6.9 Two functions with a common term $x_1x_2\bar{x}_3x_4$ in their realisation which must be separately generated (see text)

Chapter 4, Section 4.6.1), and which was shown to be appropriate in this application. A worked example may be found in Lala[15].

Further discussions on these and other related online test proposals[41–42] may be found in Abramovici et al.[17]. However the silicon area overheads required for online testing remain a serious drawback, being quoted as up to 40 % additional area for the strategy shown in Figure 6.10 and only slightly less for the previously discussed proposals. Nevertheless, the efficient layout of PLAs, particularly where a large number of inputs is involved, makes some form of online test viable in many circumstances. See further comments and additional research in References 45–47.

6.2.1.3 Built-in PLA self test

The recent emphasis on built-in self test for VLSI circuits so as to reduce or eliminate the need for complex ATPG programs and external test hardware has now influenced PLA architectures, particularly as PLAs are becoming increasingly used for random combinational logic requirements within VLSI ICs[48]. A number of proposals have now been published, and continuing research and development may be expected.

Many of the BIST concepts discussed in Chapter 5 have been considered for offline PLA built-in self test[48–59]. The early pioneering disclosure of Yajima and Aramaki[49], which predates some of the later proposals specifically using LFSR and BIST circuits, is shown in Figure 6.11. The test procedure is to scan a chosen initial pattern into the product line shift register and into the input decoder shift register, and apply a single clock pulse so as to load this data into the AND array. The resulting AND and OR parity checks and also the two final additional AND product terms are fed back to the feedback value generator, these feedback signals being known if no crosspoint faults are present. These feedback signals result in a new test pattern being fed into the shift registers and thence into the AND array so as to make a second test.

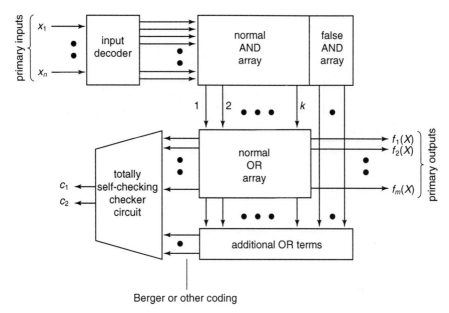

Figure 6.10   The online testable PLA of Mak, Abraham and Davidson[41]

This cycle is repeated for a total of $2n + 2k + 9$ cycles, at the completion of which the final pattern fed into the product line shift register is read out to the signature decoder.

If there is any crosspoint fault detected during a test, it will modify the future test patterns produced by the feedback value shift register, and hence at the completion of all the scheduled tests a wrong final answer will be present in the product line shift register and hence in the signature decoder. The correct (fault-free) sequence of test patterns and the final response can be determined for any given size of PLA, being independent of the functional dedication of the normal PLA outputs. The fault coverage of this strategy is shown to be all single crosspoint faults, all single stuck-at faults and all single faults in the additional self-test circuitry. However, multiple errors may be masked in the product line shift register, giving aliasing in the final output signature. Further details of this BIST method, including the derivation of the signals and signature when under test, may be found published[10,13,26,49], but the silicon area overhead is generally greater with this technique than in later BIST proposals.

Later self-test proposals include the following:

- Daehn and Mucha[50], which uses built-in logic block observation (BILBO) circuits;
- Hassan and McCluskey[51], which uses multiple linear feedback shift registers (LFSRs);
- Hua *et al.*[52], which is a BILBO development of Figure 6.7;

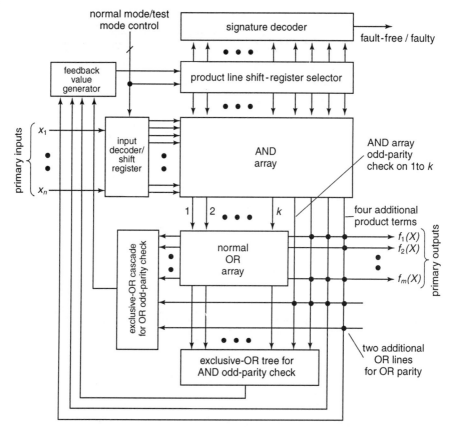

Figure 6.11 The offline built-in self-test strategy ATPLA of Yajima and Aramaki[49], omitting the clock details which load the two shift registers for clarity

- Fujiwara et al.[53], which is a development of Figure 6.11;
- Grassl and Pfleiderer[54], which employs shift registers and MISR test capture;
- Saluja et al.[35,55], which also incorporates shift registers and MISR test capture.

Details of these and further proposals may be found in the published literature[10,13,17,26,35,50-59], with comparisons of their test application length, test coverage and silicon area requirements in References 17 and 26. Here we will briefly review References 50 and 35 which will illustrate the principal developments.

Daehn and Mucha's BIST structure[50] is shown in Figure 6.12. The PLA is basically partitioned into three parts, namely the input decoder, the AND array and the OR array, with a BILBO type circuit between these three partitions. The test action is as follows:

(i) test the AND array by generating an input test set using BILBO 1, and capture and scan out the test signature using BILBO 2;
(ii) test the OR array by generating an input test set using BILBO 2, and capture and scan out the test signature using BILBO 3;
(iii) if it is possible to connect the PLA outputs and if necessary additional BILBO 3 outputs back to the PLA primary inputs, then BILBO 3 may be used to drive the output buffers/inverters and the input decoders, the test results being captured and scanned out from BILBO 1. Alternatively, if the PLA is buried, preceeding and following BIST circuits may be used to test these input and output circuits.

There are, however, certain unique features in these tests. The BILBO circuits used in autonomous test vector generation mode to supply test vectors to the AND array and to the OR array do not need to be maximum-length pseudorandom generators as discussed in Chapter 5, but instead can be nonlinear feedback shift registers (NLFSRs) as shown in Figure 6.12b and c, the output sequence generated by these circuits being as follows:

Circuit of Figure 6.12b	
outputs Q	outputs $\bar{Q}$
0 0 0 ... 0 0	1 1 1 ... 1 1
1 0 0 ... 0 0	0 1 1 ... 1 1
0 1 0 ... 0 0	1 0 1 ... 1 1
0 0 1 ... 0 0	1 1 0 ... 1 1
.	.
.	.
.	.
0 0 0 ... 1 0	1 1 1 ... 0 1
0 0 0 ... 0 1	1 1 1 ... 1 0
repeat	repeat

Circuit of Figure 6.12c	
outputs Q	outputs $\bar{Q}$
0 0 0 ... 0 0	1 1 1 ... 1 1
1 0 0 ... 0 0	0 1 1 ... 1 1
1 1 0 ... 0 0	0 0 1 ... 1 1
0 1 1 ... 0 0	1 0 0 ... 1 1
.	.
.	.
.	.
0 0 0 ... 1 1	1 1 1 ... 0 0
0 0 0 ... 0 1	1 1 1 ... 1 0
repeat	repeat

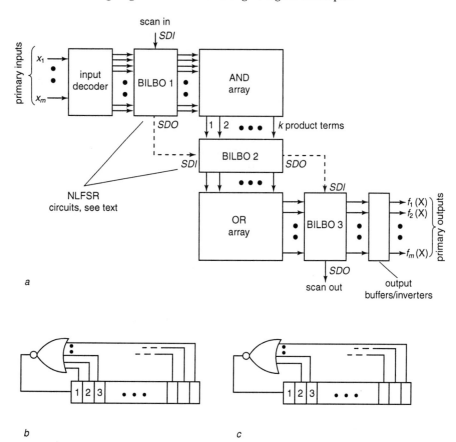

Figure 6.12   The offline built-in self-test strategy of Daehn and Mucha[50]
  a the architecture
  b the NLFSR generator circuit to test NOR/NAND structures
  c the NLFSR circuit to test input decoders

It will be appreciated that outputs Q of Figure 6.12b are the appropriate test vectors to test OR/NOR circuits, and outputs $\bar{Q}$ are the appropriate test vectors to test AND/NAND circuits. Therefore, depending upon whether the AND array of the PLA is in reality a NOR or a NAND structure, the product output lines of the array may be tested with this set of $2n+1$ test vectors. Similarly, the OR array may be tested with a set of $k+1$ test vectors The outputs of Figure 6.12b are appropriate for checking the input decoders of the PLA so as to ensure that all literals x_i and $\bar{x}_i$ are correct with no stuck-at or bridging faults.

The total number of test patterns required to test the PLA fully is therefore linearly dependent upon the number of inputs, product terms and outputs. The test signatures generated in the BILBO, however, are function dependent and have to be calculated for every PLA dedication. The signatures from each BILBO may be individually scanned out or cascaded together, but the possibility of aliasing is present. This probability, however, may be less than the theoretical value considered in Chapter 5, because all possible faults in the output functions are not equally likely in a PLA architecture as was the assumption used previously in calculating theoretical aliasing probabilities. All single crosspoint faults, single stuck-at faults and single bridging faults are claimed to be detected by these tests[17,50].

The BIST strategy of Saluja et al.[35] involves controllability of the input literals and the product lines as discussed in Section 6.2.1.1, see Figure 6.7, but now adds on-chip autonomous test vector generation and MISR signature compression to provide self test. The circuit schematic is shown in Figure 6.13. The test generator shifts a chosen sequence of 0s and 1s into the two shift registers so as to (i) individually energise the product lines leading into the array and (ii) individually control the logic values of the input data into the array, the control signal $C = 0$ being used to disable the normal x_i input data. The extra column in the AND array and the extra row in the OR array provide online function-independent parity checks if required, but the final MISR signature when under offline self test conditions will be a function-dependent test signature.

In reviewing these and the several other proposed BIST methods for PLA testing, it will be appreciated that shift registers, BILBO or modified BILBO test generators and multiple-input signature registers are commonly involved. Other means of both offline and online testing than specifically considered here may also be found, including:

- parallel-testable PLA architectures[60];
- divide and conquer testable PLA architectures[61];
- specialised layout-testable PLA architectures[37,62];
- random pattern testable PLA architectures[58,63–65].

The first three of the above may be found reviewed and evaluated in Reference 17; the final technique may be found fully discussed and referenced in Reference 58. Currently, the strategy which has the best reported fault coverage of all built-in self-test schemes is that of Fujiwara et al.[53] which uses shift registers and parity checks rather than BILBO/MISR circuit additions, and which requires a function-independent test sequence of $2k(n + 1) + 1$ test vectors. A further analysis of this proposal may be found in References 10 and 66.

To summarise, the uniform structure of PLAs has considerable advantages as far as testing is concerned, which has enhanced and will continue to enhance their attraction for many combinational logic requirements. The fault coverage problem for very large PLAs is considerably eased, with the possibility of achieving near 100 % fault coverage with minimal test generation and test verification complexity. Continued development should enable a target of, say, 10% silicon overhead to be achieved for large testable PLA designs, although there will always be a trade off between silicon

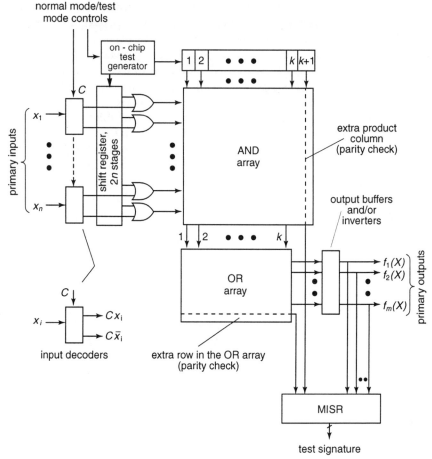

Figure 6.13 *The offline built-in self-test strategy of Saluja, Kinishita and Fujiwara*[35] *which is an extension of the test strategy shown in Figure 6.7*

overhead, time to test and other considerations such as partitioning or folding of the PLA structures.

6.2.2 Programmable gate arrays

Programmable gate arrays and particularly field-programmable gate arrays (FPGAs) have architectures which, like PLAs, are finding increasing application, most particularly in the custom IC field. Unlike PLAs, the majority of FPGAs contain storage (memory) elements as well as combinational logic, and hence the designation gate array is unfortunate. The feature of all PLAs and FPGAs, however, is a structured silicon layout consisting of a very large regular array of identical cells, sometimes referred to as tiles, the final interconnection (dedication) of which forms the custom circuit.

There has been and remains considerable development in the design and choice of the standard cell which constitutes the array architecture. The broad argument concerns whether the cell should be very simple, thus requiring the interconnection of a number of cells to make circuits such as shift registers and counters, or whether each cell should have a higher logic capability so that it may itself realise simple combinational and sequential requirements. This argument is concerned with what is now termed the granularity of the architecture[5,6,8], and is a subject of vigorous debate. A compromise that has been adopted, however, is to have two standard cells in a FPGA, one cell being entirely combinational logic gates and the other a clocked flip-flop circuit.

Publications dealing specifically with programmable gate arrays are widely available[1,5-8]; however, the original equipment manufacturer has to accept whatever choice of architecture is commercially available, and hence it is the vendor who should consider the testing implications of a particular architecture. Currently the principal vendors in this field include:

- Texas Instruments;
- Intel Corporation;
- Amtel Corporation;
- Xilinx;
- Motorola;
- GEC Plessey Semiconductors;
- Lattice Semiconductors Corporation;
- Concurrent Logic;
- QuickLogic Corporation;
- Algotronix;

and others. Takeovers and newly-formed innovative companies make this a continually evolving and dynamic field. Details of commercial products may be found in company literature such as References 67–75; Figure 6.14 illustrates typical FPGA cell designs.

Since the choice of cell available in commercial FPGAs has been chosen to enable any OEM circuit requirement to be realised, it is clearly possible to dedicate these devices to provide appropriate controllability and observability for final test purposes, or to build in scan-path or self-test mechanisms if necessary. Possibly the size of commercially-available FPGAs up until recently has not been sufficiently large to warrant full DFT mechanisms such as BILBO to be built into the dedicated product, but this may change with increasing FPGA capabilities. Some benchmark test results have been published comparing the performance and silicon area overheads building scan testing into Actel and Xilinx FPGAs, these tests using the normally-available cells in the two architectures for these comparison studies[76].

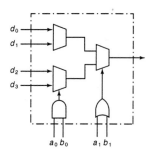

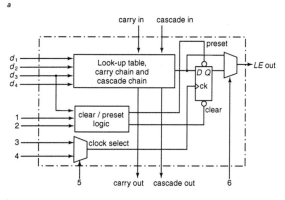

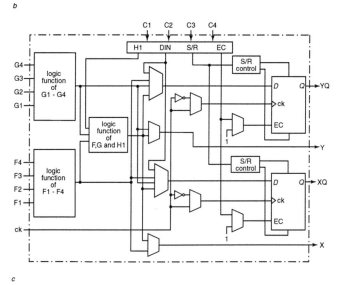

Figure 6.14　Example cells used in programmable gate arrays. The trapezoidal elements are multiplexers in all cases
　　　　　　a Actel Corporation multiplexer-based cell; a sequential cell version adds two latches to this configuration
　　　　　　b Altera Corporation FLEX 8000 cell
　　　　　　c Xilinx Corporation XC 4000 configurable logic cell (CLB)

However, very many FPGA vendors have considered the implications of scan test for completed systems (board-level testing), and as a result the primary inputs and primary outputs of many large FPGAs now have boundary scan capability built into their I/O cells[67–71,77]. The IEEE standard 1149.1, 'Standard test access port and boundary scan architecture', previously detailed in Section 5.4 of Chapter 5, is therefore available in most large FPGAs. The vendors' design of the I/O cells provides all the facilities required by standard 1149.1, typical I/O circuit details being shown in Figure 6.15.

The use of FPGAs does not therefore introduce any new dimensions or concepts into the testing requirements over and above those which we have considered in previous chapters. However, with increasing size and capabilities, the need for DFT will become more important than perhaps it has been up to now when using these versatile components.

6.2.3 Other programmable logic devices

The PLAs and FPGAs discussed in the preceeding sections together constitute the programmable devices of greatest potential for digital system design work. Leaving aside memory devices which will be considered separately in the following section, there are a number of further less complex but useful off the shelf programmable devices for OEM use, which we will briefly consider. These include the following:

- programmable array logic devices (PALs);
- programmable logic sequencers (PLSs);
- programmable diode matrix devices (PDMs), which must now be considered obsolete;
- erasable programmable logic devices (EPLDs);
- mask-programmable logic devices (MPLDs);
- gate-array logic devices (GALs);

and other possible company-specific variants or terminology. Collectively all these products may be referred to as programmable logic devices (PLDs), but the exact coverage of the term PLD is a little unsure. An alternative term, complex programmable logic device (CPLD), has recently been introduced[77], but this has generally been applied to the larger PLA and FPGA architectures rather than to the (usually smaller) devices listed above.

Details of these latter devices may be found in the literature[1,3,8,9,69,74,78–80]. In general, they are all based upon some AND/OR combinational logic structure, most often with the AND array programmable but with fixed (non-programmable) OR outputs, and with the addition of clocked D-type flip-flops or other storage/memory circuit elements in the majority of cases. Figure 6.16 illustrates a particular 12-input CMOS electrically erasable/programmable PLD, with ten primary outputs which may be programmed to be direct combinational outputs or latched/flip-flop outputs as desired.

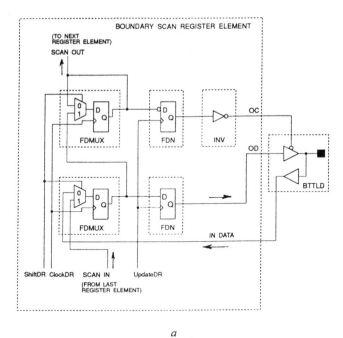

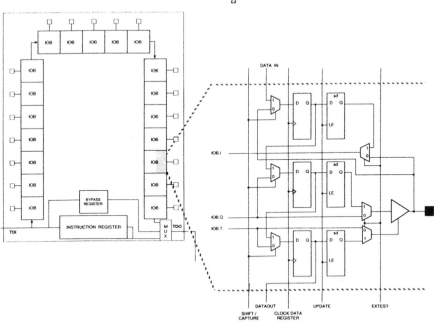

Figure 6.15 Example boundary scan provision on commercial FPGAs
  a Amtel Corporation boundary scan register with two data bits, cf. Figure 5.20b
  b Xilinx Corporation XC 4000 boundary scan cell with three data bits
  (Acknowledgements, References 69 and 70)

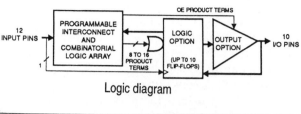

Logic diagram

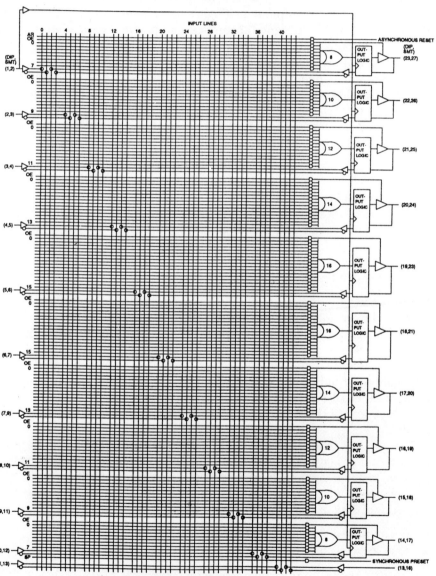

Functional logic diagram

Figure 6.16 The Amtel ATF21V 10B/BL CMOS PLD, with variable product terms ranging from eight-input to 16-input AND functions and programmable output function logic (Acknowledgements, Reference 69)

The size of this range of programmable devices is typically of the order of ten to 20 primary inputs and somewhat fewer outputs, although some devices with 60 or more I/Os are available; packing is in standard dual-inline (DIL) packages, plastic or ceramic leadless chip carriers (PLCC or LCCs), pin-grid arrays (PGAs) or other form of standard IC packaging[9]. All are normally considered to be MSI or LSI circuits rather than VLSI. No special built-in test mechanisms are therefore provided by the vendor in these architectures, except possibly in the output sequential elements—see below. Many vendors do, however, provide very good theoretical and practical design information in their data books and supporting literature, e.g. Reference 80, including high speed PCB design data and testability considerations.

However, since the majority of PLD applications are sequential, using the storage elements provided in the architectures, the flip-flop or latch circuits are often relatively complex circuits so as to provide controllability and observability for test purposes. Figure 6.17 illustrates one such design which provides the following facilities:

(i) combinational output or latched output or clocked D-type output;
(ii) power-up to $Q = 0$ state when the d.c. supply is first connected, thus providing known initialisation;
(iii) preload, allowing any arbitrary state value to be entered into each storage element, with transitions from any state to the next state being individually testable with single clock pulses;
(iv) observability of the Q output state of all storage elements by suppression of any combinational data at the outputs.

From (iii) and (iv) in particular it will be appreciated that controllability and observability of all the sequential elements is available, and hence test sequences may be shortened compared with the situation where the dedicated PLD has to sequence through all possible combinational and sequential states of the circuit. Indeed, with care it may be possible to segregate the combinational logic tests from the sequential logic tests, as is done in other DFT strategies. Hence no provision for IEEE standard 1149.1 boundary scan has yet been provided on any of these off the shelf PLD products.

The programmable logic devices now reviewed cover the specification areas shown in Figure 6.18a. Overlapping this area and extending to higher gate counts and potentially slightly higher operating speeds are the two further digital logic products, not off the shelf, which an OEM may employ for product designs. These are:

- uncommitted gate array circuits;
- standard cell circuits;

the former requiring only the interconnection mask(s) for their commitment, the latter requiring a full mask set. Both represent well established products for OEM use, usually for medium to large volume requirements

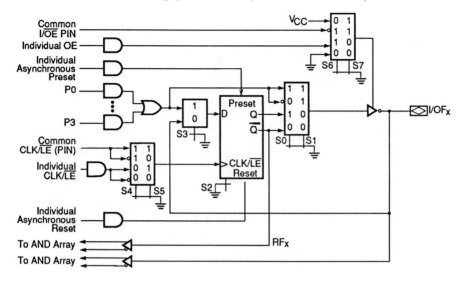

Figure 6.17   The output macrocell of Advanced Micro Devices PALCE 29MA16H-25 CMOS PLD (Acknowledgements, Reference 80)

as indicated in Figure 6.18b. Details may be found in many publications[9,11,79,81–83]*.

Although most gate array and standard cell products have strongly structured architectures, usually consisting of regular rows of active devices separated by wiring channels, unlike PLAs there is nothing in their structure which gives rise to any special test strategy, fault model or DFT consideration. The vendors of such parts usually have available in their libraries the necessary details to allow the ready incorporation of shift registers and/or higher-level macros, and therefore all the standard ingredients which may be wanted for test and DFT purposes are usually available. However, the design of buried PLA macros using conventional channelled gate array architectures is not easily possible, but more recent sea of gates devices which do not have this channelled architecture have made this and RAM and ROM structures more feasible in gate arrays. It must be appreciated that all these products are mask-committed circuits and not programmable in the sense that we have previously been considering, and therefore testing based upon crosspoint faults as in Section 6.2.1 is not at all relevant.**

---

* ASIC (application specific integrated circuit) is no longer a satisfactory term for unique OEM custom ICs, since IC vendors now market standard off the shelf ICs for specific applications which may be referred to as ASICs or ASSPs (application specific standard parts). The alternative terminology USIC (user specific integrated circuit) is preferable for describing gate array and standard cell ICs.
** A further recent confusion in terminology is the introduction of the term MPGA (mask programmable gate array) for what is usually just termed a gate array[7]. It is a pity that some internationally agreed standard designations cannot be established.

328 *VLSI testing: digital and mixed analogue/digital techniques*

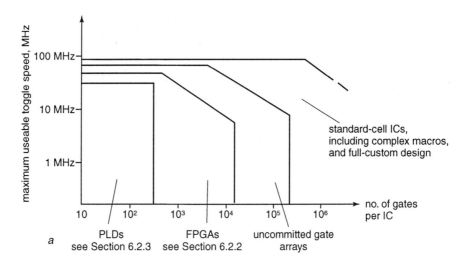

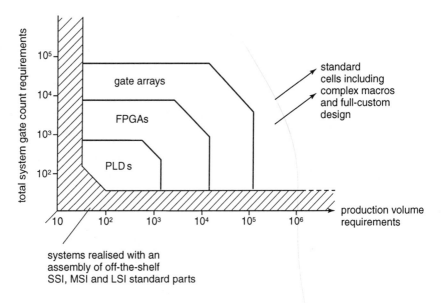

Figure 6.18 Typical technical and economic limits for the various categories of programmable and custom digital ICs. The ranges shown are subject to continuous physical and commercial evolution, and therefore represent a broad picture of these parameters
a technical maximums
b economic regions of advantage

6.3 Read-only memory testing

Reverting back to conventionally programmable devices, read-only memory (ROM) and programmable read-only memory (PROM) circuits form an essential part of all processor and microprocessor-based systems. The functional testing of these circuits therefore is of supreme importance.

The full hierarchy of memory devices, including random access memories, is shown in Figure 6.19. Details of the technology and fabrication of these devices may be found elsewhere[9], but need not concern us here since testing depends upon their dedication rather than upon their precise circuit and manufacturing details. We will, however, mention specific features when relevant to testing considerations.

Recall that in spite of its designation the read-only memory is a purely combinational circuit with n primary inputs (the address inputs), and m primary outputs (the m-bit output word), where both n and m may be tens of inputs and outputs in the largest circuits. The architecture consists of:

- input decoders, which effectively produce all 2^n combinations of the primary inputs, the minterms;
- the programmable AND array, usually termed the memory array or the memory matrix or the storage array;
- fixed output OR logic for each of the m output lines—this may be wired-OR rather than a separate n-input logic gate;
- a final enable gate so that the outputs may all be disabled to logic 0.

Each output is therefore in a sum of minterms form, e.g. output bit 1 is, say:

$$[\bar{x}_1\bar{x}_2\bar{x}_3\bar{x}_4 x_5 \bar{x}_6 x_7 x_8 + \bar{x}_1\bar{x}_2\bar{x}_3\bar{x}_4 x_5 x_6 x_7 \bar{x}_8 + \ldots\ldots]$$

if eight inputs are present. Further circuit details may be found elsewhere[4,9,84].

Whatever the precise physical mechanisms present in the ROM architecture, the final test of the dedicated memory must be to ensure that the correct literals x_i, $\bar{x}_i$ appear in each output. This involves checking not only that all required minterms appear in every output, but also that no unwanted minterms appear. The only way of ensuring this is to step through the full truthtable of the outputs, that is to apply all 2^n input test vectors and check for the correct logic 0 or logic 1 response on each test vector on all outputs. Therefore we require a generator which will provide all 2^n input test vectors; the order of application is not significant, and hence a binary counter or a LFSR circuit with the all-zero state (see Figure 5.29) can be used.

The manufacturers of ROMs which are vendor-dedicated to the requirements of a particular customer may use very expensive computer-controlled VLSI testers, running at clock speeds of several hundred megahertz. When the OEM has to do the tests, then provision for the necessary exhaustive tests must be made; with PROMs the OEM's own programming hardware resource will invariably incorporate test means which

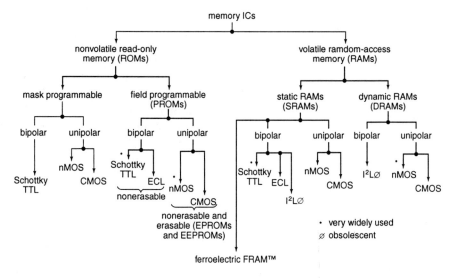

Figure 6.19 The families of commercially available memory circuits

will check the dedicated device exhaustively and functionally, but with buried memory within a custom IC appropriate controllability and observability must be provided. Notice that the normal system logic signals are unlikely to provide an exhaustive test of a ROM or PROM memory device, or enable all the outputs to be directly monitored, and therefore attention must be given to this problem in order to ensure a comprehensive check on the memory patterns.

Some consideration has been given to compressing the m-bit output response of a ROM or PROM when under exhaustive test in order not to have to store the full fault-free responses against which the outputs under test are checked. These considerations have included weighted syndrome test signatures, cyclic redundancy check signatures and parity check signatures, but to date these have largely been academic exercises and have not been adopted for mainstream memory testing[12,85-89]. However, there has been a commercial introduction of serial scan paths in memory circuits, briefly forecast in Section 5.3.2 of Chapter 5, giving what has been termed serial shadow register (SSR™) configurations. In general this has only been applied to relatively small ROMs and PROMs of MSI complexity, and not to the much larger versions which are now in widespread use. (SSR is a registered trademark of Advanced Micro Devices, Inc.)

Figure 6.20 illustrates a serial shadow register 4096 × four-bit PROM. Under normal mode conditions the four-stage shadow register *S-REG* is inoperative, and the PROM array operates normally controlling the output registers and hence the four-bit output word. However, in test mode the following resources are available:

(i) scan-in of test data to the shadow register and scan-out of the test response data;
(ii) replacement of the normal PROM array output by the data bits loaded into the shadow register;
(iii) loading of the normal PROM array outputs back into the shadow register for subsequent scan out and verification.

In total, therefore, system data may be captured and serially scanned out for test purposes, or test data scanned in to replace normal system data.

Further details of serial shadow registers may be found published[9,90-92]. The same concept has also been applied to random-access memory and PLA circuits of MSI complexity, and also to separately packaged serial shadow register ICs which may be built into PCB data buses to provide the above test facilities. Figure 6.21 illustrates such a circuit, which is packaged in a 24-pin dual-inline package.

However, the increasing use of LSI and VLSI circuits and the adoption of IEEE standard 1149.1 boundary scan has superseded these earlier MSI developments, although the principles involved remain entirely valid for the test of small assemblies containing ROM, PROM and other circuits.

6.4 Random-access memory testing

Unlike ROMs, random-access memories do not contain fixed data, but are repeatedly written to and read from during their lifetime service. There are, therefore, no initial input/output logic relationships to test for and confirm as in dedicated ROM and PROM circuits, but instead two aspects need to be considered, namely:

(i) some initial test or tests which attempt to confirm that each individual RAM cell is capable of being written to (logic 0 and logic 1) and read from;
(ii) some online test or tests to confirm that the overall RAM structure is continuing to operate correctly.

It is clearly not feasible to perform an initial test which will exhaustively confirm that any memory pattern can be stored and read correctly, since with n inputs and m outputs there are $2^{2^n} \times 2^m$ internal memory patterns theoretically possible, including trivial and identical functions.

The family of RAM circuits has been illustrated in Figure 6.19, the two classifications being static RAMs (SRAMs) and dynamic RAMs (DRAMs). Details of their specific fabrication technologies may be found elsewhere, but need not concern us here except to recall that both are characterised by the smallest possible device size and the highest possible packing density on the chip, with dynamic RAMs being dependent upon stored charges to differentiate between logic 0 and logic 1 data[9,10,14,15,89,93,94].

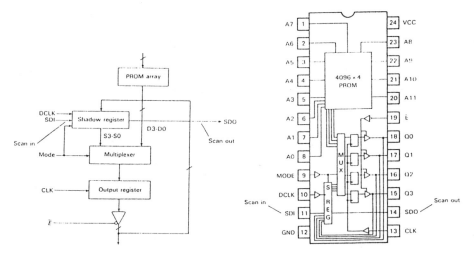

Figure 6.20 *The principle of internal scan test applied to individual memory ICs, now largely superseded by more general boundary scan techniques as memory and IC size have increased*

However, because of their structure the faults which may be experienced have certain unique effects which the testing should address. These include:

- one or more bits in the memory stuck-at 0 or 1;
- coupling between cells such that a transition from 0 to 1, or vice versa, in one cell erroneously causes a change of state in another (usually adjacent) cell—a pattern sensitive fault;
- bridging between adjacent cells, resulting in the logic AND or logic OR of their stored states;
- in dynamic RAMs unwanted charge storage which may result in a logic 0 being read as a logic 1 after a string of logic 1s has been read;
- also in dynamic RAMs, inadequate charging or discharging of memory cells to/from their logic 1 state, or undue leakage of charge between refresh periods causing loss of logic 1 data;
- temporary faults caused by external interference which result in incorrect data being generated within the memory, but with no lasting failure in the circuit.

There may also be gross faults in the input and output circuits, but these should be readily picked up by tests which are seeking to confirm the correct action of the individual memory cells, and therefore we shall have no need to consider these failings here. All these (and other) faults may be classified as hard faults if they are permanent, or soft faults if they are transitory; clearly only online testing (see later) can catch the intermittent soft faults. All these memory faults have been very extensively studied and documented, and continue to be of supreme importance as circuit geometries shrink and circuit densities increase[93–100].

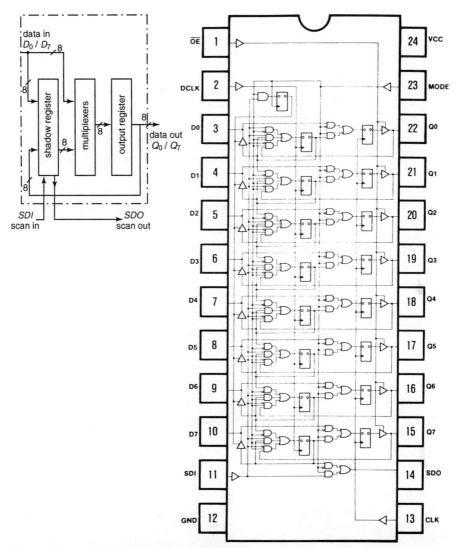

Figure 6.21 The AMD 29818 (MMI 74S818) eight-bit standalone bus-interface shadow register IC, now superseded by IEEE standard 1149-1 circuits such as shown in Figure 5.20

6.4.1 Offline RAM testing

A number of algorithms have been developed to test read-only memory arrays for hard faults. Each involves writing a 1 in each cell and performing a readout, and writing a 0 in each cell and performing a further readout, in some chosen sequence of tests which embraces each memory cell in the circuit. These tests are therefore functional tests, but they are designed to be appropriate for the hard faults that may occur. There are several basic test patterns which may be used for RAM testing, including:

- marching patterns;
- walking patterns;
- diagonal patterns;
- galloping patterns;
- nearest-neighbour patterns.

The shifting write/read patterns of the marching-one test sequence are shown in Figure 6.22a. The array is first filled with 0s, and then the first address is checked by confirming it is 0, followed by reading in a 1 and checking that the readout is now 1. The next address is then tested in the same manner, leaving the first address at 1. This procedure of progressively filling up the array with 1s continues untill the memory is full. The inverse, the marching-zero test sequence, is then performed, progressively setting the memory cells back to all 0s and checking the latest zero-valued cell at each step. Note that this procedure and the procedures immediately below relate to a single output bit of the RAM. For m outputs (an m-bit output word) the test procedure must be applied to all the memory bits, but this may be done simultaneously in parallel for all the separate m outputs if it is assumed that there is no fault coupling between outputs. This is a reasonable assumption, since each bit of the m-bit output word has its own memory array structure.

This test sequence requires $6N$ write/read cycles, where N is the total number of memory cells per output bit, and covers all individual stuck-at cell faults, most pattern-sensitive faults and address decoder faults. An additional check that all cells are at 1 at the completion of the marching-one test and all are at 0 at the completion of the marching-zero test may be done.

The walking pattern test procedure is shown in Figure 6.22b. The array is initially filled with all 0s and read out. A single 1 is then written into the first address and all locations except this one are read out to confirm that they are still 0. The single 1 cell is then verified. This action is repeated with all remaining cells being individually set to 1 against a background of 0s, and finally the whole procedure is repeated with a single 0 at each location against a background of 1s. This test is termed the WALKPAT test, and is particularly effective against recovery faults in the output sense ampifier which make it unresponsive to a logic 1 in the middle of a string of logic 0s or vice versa.

A variation on this test is the diagonal walking test shown in Figure 6.22c, which loads a complete diagonal of 1s into the array against the background of 0s, and then walks this diagonal through the array. This is repeated with a diagonal of 0s against a background of 1s. This variation requires fewer write/read cycles than WALKPAT, but does not test the sense amplifier quite as rigorously as the single 1 in all 0s or single 0 in all 1s of the WALKPAT test.

A yet further variant on WALKPAT is GALPAT, which has a so-called galloping-one and a galloping-zero test sequence. The write sequence is identical to that shown in Figure 6.22b, but the read actions read all the memory locations in order from the first one to the last one on each test.

Other variants have been proposed including galloping-row and galloping-column test sequences. However, the nearest-neighbour test sequence is slightly dissimilar, being designed specifically to detect local abnormal interactions between adjacent cells. It will be seen that in WALKPAT and GALPAT large blocks of 0 (or 1) cells remote from the single 1 (or 0) cell are repeatedly checked, and hence it would be far more economical to identify the cells in the immediate vicinity of the single 0 (or 1) bit, and verify the fault-free nature of this cluster of cells at each step. This concept is illustrated in Figure 6.22d, and clearly involves far fewer read operations than other walking or galloping proposals, but requires a more complex addressing test sequence.

Details of these and other ROM testing proposals may be found published[13,14,88-90,100-105]. However, as memory size has grown, so the time to test using these methods has become increasingly unacceptable. It may be shown that the number of read operations required by galloping tests is of the order of $O(N^2)$, where N is the number of memory cells per bit output as previously. As an example, consider a 64Mbit RAM tested by GALPAT which requires $4N^2$ test cycles[14], then with a tester speed of 100 MHz a test time of 1896 days (over five years) is involved. Such procedures are therefore only feasible for small memory circuits of kilobit capacity, or by partitioning the full RAM circuit into a number of smaller partitions which may be separately and possibly simultaneously tested.

Details of several memory partitioning concepts have been published. They include the following proposals:

- Jarwala and Pradhan's TRAM architecture as shown in Figure 6.23a, which splits the memory into 16 blocks for test purposes[106];
- Sridhar's architcture as shown in Figure 6.23b, which uses a Kbit wide parallel signature analyser (PSA) to access Kbits of memory simultaneously[107];
- Rajsuman's STD architecture as shown in Figure 6.23c, which partitions the memory address decoder into multiple levels, with the memory cells being grouped accordingly[108].

Further details of these proposals may be found pubished[14,106-108], but to date no one partitioning method has become universally accepted.

All the above test procedures are applicable to both static and dynamic RAMs, although the latter may have built-in wait perods during the tests to confirm the dynamic memory. However, if the types of faults considered likely are limited to fixed (stuck-at) cell faults, pattern sensitive faults being considered unlikely or adequately covered by the stuck-at tests, then marching tests which involve $O(N)$ rather than $O(N^2)$ tests are generally considered as adequate test strategies for very large RAMs[101].

There remains one class of static RAM listed in Figure 6.19 which to date is

336 *VLSI testing: digital and mixed analogue/digital techniques*

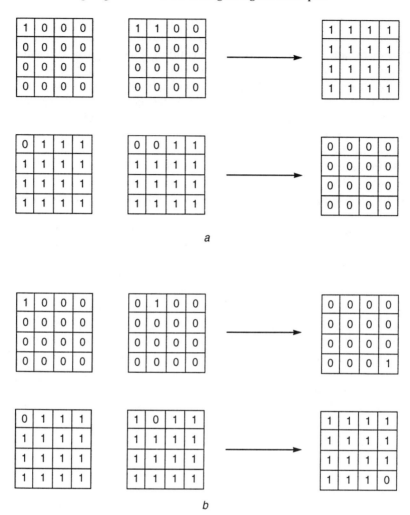

Figure 6.22 Bit test patterns for RAM memory arrays
  a marching-one/marching-zero write pattern
  b walking-one/walking zero write pattern

not represented in the literature on ROM and RAM testing. This is the non-volatile ferroelectric random-access memory (FRAM™), which employs the polarisation of a ferroelectric material such as lead-zirconate-titanate (PZT) as memory cells[9,109,110]. It is not known what the exact fault mechanism(s) may be in these memories, although it is likely that stuck-at conditions and faults in the write/read circuits will be the functional fault effects principally encountered. Test sequences which involve $O(N)$ rather than $O(N^2)$ tests are therefore probable. Dynamic faults will not be present.

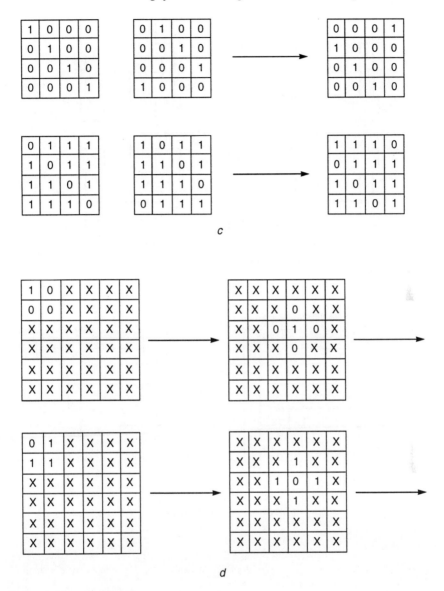

Figure 6.22   Continued
  c diagonal walking write pattern
  d nearest-neighbour write pattern, X = don't care

6.4.2 Online RAM testing

Hard faults in random-access memories have received considerable theoretical attention when considering (i) the probability of fault detection using random test patterns and (ii) the propagation of faulty data through RAMs. Details may be found in Bardell *et al.* and the references contained

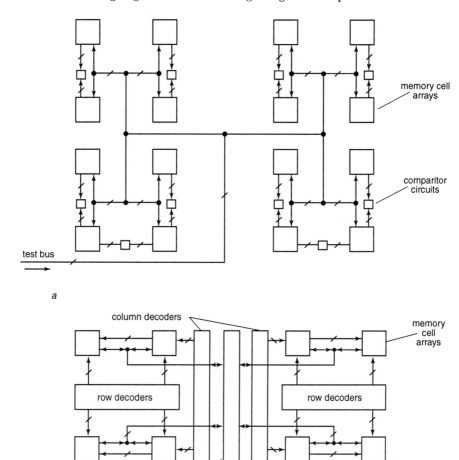

Figure 6.23 Some structured partitioning proposals for RAM testing, omitting the control details for clarity

a TRAM architecture, using a test bus for input test data and comparitors between the RAM partitions

b Sridhar's partitioning, using a parallel signature analyser for input and output test data to the partitions[107]

therein[16]. However, it generally proves impractical to estimate any worthwhile theoretical probablities for intermittent faults in memory circuits. What is known is that with the shrinkage of device geometries, the memory cells in RAM circuits become more susceptable to α-particle interference, since the

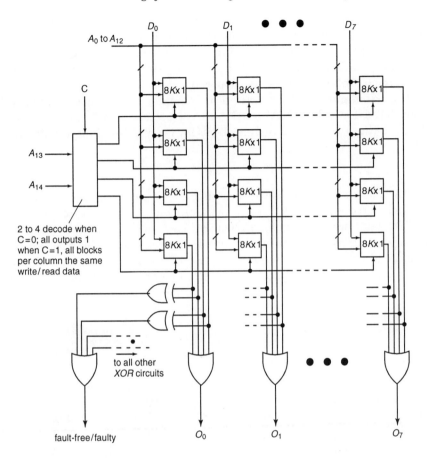

Figure 6.23   Continued
  c Rajsuman's partition of a 32K × 8 bit memory, using 2 to 4 address decoder with test override to select the 8K × 1 bit partitions

storage capacitance per cell to record a logic 1 bit becomes extremely small, and therefore the soft errors produced by this and other external interference become more frequent. An early figure of a soft error rate of about 0.1 % per 1000 hours for 64 Kbit RAMs was quoted in the early 1980s, but this is now likely to be considerably better with improvements in packaging and other fabrication developments[111-114].

Some online means of test of RAM circuits therefore becomes necessary if soft faults are to be detected and ideally corrected, and hence the coding techniques discussed in Chapter 4, Section 4.6.1, become relevant in this context. We are, therefore, considering once again the left-hand branch of Figure 4.10 and the general schematic of Figure 4.11.

340 *VLSI testing: digital and mixed analogue/digital techniques*

The additions required in RAM architectures for online testing will necessitate further output code bits as indicated in Figure 6.24. Recall from Chapter 4 that single error detection in a word can be accomplished by the addition of a simple parity check, but error correction requires the addition of more check bits for this to be achieved.

The Hamming code discussed in Chapter 4 has been extensively considered for online RAM tests[115], as was forecast in this earlier chapter. This requires the addition of k check bits for every m bits of information data, where:

$$2k \geq m + k + 1$$

plus a final overall even-parity check bit if it is desired to distinguish between a single error-correctable (SEC) bit fault and a double error-detectable (DED) fault which is not correctable. The increase in word length to incorporate these check bits was given in Table 4.5; for 16-bit data words six additional check bits are required to give SEC/DED action, for 64-bit data eight additional check bits are required. The percentage memory overhead to incorporate Hamming code correction therefore decreases as the word

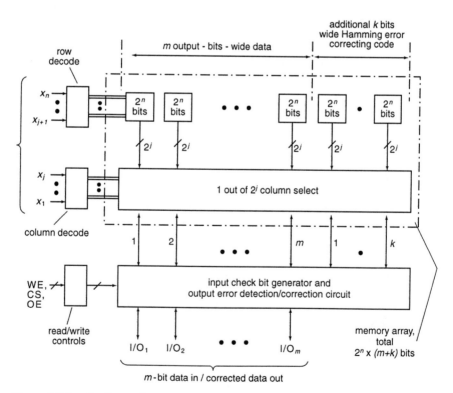

Figure 6.24 *Outline schematic of the online single error correction/double error detection (SEC/DED) Hamming code RAM circuit*

size increases, and hence is an attractive means of providing online RAM checking. It does, however, rely upon the integrity of the complex error generation/detection/correction circuitry at the data I/Os, plus of course the additional write-in and readout of the check bits on every write/read cycle.

Other proposals for RAM online test have been considered, including variants by Elkind and Siewiorck, Osman, Tanner, Su and DuCasse, and others[116-120]. Berger coding, see Chapter 4, has been used for the error detection/error correction as well as Hamming codes. Details may be found in the published literature. However, a somewhat different online test strategy was proposed by Hsiao and Bossen[121], being based upon the mathematical characteristics of Latin squares.

A Latin square is an $n \times n$ array of the elements x_1, x_2, ..., x_n, where x_i are normally integer numbers, such that each row and each column contains some permutation of all x_1, x_2, ..., x_n. A pair of Latin squares is said to be orthogonal if, when the corresponding positions in each square are considered, the ordered pair of numbers x_i, x_j in every position is unique, the same ordered pair never appearing more than once. This is illustrated in Figure 6.25a for the simple case of $n = 4$. For any n entries there exists $n - 1$ possible Latin squares.

The $n - 1$ copies of an orthogonal Latin square can be realised by the use of a maximum-length linear feedback shift register (LFSR) characterised by a primitive polynomial of degree r, where $r = \log_2 n$, the LFSR outputs being used to skew the normal address decode. The simple case for $n = 4$ ($r = 2$) is shown in Figure 6.25b, where it will be seen that the two inputs A_0 and A_1 can be changed to all the other Latin square orders by the pseudorandom output of the LFSR. The disclosed procedure for online RAM testing using this principle is therefore as follows, still considering a 4×4 square for simplicity.

A normal write/read of 16 memory cells C_0 to C_{15} is shown in Figure 6.26a, with Hamming single error correction/double error detection (SEC/DED) on the complete array. Any single bit error will be detected and corrected as in Figure 6.24. However, if two bits in a row, say C_9 and C_{11}, are in error, then detection but not correction will be present. If now this detection signal is arranged to reload the register with the LS_1 skew, see Figure 6.26b, then these two faults will be moved to different rows, and single error correction will now be possible.

However, should this shift, say, C_9 to a row which already has a single bit error, say C_3, then a double error detection will again arise. A further shift of the register to LS_2 or LS_3, see Figure 6.26c and d, will reduce this to correctable single-bit errors. If all the bit errors occur in just one row, then if detected it is always possible to disperse these errors into single error correctable rows, even if all n bits in the one row are in error.

It is clear from all the above discussions that it is possible to formulate various means for online error detection and/or error correction in RAM arrays. However, the question which must be considered is whether these

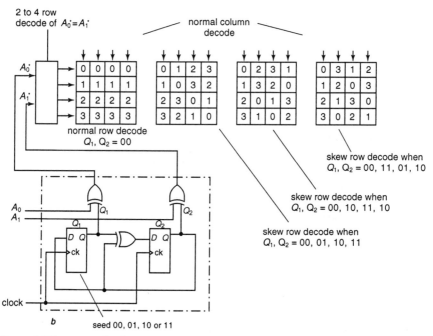

Figure 6.25 $n \times n$ Latin squares
  a the three possible Latin squares of $n = 4$
  b circuit to skew the row address decoding 0 to 3 so as to give the three Latin square positions

strategies, which fall short of, say, triple modular redundancy, truly increase system reliability. Recall our considerations of device failure rates, λ, and reliability, $R(t)$, in Chapter 4, and let us apply this theory to the problem of RAM cell failures.

Assuming that the failure rate of individual RAM cells is constant, having the value λ_i independent of any other factors, then the reliability of an individual cell, that is the probability that it remains functional over a given time t, is given by:

$$R(t)i = e^{-\lambda_i t} \qquad (6.3)$$

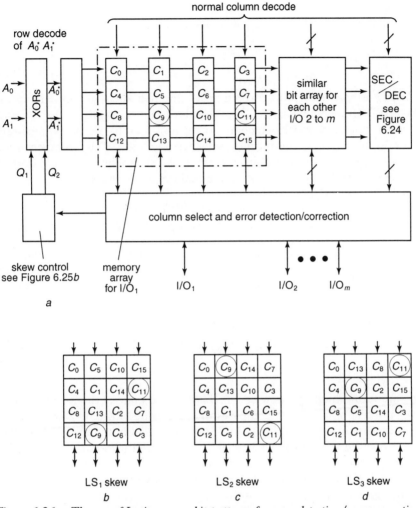

Figure 6.26 The use of Latin square bit patterns for error detection/error correction
a normal 16-bit (16 × 1 bit word) RAM storage
b skew LS_1 on a
c skew LS_2 on a
d skew LS_3 on a

where λ_i is the failure rate per hour (or 10^3 hours). The reliability of a single m-bit word is therefore:

$$R(t)_m = [R(t)_i]^m \qquad (6.4)$$

The failure rate of the read/write circuitry surrounding the memory array also has a failure rate, say λ_s, giving a reliability of:

$$R(t)_s = e^{-\lambda_s t} \qquad (6.5)$$

and hence the overall reliability of the complete RAM with 2^n m-bit words, without any error detection/error correction, is given by:

$$R(t)_{overall} = [R(t)_s].[R(t)_m]2^n$$
$$= R(t)_s\{[R(t)_i]^m\}2^n \tag{6.6}$$

When error detection/correction is applied such as the Hamming SEC/DED strategy of Figure 6.24, then the number of bits in the memory increases and additional detection/correction circuitry becomes necessary. However, a single bit failure can now be accomodated. If the number of bits per output word increases from m to $m + k$, the reliability of a single word in the memory with not more than one bit failure is now given by:

$$R(t)_{m+k} = \{[R(t)_i]^{m+k} + [m+k].[1-R(t)_i].[R(t)_i]^{m+k-1}\}$$
$$= \{[m+k].[R(t)_i]^{m+k-1} - [m+k-1].[R(t)_i]^{m+k}\} \tag{6.7}$$

The reliability of the subsidiary circuitry surrounding the array given by eqn. 6.5 now has some further circuit complexity with a reliability of, say, $R(t)_c$, and hence the overall reliability with SEC/DED checking and the correction of one bit error now becomes:

$$R(t)_{overall} = [R(t)_s].[R(t)_c].[R(t)_{m+k}]^{2^n}$$
$$= [R(t)_s].[R(t)_c].\{[m+k].[R(t)_i]^{m+k-1}$$
$$- [m+k-1].[R(t)_i]^{m+k}\}^{2^n} \tag{6.8}$$

The question which may now be asked is how the value of the reliability term in { } brackets in eqn. 6.8 compares with the value of the reliability term in { } brackets in eqn. 6.6. If the individual bit reliability $R(t)_i$ is high, then eqn. 6.8 will give an improved reliability compared with eqn. 6.6, provided that $R(t)_c$ is also reliable. However, if there is some inherent unreliability in the system then the increased circuit complexity of the SEC/DED strategy will not give good results. This is the same situation as considered in Chapter 4, where it was shown that redundancy and other strategies do not yield higher reliabilities unless the basic system has an appropriate reliability to begin with.

In the case of soft (intermittant) faults caused by external interference, then SEC/DED and similar error-correction strategies will clearly be beneficial, provided that there are no hard (permanent) faults already in the memory which have been masked by the SEC circuit. Herein lies the danger: to obtain the benefit of any single (or higher) correction circuit, the memory must be periodically checked to ensure that there are no hard faults present, without which check there is no guarantee of the continuing effectiveness of the error correction. Also if there is any hard fault present in the memory the continuing reliability of the memory is worse than it would be without any error correction/detection; for example, in a 16-bit word with six check bits, if any one bit is faulty this leaves 21 bits which may subsequently fail, which is worse than 16 bits of uncorrected memory.

The present general consensus of opinion is that error detection/error correction in random-access memories is not justified; the extra complexity of, say, the proposed strategy of Figure 6.26 is not worthwhile because of the overheads involved and the inherent risk of more random failures. Indeed, most manufacturers of memory circuits now concentrate on increasing the inherent reliablity of the device design and fabrication rather than increasing the total device complexity. However, for secure systems it now becomes necessary for the system designer to build in check resources, possibly software-based[15,122-124] in order to preserve the integrity of the overall system.

6.4.3 Buried RAMs and self test

Although standalone RAMs can be subject to marching or other functional tests which vendors consider adequate, particularly if the memory array is itself partitioned, the testing of buried RAMs in processors and custom ICs is more difficult due to limited controllability and observability of the memory I/Os. There is, of course, the attendant requirement of also testing the surrounding logic, and hence the buried memory is but one (possibly large) partition of the complete circuit. The OEM will not usually be called upon to decide the precise test sequence which will be applied to the memory; the original manufacturer will normally have a recommended test routine for its proprietary designs and technology realisations, which can be detailed to the OEM for system test[11].

Access to buried memory I/Os may be provided by multiplexing every I/O to accessible test pins and using external test generators and response checking hardware, but this entails considerable silicon area overhead and additional pins. However, if built-in self test is sought, then all the main components used in self test, namely scan paths, LFSRs, signature analysers and BILBO, may be considered, but the unique structure of the RAM will dictate the actual test vector sequences employed.

Many proposals have been published for RAM built-in self test. All entail at least two write and two read tests on all memory cells to check for stuck-at conditions, with the peripheral support circuitry also being tested by the chosen test set. The proposals of Nicolaidis[125] and of Sun and Wang[126] both employ LFSRs generating maximum-length pseudorandom sequences to drive the RAM input address decoders, the memory array being tested by marching-one and marching-zero bits on a background of solid 0s and 1s, respectively. The memory readout is captured and scanned out periodically for checking. Special precautions in these and similar BIST proposals are usually taken to ensure that the LFSR is itself fault free; also, intermediate test patterns may be scanned into the input LFSR test generator during the pseudorandom checks to test for specific memory failures.

Binary counters rather than LFSRs have been employed by Jain and Stroud[127] for embedded RAM testing, the binary counter producing not only all the memory addresses when in test mode but also the write-in bit data to the memory cells. A checkerboard pattern of memory data, see Figure 6.27, is used, which consists of writing and reading the whole memory with one checkerboard pattern of 0s and 1s, and then complementing this pattern for the following write/read sequence. The readout test results may be captured in a MISR for confirmation, or compared against the expected results which have a very simple known sequence under fault-free conditions.

A dissimilar built-in self-test strategy has been developed by Knaizuk and Hartmann[128,129], and also by Nair[130]. Here the memory addresses are divided into three groups G_0, G_1 and G_2, where

G_0 = addresses 0, 3, 6, 9, 12, 15, ...

G_1 = addresses 1, 4, 7, 10, 13, 16, ...

G_2 = addresses 2, 5, 8, 11, 14, 17, ...

The steps in the write/read test sequence are now as follows:

- step 1, write all 0s to G_1 and G_2;
- step 2, write all 1s to G_0;
- step 3, read all 0s in G_1;
- step 4, write all 1s in G_1;
- step 5, read all 0s in G_2;
- step 6, read all 1s in G_0 and G_1;
- step 7, write all 0s in G_0;
- step 8, read all 0s in G_0;
- step 9, write all 1s in G_2;
- step 10, read all 1s in G_2.

It has been shown that the above triplet grouping of write/read cycles will detect all single and multiple stuck-at faults in the memory array, the registers and the address decoders[129], but does not check that all bits in G_0 can be written to 0 or that all bits in G_1 and G_2 can be written to 1 since the memory bits in these locations may possibly take these values on power-up. However, this shortcoming can easily be overcome by adding an initialisation step[131], namely:

- step 0, write all 1s to G_1 and G_2, and all 0s to G_0.

Every bit address now experiences five write/read operations, namely:

write 0, write 1, read 1, write 0, read 0

for the G_0 address bits, and

write 1, write 0, read 0, write 1, read 1

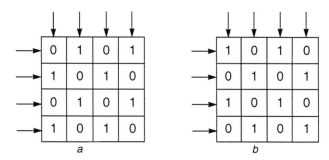

Figure 6.27 The checkerboard pattern of 0 and 1 bits for RAM testing
  a first write/read test per output bit
  b second write/read test per output bit

for the G_1 and G_2 bits, as shown in Figure 6.28a. Nair's proposals[130] are very similar, requiring four write/read cycles of all addresses, but need the additional initialisation write of all cells as above if all bits can be confirmed as being correctly written to as well as read from.

The overall schematic of this test strategy is shown in Figure 6.28b. It is reported[131] that this strategy has been succcessfully used in real-life circuits, and requires a low silicon area overhead. The test response circuit is a simple comparitor circuit which compares the read outputs against the expected fault-free response of blocks of 0s and 1s.

Several other embedded RAM test strategies have been proposed, the majority of which will be found referenced in References 13 and 88. It may be noted that in general these strategies consider only stuck-at faults in the memory arrays rather than more detailed pattern-sensitive faults. Also, the order of addressing the memory cells is largely irrelevant, the simplest possible hardware (binary counters or LFSRs or special counters) being used rather than more complex addressing sequences.

Overall, the built-in self test of buried RAMs must be considerd in the context of the test strategies for the remaining and surrounding partitions of the IC or system, and not in isolation. A further reference to RAM stuctures will therefore be found in the following section, Section 6.5, when considering microprocessor testing. The unique physical structure of RAMs will, however, usually entail very lengthy test sequences for fault-free write-to and read-from operation.

6.5 Microprocessors and other processor testing

The general-purpose microprocessor with its associated working memory is the most complex of all VLSI circuits to test, whether it is a standard off the shelf item, for example the Pentium microprocessor, or a partition within a custom VLSI integrated circuit. The broad hierarchy of all processors is illustrated in Figure 6.29; the processors for specific industrial, commercial or

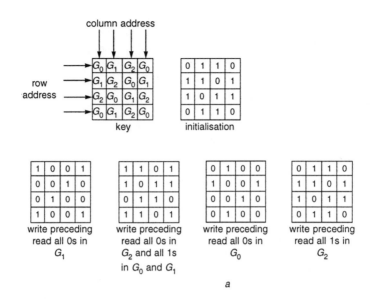
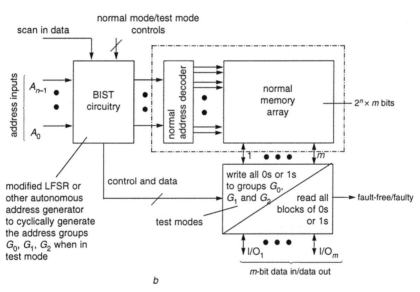

Figure 6.28 The built-in RAM testing strategy of Knaizuk and Hartman[128]
a write/read grouping G_0, G_1, and G_2 of the bit addresses for the simple case of 16 memory bits. Note the diagonal relationships
b outline schematic, omitting the normal read/write controls and clock signals, etc.

domestic applications, sometimes referred to as microcontrollers, may be very simple, possibly with only four-bit data and minimum memory requirements, but in all cases the OEM or end user will not normally know

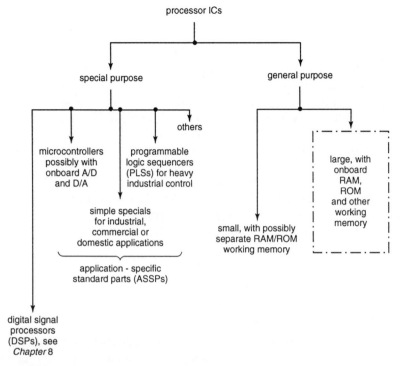

Figure 6.29   *The general processor and microprocessor hierarchies*

the detailed circuitry at the gate level and therefore testing strategies based on stuck-at fault modelling or gate-level considerations will usually be irrelevant. Some appropriate functional testing must be involved.

We must also distinguish between the tests which the vendors do on processors before dispatch, and the tests which OEM and/or end users need to do on individual processors or completed systems. Again, we do not usually know the exact test sequences which vendors do, but provision for testing will undoubtedly be built into the design of the more complex products. However, for other than very simple off the shelf products, the OEM will have vendor advice available on what tests are considered appropriate, these being nonexhaustive tests which will hopefully catch major faults in the processor circuitry.

For the DSP and special processors listed in Figure 6.29, it will usually be the case that the designer of the system in which these products are being used will formulate a series of functional tests to confirm key operations in the overall system. Incoming ICs from vendors may also be checked on receipt, using a special test set or even a final equipment as a test bed. We will have further comments on digital signal processing applications and test in Chapter 8 when we consider mixed analogue/digital test. We will, therefore, have few further comments to make here, except to note that every system will have its own unique test requirements, and that the system designer must be conversant with the needs of and strategies for testing, such as we have been considering in previous chapters of this text.

Turning therefore to the general-purpose microprocessor and considering its testing difficulties and test strategies. The general architecture which is involved is shown in Figure 6.30, all partitions being on-chip in the case of most 16, 32 and 64-bit processors. The program counter, stack pointer, index register and other operational circuits (not shown in detail in Figure 6.30) may be in PROM or PLA or other structured circuit partitions; the CPU block is the most complex, containing the arithmetic logic unit, tempory registers (accumulators) to hold the ALU data, instruction register, decode and control, and other details. A good educational introduction to microprocessors may be found in Wakerley[4] and in Hayes[84].

Neither vendor nor OEM testing at gate level is feasable with such complex architectures. Instead, key features are assumed to be covered by the following functional tests, not necessarily in this order:

- program counter test;
- scratchpad memory test;
- stack pointer and index register test;
- arithmetic logic unit and its associated register tests;
- further tests on control lines and other peripheral circuit tests.

The principal difficulty is that the normal controllability and observability of the circuit is limited to the control, address and data buses, which provide inadequate accessibility as processor size increases unless test access is built in at the design stage. Different microprocessor design teams may have dissimilar detailed test strategies[132–140], but all now involve some combination of scan-test, boundary-scan, built-in self-test and possibly I_{DDQ} tests. Earlier attempts to model complete microprocessor architectures, such as the S-graph model in which each register in the architecture is represented as a node with data flows between all the nodes in an assembled graph[14,141], did enable test patterns to be automatically generated for eight-bit microprocessors based upon functional rather than gate-level tests, but a fault coverage of only about 90 %–95 % of possible stuck-at faults was reported. Details of this early work and other similar attempts at processor modelling may be found in the literature[140–145], but currently they have little impact on VLSI processor test strategies.

6.5.1 General processor test strategies

Given that accessibility has been provided or is present in the microprocessor architecture for test purposes, then an order of test for a complete microprocessor may be as follows:

(i) isolate all devices including the CPU from the system buses using the tristate outputs or other available means;
(ii) test all buses for stuck-at and bridging faults by driving each bus to logic 0 and logic 1;

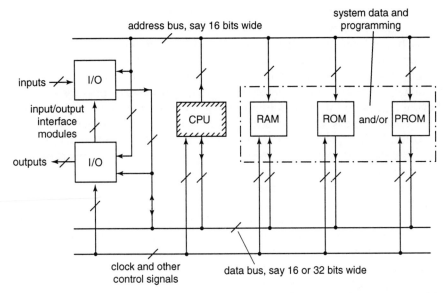

Figure 6.30 Outline schematic of a typical general-purpose microprocessor with onboard working memory in RAM and ROM, and possibly PLAs for random logic requirements

(iii) test each individual nonCPU partition by means of its own test strategy, reconnecting to the buses as necessary;
(iv) tristate again all nonCPU partitions, and then test the CPU by means of its own test strategy; this is likely to be the most complex of the partition tests, and will be considered further below;
(v) finally reconnect the CPU to normal, and carry out a final system test at maximum operational speed using some chosen set of operating instructions and known responses.

The above division of test activities may also reflect the division of design responsibilities for the complete processor, since separate small groups of design engineers are frequently responsible for particular partitions of the complete architecture. A considerable complexity in applying the final test signals and monitoring the responses and in the overall housekeeping is clearly involved in this divide and conquer test procedure. The OEM who purchases such a product, however, may not wish or be able to undertake this detailed test strategy, and hence it is possible for the vendor to specify a test program which will ensure that no major failings are present in the product. No guarantee of 100 % fault-free performance can ever be given, but the probability of correct operation can be made acceptably high.*

---

* This does not of course take into account processor design errors which may not show up until a very rare or unexpected computational exercise is being executed; on the other hand there may be manufacturing faults present which never show up because the system requirements never need or use the faulty node or nodes.

The precise functional test methods used by vendors will vary from vendor to vendor. The memory and random logic tests will not involve any new basic ideas beyond those which we have already considered; however, the key element, the core CPU, has received specific consideration which we will shortly review in Section 6.5.2. Boundary scan in accordance with IEEE standard 1149.1 is now extensively used in present-day microprocessors, having first been introduced in an off the shelf product by Motorola at the end of the 1980s in the 68040 general-purpose microprocessor. Later products by Intel, such as the 80486 onwards, and other vendors, also now incorporate boundary scan. BILBO for self test has also been incorporated, particularly for the nonmemory partitions where nonexhaustive built-in self test rather than fully exhaustive test may be used. The memory partitions which are connected directly to the data buses may be directly tested from the accessible I/Os and/or via the boundary scan, possibly by performing some chosen interchange of data between memories as an alternative to individual memory tests such as considered in Sections 6.3 and 6.4.

The large macros (megacells) provided by vendors for custom ICs may also have test provisions. For example, the Alcatel Mietec 0.7 µm CMOS standard-cell family MTC-22000 supports IEEE standard 1149.1 by providing the appropriate JTAG scan-path resources on large members of the cell library. Other application specific standard part (ASSP) processors such as the high performance Texas Instruments' DSP processor TMS.320.C50 also provide this IEEE industry standard test facility, which it is claimed provides an effective means of testing all on-chip partitions to a 99 % confidence limit[146,147]. In total, therefore, there is a considerable effort to provide for the testing of complex processor architectures, with vendor developments continuing in order to keep pace with the increase in circuit capabilities. The problem of checking for correct timing and circuit delays, however, is possibly one of the most difficult factors, but exact details of the parametric tests which vendors apply to production microprocessors are not disclosed. The OEM will not usually be concerned with such tests, as it is generally assumed that a purchased product will meet its published specification and that only functional faults may be encountered. Further details of known microprocessor testing techniques may be found in the published literature[132-146].

6.5.2 The central processing unit

The task of the CPU is to fetch instructions from the main working memory and data from the data registers, execute the required arithmetic or logical operations on the working data and feed the result back to memory or data buses. Additional control of the processor buses to connect them via the I/Os to the outside world, and all other control requirements, are also present. Figure 6.31 shows the general architecture involved.

It is not possible to perform all the arithmetic and logical operations with all possible data when testing the CPU, as the number of test vectors required

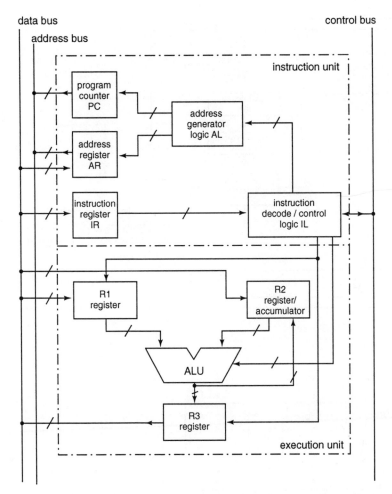

Figure 6.31 Outline schematic of the architecture of a general-purpose arithmetic logic unit, omitting exact details of the registers and multiplexing surrounding the ALU for clarity

would be impossibly large. Therefore a series of nonexhaustive tests to check that the CPU functions correctly over a range of data values is invariably performed, with appropriate checks to ensure that transfer of data from all sources to all destinations can be accomplished.

Some early work on easily-testable CPUs incorporated the shadow register concept which we previously considered in Section 6.3 in connection with memory testing. Figure 6.32a illustrates the shadow register additions to the normal CPU registers, with scan-in and scan-out being available to load in and read out data from six individual blocks of memory[92]. The three registers immediately surrounding the ALU are shown in Figure 6.32b, thus enabling any test data to be applied and the results scanned out for confirmation.

354 *VLSI testing: digital and mixed analogue/digital techniques*

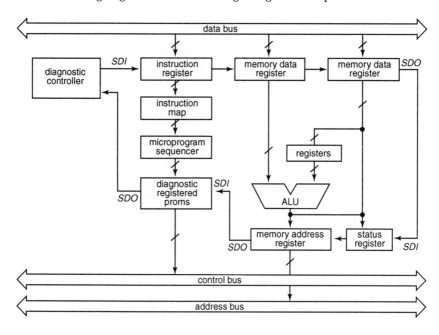

a

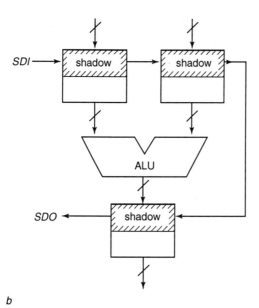

b

Figure 6.32 *The early serial shadow register concept applied to a CPU, with six shadow registers to provide controllability and observability*
  a overall schematic
  b the shadow registers around the ALU

These particular circuits have, however, been overtaken by later scan-path and built-in self-test configurations, but the principles of these early circuits still remain valid and educationally useful. CPU built-in self test using BILBO or other variations such as in Reference 148 have been pursued. Outline schematics from three publications are shown in Figure 6.33 to illustrate these developments; the three examples are:

(i) reconfigurable memory for test generation and signature capture in Reference 134, the final test data being fed to accessible (visible) data buses;
(ii) scan and test clock provision within each partition in Reference 149, which enables control and observation of all flip-flops in the system to be provided;
(iii) the weighted BIST structure used in the ALU of a communications processor[150], involving two reconfigurable registers acting as test generators and two others as MISRs.

The Intel i860 processor[140] also incorporates LSSD and boundary scan to provide test resources.

The silicon area overhead to build in full BILBO or other autonomous test vector generation and response capture remains the biggest penalty in the introduction of these strategies. However, JTAG boundary scan is now standard on all large processors and other VLSI circuits to enable not only the VLSI circuit itself but (equally important) the surrounding circuitry and interconnect to be tested. This area remains extremely dynamic, with microprocessor power and circuit density still doubling every two years or so; the recent microprocessor developments such as the Intel PentiumPRO and 80786 designs and the Motorola Power-PC604 and 620 designs will undoubtedly incorporate comprehensive built-in test strategies to cater for their millions of on-chip devices, and hence active developments in both functional and parametric testing will undoubtedly continue.

6.5.3 Fault-tolerant processors

The requirements for a circuit or system to function satisfactorily has been considered in several areas in preceeding pages of this text. Fault tolerance normally requires either additional error detection/error correction data bits, or some replication of complete modules or circuits as considered in Section 4.6.2 of Chapter 4. Triple modular redundancy (TMR) is a classic model.

Fault-tolerant processors fall into the above categories, but there are some other distinctions which are particularly associated with microprocessors and microprocessor-based systems. These are:

- fail-soft operation, which involves a distributed network of processors or equipments and which in the event of a fault can be reconfigured to cut out the faulty section to maintain system operation;

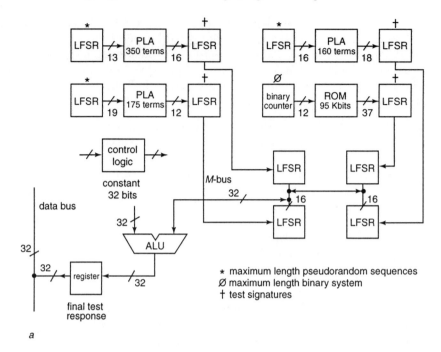

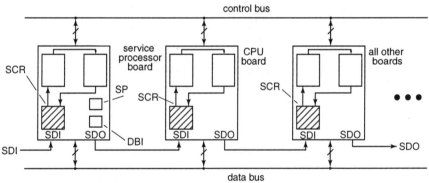

Figure 6.33 Outline schematics, showing the principal parts of some microprocessor and CPU BIST strategies

  a the 80386 μP built-in self test[134]
  b the scan subsystem of the Apollo DN10000 workstation, which provides full controllability and observability of all system flip-flops [149]

- fail-safe operation, which in the event of any fault never gives an output that is a potentially dangerous condition;
- fault-tolerant operation, which in the event of a fault still maintains correct or acceptable final output data.

There is no clear-cut distinction between the above, and a particular fail-soft network, for example, may also include fault-tolerant processor circuits. Software (programming) faults can also be involved in these considerations.

There have been many schemes for providing fault tolerance in computer and other processor systems, being particularly necessary in aerospace missions, nuclear power station control and supervisory systems, rail traffic control systems and the like. Here we will briefly review past developments to indicate the trend and scope of these activities, which continue to be necessary in similar present day applications and circumstances.

The first widely reported fault-tolerant computer system design was the STAR (self testing and repair) computer study of Avizienis *et al.* of the early 1970s[151], which investigated the basis for fault-tolerant processing hardware

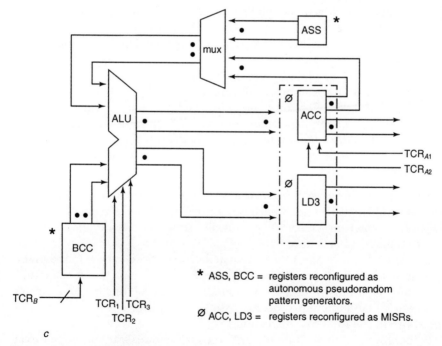

Figure 6.33 Continued

    c BIST structures in the arithmetic logic unit of a communications processor[150]

using self-checking techniques. Many of the concepts considered in this early work formed a basis for later research and development activities. These developments included the following:

(i) the PRIME architecture of 1972 by the University of California at Berkeley[152];
(ii) the FTSC (fault tolerant spaceborne computer) architecture of circa 1975[153];
(iii) the IBM space shuttle computer network of circa 1975[154];
(iv) the Pluribus fault-tolerant multiprocessor of circa 1977[155];
(v) the COPRA reconfigurable computer architectures of circa 1977[156];
(vi) the NASA FTMP (fault tolerant multiprocessor) system of circa 1977[157];
(vii) the NASA SIFT (software implemented fault tolerance) system of the same time[158];
(viii) the C.Vmp (computer-voted multiprocessor) architecture of circa 1977[159];
(ix) the TANDEM reconfigurable multiprocessor system of circa 1980 by Tandem Computers Ltd.[160];
(x) the military FTBBC (fault tolerant build block computer) systems of the Jet Propulsion Laboratory of circa 1978 onwards[161];
(xi) the Sperry Univac 1100/60 general-purpose fault-tolerant computer of circa 1979[162];
(xii) the AXE multipartition control processor for telephone exchanges of circa 1980[163];
(xiii) the STRATUS/32 hardware redundancy system of circa 1983 for financial transaction processing applications[164];
(xiv) the Bell ESS (electronic switching system) fault-tolerant computer systems for telephone exchanges from 1975 onwards[165];
(xv) the COMTRAC (computer aided traffic control system) for Japanese high-speed train control using interconnected computers, of circa 1975 onwards[166].

Following the above we have seen a proposed self-checking version of the Motorola MC68000 general-purpose controller[167], the self-checking MIL-STD-1750A microprocessor[168], plus Intel, Motorola, Hewlett-Packard and other commercial activities.

Details of the majority of the early systems listed above may be found in the special issue of the Proceedings of the IEEE, Vol. 66, No. 10, October 1978 and later surveyed in Lala[15]; in total they cover all the major concepts adopted for fault-tolerant processor sysems. We will briefly illustrate two of the above in the following discussion.

In looking at all these fault-tolerant proposals and realisations, three broad concepts may be found, namely:

(i) Local redundancy in ALUs and memory, often using Hamming SEC/DED coding[13,116,169,170]. This will cater for local single stuck-at or other faults.

(ii) A repeat of program instructions if an error is detected by appropriate check circuits. This is known as rollback, and requires the system to go back and repeat all its processing from the last fault-free check point[14,15,171,172]. This is particulary appropriate if the system has been affected by interference or suffers some transitory fault, but requires a time space for this to be executed. (This is, of course, what a human operator tends to do automatically when something goes wrong, and so has no abstruse theoretical basis!)

(iii) Multiprocessor working for reconfigurable fail-soft operation.

Consider the early Sperry Univac 1100/60 general-purpose computer, which is an example of the route which fault-tolerant processors have followed. The basic architecture is shown in Figure 6.34. Duplicated 32-bit bit-slice CPUs, cache and main working memory are present, together with duplicated data buses and duplicated system support (correction) processors. The main storage units employ SEC/DED error correction/error detection coding, and the control data storage employs simple parity checking.

In operation the input control and data information is checked by the SEC/DED and parity coding, and the output data from the two CPUs on the two output buses is compared for agreement. It is the job of the system support processors to monitor the several checks, and to call for a rewrite of the input data or for some reconfiguration of the system if final disagreement at the output is present. A considerable complexity in these attempts at error recovery is present, including a wait period before retry in case the fault is some temporary aberration. When all attempts at recovery have failed and output data is still inconsistent, then the operator is finally called and given status information to enable a repair procedure to be initiated.

Further details of this system may be found in published literature[15,162]. This strategy using modular redundancy with test and reconfiguration may be contrasted with the static redundancy strategies illustrated in Figures 4.20 and 4.22 of Chapter 4, which do not use retry or reconfiguration but instead have majority voting for the final system output.

The schematic of the computer aided traffic control system, COMTRAC, is shown in Figure 6.35a. It consists basically of three computers, each of which can be:

- online controlling;
- online standby (not controlling);
- offline (disconnected).

Normally two of the three are working in online controlling mode, with their outputs being compared by the dual system controller DSC, the third computer being online standby. Notice that there are also three DSCs, each monitoring two computers. The status register control unit, STR, monitors all three DSCs, and in the event of any disagreement being detected determines which computer is the odd man out, and where appropriate takes it off line, bringing in the previous offline computer to replace it.

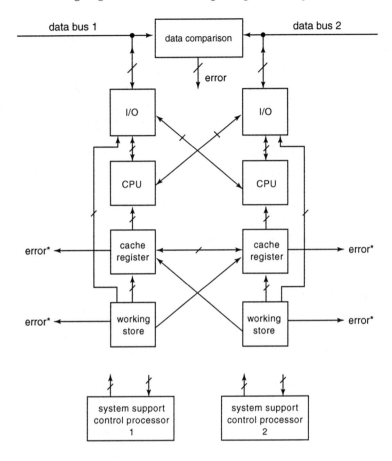

Figure 6.34 The outline schematic of the Sperry Univac 1100/60 fault-tolerant computer with duplicated resources, rewrite of data and repeat/reselection of CPU activity

However, in addition to this hardware structure, there is also a corresponding triplicated software/memory structure, as shown in Figure 6.35b. The DMS dual monitor system units contain the state control program data, and the COS configuration operating system contains the configuration/reconfiguration control program. The structure of Figure 6.35b may therefore be thought of as lying behind the hardware structure of Figure 6.35a, the whole being supervised by the status control register STR. Further details may be found in Reference 166.

There is considerable complexity in the detailed monitoring and reconfiguration circuits of this system, but it clearly illustrates the basic concept of multiple operating processors which may be switched in and out

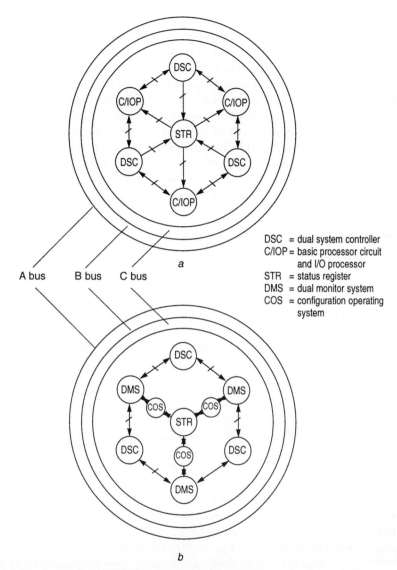

Figure 6.35 The COMTRAC train control system; bus and control connections have been omitted for clarity
  a the 3-computer hardware schematic
  b the 3-control-program software schematic

of service by appropriate checks. This and similar fail-soft processor systems will be seen to use the strategy of multiple processors effectively checking each other and disconnecting one of their members if disagreements are found.

Further details of concurrent error detection using separate coprocessors known as watchdog processors may also be found in the published literature[173,174]. However, the precise design details of fault-tolerant systems, particularly for fail-safe applications where manditory safety regulations may have to be met, is a highly specialised professional activity, with a great deal of commercial security being present.

Finally, a comment on the future possibilities of wafer scale integration (WSI) as a next phase beyond VLSI. Should WSI become a commercial proposition, then undoubtedly the provision of modular redundancy on the wafer becomes an essential part of the development, the modules possibly being as large as complete VLSI processors. Means to isolate faulty modules on the wafer at the production stage must be present, which will lead to fully configurable wafers to provide fault tolerance at the production stage and in service. The switching means which can be envisaged to isolate faulty modules may include:

- hard-configurable methods, such as electrical fuses or antifuses;
- firm-configurable methods, such as nonvolatile but alterable FET switches;
- soft-configurable volatile methods, based upon some external programming control;
- self-configurable methods, relying upon self test and automatic reconfiguration;
- triple (or higher) modular redundancy with voters, transparent to the end user.

Further comments on this future area may be found in References 175–177.

6.6 Cellular arrays

Cellular arrays are two-dimensional multiple input/multiple output architectures of combinational or sequential logic cells which provide specific signal processing or logic duties. If all the cells in the array are identical, as is usually the case, then we have an iterative cellular array.

Figure 6.36 illustrates the general structure, with data flowing in parallel through the array to generate the required outputs. A diagonal data flow between cells is also present in many applications. The particular advantage of such arrays, which are also known as systolic arrays in signal processing and arithmetic operations, is that they can usually be arranged in a very compact silicon layout, the design of which may easily be tailored to match the width of the data being processed. With minimum silicon area and hence minimum length of interconnect, maximum operating speed is also attainable.

To avoid fully exhaustive overall testing, there needs to be some means of controllability and observability of the individual cells in the array, or individual rows of cells or columns of cells, or some other subgrouping. This requirement may be met by being able to switch each cell from its normal mode to some test mode. In test mode the cell may, for example, be made

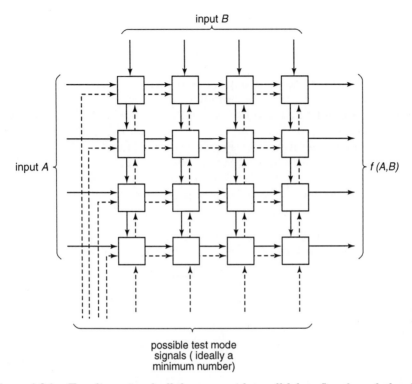

Figure 6.36 *Two-dimensional cellular array with parallel data flow through the cells. The cells may be purely combinational or contain memory (usually a D-type flip-flop). Test signal paths may be as shown dotted*

transparent so that signals may flow through unchanged to further cells, as shown dotted in Figure 6.36; other methods may be to use row and column bypassing so that test data may be concentrated upon a particular row or column at a time. The precise means of test, however, largely remains an *ad hoc* process, greatly dependent upon the cell details. The array designer, therefore, should consider a test strategy during the design phase if some minimum test procedure is required. It may be noted that scan-path techniques for giving access to all the cells in an array are not generally efficient, and some exploitation of the regularity of the array structure usually promises to provide a more efficient test strategy.

As an example, consider the case of a simple cellular array parallel multiplier with an n-bit multiplicand input A_0 to A_{n-1} and an n-bit multiplier input B_0 to B_{n-1}; this involves an iterative array of n^2 combinational cells to do the multiplication plus $n-1$ cells on the perimeter to handle carry data. Since there are $2n+1$ primary inputs (n per operand plus one for control purposes), to test the array exhaustively would require 2^{2n+1} input test vectors, which is unacceptably high for large n. However, if each cell could be separately tested, and assuming that each cell has four inputs, then to test all the cells would only require $(n^2 + n + 1) \times 2^4 = 16 \, (n^2 + n + 1)$ input test vectors.

Swartzlander and Malik[178] have, however, considered the possibility of bridging faults between adjacent cells, and have proposed a test which considers small overlapping groups of cells rather than individual cell tests. This may, for example, be a 2×2 group or a 3×3 group which walks across the whole array during a test procedure. A 2×2 group in a parallel multiplier array has a total of nine inputs and a 3×3 group a total of 14 inputs around the perimeter of the group, see Figure 6.37, and therefore 2^9 and 2^{14} test vectors respectively are required for an exhaustive group test. The statistics for the exhaustive test of a 16×16 bit parallel multiplier are given in Table 6.2 when an overlap between cells in the group tests is present.

Table 6.2 *The number of exhaustive test vectors required for group testing a 16×16 bit parallel multiplier, assuming that the test vectors can be applied to all the cell inputs (Acknowledgements, based on Reference 178)*

No. of cells per group	No. of overlapping groups per array	Inputs per group	Exhaustive test vectors per group	Exhaustive test vectors per array
1	271*	4	16	4336
1×2	256	6	64	16384
2×1	254	7	128	32512
2×2	240	9	512	122880
3×3	210	14	16384	3440640
16×16 (whole array)	1*	33	$\approx 8.6 \times 10^{10}$	$\approx 8.6 \times 10^{10}$

*cannot overlap

The practical difficulty with the above strategy is that it requires controllability and observability of every input and output on each cell from outside the array. It is suggested by Swartzlander and Malik that multiplexers can provide this accessibility, but a high silicon overhead would have to be paid for this provision.

Attempts to place the testing of iterative arrays on a more theoretical basis have been considered. Possibly the first publications in this area are those of Kautz[179] and Feng and Wu[180], followed by further considerations involving iterative arrays of processors rather than more simple digital logic cells[181-183]. Processor arrays will not be considered here, except to note that they are now increasingly considered in the context of wafer scale integration, as introduced and referenced in the preceding section. From these theoretical studies two testability objectives for iterative logic arrays have been proposed, namely:

- C-testability, in which the number of test vectors necessary to test the array is constant irrespective of the array size;

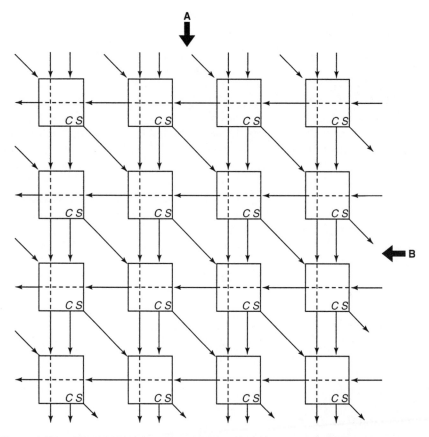

Figure 6.37 The iterative parallel multiplier array with four inputs per cell, nine inputs per group of four cells, 14 inputs per group of nine cells, etc.

- I-testability, in which the outputs of comparable cells are made identical under test conditions, with resultant ease of response comparison.

A combination of these two objectives, namely C/I–testable, has also been investigated. Details of these strategies may be found in Sridhar and Hayes[184], with further considerations in Moore *et al.*[185,186] and elsewhere. However, many of the developments of these strategies relate to single-dimensional arrays only, namely a string of cells, rather than two-dimensional $n \times n$ arrays.

Another objective test strategy is to ensure that every cell output toggles at least once from 0 to 1 and from 1 to 0 in the test procedure. Although all these concepts may set valid targets for good design and test strategies, they do not provide any direct DFT synthesis, and therefore the array designer has in practice to consider each design on an individual basis, considering:

(i) the functionality of the standard cell and how it may best be tested;
(ii) the regular flow of data through the complete array;

(iii) how test strategies may be applied and the results detected, ideally using the normal primary I/Os only.

We will look at two examples below to illustrate these ideas.

Figure 6.38a gives the outline schematic of an iterative systolic array which performs the inner product computation on input matrix **X** and a stored matrix **W**[187]. Each cell entails a clocked latch so that data progresses through the cells with a clock delay, Δt, across a cell. The flow of data across the rows is unchanged, with the delay Δt between the columns, and the results flow down the columns again with the delay Δt between rows. (The exact logic configuration need not concern us here; it is basically a full-adder configuration with the stored data bit $w_i{}^j$ per cell.)

A simple test structure for the horizontal logic flow of data is as shown in Figure 6.38b. A serial bit stream of, say, 1, 0, 1, 0, ... applied to the top right-hand test input will appear at each left-hand row output, displaced Dt between rows if all is correct. A simple comparitor circuit can therefore be used to confirm this horizontal flow of test data. The vertical logic can be checked by first clearing the array with 0 0 0 ... input data, and then clocking in a series of test vectors which flows through and exhaustively checks the logic of each cell. This test vector sequence is arranged such that every cell is simultaneously performing one of its exhaustive tests, the complete test sequence ensuring that every cell receives (at least) its own fully exhaustive test input data. Full details may be found in Reference 187. Allowing for on-chip output comparitor circuits and checking, it is stated that for a 32 × 32 array of cells a total of 249 clock periods only is necessary for an exhaustive test of the array. The number of tests for an $n \times n$ array is of length $O(n)$ rather than $O(n^2)$.

This is a good example how exhaustive testing of every cell in an array can be undertaken with a relatively small number of test input vectors, the test data flowing through the array such that every cell experiences every combination of its input data sometime during the test sequence. The formulation of such test sequences, however, remains largely intuitive.

Error correction and fault tolerance techniques may also be considered in cellular array test strategies. This often involves a spare row and/or column of cells which may be switched into service to replace a faulty row or column; column bypassing using multiplexers has been proposed for the array of Figure 6.38[187].

As a further example, consider the cellular sorting network shown in Figure 6.39a. This is one of a series of shuffle exchange networks which can sort a given input vector into some alternative output vector; feedback around the array may be present in some applications to realise the required shuffle. Each cell in Figure 6.39a consists of either a straight-through connection or a crossed-over connection, controlled by an appropriate global controller (not shown). Further details may be found in Dowd *et al.* and elsewhere[188,189]. Tests for faulty cells may be undertaken by consecutively testing each column of cells in the array, with all remaining column cells set

Testing of structured digital circuits and microprocessors 367

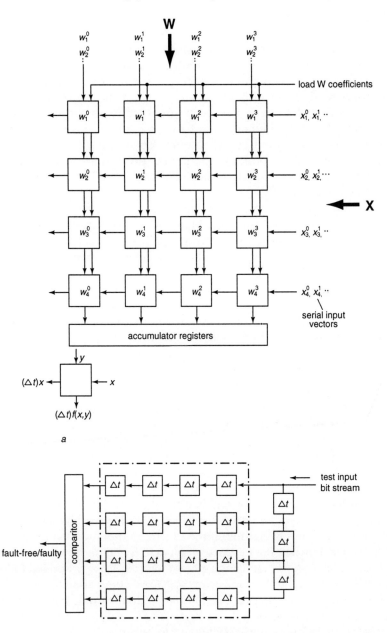

Figure 6.38 The systolic array for inner product vector computation, with input **X** flowing across the array from left to right and **W** held in the array; clock lines omitted for clarity

  a the general schematic and cell schematic
  b possible test for the horizontal data flow, with Δt time delay across each cell

to the straight-through state. The input control vectors to execute this are straightforward to determine, with, say, 0 1 0 1 0 ... and 1 0 1 0 1 ... or a single 1 on a background of 0s or a single 0 on a background of 1s being applied to the primary inputs. In general, $2 \log_2 n$ tests, where n is the number of primary inputs, is the minimum necessary for an exhaustive test.

Appropriate diagnosis of faulty test response enables any single faulty cell to be identified. Hence it has been proposed[190] that there shall be k replications of each column of cells, where $k \geq 2$, with means to bypass a faulty column using multiplexed bypass routing. If it is not possible to diagnose exactly which cell or column is faulty, then some quasi-exhaustive permutation of the column bypass facilities may be tried in order to establish a fault-free output shuffle.

In general, the simpler the cell in an iterative cellular array then the simpler the intuitive test strategy. Many other cellular structures may be found in the published literature, with certain sophisticated analyses of the data flow and test conditions for the more complex arithmetic and DSP architectures[180-185,190-197].

However, let us conclude with a research example of a general-purpose logic array.* In considering easily-testable combinational logic primitives (building blocks), attention has been given to the class of functions known as self-dual logic functions, a self-dual function being defined as:

$$f(x_1, x_2, ..., x_n) = f(\overline{\overline{x}_1, \overline{x}_2, ..., \overline{x}_n})$$

every x_i input variable in the left-hand expression being complemented in the right-hand expression together with overall complementation. This equality holds for all 2^n input combinations of the x_i input variables. A simple example of a self-dual function is $x_1 x_2 + x_1 x_3 + x_2 x_3$, as may readily be shown by compiling the truthtable for this function and its dual. All self-dual functions are also universal logic functions, since any combinational logic function may be realised using one or more self-dual functions[198].

The particular attraction of self-dual functions is that if any input vector such as $x_1 \bar{x}_2 x_3$ ... is applied to the function followed by the bit-complemented vector $\bar{x}_1 x_2 \bar{x}_3$ ... then by definition the function output will switch from 1 to 0 or vice versa. Further, if a network of self-dual functions is made then it is a simple exercise to show that if any input vector $\dot{x}_1 \dot{x}_2 ... \dot{x}_n$ is applied followed by the bit-complemented input vector $\bar{\dot{x}}_1 \bar{\dot{x}}_2 ... \bar{\dot{x}}_n$, then all self-dual function outputs throughout the network including the final outputs will switch. (The designation $\dot{x}_i$ means x_i or $\bar{x}_i$.)

Consider, therefore, a simple self-dual function such as

$$f(X) = \overline{x_1 x_2 + x_1 x_3 + x_2 x_3}$$

---

* This example could equally have been considered in the earlier section of this chapter when discussing gate array architectures, since the dividing line between gate arrays and cellular arrays is not a clear-cut distinction.

Testing of structured digital circuits and microprocessors 369

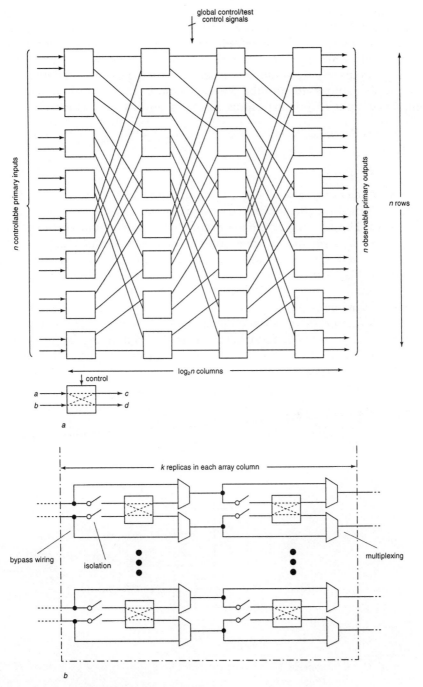

Figure 6.39　The cellular sorting network
　　　　　　a outline schematic omitting the cell control lines
　　　　　　b bypassing of faulty cells by column replication, circuit details omitted
　　　　　　　for clarity

This can be used as a universal logic building block (cell) since two-input NAND and NOR relationships can be realised by setting one of the input variables to logic 0 or 1, respectively, and hence any output function(s) may be realised by an array of such cells. In normal mode one of the network primary inputs, say x_n, would be connected to all logic cells and held at logic 0 or 1, see Figure 6.40, and the outputs would be the required functions of the remaining $n-1$ primary inputs. Applying any two test vectors:

$$\dot{x}_1 \dot{x}_2 \ldots \dot{x}_{n-1} 0$$

and

$$\bar{\dot{x}}_1 \bar{\dot{x}}_2 \ldots \bar{\dot{x}}_{n-1} 1$$

is now sufficient to cause all cell outputs to switch.

Other self-dual functions of three or more inputs may be used as universal logic cells. The advantage of this concept is that it has eliminated the need to determine test vectors to give stuck-at fault cover, since any two bit-complemented test vectors will give a test. We are effectively doubling the length of the truthtable for each output function, using one half of it for normal-mode working and the other self-dual half for test-mode working. The disadvantage is the additional silicon area penalty for the cells and the test mode connections. Futher details of this concept may be found in References 199 and 200, and is representative of continuing interest in gate arrays, cellular arrays and testability considerations for random combinational logic.

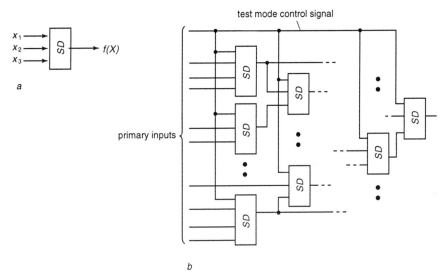

Figure 6.40 The concept of self-dual combinational functions as cells in an easily-tested logic array
        a three-input self-dual function such as $x_1 x_2 + x_1 x_3 + x_2 x_3$
        b network of self-dual cells, where the output value of all cells changes when the dual of any vector is applied

6.7 Chapter summary

This chapter has been principally concerned with products of LSI and VLSI complexity, detailed testing of which is very largely the province of the vendor of the product. The OEM and end user will not be conversant with the precise silicon details and gate-level circuitry, and therefore may have to rely upon the vendor for advice on appropriate test strategies for final system test and in-service maintenance.

Programmable logic devices, however, range from relatively small products to complex products of VLSI complexity; exhaustive testing of the MSI-size products is straightforward, but the larger products have been seen to require built-in means of easing the testing problems. The OEM has to accept whatever provisions, if any, the vendor has incorporated, or to dedicate the product so as to incorporate some chosen DFT strategy.

Microprocessors with their associated on-chip RAM and ROM memory, and often PLA partitions for random logic requirements, represent the most complex digital testing problem. There is no means of completely testing ROM and PROM structures short of fully exhaustive testing; RAM structures, however, have their own test strategies based, as we have seen, on checking the appropriate memory action of each individual cell in the array. The central processing unit requires its own test strategy which must nowadays be considered at the design stage, and which may involve scan test, built-in self test or other DFT means which the IC design team considers appropriate.

Cellular arrays, and particularly systolic arrays for digital signal processing and similar duties, have been seen to be unique in that test strategies can be formulated on an *ad hoc* intuitive basis, the data flow through the cells of the array leading to the formulation of appropriate test vector sequences which ideally exhaustively test each cell. These are purely functional tests and are not based on any fault model or other consideration. Again, it is the designer of the particular product who must consider this, and build in additional test facilities if required.

Overall, with the exception of random-access memories, fault modelling plays little part in the design and test of these structured digital products. Functional testing is the norm. Also, with the possible exception of PLAs and smaller logic partitions, automatic test pattern generation (ATPG) programs are not prominent in determining test vector sequences, exhaustive or quasi-exhaustive test sequences for system partitions being a satisfactory option.

In summary, we therefore find the following characteristics in the testing of strongly-structured digital circuits and microprcessors:

- widespread provision of IEEE standard 1149.1 boundary scan;
- increasing use of built-in self-test (BILBO or similar reconfigurable circuits) so as to minimise the volume of externally generated and inspected test data;
- exhaustive or quasiexhaustive testing of partitions;

- replicated processor systems where critical and safety applications are involved;
- intuitive test strategies for systolic and other specialised arrays, based upon the normal functionality of the standard cells in the array;
- functional testing of all partitions rather than any other basis of determining the necessary test vector sequences.

6.8 References

1. WHITE, D.E.: 'Logic design for array-based circuits: a structured approach' (Academic Press, 1992)
2. BROWN, S.D., FRANCIS, R.J., ROSE, J., and VRANESIC, Z.G.: 'Field programmable gate arrays' (Kluwer Academic Publishers, 1992)
3. BOLTON, M.: 'Digital system design with programmable logic' (Addison-Wesley, 1990)
4. WAKERLEY, J.F.: 'Digital design, principles and practice' (Prentice Hall, 1994)
5. MOORE, W., and LUK, W. (Eds.): 'FPGAs' (Abingdon EE and CS Books, 1991)
6. MOORE, W., and LUK, W. (Eds.): 'More FPGAs' (Abingdon EE and CS Books, 1994)
7. TRIMBERGER, S.M. (Ed.): 'Field-programmable gate array technology' (Kluwer Academic Publishers, 1994)
8. OLDFIELD, J.V., and DORF, R.C.: 'Field-programmable gate arrays' (Wiley, 1995)
9. HURST, S.L.: 'VLSI custom microelectronics: digital, analog and mixed signal' (Marcel-Dekker, 1998)
10. WILLIAMS, T.W.: 'VLSI testing' (North-Holland, 1986)
11. NEEDHAM, W.M.: 'Designer's guide to testable ASIC design' (Van Nostrand Reinhold, 1991)
12. WILKINS, B.R.: 'Testing digital circuits, an introduction' (Van Nostrand Reinhold, 1986)
13. RUSSELL, G., and SAYERS, I.L.: 'Advanced simulation and test methodologies for VLSI design' (Van Nostrand Reinhold, 1989)
14. RAJSUMAN, R.: 'Digital hardware testing: transistor level fault modeling and testing' (Artech House, 1992)
15. LALA, P.K.: 'Fault tolerant testable hardware design' (Prentice Hall, 1985)
16. BARDELL, P.H., McANNEY, W.H., and SAVIR, J.: 'Built-in test for VLSI: pseudorandom techniques' (Wiley, 1987)
17. ABRAMOVICI, M., BREUER, M.A., and FREIDMAN, A.D.: 'Digital systems testing and testable design' (Computer Science Press, 1990)
18. SMITH, J.E.: 'Detection of faults in programmable logic arrays', *IEEE Trans.*, 1979, **C-28**, pp. 845–853
19. OSTAPKA, D.L., and HONG, S.J.: 'Fault analysis and test generation for programmable logic arrays (PLAs)'. Proceedings of IEEE conference on *Fault tolerant computing*, 1978, pp. 83–89, reprinted in *IEEE Trans.*, 1979, **C-28**, pp. 617–627
20. CHA, C.W.: 'A testing strategy for PLAs'. Proceedings of IEEE 15th conference on *Design automation*, 1978, pp. 326–331
21. EICHELBERGER, E.B., and Lindbloom, E.: 'Heuristic test-pattern generation for programmable logic array', *IBM J. Res. and Dev.*, **24** (1), pp. 15–22
22. MIN, Y.: 'A unified fault model for programmable logic arrays'. CRC technical report 83–5, Stanford University, CA, 1983

23 MIN, Y.: 'Generating a complete test set for PLAs'. CRC technical report 83-4, Stanford University, 1983
24 AGRAWAL, V.K.: 'Multiple fault detection in programmable logic arrays', *IEEE Trans.*, 1980, **C-29**, pp. 518-522
25 BOSE, P., and ABRAHAM, J.A.: 'Test generation for programmable logic arrays'. Proceedings of IEEE conference on *Design automation*, 1982, pp. 574-580
26 ZHU, X., and BREUER, M.A.: 'Analysis of testable PLA designs', *IEEE Des. Test Comput.*, 1988, **5** (4), pp. 14-27
27 SERRA, M., and MUZIO, J.C.: 'Testing programmable logic arrays by the sum of syndromes', *IEEE Trans.*, 1987, **C-36**, pp. 1097-1101
28 YAMADA, T.: 'Syndrome-testable design of programmable logic arrays'. Proceedings of IEEE international conferrence on *Test*, 1983, pp. 453-458
29 SERRA, M., and MUZIO, J.C.: 'Space compaction for multiple-output circuits', *IEEE Trans.*, 1988, **CAD-7**, pp. 1105-1113
30 FUJIWARA, H., and KINOSHITA, K.: 'A design of programmable logic arrays with universal tests', *IEEE Trans.*, 1981, **C-30**, pp. 823-828
31 HONG, S.J., and OSTAPKO, D.L.: 'FITPLA: a programmable logic array for function independent testing'. Proceedings of IEEE 10th international symposium on *Fault tolerant computing*, 1980, pp. 131-136
32 KHAKBAZ, J.: 'A testable PLA design with low overhead and high fault coverage', *IEEE Trans.*, 1984, **C-33**, pp. 743-745
33 RAJSKI, J., and TYSZER, J.: 'Easily testable PLA design'. Proceedings of *EUROMICRO*, 1984, pp. 139-146
34 REDDY, M., and HA, D.S.: 'A new approach to the design of testable PLAs', *IEEE Trans.*, 1987, **C-36**, pp. 201-211
35 SALUJA, K.K., KINISHITA, K., and FUJIWARA, H.: 'An easily testable design of programmable logic arrays for multiple faults', *IEEE Trans.*, 1983, **C-32**, pp. 1038-1046
36 PRADHAN, D.H., and SON, K.: 'The effect of untestable faults in PLAs and a design for testabity'. Proceedings of IEEE international conference on *Test*, 1980, pp. 359-367
37 BOZORQUI-NESBAT, S., and McCLUSKEY, E.J.: 'Lower overhead design for testability of programmable logic arrays'. Proceedings of IEEE international conference on *Test*, 1984, pp. 856-865
38 MIN, Y.: 'A PLA design for ease of test generation'. Digest of IEEE 14th international symposium on *Fault-tolerant computing*, 1984, pp. 436-444
39 KHAKBAZ, J., and McCLUSKEY, E.J.: 'Concurrent error detection and testing for large PLAs', *IEEE Trans.*, 1982, **ED-29**, pp. 756-764
40 CHEN, C.Y., FUCHS, W.K., and ABRAHAM, J.A.: 'Efficient concurrent error detection in PLAs and ROMs'. Proceedings of IEEE *ICCD*, 1985, pp. 525-529
41 MAK, G.P., ABRAHAM, J.A., and DAVIDSON, E.S.: 'The design of PLAs with concurrent error detection'. Proceedings of IEEE 12th symposium on *Fault tolerant computing*, 1982, pp.303-310
42 DONG, H.: 'Modified Berger codes for the detection of unidirectional errors'. Proceedings of IEEE 12th symposium on *Fault tolerant computing*, 1982, pp. 317-320
43 WANG, S.L., and AVIZIENES, A.: 'The design of totally self-checking circuits using programmable logic arrays'. Proceedings of IEEE 12th symposium on *Fault tolerant computing*, 1979, pp. 173-180
44 PIESTRAK, S.J.: 'PLA implementations of totally self-checking circuits using m-out-of-n codes'. Proceedings of IEEE *ICCD*, 1985, pp. 777-781
45 SHEN, Y.-N., and LOMBARDI, F.: 'On a testable PLA design for production'. Proceedings of 4th technical workshop on New directions for testing, Victoria, BC, Canada, 1989, pp. 89-134

46 WESSELS, D., and MUZIO, J.C.: 'Adding primary input coverage to concurrent checking for PLAs'. Proceedings of 4th technical workshop on *New directions for testing*, Victoria, BC, Canada, 1989, pp. 135–154
47 SUN, X., and SERRA, M.: 'Concurrent checking and off-line data compaction testing with shared resources in PLAs'. Proceedings of 4th technical workshop on *New directions for testing*, Victoria, BC, Canada, 1989, pp. 155–166
48 LUI, D., and McCLUSKEY, E.J.: 'Design of large embedded CMOS PLAs for BIST', *IEEE Trans.*, 1988, **CAD-7**, pp. 50–59
49 YAJIMA, S., and ARAMAKI, T.: 'Autonomous testable programmable arrays'. Proceedings of IEEE 11th symposium on *Fault tolerant computing*, 1981, pp. 42–43
50 DAEHN, W., and MUCHA, J.: 'A hardware approach to self-testing of large PLAs', *IEEE Trans.*, 1981, **C-30**, pp. 829–831
51 HASSAN, S.Z., and McCLUSKEY, E.J.: 'Testing PLAs using multiple parallel signature analysers'. Proceedings of IEEE 13th international symposium on *Fault tolerant computing*, 1983, pp. 422–425
52 HUA, K. A., JOU, J.-Y., and ABRAHAM, J.A.: 'Built-in tests for VLSI finite state machines'. Digest of IEEE 14th international symposium on *Fault tolerant computing*, 1984, pp. 292–297
53 FUIJIWARA, H., TREUER, R., and AGRAWAL, V.K.: 'A low overhead, high coverage built-in self-test PLA design'. Digest of IEEE 15th international symposium on *Fault tolerant computing*, 1985, pp. 112–117
54 GRASSL, G., and PFLEIDERER, H.J.: 'A function independent test for large programmable logic arrays', *Integ. VLSI J.*, 1983, **1** (1), pp. 71–80
55 SALUJA, K.K., KINOSHITA, K., and FUJIWARA, H.: 'A multiple fault testable design of programmable logic arrays'. Proceedings of IEEE 11th international symposium on *Fault tolerant computing*, 1981, p. 44–46
56 BREUER, M.A., and SCHABAN, F.: 'Built-in test for folded programmable logic arrays', *Microprocess. Microsyst.*, 1987, **11**, pp. 319–329
57 WANG, S.L., and SANGIOVANNI-VINCENTELLI, A.: 'PLATYPUS: a PLA test pattern generation tool'. Proceedings of IEEE 22nd conference on *Design automation*, 1985, pp. 192–203
58 HA, D.S., and REDDY, S.M.: 'On the design of random pattern testable PLAs'. Proceedings of IEEE international conference on *Test*, 1986, pp. 688–695
59 HA, D.S., and REDDY, S.M.: 'On BIST PLAs'. Proceedings of IEEE international conference on *Test*, 1987, pp. 342–351
60 BOSWELL, C., SALUJA, K., and KINOSHITA, K.: 'A design of programmable logic arrays for parallel testing', *J. Comput. Syst. Sci. and Eng.*, 1985, **1**, October, pp. 5–16
61 UPADHYAYA, J.S., and SALUJA, K.: 'A new approach to the design of built-in self-testing PLAs for high fault coverage', *IEEE Trans.*, 1988, **CAD-7**, pp. 60–67
62 SON, K., and PRADHAN, D.K.: 'Design of programmable logic arrays for testability'. Proceedings of IEEE international conference on *Test*, 1980, pp. 163–166
63 HA, D.S., and REDDY, S.M.: 'On the design of pseudo-exhaustive testable PLAs', *IEEE Trans.*, 1988, **C-37**, pp. 468–472
64 FUJIWARA, H.: 'Design of PLAs with random pattern testability', *IEEE Trans.*, 1988, **CAD-7**, pp. 5–10
65 FUJIWARA, H.: 'Enhancing random pattern coverage of programmable logic arrays via a masking technique', *IEEE Trans.*, 1989, **CAD-8**, pp. 1022–1025
66 TREUER, R., FUJIWARa, H., and AGRAWAL, V.K.: 'Impementing a built-in self-test PLA design', *IEEE Des. and Test Comput.*, 1985, **2**, April, pp. 37–48
67 'FPGA applications handbook'. Texas Instruments, Inc., TX, publication SRFA.001, 1993

68 'JTAG 1149.1 specifications for in-circuit reconfiguration and programming of FLEXlogic FFGAs'. Intel Corporation, CA, application note AP-390, 1993
69 'Configurable logic design and application book'. Amtel Corporation, CA, publication 0266B, 1995
70 'The programmable logic data book'. Xilinx, Inc., CA, publication PN.0401224, 1994
71 'FPGA data, field programmable gate arrays'. Motorola, Inc., AZ, publication DL 201/D,1994
72 'Configurable array logic user manual'. Algotronix Ltd., Edinburgh, Scotland, 1992
73 'Programmable interconnect data book'. Aptix Corporation, CA, 1993
74 'Programmable data book'. Altera Corporation, CA, 1994
75 'The FPGA designer's guide'. Actel Corporation, CA, 1993
76 WEINMANN, U., KUNZMANN, A., and StROHMEIER, U.: 'Evaluation of FPGA architectures', in [5], pp. 147–156
77 JENKINS, J.H.: 'Designing with FPGAs and CPLDs' (Prentice Hall, 1994)
78 HURST, S.L.: 'Custom-specific integrated circuits' (Marcel Dekker, 1985)
79 'pASIC family very high speed CMOS FPGAs'.QuickLogic, Inc., CA, 1991
80 'Device data book and design guide'. AMD, PAL®s, CA, 1995
81 GIACOMO, J.D.: 'Designing with high performance ASICs' (Prentice Hall, 1992)
82 Johansson, J., and Forskitt, J.S.: 'System designs into silicon' (Institute of Physics Publishing, 1993)
83 HUBER, J.P., and ROSNECK, M.W.: 'Succssful ASIC design the first time through' (Van Nostrand Reinhold, 1991)
84 HAYES, J.P.: 'Introduction to digital logic design' (Addison-Wesley, 1993)
85 Goldberg, J., Lewitt, K.N., and Wensley, J.H.: 'An organisation for a highly survivable memory', *IEEE Trans.*, 1974, **C-23**, pp. 693–705
86 ZORIAN, Y., and IVANOV, A.: 'An effective BIST scheme for ROMs', *IEEE Trans.*, 1992, **C-41**, pp. 646–653
87 AKERS, S.B.: 'A parity bit signature for exhaustive testing'. Proceedings of IEEE international conference on *Test*, 1986, pp. 48–52
88 SAVIR, J., and BARDELL, P.H.: 'Built-in self-test: milestones and challenges', *VLSI Des.*, 1993, **1** (1), pp. 23–44
89 TUSZYSKI, A.: 'Memory testing', in [10], pp. 161–228
90 BENNETTS, R.G.: 'Design of testable logic circuits' (Addison-Wesley, 1984)
91 LEE, F., COLI, V., and MILLER, W.: 'On-chip circuitry reveals system's logic states', *Electron. Des.*, 1983, **31** (8), pp. 109–124
92 BIRKNER, J., COLI, V., and LEE, F.: 'Shadow register architecture simplifies digital diagnosis'. Monolithic Memories, Inc., CA, Application note AN.123 1983
93 DILLINGER, T.E.: 'VLSI engineering' (Prentice Hall, 1988)
94 PRINCE, B.: 'Semiconductor memories' (Wiley, 1995, 2nd edn.)
95 McPARTLAND, R.J.: 'Circuit simulation of alpha-particle-induced soft errors in MOS dynamic RAMs', *IEEE J. Solid-State Circuits*, 1981, **SC-16**, pp. 31–34
96 ZACHAN, M.P.: 'Hard, soft and transient errors in dynamic RAMs', *Electron. Test*, March 1992, pp. 42–54
97 WESTERFIELD, E.C.: 'Memory system strategies for soft errors and hard errors'. Proceedings of IEEE *WESCON*, 1979, Paper 9-1, pp. 1–5
98 GEILHUFE, M.: 'Soft errors in semiconductor memories'. Proceedings of *IEEE COMPCON*, 1979, pp. 210–216
99 MAY, T., and WOODS, M.: 'Alpha-particle-induced soft errors in dynamic memories', *IEEE Trans.*, 1979, **ED-26**, pp. 2–9
100 DEKKER, R., BEENKER, F., and THIJSSEN, L.: 'A realistic fault model and test algorithm for static random access memories', *IEEE Trans.*, 1990, **CAD-9**, pp. 567–572

101 NAIR, R., THATTE, S.M., and ABRAHAM, J.A.: 'Efficient algorithms for testing semiconductor random access memories', *IEEE Trans.*, 1978, **C-27**, pp. 572–576
102 HAYES, J.P.: 'Detection of pattern sensitive faults in random access memories', *IEEE Trans.*, 1975, **C-24**, pp. 150–157
103 SUK, D.S., and REDDY, S.M.: 'A march test for functional faults in semiconductor random access memories', *IEEE Trans.*, 1981, **C-30**, pp. 982–985
104 MARTINESCU, M.: 'Simplified and efficient algorithms for functional RAM testing'. Proceedings of IEEE international conference on *Test*, 1992, pp. 236–239
105 SAVIR, J., McANNEY, W.H., and VECCHIO, S.: 'Testing for coupled cells in random access memories'. Proceedings of IEEE international conference on *Test*, 1989, p. 439–451
106 JARWALA, N.T., and PRADHAN, D.K.: 'TRAM: a design method for high-performance easily testable multimegabit RAMs', *IEEE Trans.*, 1988, **C-37**, pp. 1235–1250
107 SRIDHAR, T.: 'A new parallel test approach for large memories', *IEEE Des. and Test Comput.*, 1986, **3** (4), pp. 15–22
108 RAJSUMAN, R.: 'An apparatus and method for testing random access memories'. US Patent Application 620359 1990
109 'Technical report, nonvolatile ferroelectric technology and product'. Ramtron Corporation, CO, 1988
110 WEBBER, S.: 'A new memory technology', *Electronics*, 1988, 18 February, pp. 91–94
111 TUMMALA, R., and RYMASZEWSKI, E.: 'Microelectronics packaging handbook' (Van Nostrand Reinhold, 1989)
112 SINNADURAI, F.N.: 'Handbook of microelectronic packaging and interconnect technologies,' (Electrochemical Publications, Scotland, 1985)
113 MAY, T.C., and WOODS, M.H.: 'A new physical mechanism for soft faults in dynamic memories'. Proceedings of IEEE 16th symposium on *Reliability physics*, 1978, pp. 33–40
114 BRODSKY, M.: 'Hardening RAMs against soft errors', *Electronics International*, **53** (8), pp. 117–122
115 SARRAZIN, D.B., and MALIK, M.: 'Fault tolerant semiconductor memories', *IEEE Computer*, 1984, **17** (8), pp. 49–56
116 ELKIND, S.A., and SIEWIORCK, D.P.: 'Reliability and performance of error-correcting memory and register arrays'. Proceedings of IEEE international conference on *Test*, 1980, **C-29**, pp. 920–927
117 OSMAN, F.: 'Error correcting techniques for random access memories', *IEEE J. Solid State Circuits*, 1982, **SC-27**, pp. 877–881
118 TANNER, R.M.: 'Fault-tolerant 256K memory designs', *IEEE Trans.*, 1984, **C-33**, pp. 314–322
119 SU, S.Y.H., and DuCASSE, E.: 'A hardware redundancy reconfiguration scheme for tolerating multiple module failures', *IEEE Trans.*, 1980, **C-29**, pp. 254–257
120 PRADHAN, D.K.: 'A new class of error correcting/detecting codes for fault-tolerant computer applications', *IEEE Trans.*, 1980, **C-29**, pp. 471–481
121 HSIAO, M.Y., and BOSSEN, D.C.: 'Orthogonal Latin square configuration for LSI memory yield and reliability enhancement', IEEE Trans., 1975, **C-24**, pp. 512–516
122 HECHT, H.: 'Fault-tolerant software', *IEEE Trans.*, 1979, **R-28**, pp. 227–232
123 FERIDUN, A.M., and SHIN, K. G.: 'A fault-tolerant microprocessor system with rollback recovery capabilities'. Proceedings of 2nd international conference on *Distributed computer systems*, 1981, pp. 283–298
124 ANDERSON, A., and LEE, P.: 'Fault tolerance: principles and practice' (Prentice Hall, 1980)

125 NICOLAIDIS, M.: 'An efficient built-in self-test scheme for functional test of embedded RAMs'. Proceedings of IEEE 15th symposium on *Fault tolerant computing*, 1982, pp. 118–123
126 SUN, Z., and WANG, L.-T.: 'Self-testing embedded RAMs'. Proceedings of IEEE international conference on *Test*, 1994, pp. 148–156
127 JAIN, S.K., and STROUD, C.E. 'Built-in self testing of embedded memories', *IEEE Des. Test of Comput.*, 1986, **3** (5), pp. 27–37
128 KNAIZUK, J., and HARTMANN, C.: 'An algorithm for testing random access memories', *IEEE Trans.*, 1977, **C-26**, pp. 414–416
129 KNAIZUK, J., and HARTMANN, C.: 'An optimum algorithm for testing stuck-at faults in random access memories', *IEEE Trans.*, 1977, **C-26**, pp. 1141–1144
130 NAIR, R.: 'Comments on "An optimum algorirthm for testing stuck-at faults in random access memories"', *IEEE Trans.*, **C-28**, 1979, pp. 256–261
131 BARDELL, P.H., and McANNEY, W.H.: 'Built-in test for RAMs', *IEEE Des. Test Comput.*, 1988, **5** (4), pp. 29–36
132 JANI, D., and ACKEN, J.M.: 'Test synthesis for microprocessors', Digest of IEEE international conference on *Test*, Test synthesis seminar, 1994, pp. 4.1.1–4.1.8
133 SHIH, F.W., CHAO, H.H., ONG, S., DIAMOND, A.L., TANG, J.Y.-F., and TREMPEL, C.A., 'Testability design for Micro/370, a system 370 single-chip microprocessor'. Proceedings of IEEE international conference on *Test*, 1986, pp. 412–418
134 GELSINGER. P.: 'Design and test of the 80386', *IEEE Des. Test Comput.*,1987, **4** (3), pp. 42–50
135 BHAVSAR, D.K., and MINER, D.G.: 'Testability strategy for a second generation VAX computer'. Proceedings of IEEE international conference on *Test*, 1987, pp. 818–825
136 NOZUMA, A., NISHIMURA, A., and IWAMURA, J.: 'Design for testability of a 32-bit microprocessor, the TXI'. Proceedings of IEEE international conference on *Test*, 1988, pp. 172–182
137 GALLUP, M.G., LEDBETTEr, W., McGARITY, R., McMAHAN, S., SCHEUR, K.C., SHEPHERD, C.G., and GOOD, L.: 'Testability features of the 68000'. Proceedings of IEEE international conference on *Test*, 1990, pp. 749–757
138 JOSEPHESON, D.D., DIXON, D.J., and ARNOLD, B.J.: 'Test features for the HP PA77100LC processor'. Proceedings of IEEE international conference on *Test*, 1993, pp. 764–772
139 DANIELS, R.G., and BRUCE, W.C.: 'Built-in self-test trends in Motorola microprocessors', *IEEE Des. Test Comput.*, 1985, **1** (2), pp. 64–71
140 PERRY, T.S.: 'Intel's secret is out', *IEEE Spectr.*, 1986, **26** (4), pp. 21–27
141 THATTE, S.M., and ABRAHAM, J.A.: 'Test generation for microprocessors', *IEEE Trans.*, 1980, **C-29**, pp. 429–441
142 BRAHME, D., and ABRAHAM, J.A.: 'Functional testing of microprocessors', *IEEE Trans.*, 1984, **C-33,** pp. 475–485
143 BELLON, C., LIOTHIN, A., SADIER, S., SAUCIER, G., VELAZCO, R., GRILLOT, F., and ISSENMANN, M.: 'Automatic generation of microprocessor test programmes'. Proceedings of IEEE 19th conference on *Design automation*, 1982, pp. 566–573
144 BELLON, C.: 'Automatic generation of microprocessor test patterns', *IEEE Des.Test Comput.*, 1984, **1** (1), pp. 83–93
145 SHIDAR, T., and HAYES, J.P.: 'A functional approach to testiing bit-sliced microprocessors', *IEEE Trans.*, 1981, **C-30**, pp. 563–571
146 'Testability primer'. Texas Instruments Inc., TX, publication SSYA.002A, 1991
147 'TMS 320 C50 DSP preview bulletin'.Texas Instrument Inc., TX, publication MPL 268/GB-2589 Pb, 1989

148 CERNY, E., ABOULHAMID, M., BOIS, G., and CLOUTIER, J.: 'Built-in self-test of a CMOS ALU', *IEEE Des. Test Comput.*, 1988, **5** (4), pp. 38–48
149 DERVISOGLU, B.I.: 'Scan path architecture for pseudorandom testing', *IEEE Des. Test Comput.*, 1988, **6** (4), pp. 32–48
150 MARTINEZ, M., and BRACHO, S.: 'Weighted BIST structures in the arithmetic unit of a communications processor', *IEE Proc., Comput. Digit. Tech.*, 1995, **142 CDT**, pp. 360–366
151 AVIZIENIS, A., GILLEY, G.C., MATHUR, F.P., RENNELS, D.A., ROHR, J.A., and RUBIN, D.K.: 'The STAR (self-testing and repair) computer: an investigation into the theory and practice of fault-tolerant computer design', *IEEE Trans.*, 1971, **C-20**, pp. 1312–1321
152 BASKIN, H.B., BORGERSON, B.R., and ROBERTS, R.: 'PRIME: a modular architecture for terminally oriented system'. Proceedings of IEEE spring joint conference on *Computers*, 1972, pp. 431–437
153 STIFFLER, J.J.: 'Architectural design for near 100% fault coverage'. Proceedings of IEEE international symposium on *Fault tolerant computing*, 1976, pp. 134–137
154 SKLAROFF, J.R.: 'Redundancy management techniques for space shuttle computers', *IBM J. Res. Dev.*, 1976, **20** (1), pp. 20–27
155 KATSUKI, D., ELSAN, E.S., MANN, W.F., ROBERTS, E.S., ROBINSON, J.G., SKOWRONSKI, F.S., and WOLF, E.W.: 'Pluribus—an operational fault-tolerant microprocessor', *Proc. IEEE*, 1978, **66**, pp. 1146–1157
156 MERAUD, C., and LLORET, P.: 'COPRA: a modular family of reconfigurable computers'. Proceedings of IEEE conference on *national aerospace and electronics*, 1978, pp. 822–827
157 HOPKINS, A.L., SMITH, T.B., and LALA, J.H., 'FTMP—a highly reliable fault-tolerant microprocessor for aircraft', *Proc. IEEE*, 1978, **66**, pp. 1221–1239
158 WENSLEY, J.H., LAMPORT, L., GOLDBERG, J., GREEN, M.W., LEVITT, K.N., SMITH, P.M.M., SHOSTAK, R.E., and WEINSTOCK, C.B.: 'SIFT: design and analysis of fault-tolerant computer for aircraft control', *Proc. IEEE*, 1978, **66**, pp. 1240–1255
159 SEIWIOREK, D.P., KINI, V., MASHBURN, H., McCONNEL, S., and TSAO, M.: 'A case study of C.MMP, Cm and C.Vmp: experiences with fault-tolerance in multiprocessor systems', *Proc. IEEE*, 1978, **66**, pp. 1178–1199
160 KATZMAN, J.A.: 'A fault-tolerant computing system', Proceedings of IEEE international conference on *System sciences*, 1978, pp. 85–102
161 RENNELS, D.A.: 'Architectures for fault-tolerant spacecraft computers', *Proc. IEEE*, 1978, **66**, pp. 1255–1268
162 BOONE, L.A., LEIBERGOt, H.L., and SEDMAK, R.M.: 'Availability, reliability and maintainability aspects of the Sperry Univac 1100/60'. Proceedings of international IEEE international symposium on *Fault tolerant computing*, 1980, pp. 3–8
163 OSSFIELD, B.E., and JONSSON, I.: 'Recovery and diagnosis in the central control of the AXE switching system', *IEEE Trans.*, 1980, **C-29**, pp. 482–491
164 HERBERT, E.: 'Computers: minis and mainframes', *IEEE Spectr.*, 1983, **20** (1), pp. 28–33
165 TOY, W.N.: 'Fault-tolerant design of local ESS processors', *Proc. IEEE*, 1978, **66**, pp. 1126–1145
166 IHARA, H., FUKNOKA, K., KUBO, Y., and YOKOTA, S.: 'Fault tolerant computer system with three symmetrical computers', *Proc. IEEE*, 1978, **66**, pp. 1160–1177
167 MARCHAL, P., NICOLAIDIS, M., and COURTOIS, B., 'Microarchitecture of the MC 86000 and the evaluation of a self-checking version'. Proceedings of NATO Advanced Study Institute, Microarchitectures of VLSI Computers, 1984
168 HALBERT, M.P., and BOSE, S.M.: 'Design approach for a VLSI self-checking

MIL-STD-1750A microprocessor'. Proceedings of IEEE 14th symposium on *Fault tolerant computing*, 1984, pp. 254–259
169 HWANG, S., RAJSUMAN, R., and MALAIYA, Y.K.: 'On the testing of microprogrammed processors'. Proceedings of IEEE international symposium on *Microprogramming and microarchitectures*, 1990, pp. 260–266
170 HOPKINS, A.L.: 'A fault-tolerant information processing concept for space vehicles', *IEEE Trans.*, 1971, **C-20**, pp 1394–1403
171 MERAUD, C., and BROWAEYS, F. 'Automatic rollback techniques of the COPRA computer'. Proceedings of IEEE international symposium on *Fault tolerant computing*, 1976, p. 23–29
172 LEE, Y.H., and SHIN, K.G.: 'Rollback propagation, detection and performance evaluation of FTMR2M–a fault-tolerant microprocessor', *Proc. Comput. Archit.*, 1982, pp. 171–180
173 MAHMOOD, A., and McCLUSKEY, E.J.: 'Concurrent error detection using watchdog processors', *IEEE Trans.*, 1988, **C-37**, pp. 160–174
174 SAXENA, N.R., and McCLUSKEY, E.J.: 'Control flow checking using watchdog assist and extended-precision checksums'. Proceedings of IEEE conference on *Fault tolerant comput.*, 1989, pp. 428–435
175 EVANS, R.A.: 'Wafer scale integration', in MASSARA, R.E. (Ed.): Design and test techniques for VLSI and WSI circuits' (Peter Peregrinus, 1989), pp. 261–289
176 JESSOPE, C.R., and MOORE, W.R. (Eds.): 'Wafer Scale Integration' (Adam Hilger, 1986)
177 KATEVENIS, M.G.H., and BLATT, M.G.: 'Switch design for soft-configurable WSI systems'. Proceedings of conference on VLSI, 1985, pp. 197–219
178 SWARTZLANDER, E.E., and MALIK, M.: 'Overlapped subarray segmentation: an efficient test for cellular arrays', *VLSI Des.*, 1993, **1** (1), pp. 1–7
179 KAUTZ, W.H.: 'Testing for faults in combinational cellular logic arrays'. Proceedings of 8th annual symposium on *Switching and automation theory*, 1967, pp. 161–174
180 FENG, T.Y., and WU, C.L.: 'Fault diagnosis for a class of multistage interconnected networks', *IEEE Trans.*, 1981, **C-30**, pp. 743–758
181 PEASE, M.C.: 'The indirect binary n-cube microprocessor array', *IEEE Trans.*, **C-26**, pp. 458–473
182 FALAVARJANI, K.M., and PRADHAN, D.K.: 'Fault diagnosis of parallel processor interconnections'. Proceedings of IEEE 11th symposium on *Fault tolerant computing*, 1981, pp. 209–212
183 SIEGEL, H.J., HSU, W.T., and JENG, M.: 'An introduction to the multistage cube family of interconnected networks', *J. Superconducting*, 1987, **5** (1), pp. 13–42
184 SRIDHAR, T., and HAYES, J.P.: 'Design of easily-tested bit-sliced systems', *IEEE Trans.*, 1981, **C-30**, pp. 842–854
185 MOORE, W.R., McCABE, A.P.H., and URQUHART, R.B., (Eds.): 'Systolic arrays' (Adam Hilger, 1987)
186 MARNANE, W.P., and MOORE, W.R.: 'Testing of VLSI regular arrays'. Proceedings of IEEE international conference on *Computer design*, 1988, pp. 145–148
187 MOORE, W.R.: 'Testability and diagnosability of a VLSI systolic array', in MASSARA, R.G. (Ed.): 'Design and test techniques for VLSI and WSI Circuits' (Peter Peregrinus, 1989), pp. 145–157
188 DOWD, M., PERL, Y., RUDOLF, L., and SAKS, M.: 'The balanced sorting network', *Proc. ACM Distributed Computing*, 1983, pp. 161–172
189 RUDOLF, L.: 'A robust sorting network', *IEEE Trans.*, 1985, **C-34**, pp. 326–335
190 SUN, J., CERNY, E., and GECSEI, J.: 'Design of a fault-tolerant sorting network'.

Proceedings of Canadian conference on *VLSI* (CCVLSI'89), 1989, pp. 235–242
191 LI, F., ZHANG, C.N., and JAYAKUMAS, R.: 'Latency of computational data flow and concurrent error detection in systolic arrays'. Proceedings of Canadian conference on *VLSI* (CCVLSI'89), 1989, pp. 251–258
192 JACOB, A., BANERJEE, P., CHEN, C.-Y., FUCHS, W.K., KUO, S.-Y., and REDDY, A.L.N.: 'A fault tolerance technique for systolic arrays', *IEEE Computer*, 1987, **20** (7), pp. 65–74
193 COSENTINO, R.J.: 'Concurrent error correction in systolic architectures', *IEEE Trans.*, 1988, **CAD-7**, pp. 117–125
194 GULATI, R.K., and REDDY, S.M.: 'Concurrent error detection in VLSI array structures'. Proceedings of IEEE international conference on *Computer design*, 1986, pp. 488–491
195 CHEN, T.-H., LEE, Y.-P., and CHEN, L.-G.: 'Concurrent error detection in array multipliers by BIDO ', *IEE Proc., Comput. Digit. Tech.*, 1995, **142**, pp. 425–430
196 WEY, C.-L.: 'Design and test of C-testable high-speed carry-free dividers', *IEE Proc., Comput. Digit. Tech.*, 1995, **142**, pp.193–200
197 WOODS, R.F., FLOYD, G., WOOD, K., EVANS, R., and McCANNY, J.V.: 'Programmable high performance IIR filter chip', *IEE Proc., Circuit Devices Syst.*, 1995, **142**, pp. 179–185
198 HURST, S.L., MILLER, D.M., and MUZIO, J.C.: 'Spectral techniques in digital logic' (Academic Press, 1985)
199 TAYLOR, D.: 'Design of certain customised structures incorporating self-test'. PhD. thesis, Huddersfield University, UK,1990
200 TAYLOR, D., and PRITCHARD, T.I.: 'Testable logic design based upon self-dual primitives', *J. Semicust. ICs*, 1991, **9** (1), pp. 29–35

Chapter 7
Analogue testing

7.1 Introduction

In contrast to the vast amount of research and published information on digital logic testing, the volume of information concerning analogue testing is very small. Indeed, most available textbooks that deal with electronic testing make no mention of analogue circuits, concentrating instead on the more pervasive digital testing theory and practice. This text also will be seen to be predominately digital, which, unfortunately, reflects the historical emphasis in this subject area in spite of the world being analogue in nature.

As introduced in Chapter 1, analogue-only systems are not generally characterised by having a large number of circuit elements (macros), but instead have relatively few but individually more complex elements to build the required system. The basic circuit design of such elements (amplifiers, filter networks, comparitors, etc.) is covered in many excellent teaching textbooks[1-8], but to optimise the design of monolithic macros for production purposes is much more specialised, requiring a great deal of experience and in-house knowledge. Most OEM designers, therefore, use wherever possible predesigned commercially-available circuits, the system design being one or other of the following techniques:

(i) PCB assemblies of commercial off the shelf items such as operational amplifiers, etc.;
(ii) uncommitted analogue arrays, dedicated to the required system specification by the final custom metalisation mask(s);
(iii) full-custom IC designs, invariably designed by a vendor or specialist design team.

Details of (ii) and (iii) may be found in the published literature dealing with custom microelectronics (ASICs or USICs), but this literature contains relatively little detail on testing requirements[8-11].

Analogue-only systems cannot generally be classified as VLSI, being much more MSI/LSI in scale if a count of transistors is made. However, when we turn to mixed-mode circuits, that is circuits containing both digital and analogue elements, which we will defer until the following chapter, then we are back again in the field of VLSI complexities in both design and test. Figure 7.1 therefore indicates our present and forthcoming interests, with more detailed considerations arising in mixed-mode circuits compared with the analogue-only case.

7.2 Controllability and observability

The requirements for controllability and observability apply equally to analogue circuits as they do to digital circuits, and hence if a particular analogue element is buried within several others it may be appropriate to provide individual test access to the I/Os of the element. For small analogue assemblies, the system I/Os may provide adequate access, but if, for example, there is an operational amplifier not directly accessible from the primary inputs then direct test access may be appropriate in order that parametric tests can be undertaken. This is illustrated in Figure 7.2, where the individual elements may possibly be unpackaged (naked) ICs assembled in a thick-film or thin-film hybrid construction package. Notice that it may be necessary to apply signals to test the individual item which can never be applied when the circuit is in normal working mode, in order that its full functional specification may be checked. This is a somewhat philosophical situation, since it may be argued that it should not be necessary to test any component beyond the range of signals it receives in normal operation. However, it is usually considered desirable to test analogue circuits to their full design specification and tolerance if at all possible, rather than to some minimum necessary for system operation.

A possible way of providing this increased controllability and observability can be via analogue multiplexers, as shown in Figure 7.3. The normal mode/test mode controls must be digital, since the multiplexers are conducting/nonconducting paths consisting of FET transmission circuits or buffer amplifiers switched in and out of service. However, it is inevitable that there will be some degradation of the analogue signals passing through them, which may have to be taken into consideration in both the normal mode and the test mode of operation.

As far as the mechanical aspects of providing controllability and observability are concerned, provided that the required nodes are brought out to accessible pins, or in the case of a PCB assembly of standard off the shelf parts provided all I/O connections are still visible, then there are usually no difficulties. A test socket or a bed of nails probe assembly is appropriate to provide the test connections, with the addition of individual hand-held or hand-controlled probes if necessary.

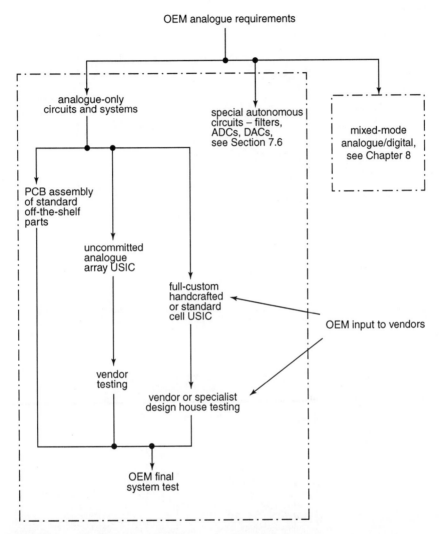

Figure 7.1   The hierarchy of analogue circuit and testing areas

Further details of test hardware will be given in Section 7.4. As far as the OEM is concerned, the system test objective will be to confirm that all components, be they off the shelf ICs or other items, are fault free, and to identify and replace any faulty replaceable component on the PC board or other system assembly. As far as the vendor of an analogue IC is concerned, be it an off the shelf or a custom component, the objective will be to confirm that the device fully meets its published specification, with no diagnostics (on-chip fault location) normally being required on a production run unless serious yield problems are encountered.

384 *VLSI testing: digital and mixed analogue/digital techniques*

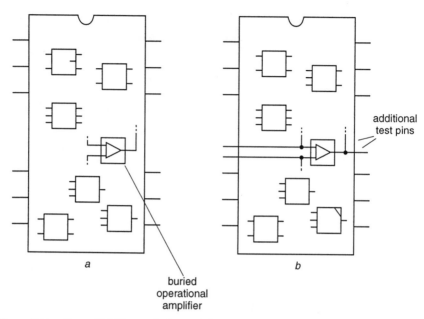

Figure 7.2 *Controllability and observability of buried analogue macros*
  a buried operational amplifier
  b additional test pins

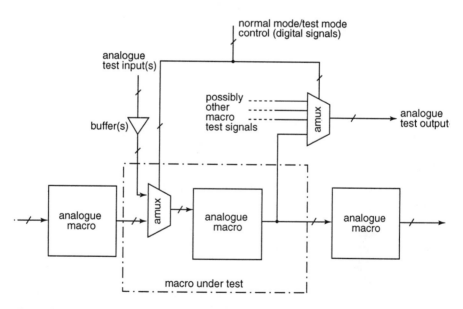

Figure 7.3 *The inclusion of analogue multiplexers to give controllability and observability but at the expense of an increase in circuit complexity and perhaps some loss of circuit performance and tolerancing*

7.3 Fault models

The stuck-at fault model so widely used in digital simulation and test pattern generation has no direct counterpart in the analogue area. Unlike the digital case where the majority of faults logically result in one or more stuck-at nodes, in analogue circuits the types of faults may be:

(i)   deviation faults, where the value of a parameter deviates or drifts in some way from its nominal value;
(ii) catastrophic faults, where the value of a parameter or parameters deviates widely from nominal (i.e. open circuit, short circuit, breakdown, etc.);
(iii) dependent faults, where the occurrence of one fault causes the occurrence of other parametric fault(s);
(iv) environmental faults, where some environmental condition (temperature, humidity, etc.) causes the occurrence of circuit faults;
(v)   intermittent faults, usually catastrophic when they occur but only temporary (e.g. bad connections within resistors or capacitors, etc.).

Faults (i), (iii) and (iv), therefore, are the types of faults in analogue circuits that can cause the circuit to go outside its design specification, but possibly not sufficiently to disable the whole system completely, unlike digital stuck-at faults which result in either the system continuing to work correctly when the stuck-at node is not involved, or to give completely wrong answers in the other situation.

Overall, the various types of fault which can be experienced in analogue circuits due to fabrication imperfections may be considered to be[12,13]:

- hard faults, that is faults which are irreversible and give a circuit performance catastrophically different from normal;
- soft faults, where the circuit continues to function, but outside an acceptable specification.

The actual faults which may be experienced and which affect the analogue performance include:

- MOSFET short circuits between drain-source, drain-gate, and gate-source;
- MOSFET floating gate faults;
- open-circuit or short-circuit capacitor and resistor faults;
- bipolar transistor short circuits between collector-base, base-emitter and emitter-collector, and to-substrate faults;
- process gradients across the chip which cause mismatch between ideally balanced components;
- leaky junctions in both diodes and transistors;
- pin holes and other local fabrication defects;
- mask misalignments and general out of specification wafer fabrication, although this should be detected by the wafer drop-ins (see Figure 1.1) before wafer slicing is undertaken;

and other hard and (particularly) soft faults[14-16]. Hence, the simple stuck-at fault model of the digital circuit is inadequate for applying to analogue circuits in order to cover all possible parametric changes in circuit performance.

Nevertheless, attempts have been made to fault model analogue circuits, on the lines indicated in Figure 7.4. One argument which has been put forward is that although in the digital case the stuck-at fault model does not cover all the possible faults within a digital circuit or system, in practice it proves to be able to cover a very large number of other faults; therefore it is argued that if a given hard fault condition is chosen as the fault model for an analogue circuit, its use will also cover a large number of other possible hard and soft faults.

Details of this development may be found in the published literature[12,17-20]. This fault model strategy for analogue circuits has been justified by reports which suggest that up to 83.5 % of the observed faults in analogue circuits are hard (catastrophic) faults[21], but even so this leaves 16.5 % (or more) faults which would not necessarily be detected by fault modelling based upon some open-circuit or short-circuit fault modelling strategy. Other published data quotes only 75 % hard faults in production analogue circuits[22].

Even if analogue fault modelling is adopted, there is still the requirement to simulate the circuit with each selected fault. This normally involves a SPICE simulation[7,8,23-25], which itself becomes prohibitively costly in time and effort when very large analogue circuits and systems are involved unless some partitioning and controllability and observability provisions are built into the design. Higher level commercial simulators such as HELIX™[25-27] may ease this problem, but the higher the level of simulation, the less fine detail will be present in the simulation results. Temperature and other physical parameter variations may not be possible unless SPICE or some SPICE-derivative simulation is employed[24,28].

Fault modelling of analogue circuits is therefore no longer considered to be a main-stream strategy for analogue testing; instead, detailed fault-free simulation characterisation of the circuit and functional tests to detect any unacceptable deviation from the norm currently represents the generally accepted method of test. In a sense this mirrors the digital situation, where stuck-at fault modelling and ATPG have largely been superseded for VLSI by partitioning and functional tests.

7.4 Parametric tests and instrumentation

Digital circuit testing involves only the detection of logic 0 and 1 signals on the primary outputs of the circuit or network under test, except perhaps at prototype stage and at selected times in a production run when the vendor may undertake some more detailed timing or other parametric tests to

Analogue testing 387

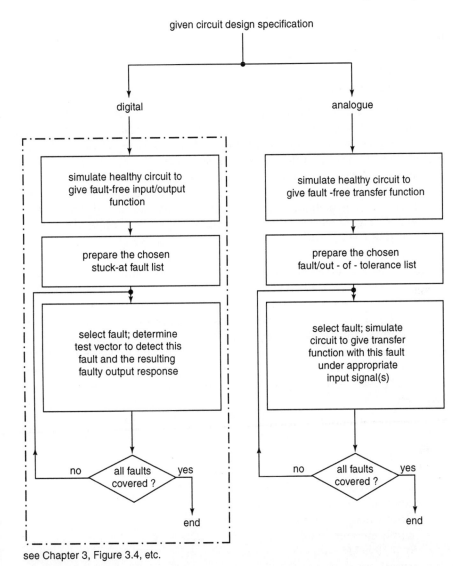

see Chapter 3, Figure 3.4, etc.

Figure 7.4 The attempt to fault model analogue circuits corresponding to the fault modelling of digital circuits

confirm the stability of the production process. Analogue circuit testing, however, always involves more detailed testing to ensure fault-free system performance.

The range of analogue-only circuits which are widely encountered include:

- amplifiers, particularly operational amplifiers;
- regulators;
- oscillators, and particularly voltage-controlled oscillators;

- comparitors;
- phase-locked loops;
- sample-and-hold circuits;
- analogue multipliers;
- analogue filters;

and others. Analogue to digital and digital to analogue converters must also be considered, although it is debatable whether these should be classified under analogue-only or mixed analogue/digital devices. Because of their special importance we will consider them separately, see Section 7.6. In addition to all these dissimilar circuit functions must be added the different applications in which they may be used, which in its turn introduces further testing requirements. Figure 7.5 is a well known illustration of this diversity of analogue macros[29], which precludes having a simple range of building blocks and standard test methodologies for all possible applications.

Manufacturers/vendors of off the shelf analogue parts and of semicustom analogue arrays and standard cells give parametric data on all standard circuit designs. In the case of analogue arrays using individual FETs or bipolar transistors, the full range of data required for SPICE simulation by the OEM may not be available, and therefore default values will have to be used in design simulations. However for circuit macros, the vendors' data sheets will give appropriate transfer characteristics to enable the OEM to build up the necessary test values and fault-free tolerances for system test purposes. Example vendor's data is as follows, additional data being available if required:

- Operational amplifier macro
  L-level output voltage 1.2 V
  H-level output voltage 4.8 V
  small-signal differential output voltage gain 85 dB
  unity gain bandwidth 75 kHz
  slew rate at unity gain 0.05 V/µs
  offset voltage < 10 mV
  supply current 10 µA
  load 1000 kΩ/100 pF
  typical phase margin 60°

  All parameters specified at V_{DD} = 5 V, T_{amb} = 25° C

- Analogue comparitor macro
  input common-mode range 0.5–3.5 V
  propagation delay time for 5 mV overdrive 85 ns
  propagation delay time for 500 mV overdrive 30 ns
  offset voltage 10 mV
  supply current 200 µA

  All parameters specified at V_{DD} = 5 V, T_{amb} = 25° C

Analogue testing 389

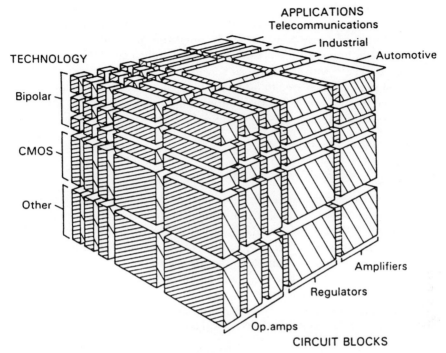

Figure 7.5 A graphical representation of the diversity of analogue circuits, each with their own unique requirements (Based upon Bray and Irissou[29])

- RC oscillator with external RCs macro
  frequency with R = 80 kΩ, C = 100 pF, = 67 kHz, supply current 55 µA typical, 110 µA max
  frequency with R = 50 kΩ, C = 100 pF, = 83 kHz, supply current 75 µA typical, 150 µA max
  All parameters specified at V_{DD} = 5 V, T_{amb} = 25° C

Note that a supply current check is always an appropriate first check to do on all completed assemblies, although it may be difficult to determine the minimum and maximum fault-free values for a new design—gross faults will of course be obvious.

To summarise, the parametric data for device or system test must come from the vendors of the standard parts or designers of special circuits; the precise range of tests which vendors may perform before dispatch and the tests which OEM designers do on their in-house system designs must perforce be decided on an individual basis, but will inevitably include functional tests measured against fault-free response data.

Turning to the instrumentation requirements, we have seen in Chapter 1 how standard instrumentation may be assembled into a rack and stack assembly to provide a general-purpose signal source and output measurement facility, see Figure 1.5 for example. This is possibly the most common type of facility that an OEM may use in the development and prototype stage of a new design, possibly using a more simple specific test rig for production line testing. Further mechanical and electrical details may be found in Parker[30], with the appreciation that analogue test systems such as illustrated in Figure 1.5 are almost invariably custom-built assemblies.

However, in addition to custom-built rack and stack assemblies, there are three other types of analogue test resources commercially available, namely:

- complex general-purpose testers;
- design verification testers;
- individual bench-top instruments.

The first of these categories covers extremely sophisticated equipment often costing several million dollars, examples being the Sentry Series 80, Teradyne Series A500, Avantest Series T3700 and LTX Series 7 testers. Figure 7.6 illustrates one such product, from which it will be appreciated that such resources are similar in complexity and cost to the VLSI digital tester illustrated in Figure 1.4. These analogue testers allow measurement of voltage down to a few microvolts, voltage gains up to or exceeding 120 dB, a dynamic range of more than 140 dB, voltage noise densities in the nanovolt per Hz region, bias measurements down to a few picoamps and frequencies from d.c. to hundreds of megahertz.

The time taken for comprehensive analogue tests using purely analogue instrumentation tends to be very long, and hence faster methods using some form of digital signal processing and digital readout in the test resource is desirable for more rapid production test purposes. The second of the categories of test systems listed above, namely the design verification testers, includes digital signal processing capability. Commercial products include the IMS Logic Master XL, the Hilevel-Topaz systems, Hewlett-Packard 80805 testers, ASIX-1 and ASIX-2 testers and others, with a cost of about one tenth that of the previous category of tester. However, these are all basically digital logic testers with limited analogue testing capability; as such they may be appropriate and relevant for an OEM who has analogue, digital and mixed analogue/digital testing needs, but the analogue test resource may be too limited for certain testing duties.

Thirdly, the bench-top resources are normal individual instruments for specific purposes, such as waveform analysers, synthesisers, data loggers, down to filter units and voltmeters, etc. Such instrumentation may be assembled to provide a test bench for a particular OEM product, but the distinction between a rack and stack resource and a bench-top assembly of standard instruments is becoming blurred since most standard instruments of any standing now have a backplane general purpose instrumentation bus (GPIB or HPIB) built into them to enable them to be jointly controlled under PC supervision. All these categories therefore may now be referred to as automated analogue test equipments, AATEs. The use of a commercial or

Analogue testing 391

Figure 7.6 A comprehensive analogue test system, normally for vendors' use; cf. Figures. 1.4 and 8.3 (Photo courtesy Avantest Corporation, Japan)

in-house test programming language (TPL) to exercise the various generic instruments linked by the instrument bus is normally used, linking simulation to test as follows:

design simulation data
↓
TPL file (inputs, outputs, instrumentation set-ups, etc.)
↓
executable test program for the device under test
↓
compare test results with design simulation data.

The design simulation for relatively small analogue circuits can be SPICE or one of its many commercial derivatives such as CDSPICE-Verilog, or other available software which caters for larger circuits and systems. However, in order that the analogue designer does not have to write detailed software programs for the test of every new circuit or system, a longer term objective which both vendors and OEMs would appreciate would be the availability of macro library test data, corresponding to the vendors' library of design data, thus giving the design engineer both the circuit data and relevant test data information. This would allow the possibility of readily entering specific operating voltages, frequencies, bandwidths, etc. into the TPL test data program(s), thus automatically generating the required instrumentation instructions for the circuit or system test.

7.5 Special signal processing test strategies

In the foregoing discussions it was implicitly assumed that the a.c. test signals used in test mode to energise the analogue circuit or system under test are as far as possible the same as those which the circuit receives in real life, and the resulting circuit output signals are therefore also normal circuit response signals. However, the increasing adoption of digital signal processing strategies allows nonstandard test signals to be considered for test purposes in addition to the normal mode waveforms.

The disadvantages of using conventional analogue instrumentation include:

- the difficulty of setting input parameters to exact or repeatable values;
- the inaccuracy in measuring and/or reading output parameters;
- the difficulty of synchronising two or more test inputs which may be necessary in some analogue tests;
- the slow response of all measuring instruments which rely upon mechanical movements to give their output reading.

The last is particularly significant, since the settling time of nondigital instrumentation places a limit on the speed with which the circuit may be checked[31]. Also, it is possible to apply other than conventional a.c. signals to the circuit under test; for example if a pseudorandom binary sequence from an autonomous LFSR generator was fed into an amplifier or filter or other analogue circuit, something is likely to appear at the circuit output(s) which may be an acceptable signature for the circuit under test, but conventional a.c. instruments are unlikely to appreciate such unusual output waveforms.

There is, therefore, growing significance in specifying test data for analogue circuits which involve the digital domain; signal processing techniques and detailed mathematical analysis may be involved in many of these test strategies. We will consider some of these developments below, all of which apply equally to the mixed analogue/digital test methodologies of the next chapter as to the analogue-only case being considered here.

7.5.1 DSP test data

A major advantage of using digital signal processing techniques to test analogue circuits is that the final test response is analysed in the digital domain rather than in the analogue domain, which considerably eases the problems of deciding when the response data is acceptable. Notice that in digital circuit testing if any output bit differs from the fault-free response then a fault is recorded, but in analogue testing there is always some acceptable variation in output signal values, which may be considerably easier to define and measure in the digital domain than in the analogue domain. Also, as previously noted, testing times can be appreciably reduced, a factor of ten to

20 times faster being quoted for analogue filters by analysing magnitudes and phases via a fast Fourier transform (FFT) obtained from a single-pass measurement[8].

Perhaps the most powerful attraction of using DSP techniques for analogue testing is for built-in self-test (BIST) strategies. However, this is invariably associated with mixed analogue/digital systems, where the digital part also has a built-in self-test resource, and therefore we will defer consideration of this until the following chapter.

The DSP emulation of analogue test instrumentation is generally of the form shown in Figure 7.7a, where the PC or workstation contains the software program to generate the required input stimuli and to analyse and display the resulting output test data. The particular input stimuli and output test signature(s) are specific for the particular analogue circuit under test.

In most cases the input stimuli will be generated by a digital to analogue converter whose input is the sequence of words provided by the preceding memory. Therefore any waveform and frequency input can in theory be applied to the circuit under test, limited only by the performance of the DAC and its input data. Similarly, the output will generally be processed by an analogue to digital converter, the digital output of which is sampled and captured in memory for test acceptance/rejection. This is illustrated in Figure 7.7b. It may be necessary to include appropriate bandpass filters in the analogue input and output paths to eliminate the possibility of any high-frequency components in the input and aliasing in the output, together with some overall sequencing and synchronisation to ensure correct input/output relationships of the digital data. In general the maximum sampling frequency in the digital domain should be, say, at least ten times the maximum analogue frequency, which places a limit on this type of testing for very high speed applications. Further details may be found in the published literature[8,31-35]. Sampling speed requirements and aliasing (the Nyquist criteria) may be found in Reference 31 and elsewhere[36].

There are commercially available testers which employ such digital domain methods for analogue circuit testing, such as the LTX workstation-based Hi.T linear test system illustrated in Figure 7.8. This computer-aided test development and test application system is based upon an Apollo DOMAIN network, with a multiple windowing and graphics capability to plot the results of (i) initial simulation, (ii) test development procedures, and (iii) final active test measurements, in write—edit—run procedures.

The Hi.T test system architecture is generally as shown in Figure 7.7a, and provides both high-accuracy synthesis and high-accuracy measurement of a.c. and d.c. signals. Waveforms with better than 100 dB signal to noise performance are available, with the particular feature that each pin connected to the circuit under test can be made:

(i) an independent a.c. waveform source;
(ii) an independent a.c. waveform measurement;

394 VLSI testing: digital and mixed analogue/digital techniques

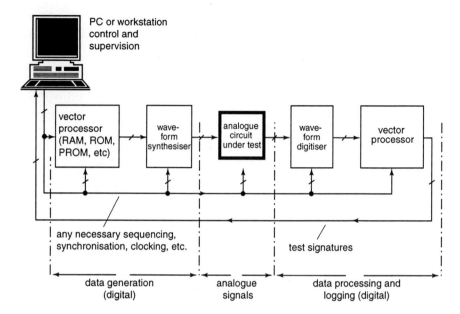

Figure 7.7 DSP testing strategies for analogue circuits
   a general methodology with digital domain input and output data
   b input DAC/output ADC configuration

(iii) an independent d.c. source;
(iv) an independent d.c. measurement;

for up to 96 pins maximum. This DSP per pin feature, each pin with its own small amount of local memory, is said to offer extreme flexibility to the user to test any analogue or analogue-with-digital circuit at maximum speed, the multiple processors and local pin memory eliminating the need to distribute continuously any large amounts of DSP data around the system during test. Synchronisation of all the per-pin DSP resources is precisely controlled.

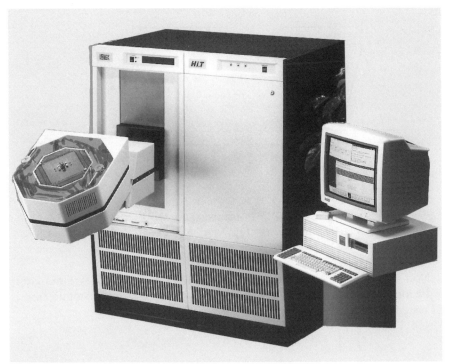

Figure 7.8 The LTX Hi.T linear IC test system (Photo courtesy LTX Corporation, MA)

7.5.2 Pseudorandom d.c. test data

The use of pseudorandom sequences generated by autonomous linear feedback shift registers has been proposed for the testing of analogue circuits, particularly using maximum-length pseudorandom sequences (M-sequences). The testing method is therefore similar to that previously shown in Figure 3.16, but the circuit under test is now an analogue circuit rather than a logic circuit or system.

Recall from Chapter 3 and Figures 3.17 and 3.20 that when the feedback taps are appropriately chosen, an n-stage LFSR will act as an autonomous test vector generator with a maximum length sequence of $2^n - 1$ vectors before the sequence repeats. The characteristics of the sequence are pseudorandom, and it has the autocorrelation function given in eqn. 3.15. This autocorrelation function gives maximum correlation when any vector is correlated with itself, and minimum correlation when correlated with any other time-shifted vector in the M-sequence. It therefore approximates to a unit impulse, as shown in Figure 7.9. There is the small d.c. offset $(1/2^M)\, a^2$ in this near unit impulse, where $M = 2^n - 1$, so that the total area under the correlation function remains zero.

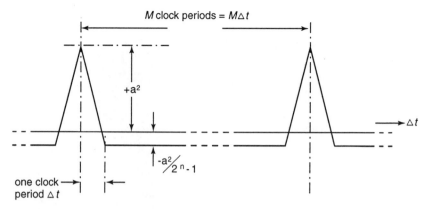

Figure 7.9 *The autocorrelation of a maximum-length M-sequence, where* n *is the number of stages in the LFSR and* $M = 2^n - 1$ *is the maximum length vector sequence*

Now if there was available a true unit impulse input, the Weiner-Hopf equation given below gives the impulse response of a given analogue circuit or system[36-40]:

$$\Phi_{XY}(\tau) = \int_{\tau=0}^{T} h(\tau)\Phi_{XX}(\tau - \tau_1) \, d\tau_1 \qquad (7.1)$$

where

$\Phi_{XX}(\tau)$ = the unit impulse input

$\Phi_{XY}(\tau)$ = the impulse response of the circuit.

Therefore, a test strategy as shown in Figure 7.10a can be proposed. The output of the circuit under test is multiplied by the M-sequence delayed by $0\Delta t, 1\Delta t, ..., (M-1)\Delta t$, and then integrated and sampled to provide discrete points from the circuit's impulse response.

However, it is extremely costly in test circuit overheads to provide the $M-1$ delayed sequences, multiplication and integration, and therefore experiments have been made to use a very much reduced set, possibly only two delayed sequences, in the test structure. This is illustrated in Figure 7.10b, where only two delays are incorporated[40], together with possibly simplified test signature sampling at the output. Notice that provided the maximum-length autonomous LFSR generator is a type A generator and not a type B, see Figure 3.20, and if the output sequence from the first stage is used to drive the circuit under test, then the delayed sequences $1\Delta t, 2\Delta t$, up to $(n-1)\Delta t$ are freely available at the 2nd, 3rd, up to the the final nth stage output of the register. Another factor which should be considered in attempting any form of impulse-based test is that the period of the M-sequence must be greater than the impulse decay of the circuit under test, or otherwise the impulse response of the circuit will be overlaid with subsequent circuit responses[40].

Analogue testing 397

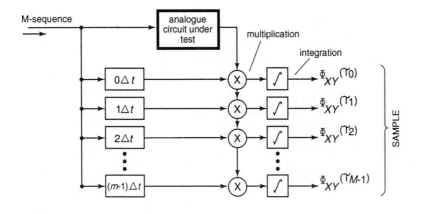

b

Figure 7.10 Published concepts of using pseudorandom LFSR M-sequences for analogue testing
a unit impulse response testing
b a greatly simplified pseudorandom test procedure[40]

It is not apparent that using pseudorandom M-sequences offers any strong advantages for the testing of analogue-only circuits. There may be a better case to make with mixed analogue/digital circuits, where the autonomous LFSR generator may already be available for d.c. (logic) test purposes. However, in any case, the full mathematical implementation of eqn. 7.1 is unlikely to be an economical proposition, and hence it is more realistic to suggest that every analogue circuit should be considered individually, attempting to show what the circuit response would be to some suggested pseudorandom d.c. input sequence either by simulation or by the use of a gold circuit to determine the fault-free response and the potential fault coverage.

Some further research work has also been pursued wherein the d.c. supply current to operational amplifiers, comparators and other simple analogue circuits is monitored while a pseudorandom d.c. test input is applied[41-44]. It is claimed that this I_{DDQ} type test will detect a high percentage of internal circuit faults, with no necessity to monitor any internal nodes or primary I/Os. It was also reported to be advantageous to ramp the d.c. supply voltage during test rather than to maintain the supply always at a nominal value. However, this and the previous types of proposed pseudorandom test strategies are unlikely to provide a good test methodology for any soft faults in analogue circuits.

7.6 The testing of particular analogue circuits

The testing of particular types of analogue circuit other than straightforward operational amplifiers, comparators, etc., has received very specific attention, and test strategies have been proposed or developed to cater for these analogue elements. Here we will introduce work which has been published on two types, namely filters and analogue to digital/digital to analogue converters. These strategies are alternatives to applying normal-mode input signals and confirming that the outputs are within specified test limits.

7.6.1 Filters

Filter design is one of the few area of microelectronic design which may be fully automated. Given a required filter specification and the designer's choice of type of filter configuration, then component values may be automatically generated by CAD software programs. This is the nearest to true silicon compilation that is generally available to an OEM system designer, since the outline circuit configurations are known and only component evaluation and overall simulation is involved[8,45,46].

The full hierarchy of filters is illustrated in Figure 7.11. Passive filters, usually composed of discrete resistors, capacitors and inductors, need not concern us here; almost without exception in microelectronic-based systems the required filters will be active, based upon the use of operational amplifiers to provide circuit gain and the required transfer functions. CAD filter software is invariably used for active filter design, with more basic simple circuit simulation or SPICE being sufficient for any required passive filter design activity[11,24,25,47].

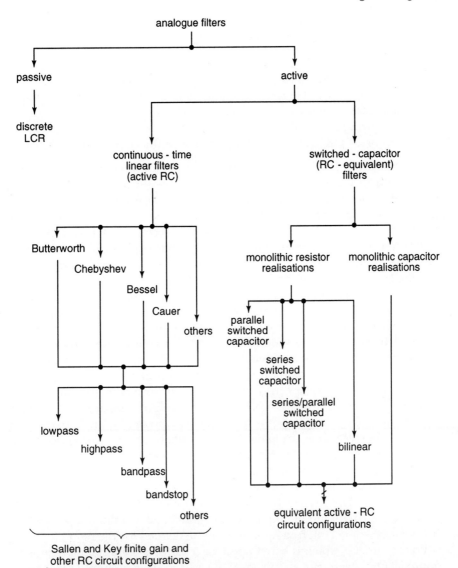

Figure 7.11 The hierarchy of analogue filters. Note that in switched-capacitor filters the internal switching frequency must be $\gg$ than the frequency of use of the filter. Finite impulse response (FIR) and infinite impulse response (IIR) filters have been omitted from this table as they are digital-only devices

As shown in Figure 7.11, active filters may be divided into continuous-time linear filters and switched-capacitor (SC) filters. Continuous-time linear filters mirror the characteristics which LCR passive filters can provide, but with generally improved characteristics. Because inductors are almost impossible to realise in monolithic form, operational amplifier configurations are necessary in order to realise inductance. The conventional types of filter, namely:

- Butterworth, giving maximally-flat passband response;
- Chebyshev, giving steepest transition from passband to stopband but at the expense of ripple in the passband;
- Bessel, giving maximally-flat time delay;
- Cauer, or elliptic, giving the sharpest knee at the edges of the bandpass;

and variants are freely possible. However, since capacitors are particularly convenient to make in monolithic form, with closely matched rather than absolute values being readily available, switched-capacitor filters are now much more attractive for microelectronic realisation than continuous-time filters. MOS technology can provide the necessary matched capacitors, efficient switches and adequate operational amplifier performance for almost all applications. Figure 7.12 shows a fifth-order filter as an illustration of switched-capacitor circuit design.

We have no need to consider any of the design details of filters here, since all test procedures are undertaken using the normal primary input and output terminals, with no reference to any internal nodes. Many excellent texts on filter design are widely available[48–55], which may be consulted for detailed design theory and practice. Testing is conventionally straightforward, using standard commercial instrumentation such as sweep-frequency oscillators, spectrum analysers, vector voltmeters and other instrumentation, see for example Figure 1.5. Some standard digital signal processing of the filter output response may be present in order to quantify particular performance parameters such as phase response, ringing, etc.[32], with data logging to provide recorded test information.

However, there has been research into nonconventional ways of testing analogue filters, such as impulse and transient testing, which we will briefly consider here. The strategy behind many of these developments has been to simulate the circuit under fault-free and chosen fault conditions, from which information some fault-free signature can be defined for the filter.

One approach, which is based upon hard faults, is the d.c. fault dictionary approach[54,56]. Here the circuit response to a given d.c. input waveform or impulse is simulated (i) under fault-free conditons and (ii) under some set of faults which includes internal nodes individually set at voltages outside their normal fault-free tolerance band. From these simulations it may be possible to specify an output response which will act as a signature for some or all the faults in the chosen faults list[18,31,54–61]. This technique has been applied to continuous-time filters, but does not appear to be able to provide a 100 % fault cover of the faults list when only the input and output nodes are accessible. It may, however, provide a very rapid and economical first test before more detailed parametric tests on the output response are undertaken, for example as an initial wafer-probe test.

Analogue testing 401

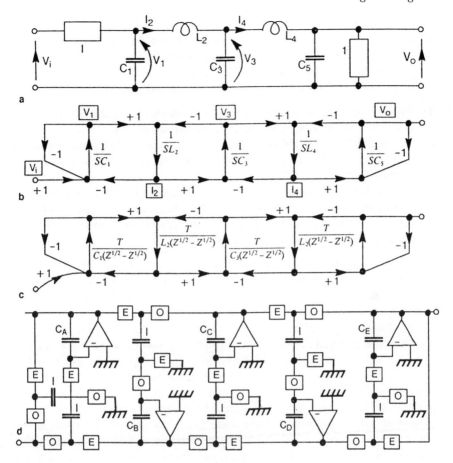

Figure 7.12 Fifth-order switched-capacitor design illustration
a prototype LCR filter schematic
b signal flow graph
c discrete time signal flow graph
d final schematic design with differential-input operational amplifier integrators; E = even phase, O = odd phase switching elements
(Acknowledgement, based upon Reference 67)

Instead of monitoring the output voltage as is done in the d.c. fault dictionary approach, supply current monitoring has been considered as an alternative overall test strategy[57–62], but again such tests are unlikely to be able to be sensitive enough to cater for just out of specification soft failures. Details of certain tests on continuous-time filters with built-in current sensing have been published[18,61]. The results shown in Binns *et al.*[61] relate to a second-order bandpass filter using a test structure as shown in Figure 7.13a, with the test register connecting a test load to the filter under test and triggering the pulse generator so as to apply d.c. input pulses to the filter. The transient current measurement is made by means of the built-in current monitoring circuit shown in Figure 7.13*b*. The principal suggested application of this test strategy, however, is not for the test of separate standalone circuits, but for buried macros where the primary inputs and outputs of the macro are not freely available.

Other test strategies which have been considered include:

- digital modelling, in which hard faults in analogue circuits are considered as stuck-at faults in an equivalent digital circuit model;
- functional K-map modelling, which is a variation on digital modelling but uses a Karnaugh-map type representation to plot the faults and their commonality;
- logical decomposition, which consists of decomposing a (usually large) analogue circuit or system into subcircuits, with the summation of simulated and measured currents at the connecting nodes of the subcircuits being used as the test parameter.

Details of these proposed strategies may be found published[31,55,63,64], but are more relevant to consider for mixed-signal analogue/digital systems than for relatively simple standalone analogue macros. They also tend to cater for catastrophic (hard) faults rather than for out of tolerance testing.

The use of pseudorandom sequences and the impulse response relationships given in Secton 7.5.2, see eqn. 7.1, has also been considered for filter testing. In particular, the reduced set of correlations illustrated in Figure 7.10*b* has been claimed to be relevant for the testing of continuous-time filters[40], but no detailed substantiation to show that all the filter parameters are comprehensively checked is currently known.

Also as far as is known, there have been no published developments concerning test strategies which are specific to switched-capacitor filter designs. There may well be, indeed are, considerably dissimilar internal faults that can occur in switched-capacitor filters compared with continuous-time filters, and perhaps the fault modelling of the former may be easier, but this does not yet appear to have been reflected in any generally known published studies of test strategies specifically for SC filters. Intuitively it ought to be easier to test an SC filter than a continuous-time filter, particularly if the internal switching waveforms are accessible, and hence further developments in this particular area may be expected.

In considering all these possibilities for the test of filter circuits, it is clear that if the input of the circuit is given an appropriate impulse or other d.c. waveform, then some transitory response will appear at the filter output, provided that the input signal has frequency components within the passband of the filter. Transient response, therefore, is a theoretically valid test strategy to consider, provided that the output response contains sufficient information to provide an acceptable test signature[65]. However on balance it would seem that conventional testing of standalone filters, where there are no problems of controllability and observability, at present remains the preferred means of testing. This may not be the case for other more complex

Analogue testing

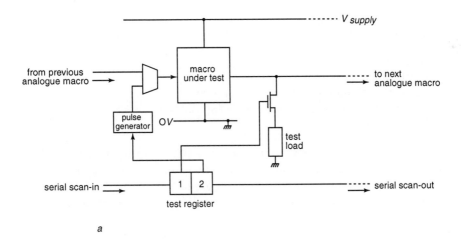

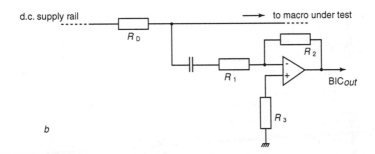

Figure 7.13   *Current signature monitoring*
    a outline test strategy with a test load applied to the a.c. macro under test
    b the built-in current monitoring, where R_d is a small-value dropping resistance in the supply rail

types of analogue circuit, or if speed of testing is a critical consideration, or for built-in analogue circuits, for example see the next section covering A-to-D and D-to-A converters and Chapter 8.

7.6.2 Analogue to digital converters

As previously mentioned, analogue to digital and digital to analogue converters (ADCs and DACs) should strictly be classified as mixed-signal analogue/digital devices, but for convenience we will consider them here rather than in the following chapter since they are well-defined autonomous devices.

Standard off the shelf ADCs and DACs are usually tested by their vendors using extremely sophisticated testers tailor made for the purpose, and costing in the region of \$10^6 to \$10^7. The test procedures would largely be functional in accordance with the device specification, the emphasis being upon:

- throughput and speed of test;
- purity of the analogue waveforms;
- equal quantisation of the digital increments;
- temperature coefficients and conversion stability;
- settling time;

and possibly other detailed parameters. The OEM testing of such standard parts, when required, would also be functional, and in the case of relatively small converters would not involve a very great deal of detail or effort. One possible OEM strategy when both A-to-D and D-to-A converters are available is to back to back test them, feeding the ADC output into the DAC, and comparing the DAC output with an a.c. test input fed into the ADC[66]. Allowing for the quantisation in the digital signals, see below, the a.c. output should be the same as the a.c. test input. The opposite arrangement of a DAC feeding an ADC may also be used, see Figure 7.21 later. In these situations it may be advantageous to have a higher performance DAC available than the ADC under test, and conversely a higher performance ADC available than the DAC under test, if possible. In a further situation, if an OEM has a relatively simple system design which involves an input ADC, some internal digital signal processing, and an output DAC, a possible test procedure may be an overall functional test using the primary analogue I/Os only, not observing the internal digital interfaces unless necessary.

However, when ADCs and DACs are buried within more complex systems, often with restricted controllability and observability, then it may be advantageous to consider some alternative possibilities for the testing of such macros. Here we will briefly look at some of the concepts which have been considered.

Considering in this section analogue to digital converters: many detailed forms of ADCs are available[1,48,66,67], all of which have quantisation steps as illustrated in Figure 7.14a. The number of steps in the digital code is the resolution of the converter; a 16-bit converter for example has a resolution of 1 bit in 2^{16}, which is 0.00152 % (15.2 p.p.m.) of the full-scale reading. Because one digital code word represents a discrete interval of the analogue input signal, this interval is usually referred to as the quantisation step, QS, its amplitude being given by:

$$QS = \frac{1}{2^n} FS \qquad (7.2)$$

where n is the number of bits in the conversion and FS is the full-scale reading of the analogue signal. There is therefore a maximum quantisation error of $QE_{max} = \pm\frac{1}{2}QS$ in the output voltage as indicated in Figure 7.14b, assuming a perfect converter.

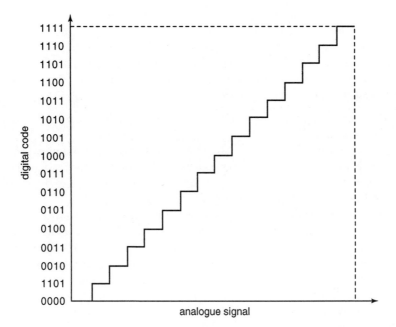

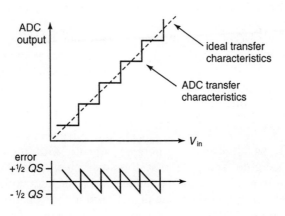

Figure 7.14 Analogue to digital conversion characteristics
  a transfer characteristics of an ideal four-bit converter
  b the quantisation error of an ADC conversion

The quantisation error for a converter with a perfect sinusoidal input waveform and four-bit quantisation ($QS = 1/16\ FS$, $QE_{max} = \pm 1/32\ FS$) is illustrated in Figure 7.15. However, this assumes that the analogue to digital conversion steps are all perfect, and not subject to any of the possible errors, variations or tolerancing which may occur in real life. Among these imperfections may be:

- offset error, which is a constant shift of the complete transfer characteristic;
- gain error, which gives rise to an incorrect full-scale value—a negative gain error is when the analogue full-scale value is associated with a digital code smaller than 1 1 ... 1 1, and a positive gain error is when the digital code of 1 1 ... 1 1 is with a less than full-scale analogue signal;
- nonlinearity, which may be divided into integral nonlinearity and differential nonlinearity; integral nonlinearity is a global measure of the nonlinearity, and is defined as the maximum deviation of the actual transfer characteristic from a straight line drawn between zero and the full-scale point of the ideal converter, differential nonlinearity is defined as the maximum deviation of any of the analogue voltage changes for one bit change in the digital representation from the ideal analogue change of $1/2^n\ FS$.

All these error characteristics are conventionally referred to the analogue voltage input, and as such hold equally for D-to-A as well as A-to-D converters. (There is a further error which is particularly associated with D-to-A converters, namely nonmonotonicity error, which we will include in Section 7.6.3.)

The above types of ADC error are illustrated in Figure 7.16. In general the vendor will primarily be concerned to ensure that all these possible errors, plus others such as noise, settling time and temperature coefficients affecting gain and linearity[66–72] lie within specified limits, and the OEM should have little need to test for other than gross faults in commercial off the shelf products.

The functional testing of ADCs is usually undertaken with sinusoidal voltage waveforms, which can be generated off-chip with high purity. However, if very accurate triangular or trapisoidal voltage waveforms are available, giving constant dV/dt voltage gradients, using such an input to test an A-to-D converter should result in the digital output code changing with exactly equal time intervals between each code word increment. Therefore a possible test for gain and linearity is to check that the digital output code is correct at two or more points, say zero, mid range and full scale, and that the increments in the digital output code occur at equally spaced time intervals. There may be difficulties in instrumenting such a test, but it would cover most of the potential imperfections in the analogue to digital conversion. However, other indirect tests have been proposed which may be more practical, such as considered below.

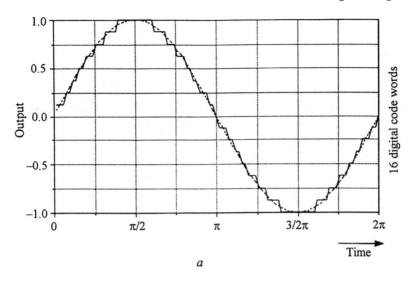

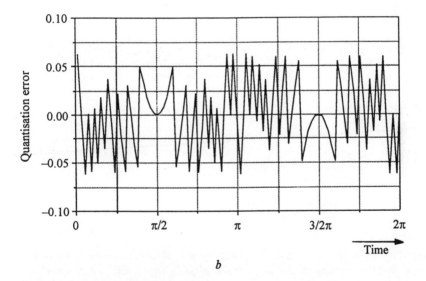

Figure 7.15 A-to-D conversion of a perfect sinewave with four-bit quantisation
  a the original sinewave and the four-bit quantisation
  b the quantisation error (Acknowledgement, based upon Reference 66)

Before we consider some of this published information, let us first summarise the important relationships that arise between (i) the frequency of an ADC input signal and (ii) the internal sampling frequency of the converter which controls when the individual (discrete) output words are generated. If the a.c. input signal is an extremely low frequency signal

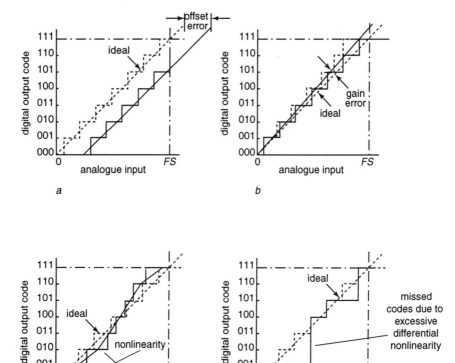

Figure 7.16 Errors in analogue to digital converters, three-bit (eight digital code words) shown for clarity
 a offset error
 b gain error
 c integral nonlinearity error
 d differential nonlinearity error

compared with the internal sampling speed of the converter, then it is intuitively obvious that all the steps in the digital output will be exercised, and therefore in theory one cycle of this low input frequency of the correct amplitude should be a sufficient test for the converter. However, as the input frequency rises then one cycle of the input waveform will increasingly be inappropriate to check the A-to-D conversion fully.

Considering an eight-bit ADC with 256 output code words (0 to 255) with, say, a 5 kHz pure sinusoidal input signal, and assume that the internal sampling frequency of the converter is 25 000 samples per second, which is above the limit at which aliasing of 5 kHz can occur. Suppose also we apply this sinusoidal input for 200 complete cycles. Therefore we have:

- input frequency, $f_{in} = 5$ kHz;
- no. of input cycles, $M = 200$;
- unit test period, $UTP = 200$ cycles at 5 kHz,
  $= M(1/f_{in})$, $= 40$ ms;
- sampling frequency, $f_{sm} = 25$ kHz;
- no. of samples N per test period $= f_{sm} \times UTP$
  $= 1000$.

Thus, assuming that the input voltage is appropriate, it would appear that there are a sufficient number of samples taken to be a test of the 256-word eight-bit converter.

However, if we look more closely at this situation of $f_{in} = 5$ kHz and $f_{sm} = 25$ kHz, it will be appreciated that if these frequencies are exact then the same values on the a.c. input waveform will be sampled on each input cycle, and that only five sample points (i.e. only five different output code words) will be present. Therefore, for this test procedure to be satisfactory, we must break this exact integer relationship between f_{in} and f_{sm} so that the input signal drifts with respect to the instants of sampling over the test period.

Two possibilities now arise, namely coherent sampling and noncoherent sampling. In coherent sampling the integer relationship between f_{in} and f_{sm} is no longer present, but a form of synchronisation between the two over the unit test period is maintained as follows.

Starting from the desired number of samples, N, per test period and the given sampling frequency, f_{sm}, we have:

$$N = f_{sm} \times UTP$$

$$f_{sm} = N(1/UTP)$$

$$= N\delta$$

where δ is termed the primitive frequency of the test. The input frequency f_{in} should now be:

$$f_{in} = M\delta$$

where M is the number of input cycles of the input frequency, but where M and N now have no common factor other than unity. Taking the above example, we have:

$$\delta = \frac{f_{sm}}{N} = \frac{25 \text{ kHz}}{1000}$$
$$= 25 \text{ Hz}$$

allowing us to choose frequencies near our original input frequency of 5 kHz as follows:

M	$f_{in} = M\delta$
197	4925 Hz
199	4975 Hz
200	Not allowed
201	5025 Hz
203	5075 Hz
⋮	⋮

Choosing, say, the input frequency 5025 Hz, then the complete coherent test specification is now:

- input frequency, f_{in} = 5.025 kHz;
- no. of input cycles, M = 201;
- unit test period, UTP = 201 cycles at 5.025 kHz
  = 40 ms;
- sampling frequency, f_{sm} = 25 kHz;
- no. of samples, N, per test period = 1000.

The 1000 test samples are now distributed throughout the range of the digital code words, with no exact point on the a.c. sinewave input signal being sampled more than once in the 201 cycles—the finite resolution of the converter will however mean that the same output digital code is of course given more than once.

If the sampling of the input waveform is noncoherent, then f_{in} is offset from the value where there is an exact integer relationship between f_{in} and the sampling frequency f_{sm}, but with no other specific relationship between them such as illustrated above. Since there is now no longer any recognised form of synchronisation between the two, the sampling process effectively becomes a random sampling process, and a very large number of a.c. input cycles may be necessary to ensure that the device test is fully comprehensive. Both coherent sampling and noncoherent sampling (sometimes referred to as correlated and uncorrelated, respectively) will be found in the following discussions.

Continuing, therefore, with alternative published ADC test strategies, one possibility is to determine the frequency of occurrence of each digital output code word with a given test input simulation. A single constant dV/dt ramp input voltage to an A-to-D converter should result in the equal occurrence of every digital output code, provided that the dV/dt rate of change of the input signal and the internal ADC sampling rate are appropriately correlated; if the input ramp and the sampling rate are uncorrelated then the cumulative occurrence of each output code over very many cycles of the input ramp is necessary. However, this method of testing, known as histogram testing, conventionally uses a pure sinewave rather than any other input waveform, the frequency being selected such that the testing is noncoherent. This gives rise to a cusp-shaped distribution of the output code words, as illustrated in Figure 7.17. A very large number of cycles of the sinusoidal input is applied in order to establish a mean response. The fault-free probability function is given by

$$p(V) = \{\pi(A^2 - V^2)^{1/2}\}^{-1} \tag{7.3}$$

where A is the peak amplitude of the sinewave, input = $\frac{1}{2}FS$, and $p(V)$ is the probability of occurrence of a particular digital output code word. (A somewhat longer equation in terms of a sinusoidal input voltage $A \sin \omega t$ has been given by Doernberg et al.[72].) With this representation, permanent non-linear behaviour in the converter will result in spikes appearing on this ideal probability histogram, and missing codes will cause gaps (zero values). Offset errors will result in the histogram being asymmetrical about the centre, and gain errors will result in a change of width of the histogram[69,72,73].

Unfortunately, the considerable number of cycles of the input waveform necessary in order to average out the occurrence of each digital output word is a serious problem, particularly if this is a test of a buried A-to-D converter. A figure of 2×10^5 output samples for an eight-bit ADC has been quoted[72], which results in long test times to obtain accurate test data, and a considerable volume of data to be shifted off-chip if the DAC is buried. The testing time is to a first degree independent of the frequency of the a.c. waveform; if a low frequency is used then more samples are recorded per cycle; if a higher frequency is used then more cycles are necessary in order to obtain the total sample count. In general a high frequency, approaching the upper limit of the ADC specification, is a better overall test.

An alternative test method is to employ a fast Fourier transform (FFT) on the converter output[48,69,71,72]. A displaced pure sinusoidal input voltage of $\{\frac{1}{2}FS + FS \sin \omega t\}$ is applied to the converter, and 2^n samples are taken at maximum sampling speed. These output samples are converted to the frequency domain by the FFT, and the spectral content of this output is examined. Harmonics of the pure input sinewave frequency are present in the FFT output spectrum, caused by the nonlinearity of the converter, which may be used to confirm that the converter performance is within its specification. The frequency of the input sinewave must be chosen such that the harmonics in the FFT output do not coincide with the sampling frequency of the ADC. It is found that if the ratio of the fundamental to the highest value harmonic is greater than $6n$ dB, the nonlinearity of the converter is negligible compared with the inherent quantisation noise of the conversion.

Looking at these two test strategies, neither is in practice ideal. The conventional histogram method requires the large number of a.c. input cycles to build up a smooth plot, and the FFT method is highly sensitive to the purity of the a.c. input signal and input noise, particularly when n is large. However, there has been further research based upon a histogram method which attempts (i) to extract signal to noise information from the histogram

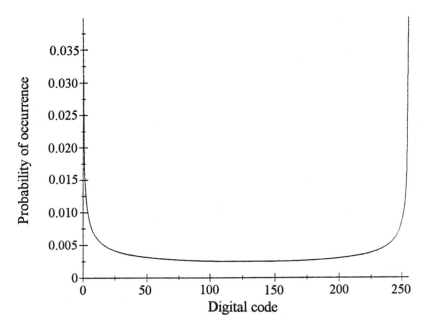

Figure 7.17   The ideal probability of occurrence of the output code words of an eight-bit ADC with many cycles of a pure sinewave input

measurements so as to eliminate the need to undertake a FFT analysis, and (ii) to use coherent sampling in order to reduce the large number of cycles necessary.

Consider again the histogram plot: the effect of a large number of cycles is to smooth out any quantisation and background noise, but if the number of cycles is restricted then these factors will not be smoothed out, and a plot such as shown in Figure 7.18a will result. However, if the number of samples is now limited, then the sinewave input should be made coherent with the ADC sampling in order to maximise the amount of information obtained from each sample and spread the noise uniformly throughout the spectral range, and to obtain repeatable results on each test[73]. The number of samples required to produce the simulated plot of Figure 7.18a is given as 2048, using an eight-bit 256 output code word DAC[74,75].

To establish a link between the signal to noise output characteristic of the ADC and this histogram data, a new parameter Δj has been introduced which defines the difference between the histogram plot obtained from an ideal device and that obtained from a real-life device, where Δj is defined as:

$$\Delta j = \frac{1}{S} \left\{ \sum_{i=0}^{2n-1} d_i \right\} \tag{7.4}$$

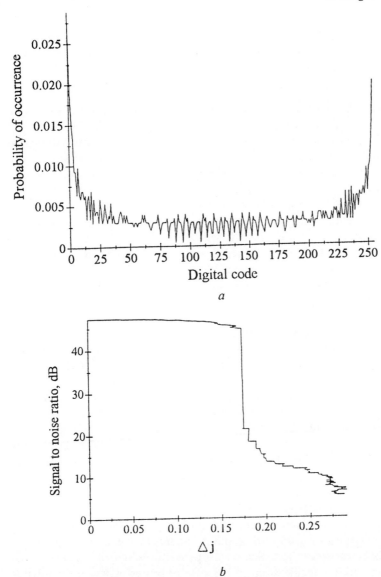

Figure 7.18 Simulation results for an eight-bit DAC with a reduced number of sinusoidal input cycles, the input being made coherent with the ADC sampling frequency
 a simulated results of perfect converter (ideal quantisation)
 b simulated results for signal to noise ratio SNR versus Δj
  (Acknowledgements, References 74, 75, 77)

where S = total number of samples in the histogram
 n = number of bits in the DAC output (as previously)
 i = the output code word number

and

d = the difference in the number of occurrences of each code word i in the ideal simulated case and the actual measured number of occurrences under test

= |(ideal no. of occurrences) − (actual no. of occurrences)|$_i$

Δj is therefore an averaged summation of the difference between the actual and the perfect number of occurrences of every output code word from the converter. The more imperfect the ADC output becomes, the more the signal to noise ratio of the output should decrease, but Δj should increase.

The relationship between the parameter Δj and the signal to noise ratio of the ADC output has been investigated by adding noise to the output simulation of Figure 7.18a, and determining the signal to noise ratio and the value of Δj on each simulation. This gives the SNR versus Δj characteristic shown in Figure 7.18b, from which it will be seen that there is a very sharp knee in the characteristic as the SNR value deteriorates. Repeating this procedure for a number of other possible output errors also shows this very sharp change associated with some critical value of Δj, making Δj a powerful parameter to consider as the ADC test signature[74,75].

The proposed method of test for a given type of A-to-D converter therefore initially involves a very computer-intensive simulation procedure to determine the correct graph of SNR versus Δj for the particular device. The actual test procedure may then be as shown in Figure 7.19a where the RAM memory is addressed by each individual output code word of the ADC, every time a memory is addressed its output word being increased by one. For example, if output code word 1 1 0 0 1 1 0 0 occurs eleven times during the test, the final m-bit output word of the memory on this address will be ... 0 1 0 1 1. Figure 7.19b shows the graph of SNR versus Δj for the Analog Devices' AD7575 eight-bit ADC; a test of this device is quoted as giving a Δj value of 0.286, which corresponds to an SNR value of 46.8 dB; the value obtained by conventional FFT analysis was 45.8 dB, and the manufacturer's minimum pass value is given as 45 dB.

From the above discussions, there are clearly many ways of approaching analogue to digital converter testing, but it remains the vendor who generally has to perform the more sophisticated test procedures in order to ensure that products are within specification. In the case of buried ADCs, it is probable that it will be a mixed analogue/digital circuit or system under consideration, possibly with some built-in self-test strategy, and hence an isolated test of the ADC may not be relevant. We will consider mixed-signal BIST and other test strategies in the following chapter.

7.6.3 Digital to analogue converters

Turning from the testing of individual A-to-D converters to D-to-A converters, again a number of test strategies have been proposed. The error

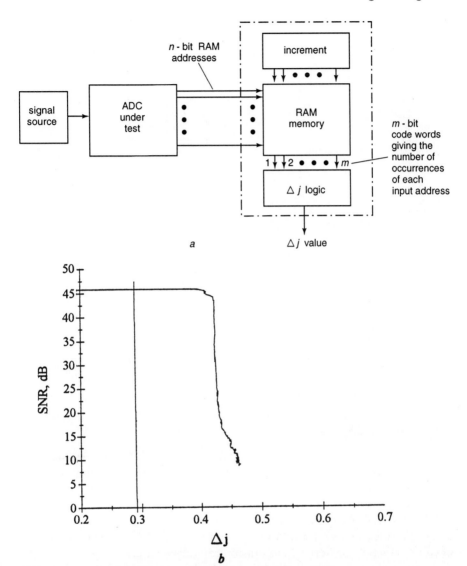

Figure 7.19 *The use of Δj for ADC testing*
  a simplified schematic of the hardware for generating Δj, omitting any coherence and other control signals
  b the graph of SNR versus Δj for the AD7575 A-to-D converter (Acknowledgements, References 74, 75, 77)

characteristics in Figure 7.16 for A-to-D converters apply equally to D-to-A converters, although they may be plotted with interchanged axes[48], but there is one further error characteristic which we must include here for completeness, this being nonmonotonicity error. This is shown in Figure 7.20, and is defined as an error whereby an increasing digital output code

416 *VLSI testing: digital and mixed analogue/digital techniques*

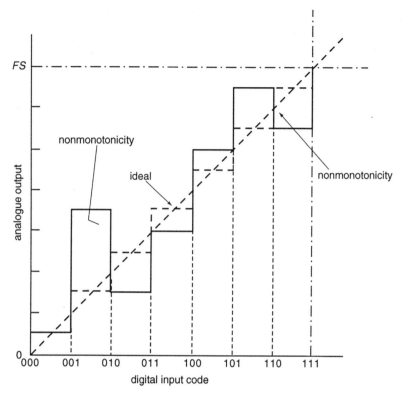

Figure 7.20   *Nonmonotonicity errors in D-to-A converters, two occurrences of nonmonotonicity shown*

does not result in an increasing analogue output voltage. (When the converter is monotonic, then the output voltage always increases with an increasing digital input code, although possibly still subject to the various errors and imperfections previously defined.)

Like ADCs, there are a variety of different DAC circuit implementations, details of which may be found in the literature[1,11,48,70,76]. Such circuit details need not concern us here where our interests are restricted to test strategies, since once again the principal testing methods are functional, looking at the overall input/output relationships rather than at device level. However, DAC testing does not have such a wide range of possible strategies as the ADC case, and very largely is confined to:

- applying each of the digital input code words in sequence;
- confirming that the analogue output on each code word is within its acceptable voltage limits;
- confirming that this conversion accuracy is maintained at the maximum specified bit rate of the converter.

This testing procedure may be undertaken with appropriate general-purpose or specialist instrumentation. Alternatively, the use of known-good converters may be possible as illustrated in Figure 7.21, but this may only be realistic for relatively small DACs, say eight-bit maximum, due to the very small quantising steps as bit size increases. Notice that it may be relevant to apply a subset of the digital input code words rather than the full 2^n set, say code words 0, 1, 3, 7, 15, ..., $2^n - 1$, which would exercise all the individual bit paths within the converter, although this would not cover all possible internal out of specification faults.

If the digital code word test generator (counter) could be made to increment and decrement continuously from 0 0 ... 0 0 0 to 1 1 ... 1 1 1 and back to 0 0 ... 0 0 0, then the analogue output from the DAC should be a stepped triangular waveform. A Fourier analysis of this output may now be a test signature for the converter, since any out of specification irregularities or missing conversions will result in abnormal harmonics in the output. However, the sensitivity of this method for DACs with a high number of digital inputs may be poor, and if the DAC is buried within other circuitry it may not be a convenient technique to implement—an on-chip high-speed synchronous reversible binary counter for the input test data is a difficult requirement to provide and the output would need to be brought off-chip for FFT analysis, thus making both controllability and observability problematic. Notice also that because of the possible error of $\pm\frac{1}{2}$ (least significant bit) in the a.c. output voltage, an eight-bit DAC will have an uncertainty of $1/256 = 0.4\%$ in the a.c. output, and a 12-bit DAC an uncertainty of $1/4096 = 0.024\%$, and this will give rise to a natural signal to noise ratio of about −48 dB for the eight-bit converter and about −72 dB for the 12-bit converter.

There has been certain published literature on other test strategies for D-to-A converters, including transient response analysis[77-79]. One proposal is somewhat akin to the Δj test strategy described above for DAC testing (see eqn. 7.4), in that an index of functionality, I_F, is determined which is a summation of a difference between a fault-free (simulated) parameter for the device under test and the actual measured parameter, taken over a selected range of output values. The strategy is as follows:

(i) consider a set of S data samples taken from the output voltage of the D-to-A converter, taken at equal time intervals Δt between samples;
(ii) let X be this set of S data samples from the converter;
(iii) let Y be the same set of S data samples from a perfect converter, obtained by simulation or other means.

For example, Y may be, say, 0, 0.1, 0.2, 0.4, ..., 5.0 volts, and X may be 0, 0.1, 0.18, 0.36, ..., 4.8 volts, or some other difference from perfect. Then the cross-correlation function R_{XY} is given by:

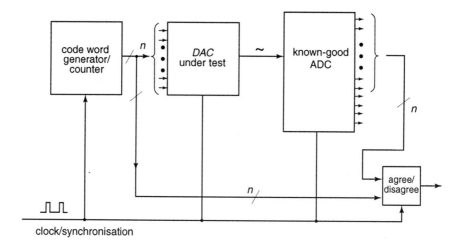

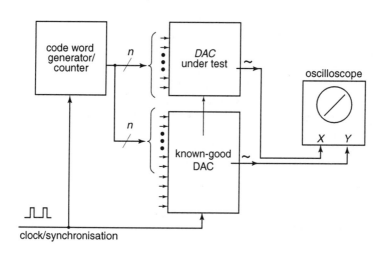

Figure 7.21 Possible OEM testing of D-to-A converters using ideally higher specification supporting converters
 a comparing digital inputs and outputs
 b Lissajous-type oscillograph plot of the analogue outputs

$$R_{XY}(\tau) = \frac{1}{S}\left\{\sum_{k=1}^{S} x(k\,\Delta t - \tau) \cdot y(k\,\Delta t)\right\} \qquad (7.5)$$

and the auto-correlation function R_{YY} is given by

$$R_{YY}(\tau) = \frac{1}{S}\left\{\sum_{k=1}^{S} y(k\,\Delta t - \tau) \cdot y(k\,\Delta t)\right\} \tag{7.6}$$

where τ is the relative time offset between the data being correlated, from which the index of functionality I_F is given by the difference of all correlation values:

$$I_F = \left\{1 - \frac{1}{c}\left(\sum_{\tau=0}^{S-1}|R_{XY} - R_{YY}|\right)\right\} \tag{7.7}$$

where c is some constant appropriate to normalise the range of values taken by I_F for a specific device under test.

A digital input test consisting of the code words 0, 1, 3, 7, 15, 31, ..., $2^n = 1$ is employed, which will detect all stuck-at and bridging faults in the DAC input lines and the weighting of each line in the analogue output. The value of I_F under this test is then determined from the set of output voltages sampled on these input words. The problem remaining, however, is to define go/no-go limits on I_F which reflect within-specification and outside-specification DAC performance. As with the previous Δj test strategy, this has been considered by simulating possible converter errors and determining the resulting value of I_F, from which computer-intensive calculations the limits on the acceptable/nonacceptable I_F test values for a given DAC may be determined. It is also claimed to be possible to scale the value of I_F such that $I_F = 100$ represents a perfect device and $I_F = 0$ represents a device just within specification, and then all negative values represent failures; catastrophic faults such as stuck-at input and output lines can give rise to negative numbers of the order of -10^3 or -10^4. Further details of this and similar strategies may be found published[11,77,80]. An attraction of the above DAC test method is that the digital input code words are extremely easy to produce, only a shift register filling with 1s being required, and also the number of analogue samples brought off-chip for processing is relatively small.

Digital to analogue converters, therefore, will be seen to be generally easier to test than analogue to digital converters. However, we will briefly consider both of them again in the following chapter in the context of mixed analogue/digital circuits and systems, particularly where built-in self test is present.

7.6.4 Complete A/D subsystems

There are commercially available a number of relatively complex converter units (subsystems) for general-purpose and specific application use, all of which involve some internal A-to-D and/or D-to-A conversion circuitry. Among such products are:

- combined controllable DAC/ADC monolithic circuits, often used with microprocessor and microcontroller systems;
- video ADC circuits for the very high speed digitising and formatting of video signals;
- video ADC circuits to reconstruct analogue video data from digitised data with blanking, composite synchronisation and other video information;
- modular ADC systems which provide multiple inputs/outputs.

An example of the last is illustrated in Figure 7.22, providing an entirely self-contained subsystem which gives professional isolation of the analogue and digital sections, programmable analogue voltage gain, stable references and other desirable circuit features.

The testing of these complex off the shelf units is the province of the vendor. No known discussions have been published concerning the detailed testing of such circuits, and it is therefore to be assumed that a series of appropriate parametric and functional tests is applied before dispatch. Although the circuit complexity (gate count) is high, it falls short of that usually considered necessary for building in scan test or BIST if all components are on one single chip, as is necessary with the latest state of the art microprocessors.

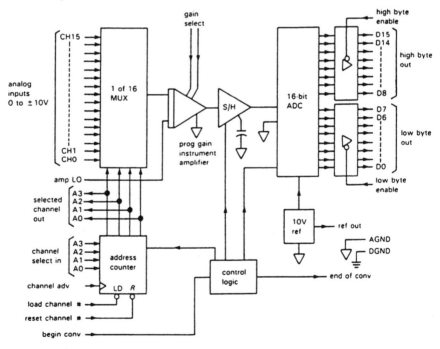

Figure 7.22   The DT5716 modular ADC system, providing 16 single-input or eight differential-input a.c. inputs, sample and hold, variable gain and 16-bit A-to-D conversion at 20 kHz throughput rate (Acknowledgements, Data Translation Inc., MA)

The OEM testing of such circuits must be an overall functional test when required. No precise knowledge of the internal circuitry down to gate level is generally available, and hence primary input to primary output tests to confirm the device specification is all that can realistically be undertaken.

7.7 Chapter summary

This chapter has ranged rather summarily over the testing requirements of autonomous analogue-only circuits and A-to-D and D-to-A converters, the vast majority of which will be off the shelf standard parts or standard designs. The difficulty is the very wide diversity of circuits and applications involved, see Figure 7.5, and the detailed specialist knowledge of analogue circuit design which is necessary to discuss the specification, the design and the tolerancing of most circuits in any depth. Also, all analogue circuits have an acceptable tolerance of performance in the time and frequency domain which must be part of the testing specification for the circuit, and hence simple deterministic go/no-go tests are difficult to formulate.

As we have seen, functional testing of such circuits is the norm, using commercial instrumentation to generate test inputs and monitor test response. The vendor will inevitably be highly involved in the production testing of standard parts, using considerable resources and in-house expertise; the OEM will generally be involved in less comprehensive functional tests of individual components or parts.

However, the most significant factor to appreciate is that digital signal processing is becoming the most successful way of testing other than simple analogue circuits, since DSP gives the particular advantages of:

- accurate generation of analogue functions (frequency, waveform and magnitude values);
- accurate measurement of output test data;
- accurate timing and/or synchronisation of any critical test data;
- repeatability of all test data and measurements, free from drift, thermal effects, etc. to a very high degree;
- no long settling times as are unavoidably present with conventional analogue instruments;
- the ability to perform high-speed processing of test data for specific test strategies.

In attempting to summarise the many ways which have been suggested for processing analogue output test data, we may broadly say that all seek to measure how like the actual test response is to the within-specification test response, by correlation, sampling or other means. A possible irony in this analogue test situaion is that it is the D-to-A and A-to-D converters which are the key elements within any DSP tester, and hence they must be of the highest possible performance for fast and accurate analogue testing.

Supply current monitoring of analogue circuits, corresponding to the I_{DDQ} tests for CMOS digital circuits, may find a place in the test strategies for some CMOS analogue circuits, and to a lesser extent bipolar, although this may only be relevant as a test for gross faults and not parametric out of tolerance drifting.

For further details of basic analogue circuits reference should be made to the many excellent graduate texts previously referenced[1-6,48,69,70,76] and others; for further information on DSP theory and practice and on DSP testing see References 32, 81. In the following chapter dealing with mixed-signal analogue/digital testing we will build upon the information covered in this chapter, but we will have no need to introduce any new testing concept for either the analogue or digital parts over and above that which we have now considered in previous pages.

7.8 References

1 HOROWITZ, P., and HILL, W.: 'The art of electronics' (Cambridge U.P., 1991)
2 SEDRA, A.S., and SMITH, K.C.: 'Microelectronic circuits' (Holt, Reinhart and Winston, 1987)
3 BOBROW, L.S.: 'Elementry linear circuit analysis' (Holt, Reinhart and Winston, 1987)
4 SAVANT, E.J., RODEN, M.S., and CARPENTER, G.L., 'Electronic design: circuits and systems' (Benjamin/Cummings, 1991)
5 ALLEN, P.E., and SANCHEZ-SINECIO, E.: 'Switched capacitor circuits' (Van Nostrand Reinhold, 1984)
6 VAN VALKENBURG, M.E.: 'Analog filter design' (Holt, Reinhart and Winston, 1981)
7 TOUMAZOU, C., LIDGEY, F.C., and HAIGH, D.G. (Eds.): 'Analogue IC design: the current mode approach' (IEE Peter Peregrinus, 1993)
8 TRONTELJ, J.,TRONTELJ, L., and SHENTON, G.: 'Analog digital ASIC design' (McGraw-Hill, 1989)
9 HURST, S.L.: 'Custom VLSI microelectronics' (Prentice Hall, 1989)
10 HURST, S.L.: 'Custom-specific integrated circuits' (Marcel Dekker, 1985)
11 SOIN, R., MALOBERTI, F., and FRANCA, J. (Eds.): 'Analogue—digital ASICs: circuit techniques, design tools and applications' (IEE Peter Peregrinus, 1991)
12 DUHAMEL, P., and RAULT, J.-C.: 'Automatic test generation techniques for analog systems: a review', *IEEE Trans.*, 1979, **CAS-26**, pp. 411–440
13 BANDLER, J.W., and SALAMA, A.E.: 'Fault diagnosis of analog circuits,' *Proc. IEEE*, 1985, **73**, pp. 1279–1325
14 FANTINI, F., and MORANDI, C.: 'Failure modes and mechanisms for VLSI ICs— a review', *Proc. IEE*, 1985, **132**, pp. 74–81
15 FANTINI, F., and VANZI, M.: 'VLSI failure mechanisms'. Proceedings of *CompEuro* '87, 1987, pp. 937–943
16 AMERASEKERA, E.A., and CAMBELL, D.S.: 'Failure mechanisms in semiconductor devices' (Wiley, 1987)
17 OHLETZ, M.J.: 'Hybrid built-in self-test (HBIST) for mixed analogue/digital integrated circuits'. Proceedings of European conference on *Test*, ETC '90, 1990, pp. 307-316
18 HARVEY, R.J.A., BRATT, A.H., and DOREY, A.P.: 'A review of testing methods for mixed-signal ICs', *Microelectronics J.*, 1993, **24**, pp. 663–674

19 WEY, C.L., and SAEKS, R.: 'On the implementation of an analog ATPG, the nonlinear case', *IEEE Trans.*, 1988, **IM-37**, pp. 252–258
20 MIJOR, L., and VISVANATHAN, V.: 'Efficient go/no-go testing of analog circuts'. Proceedings of IEEE international symposium on *Circuits and systems*, 1987, pp. 414–417
21 STAPPER, C. H., ARMSTRONG, F.M., and SAJI, K.: 'Integrated circuit yield statistics', *Proc. IEEE*, 1983, **71**, pp. 453–470
22 GALIAY, J., CROUZET, Y., and VERGINAULT. M.: 'Physical versus logical fault models for MOS LSI circuits: impact on their testability', *IEEE Trans.*, 1980, **C-29**, pp. 527–531
23 ANGONETTI, P. (Ed.): 'Semiconductor device modelling with SPICE' (McGraw-Hill, 1994)
24 VLADIMIRESCU, A.: 'The SPICE Book' (Wiley, 1994)
25 BERSFORD, R.: 'Circuit simulators at a glance', *VLSI Syst. Des.*, 1987, **8** (9), pp. 76, 77
26 MASSARA, R.E. (Ed.): 'Design and test techniques for VLSI and WSI circuits' (IEE Peter Peregrinus, 1989)
27 'HELIX command reference manuals, vols. 1 and 2' (Silvar-Lisco, 1986)
28 WALSH, K., and WOLFE, B.: 'Mixed-domain analysis for circuit simulation', *VLSI Syst. Des.*, 1987, **8** (9), pp. 44–49
29 BRAY, D., and IRISSOU, P.: 'A new gridded bipolar linear semicustom array family with CAD support', *J. Semicust. ICs*, 1986, **3**, June, pp. 13–20
30 PARKER, K.P.: 'Integrating design and test: using CAE tools for ATE programming' (IEEE Computer Science Press, 1987)
31 SHEPHERD, P.: 'Testing mixed analogue and digital circuits', in [11], pp.334–359
32 MAHONEY, M.: 'DSP-based testing of analog and mixed-signal circuits' (IEEE Computer Science Press, 1987)
33 SHANNON, C.E.: 'Communications in the presence of noise', *Proc. IRE*, 1949, **37**, pp. 10–20
34 BROWN, D., and DAMIANOS, J.: 'Method for simulation and testing of analog/digital circuits', *IBM Tech. Discl. Bull.*, 1983, **25**, pp. 636–638
35 HOCHWALD, W., and BASTIAN, J.: 'A d.c. approach for analog fault dictionary determination', *IEEE Trans.*, 1979, **CAS-26**, pp. 523–529
36 OPPENHEIM, A.V., and SCHAFER, R. W.: 'Discrete-time signal processing' (Prentice Hall, 1989)
37 BANDLER, J.W., and SALAMA, A.E.: 'Fault diagnosis of analog circuits', *Proc. IEEE*, 1988, **73**, pp. 1279–1325
38 TOWILL, D.R.: 'Dynamic testing of control systems', *Radio Electron. Eng.*, 1977, **47**, pp. 501–521
39 TSAO, S.H.: 'Generation of delayed replicas of maximal-length linear binary sequences', *Proc. IEE*, 1964, **111**, pp. 1803–1806
40 ROBSON, M., and RUSSELL, G.: 'Digital techniques for testing analogue functions'. Proceedings of IEE colloquium on *Systems design for testability*, 1995, pp. 6.1–6.4
41 BELL, I.M., CAMPLIN, D.A., TAYLOR, G.E., and BANNISTER, B.R.: 'Supply current testing of mixed analogue and digital ICs', *Electron. Lett.*, 1991, **27**, pp. 1581–1583
42 DA SILVA, J.M., MATOS, J.S., BELL, I.M., and TAYLOR, G.E.: 'Cross-correlation between i_{dd} and v_{out} signals for testing analogue circuits', *Electron. Lett.*, 1995, **31**, pp. 1617–1618
43 BELL, I.M., CAMPLIN, D.A., TAYLOR, G.E., and BANNISTER, B.R.: 'Can supply current monitoring be applied to testing of analogue as well as digital portions of mixed ASICs?'. Proceedings of European conference on *Design and automation*, 1992, pp. 538–542

44 ECKERSALL, K.R., TAYLOR, G.E., BELL, I.M., and TOUMAZOU, C.: 'Current monitoring for test'. Proceedings of IEEE international symposium on *Circuits and systems*, tutorial, 1994, pp. 551–568
45 WEDER, U., and MÖSCHWITZER, A.: 'SCF: a gate-array switched capacitor filter design tool', *J. Semicust. ICs*, 1990, **7**, March, pp. 15–21
46 RABAEY, J., GOOSENS, G., CATTHOOR, F., and DE MAN, H.: 'CATHEDRAL computer-aided synthesis of digital signal processing systems'. Proceedings of IEEE conference on *Custom integrated circuits*, 1987, pp. 17–19
47 ACUNA, A., DERVENIS, J., PAGONES, A., and SALAH. R.: 'iSPLICE.3: a new simulator for mixed analog/digital circuits'. Proceedings of IEEE conference on *Custom integrated circuits*, 1989, pp. 13.1.1–13.1.4
48 GEIGER, R.L., ALLEN, P.E., and STRADER, N.R.: 'VLSI designs for analog and digital circuits' (McGraw-Hill, 1990)
49 ALLEN, P.E., and SÁNCHEZ-SINENCIO, E.: 'Switched capacitor circuits' (Van Nostrand Reinhold, 1990)
50 SCHAUMAN, R., GHAUSI, M.S., and LAKER, K. R.: 'Design of analog filters: passive active RC and switched capacitor' (Prentice Hall, 1990)
51 HAIGH, D.H., TOUMAZOU, C., FRANCA, J.E., and SINGH, B.: 'Switched capacitor filters', *in* [11]
52 VAN VALKENBURG, M. E.: 'Analog filter design' (Holt, Reinhart and Winston, 1981)
53 SCHAUMANN, R., SODERSTRAND, M., and LAKER, K. (Eds.): 'Modern active filter design' (IEEE Press, 1981)
54 HOCHWALD, W., and BASTIAN, J.: 'A DC approach for analog fault dictionary determination', *IEEE Trans.*, 1990, **CAS-26**, pp. 523–529
55 AL-QUTAYRI, M.A., and SHEPHERD, P.R.: 'On the testing of mixed-mode integrated circuits', *J. Semicust. ICs*, 1990, **7** (4), pp. 32–39
56 LIN, P.M., and ELCHERIF, Y.F.: 'Analog circuits faults dictionary—new approaches and implementation', *Int. J. Circuit Theory Appl.*, 1985, **13**, pp. 149–172
57 MALY, W., and NIGH, P.: 'Built-in current testing—a feasibility study'. Proceedings of IEEE international conference on *Test*, 1988, pp. 340–343
58 HAWKINS, C.F., and SODEN, J.M.: 'Electrical characteristics and testing considerations for gate oxide faults in CMOS ICs', Proceedings of IEEE international conference on *Test*, 1985, pp. 544–555
59 KEATING, A.M., and MEYER, D.: 'A new approach to dynamic IDD testing'. Proceedings of IEEE international conference on *Test*, 1987, pp. 148–157
60 JENSON, F., and MOLTOFT, J.: 'Reliability indicators', *Qual. and Reliab. Eng. Int.*, 1986, **2**, pp. 39–44
61 BINNS, R.J., TAYLOR, D., and PRITCHARD, T.I.: 'Testing linear macros in mixed-signal systems using transient response testing and dynamic supply current monitoring', *Electron. Lett.*, 1994, **30**, pp. 1216–1217
62 PATYRA, M., and MALY, W.: 'Circuit design for built-in current testing'. Proceedings of conference on *Custom integrated circuit*, 1991, pp. 13.4.1–13.4.5
63 BROWN, D., and DAMIANOS, J.: 'Method for simulation and testing of analog/digital circuits', *IBM Tech. Discl. Bull.*, 1983, **25**, pp. 6367–6368
64 SALAMS, A., STARZYK, J., and BANDLER, J.: 'A unified decomposition approach for fault location in large analog circuits', *IEEE Trans.* 1984, **CAS-31**, pp. 609–622
65 BUTLER, I.C., TAYLOR, D., and PRITCHARD, T.I.: 'Effect of response quantisation on the accuracy of transient response results', *IEE Proc. Circuits Devices and Syst.*, 1995, **142**, pp. 334–338
66 MALOBERTI, F.: 'Data converters', *in* [11], pp. 117–142
67 MALOBERTI, F., and O'LEARY, P.: 'Oversampling converters', *in* [11], pp. 143–172
68 GERSHO, A.: 'Principles of quantisation', *IEEE Trans.*, 1978, **CAS-25**, pp. 427–436

69 SHEINGOLD, D.H. (Ed.): 'Analog-digital conversion handbook' (Prentice Hall, 1986)
70 HNATEK, E.R.: 'A users handbook of D/A and A/D converters' (Wiley, 1986)
71 NAYLOR, J.R.: 'Testing digital/analog and analog/digital converters', *IEEE Trans.*, 1978, **CAS-25**, pp. 527–538
72 DOERNBERG, J., LEE, H.S., and HODGES, D.A.: 'Full-speed testing of A/D converters', *IEEE J. Solid-State Circuits*, 1984, **SC-19**, pp. 820–827
73 BOBBA, R., and STEVENS, B.: 'Fast embedded A/D converter testing using the microcontroller's resources'. Proceedings of IEEE international conference on *Test*, 1990, pp. 598–604
74 RACZKOWYCZ, J., and ALLOTT, S.: 'Embedded ADC characterisation techniques', *IEE Proc., Comput. Digit. Tech.*, 1995, **142**, pp. 145–152
75 ALLOTT, S., and RACZKOWYCZ, J.: 'The characterisation of embedded analogue to digital converters'. Proceedings of 2nd international conference on *Advanced A-D and D-A conversion techniques and their applications*, Cambridge, UK, 1994, pp. 163–168
76 CANDY, J.C., and TEMES, G.C. (Eds.): 'Oversampling delta-sigma converters' (IEEE Press, 1991)
77 TAYLOR, D., EVANS, P.S.A., and PRITCHARD, T.I.: 'Testing for functional defects in embedded digital-to-analogue converters using dynamic stimuli and transient response', *Microelectron. J.*, 1994, **25**, pp. 415–424
78 TAYLOR, D., EVANS, P.S.A., and PRITCHARD, T.I.: 'Testing of mixed-signal systems using dynamic stimuli', *Electron. Lett.*, 1993, **29**, pp. 811–813
79 EVANS, P.S.A., AL-QUTAYRI, M.A., and SHEPHERD, P.R.: 'A novel technique for testing mixed-signal ICs'. Proceedings of European conference on *Test*, ETC '91, 1991, pp. 301–306
80 SCHREIBER, H.H.: 'Fault dictionary based upon stimulus design', *IEEE Trans.*, 1979, **CAS-26**, pp.529–537
81 MITRA, S.K.: 'Handbook for digital signal processing' (Wiley, 1993)

Chapter 8
Mixed analogue/digital system test

8.1 Introduction

Outside the restricted world of digital computation, the majority of industrial, commercial and scientific microelectronic applications involve some analogue signals at input(s) and/or output(s). The analogue input signals may be converted into digital signals entirely within specific transducers (sensors), and therefore from the OEM point of view may not be regarded as mixed-signal designs, the OEM designer accepting such items as self-contained off-the-shelf components with digital outputs. Similarly, it may be possible to have self-contained output components to translate appropriate digital output signals directly into some analogue output display or other facility.

We will not, therefore, be concerned here with systems which use separate self-contained analogue to digital sensors and digital to analogue output devices; such systems have full accessibility of the digital interfaces, and testing can therefore be a digital-only test of the circuit design, plus an overall functional system test as appropriate. Instead we will here consider the more complex situation where there may be some analogue signal processing, some A-to-D and D-to-A conversion and/or some signal processing, the whole representing some fairly large design of VLSI complexity.

Figure 8.1 illustrates the possible complexity that may be encountered. In many situations the analogue processing may not be as complex as indicated here, all the required processing being done in the digital domain. However, any filtering may involve operational amplifiers, and hence there may not be a simple A-to-D conversion from the analogue input(s) directly to the digital logic core in all cases.

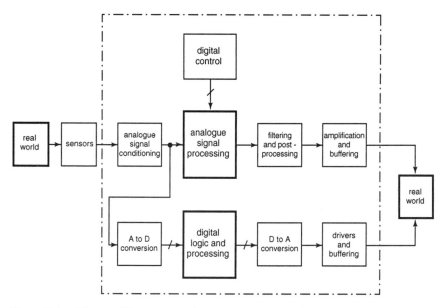

Figure 8.1   The possible structure of a complex analogue/digital system

The applications for mixed analogue/digital systems are extremely diverse; they include:

- telecommunications;
- medical;
- automotive;
- military;
- industrial control and supervision;
- consumer products;

and others. In the last area we have the every-day complexity of video recorders, compact disc players, electronic music synthesisers and organs, telephone and fax equipment, and so on. Increasingly manufacturers are attempting to place the whole circuit on one VLSI chip, a systems-on-a-chip design solution, which must involve a detailed consideration of how-shall-we-test-it at the design stage. In the more general-purpose area, we have off-the-shelf products such as the Motorola 68705 microcontroller shown in Figure 8.2, which incorporates an analogue input port that can accept up to eight analogue inputs, these analogue inputs being routed to a shared A-to-D converter which quantises each input into eight-bit 256-level digital data. These and other off-the-shelf products must be comprehensively vendor tested, and may include some scan-test or built-in self-test provisions, see later.

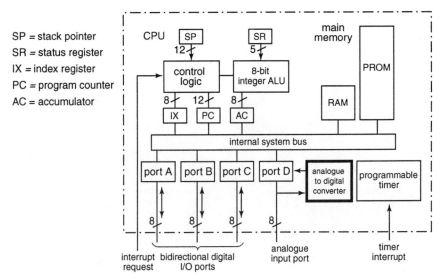

Figure 8.2 Schematic of the Motorola 68705 microcontroller, which has analogue input capability

The frequency range of operation of mixed analogue/digital systems is also extremely wide, for example:

- seismic, from fractions of a Hz up to hundreds of Hz;
- audio, from tens to thousands of Hz;
- sonar, from tens to thousands of Hz;
- telecoms, from hundreds of Hz to MHz and GHz;
- video, radio, TV, from kHz to GHz.

In general, the test strategies that we will be considering in the following pages will not involve the extreme frequencies, and hence we will not have any specific need to refer to frequency of operation in our developments. Further details and discussion on the design and applications of mixed analogue/digital circuits and systems may be found in Soin *et al.*[1], and in Trontelj *et al.*[2], and the further references contained therein.

Many of the complex VLSI mixed analogue/digital ICs are application-specific standard parts (ASSPs), designed by professional design houses or vendors as a standard off-the-shelf part for any customer's use. This is particularly so for entertainment applications and commonly-encountered commercial and industrial applications. The vendor may employ very costly test equipment such as illustrated in Figure 8.3 for prototype and production testing; the purchaser, on the other hand, is more likely to use some specific test bench or rack and stack assembly for functional testing. However, as will be seen later, built-in test facilities may be provided by the vendor on increasingly complex ASSPs.

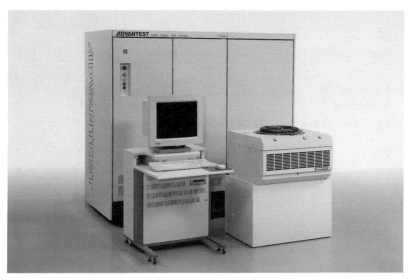

Figure 8.3 Avantest T7341 mixed-signal test system, featuring two-channel 200 MHz function generation, digitiser, digital-controlled test units and other facilities (Photo courtesy Avantest Corporation, Japan)

Thus, in considering mixed analogue/digital circuits and systems, we may distinguish between:

(i) complex off-the-shelf standard parts, with vendor testing and with the OEM having functional information but no detailed information down to gate or transistor level;
(ii) complex OEM-designed systems using standard parts in a PCB or other appropriate assembly;
(iii) mixed analogue/digital custom ICs providing a single-chip solution for a specific system requirement.

It is therefore the latter two categories which will be of principal relevance in our later discussions.

8.2 Mixed-signal user-specific ICs

Mixed-signal user-specific ICs (USICs or ASICs) are available in uncommited-array form, requiring only the metalisation mask(s) for custom dedication, and in standard-cell form, requiring a full mask set to fabricate the on-silicon assembly of the analogue and digital macros. Details of such products may be found in the published literature[3–11]; Figure 8.4 indicates the range of all types of custom ICs, from which it will be seen that mixed analogue/digital capability is a contribution from both sides of the analogue-digital divide on one chip assembly.

Mixed analogue/digital system test 431

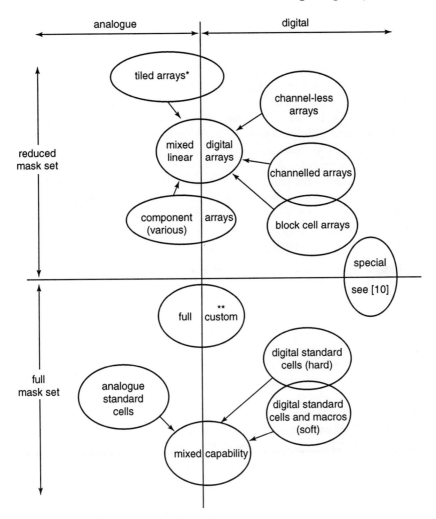

* Some (usually inefficient) digital capability
** Anything possible within the limits of the technology

Figure 8.4 *A survey of the principal design methodologies for custom ICs, with considerable blurring of the precise boundaries between types. Architectures which span the divide between a full mask set and a reduced mask[10] are uncommon*

The design phase of such circuits (and indeed of all mixed analogue/ digital circuits) inevitably partitions into an initial consideration of the analogue circuits and of the digital circuits, followed by the interface requirements between them. This initial partitioning of duties continues through the simulation phase right down to final test, since it is not possible to consider the complete circuit as one homogeneous whole from either the design or the testing point of view. The interface between the two, however, remains the critical area, and it is here where existing CAD and simulation resources are at their weakest.

The overall schematic for the design and fabrication of a mixed-signal USIC is generally as shown in Figure 8.5. If the overall system is small, using possibly component-level uncommitted arrays (tiles and mosaics, etc.[3,6]) with a reduced mask set, then an overall SPICE or SPICE-derivative simulation may be possible. This is, however, only feasible if the total number of transistors and other components involved is a few hundred and not thousands; for example, it has been reported that to simulate just one analogue to digital conversion in a high-speed 10-bit A-to-D converter at the transistor level would take over twenty hours of workstation time[12].

As system requirements grow, then the need to use predesigned library macros becomes increasingly necessary. Among the analogue macros which vendors may have available are:

- operational amplifiers—general purpose, low power, high frequency, high drive, low noise and other variants;
- comparitors—general purpose, high speed, low power and other variants;
- programmable gain amplifiers;
- filters—switched-capacitor and continuous-time;
- bandgap voltage references;
- current sources, current references and current mirrors;
- analogue switches and multiplexers;
- transconductance amplifiers;
- A-to-D and D-to-A converters;
- RC and crystal oscillators;
- nonlinear expanders, compressors and limiters;

with both bipolar, MOS and BiCMOS being represented in the market place. Again, as noted in the previous chapter, it is the variety of analogue elements each with their own tolerance bands which is the problem in analogue and mixed analogue/digital circuit design and test. Figure 8.6 shows a representative mixed-signal IC which illustrates a typical mix of circuits.

Simulation of LSI/VLSI mixed-signal ICs is therefore forced to use other than just SPICE-based simulation. This may be:

(i) using SPICE or some alternative analogue simulation package for the analogue elements*, with a separate digital simulator such as HILO™, VHDLsim™, Verilog-XL™, Lsim™ or other commercially-available software package for the digital elements;
(ii) using mixed-signal simulation, see later, which can cater for both analogue and digital elements, but not in such fine detail as SPICE.

---

* As noted in the previous chapter, particular CAD software is available for switched-capacitor filters and other specific analogue elements, but they are usually standalone packages which are difficult to link together with other software.

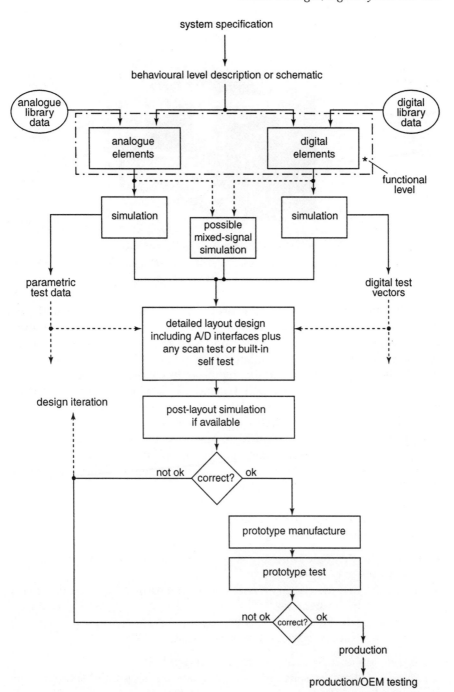

Figure 8.5 *The hierarchical design route for the majority of mixed-signal IC designs, with the basic analogue and digital elements being predesigned library data*

434 VLSI testing: digital and mixed analogue/digital techniques

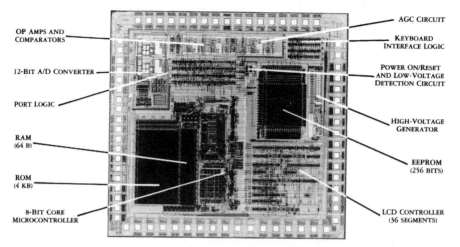

Figure 8.6   An example combined analogue/digital IC, chip size 215×215 mil^2 (Photo courtesy Sierra Semiconductor Corporation, CA)

With (i) above there remains the problem of linking the two simulators, which are run separately, to interact together. This has been termed a glued simulator approach, and involves feeding each simulator's output data to the other. A true mixed-signal simulator, however, is a single simulator which handles both the analogue and the digital components at a behavioural level. However, the distinction between (i) and (ii) is blurred, since there is a growing significance in the development and use of systems integration software which can take the best from the analogue simulation area and the best from the digital simulation area, and weld them together in a highly integrated (seamless) design package. Figure 8.7 illustrates a mixed-signal simulation resource handling both analogue and digital library elements in the design of an analogue/digital system.

This is still a highly volatile CAD area, and from our interests here does not yet have a clear path through to a final testing strategy. There are still reported problems in mixed-signal simulation, including:

- noise from the logic switching action induced into the analogue circuits, which cannot easily be modelled;
- the presence of high-power analogue elements near the vicinity of very low power logic elements;
- tolerancing the analogue components for parametric test purposes, which is generally unnecessary for the digital parts.

Further details of these aspects may be found published[12-14].

Because of these present difficuties in both the design and simulation of analogue/digital ICs, there are breadboarding techniques commercially available which exploit the capabilities of digital simulation while providing hardware resources for the analogue side. This is illustrated in Figure 8.8,

Mixed analogue/digital system test 435

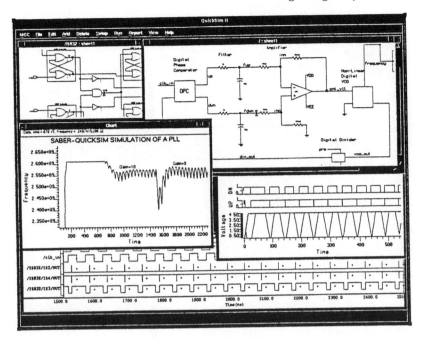

Figure 8.7 Mixed-signal simulation using Saber™/QuickSim™ CAD systems integrated software (Photo courtesy Analogy, Inc., OR)

from which it will be noted that the digital CAD software physically interfaces to real-life analogue elements, these analogue elements being individually packaged library elements identical to that which will be incorporated in the final mixed-signal IC. This Lego-type assembly of the analogue elements reflects the way in which many designers have conventionally undertaken prototype system designs; normal bench-top instrumentation can readily be used to monitor the analogue circuits, giving a system evaluation prior to integration. However, the interconnections between such analogue circuit elements do not exactly represent the final IC interconnect impedance, and the interface between the analogue assembly and the digital software is also not exact. Nevertheless, for frequencies up to possibly tens or hundreds of kHz, this design methodology has proved to be useful to the OEM where efficient, affordable and user-friendly CAD software is not otherwise available. A further variant which may be used during an initial design phase is to partition the digital logic so it may be realised in reconfigurable static RAM-based field-programmable gate arrays (FPGAs)[15]; indeed, we may well find the use of SRAM-based FPGAs and CPLDs increasing in both the design and the test formulation stages of complex IC designs.

The information in this section has not directly involved the primary subject of this text, namely testing, but has been included to illustrate the problems of design and simulation when mixed-mode signals are present. Details of circuit and system design may be found in the published literature,

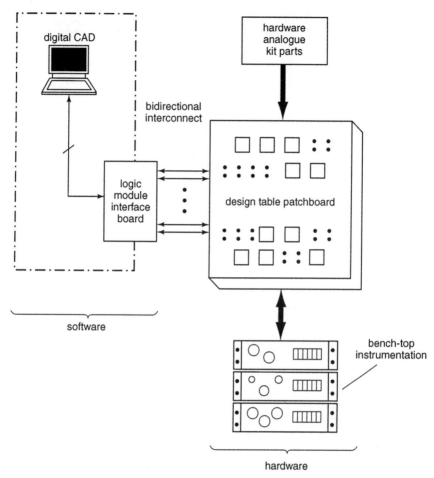

Figure 8.8   The concept of combining digital CAD software with analogue hardware elements for mixed-signal IC design purposes

but testing does not feature prominently in many publications[1,2,16–24]. However, having examined these design problems, we will continue below with the testing aspects and strategies.

8.3 Partitioning, controllability and observability

There is no single comprehensive test strategy for a complete analogue/digital IC or system. Instead, the generally accepted approach up to now has been to consider the digital part and the analogue part separately for test purposes, which is a continuation of the divide which as we have just seen exists in the specification and design phase. This classic testing strategy is therefore as shown in Figure 8.9, with (as far as possible) digital testing

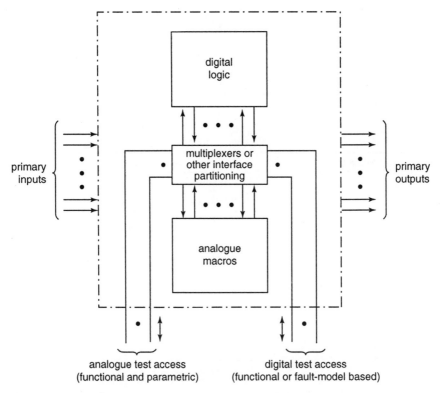

Figure 8.9 *The classic test strategy for mixed analogue/digital circuits and systems, wherein the two halves are considered separately for both design and test purposes*

methods being applied to the digital parts and analogue testing methods being applied to the analogue parts. The exception to this is, of course, when the complete system is fairly small and uncomplicated, in which case overall functional testing using the primary inputs and primary outputs may be sufficient. This will not, however, test the parametric tolerances of all the analogue elements.

This divide and conquer procedure for test therefore usually follows the routes shown in Figure 8.10. This has invariably been the (intuitive) procedure that an OEM follows when designing a mixed-signal hybrid or PCB system, where full controllability and observability of the primary inputs and outputs is freely available, and to a lesser extent when a mixed-signal IC design is being considered[2,25-28]. The presence of D-to-A and A-to-D converters and switched-capacitor filters to some extent blurs this division, and this is why they were considered as autonomous analogue elements in the previous chapter. However, recall from this chapter that there are no accepted fault models for the analogue elements, corresponding to the stuck-at model which may be used for the digital elements, and therefore in

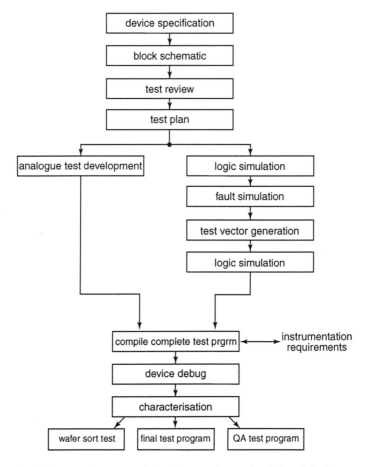

Figure 8.10 The test programme flow diagram for a mixed-signal design, assuming that the analogue and digital elements are considered independently and that accessibilty of all analogue/digital interfaces is available

addition to controllability and observability a third issue arises when considering the analogue part. This is the completeness of the a.c. tests [29].

Completeness of the a.c. tests is concerned with how much time should be spent in performing the analogue testing—how many parameters need be measured on each macro and to what accuracy they should be measured in a production run. There is no ready answer to this, and each analogue/digital designer will have to make his or her own decisions on what tests should be done for a particular design. Completeness is in some ways analogous to fault coverage in digital circuits, since in both the analogue and digital cases it is not usually feasible to spend time checking for every possible out of tolerance parameter or circuit fault. We should, however, distinguish here between:

(i) simulation of the design, which should be as comprehensive as possible;
(ii) prototype testing, which again should be as comprehensive as possible, with no immediate time constraints;
(iii) final production testing, which brings in the consideration of completeness of test and the minimum number of necessary and sufficient parameters to be tested.

The latter will influence the number of interface nodes which need to be made accessible for test, see Figure 8.9, and the test signals present when in test mode. In the case of mixed-signal custom ICs, where a vendor is manufacturing a USIC for a particular OEM, the vendor should be able to detail a subset of parametric tests for each analogue library macro, and it is therefore part of the final testing procedure to execute these requirements if at all possible.

In general, it has been found that in the majority of mixed-signal ICs most of the circuitry is digital, with possibly 20–25 % analogue. It is usual for the analogue tests to be undertaken before the digital tests, although the actual time to test the analogue elements may be comparable to that necessary to test the digital elements—the cost of testing may even be greater.

With an effective division for test purposes of the complete mixed-signal IC or system into its two disparate halves, the overall test schedule becomes:

(i) analogue-only testing of the analogue elements, as discussed in Chapter 7;
(ii) digital-only testing of the digital elements, as discussed in earlier chapters;
(iii) some final overall functional check following satisfactory completion of (i) and (ii) to ensure that the normal-mode connections between the analogue and digital parts are intact.

Figures 7.2, 7.3 and 7.10 of the previous chapter introduced the concept of providing analogue multiplexers (AMUXs) to give access to buried analogue macros—the same requirements hold for access to the analogue/digital interfaces in a mixed-signal environment. There may, of course, be a need to give accesss to the primary I/Os of particularly important analogue macros within the analogue half of the circuit as well as the A-to-D and D-to-A interfaces. The AMUXs must, however, handle the bandwith of the analogue signals, and (ideally) introduce no attenuation or nonlinear distortion in the signal paths, principles which may be found detailed by Wagner and Williams[30].

If any scan-path testing of the digital part of the circuit is in place, then it is possible to use the scan chain to control the analogue multiplexers, rather than having to use further primary inputs for AMUX control. This is illustrated in Figure 8.11. We may also wish the scan-path latches to provide selective disabling of local feedback paths, if this should be necessary to ease the testing. Notice that with the schematic arrangement shown in Figure 8.11 we are still enabling the analogue elements to be tested on their own by any

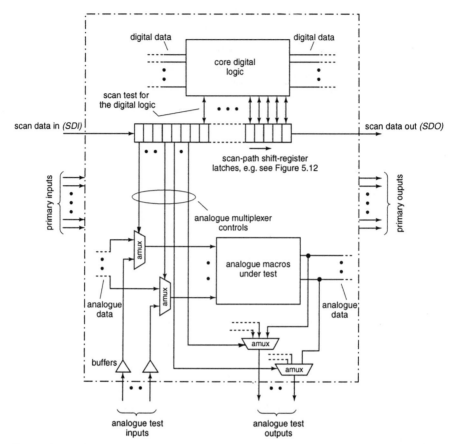

Figure 8.11 An extension of Figure 7.3 to utilise a digital scan-path resource for selectively controlling the analogue multiplexers; the normal-mode analogue/digital interfaces of the complete circuit have been omitted for clarity. Note the scan path is the AMUX control, and does not handle any analogue test data

of the means considered in Chapter 7, with the digital part being a separate consideration. Note also that on a mixed-signal IC the d.c. supply rails for the digital logic and the analogue elements are invariably kept separate or well decoupled from each other, and hence it is still theoretically possible on a mixed-signal IC to consider current monitoring tests, such as illustrated in Figure 7.13, for the analogue macros.

With complete analogue/digital segregation as in Figures 8.9 and 8.11, all our previous discussions on analogue testing using conventional instrumentation or DSP techniques, etc., therefore remain relevant [2,21,30–35]. Note that Duhamel and Rault[31] contains an extensive listing of about 500 references to early pioneering work in analogue test generation techniques. However, there remains the possibility of some form of scan test or built-in self test to cover the analogue half, which we will consider in the following sections. Unlike the digital-only situation considered earlier, the following designed for test strategies are all offline, as it is currently not realistic to consider any form of online monitoring of the analogue signals, corresponding to the online digital code checkers and other DFT techniques considered in Chapter 4—perhaps this may need reconsideration in the future?

8.4 Offline mixed-signal test strategies

The published literature on scan test and built-in self test is predominately concerned with digital-only circuits and systems; for example, the very comprehensive survey by Savir and Bardell of 1993, which describes BIST developments since its inception and provides an extensive bibliography[37], does not cover any mixed-signal concepts or developements. A later publication[38] also does not feature analogue considerations in its many overviews and perspectives of test synthesis. The dichotomy between these two areas remains to be acceptably bridged in the field of design and test.

Nevertheless there have been noticeable research activities in this area, and these will continue and grow with the increasing importance of mixed-signal systems on a chip. Here we will review some of the published information which is currently available, which indicates the train of on-going developments. As will be seen below, the evolving IEEE boundary-scan standard 1149.4 will be prominent in the future, as standard 1149.1 is now for the digital area. There is, however, unlikely to be any very novel breakthrough in completely new concepts, but rather a continuous refining of ideas and the application of specific DSP techniques, plus acceptance that there must be an appreciable silicon penalty for incorporating on-chip analogue DFT strategies.

8.4.1 Scan-path testing

If the inputs to the analogue macros are primary inputs, such that they may be exercised with known analogue test input data, then it is possible to incorporate a digital scan chain specifically to scan out data relating to the response of the analogue circuits. The digital logic will have its separate scan chain to monitor the operation of the digital circuits. This concept is illustrated in Figure 8.12

Notice that in conventional scan chains for digital logic, test data may be loaded into the scan chain and the output response captured, as was discussed in Section 5.3 of Chapter 5; however, for the analogue tests shown here, the primary inputs provide the test input data, and the scan path is only used to capture the resultant a.c. test response and then feed it off-chip. The scan test relies upon the on-chip A-to-D converters to provide the digital

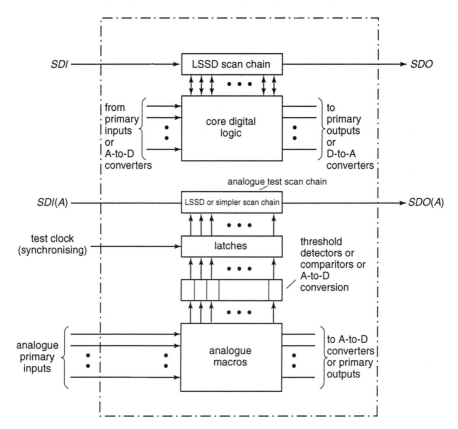

Figure 8.12 *The use of a dedicated scan chain to shift a.c. data off-chip in addition to the conventional digital logic scan chain; the normal-mode analogue/digital interfaces of the complete circuit have been omitted for clarity*

inputs to the analogue test scan chain, which may be the existing A-to-D interface converters of the circuit or additional converters to monitor critical nodes in the analogue circuits. Rather than using A-to-D conversion to load the scan chain it may be possible to employ more simple threshold detection circuits, each of which will give a binary output indication if its analogue signal is within acceptable limits when measured.

There is clearly a very significant silicon area overhead present in this proposed test strategy. If the analogue inputs are not primary inputs, or if additional a.c. test inputs are required, then there is the further complication of having to provide test access such as shown in Figures 8.9 and 8.11. Additionally, it is necessary to control the point at which the data is loaded into the scan chain from the analogue macros, which involves appropriate clocking of the latches synchronised to the phases of the a.c. test waveforms.

As an alternative to LSSD-type scan paths with some form of A-to-D conversion, analogue scan paths have been proposed which effectively scan through analogue data rather than binary data. These proposals employ sample and hold operational amplifiers, chained together with transmission switches, the whole being under the control of some appropriate scan-path clocking.

The disclosures of Wey[39–41] are shown in Figure 8.13. Each analogue node which is to be monitored feeds an analogue shift register (ASR) circuit, these circuits being chained together to form the analogue scan path. Each ASR circuit is buffered from the node under test so as not to load the normal analogue circuit elements.

The proposed ASR circuit shown in Figure 8.13b consists of two sample and hold operational amplifiers, with tranmission switch S1 controlling the sampling and switches S2 and S3 controlling the scan action. The digital signals to control the sample and hold and the serial scan action are shown in Figure 8.13c. All the ASR stages may be loaded in parallel with the sample and hold a.c. data, the values being held in C1, then being shifted through from C1 to C2 and thence to the following ASR stage by the interleaved clocks which control switches S2 and S3. Serial scan out of all sample data therefore results. If the a.c. data required to be monitored is not simultaneously available, then more than one sample, hold and scan sequence becomes necessary.

It will be seen that the two sample and hold operational amplifiers per ASR circuit are analogous to the two latches used in the digital scan-path circuits shown in Figures 5.8 and 5.10, data being shifted from one to the other during scan-out. However, Wey has suggested an alternative analogue shift register circuit which employs only one sample and hold amplifier per stage, as shown in Figure 8.14a, but if all the stages are simultaneously loaded by S1 then it is not possible to scan out the stored data in exactly the same manner as previously. Instead a separate control of the S2 transmission switches is necessary as shown in the Figure. The data sampled by ASR1 will be the final sample output, cascading through all the following conducting ASR stages.

Two critical factors are clearly necessary in all such arrangements, namely:

(i) highly accurate sample and hold voltages;
(ii) a lossless scan path, that is all operational amplifiers provide a stage gain of exactly unity.

The usual form of the analogue switches S1, S2, S3 is shown in Figure 8.14b, being the conventional n-channel/p-channel FET pair. This requires that the peak to peak a.c. signal voltage shall be less than the d.c. voltage controlling the FETs in order that gate to source voltages will always be above their threshold cut-off value, but in any case there will always be some attenuation through the switches, slightly nonlinear, which will impair the ideal transmission of the sampled a.c. data. The small-signal a.c. equivalent circuit of an ASR stage from one storage capacitor, C1, to the following capacitor is shown in Figure 8.14c, where:

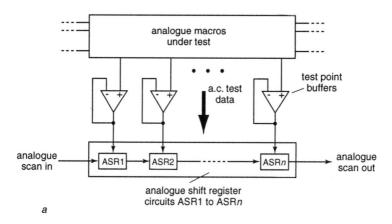

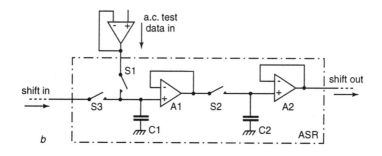

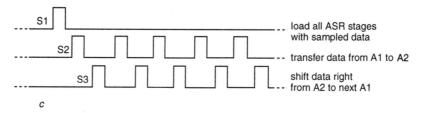

Figure 8.13   *The analogue shift register circuit*
       a outline schematic, omitting any test signal inputs to the analogue macros
       b the ASR operational amplifier circuit
       c control signals (clocks) for the transmission switches S1, S2 and S3

C_1, C_2 = the lumped stray capacitances of the circuit

R_{in} = the input impedance of the circuit, ideally infinite

R_{out} = the output impedance of the amplifier, ideally zero

R_{sw} = the series resistance of the FET transmission switch S2

A = the voltage gain of the operational amplifier, ideally exactly +1.0 +

Mixed analogue/digital system test 445

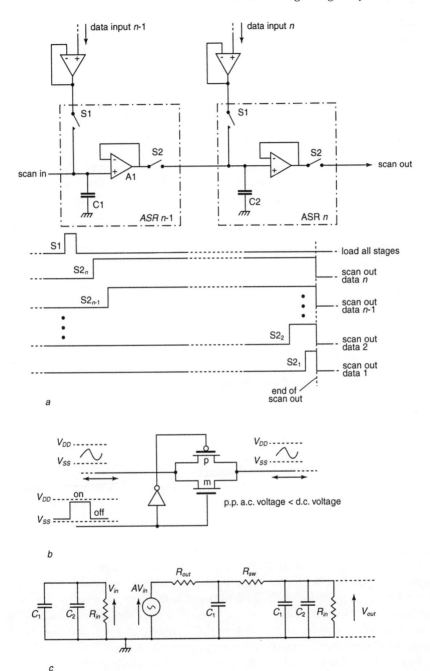

Figure 8.14 Simplified analogue shift register circuit
            a outline schematic and control signals
            b the classic CMOS tranmission switch circuit used in ASR circuits
            c the a.c. small-signal equivalent circuit of one ASR transfer stage

Clearly, the operational amplifier parameters are key factors in this transfer of sampled data. In theory the operational amplifiers could be designed to provide a small compensating gain for the imperfections of the circuit, a monolithic realisation of the parts aiding this matching, but published literature does not mention whether this is necessary or has been incorporated.

The silicon area of these proposals is appreciable. The ASR circuit shown in Figure 8.13a is not unduly different in silicon area from the digital L1/L2 circuit shown in Figure 5.8, consisting of about ten FETs per operational amplifier plus the two FETs per switch, but the point to note is that this a.c. scan path is additional to the normal-mode analogue circuits, whereas in the digital logic case the L1/L2 latches are part of the working digital logic network. It is not known whether this a.c. test strategy has been used in real-life mixed-signal ICs, but it does constitute a possible means of providing observability of internal nodes with the minimum number of additional I/Os. The number of ASR circuits that can be cascaded to form an analogue scan chain must however be limited by the imperfections that accrue with a large number of registers in series. Finally, it may not always be appropriate to sample all the analogue test nodes at the same instant, and hence more than one S1 clock may become necessary.

8.4.2 Built-in self-test strategies

Because of the proven advantages of built-in self test for VLSI digital ICs, namely:

(i) minimising the need for complex off-chip test hardware;
(ii) enabling the internal logic to be tested at full working speed;

see Chapter 5, there have been corresponding considerations for on-chip testing of analogue macros. The problem is clearly not as straightforward as the digital case, since we are interested in parametric test strategies rather than normal-mode functional operation. However, functional testing of the analogue macros is more readily attainable than full parametric tests, and features in most of the analogue BIST developments to date.

Figure 8.15a shows the general strategy that has been pursued. The digital logic is considered to be self-tested by BILBO or some equivalent methodology, with the digital elements used in this logic test being shared as far as possible with the proposed a.c. test. However, the position of the on-chip analogue macros and whether their inputs and outputs are freely available clearly influences the relevant requirements.

Figure 8.15b illustrates the variants which may be encountered, and hence the a.c. controllability and observability needs for the built-in self test. However, it is currently found that in most mixed-signal ICs the digital logic part usually takes up about 80 % of the active area of the chip, and that the analogue elements are peripheral elements; the digital logic, the core logic, is therefore the main buried part of the circuit layout[41-43]. The observability of the response of the input analogue macros and the controllability of the test data into the output analogue macros therefore constitute the principle difficulties for the a.c. tests. Note that we are not interested here at all in fault location, but only in fault detection.

The disclosures of Ohletz covering his hybrid built-in self-test (HBIST) strategy assumes the latter topology, being a digital core with primary input and primary output analogue macros[42]. The chip architecture without built-in self test is therefore as shown in Figure 8.16a, appropriate A-to-D and D-to-A interfaces being present between the analogue and digital parts. The test strategy for incorporating the analogue elements into a built-in self-test environment has the following universal guide lines:

(i) established BIST techniques for digital-only circuits shall be maintained for the digital elements in all mixed-signal circuits;
(ii) the self-test implementation for the analogue macros shall be compatable with the digital BIST circuits, or incorporate them as far as possible;
(iii) all the normal analogue to digital interfaces shall be tested at least once during the tests;
(iv) reconfiguration of the normal-mode circuits into a test mode shall as far as possible be done in preference to switching in entirely separate and additional test mode only circuits.

Figure 8.16b therefore shows the disclosed BIST strategy for the mixed-signal topology of Figure 8.16a, the action of which is as follows.

The IC is prepared for self test by the normal mode/test mode control signal NTC, which in test mode disconnects the primary a.c. input pins from the input analogue macros, connecting instead the a.c. test response data ACTR, and also disconnects the a.c. outputs from the primary a.c. output pins. The complete circuit is therefore isolated from its normal peripheral a.c. inputs and outputs. The self test of the digital core logic is first undertaken in the conventional BIST way, all the sequential storage elements being reconfigurable into BILBO LFSR configurations which can operate as autonomous test vector generators or as multiple-input signature registers (MISRs), as detailed in Chapter 5—see in particular Figure 5.24. During self test, the left-hand BILBO register of Figure 8.16b, BILBO 1, is run as the autonomous test vector generator, providing a maximum-length pseudo-random test sequence to exercise the combinational logic. Recall that in this mode the normal inputs to the storage elements which constitute the BILBO register are conventionally isolated, see Figure 5.28, and therefore here the A-to-D data has no influence. At the same time the right-hand BILBO register, BILBO 2, is run as an MISR to capture the test signature, which is subsequently scanned out in the usual manner for fault-free/faulty checking.

Following the self test of the digital logic, BILBO 2 of Figure 8.16b is reconfigured as an autonomous test vector generator, and BILBO 1 as an

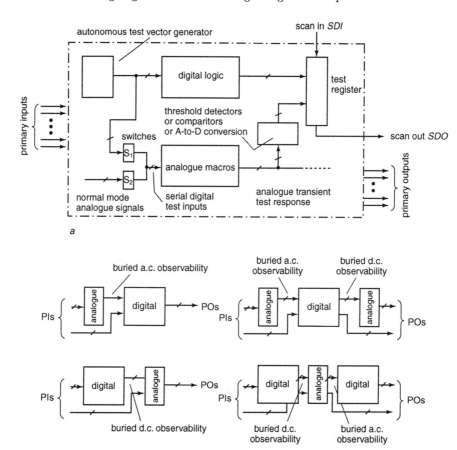

Figure 8.15 Sharing of digital built-in self-test resources for analogue self test
a general schematic
b the topological variants which may be encountered, omitting the internal A-to-D and D-to-A converters

MISR. The pseudorandom test vector sequence now exercises the output analogue macros via the normal D-to-A converters, the a.c. test response ATCR being fed back to the input analogue macros and thence via the A-to-D converters to the BILBO 1 MISR circuit. Notice two points, namely:

(i) all the signal paths between the analogue and the digital partitions are tested in their normal working condition in this a.c. self test;

(ii) the final content of the MISR register of the logic test can act as a seed for the same BILBO register when switched to become the test generator for the a.c. tests.

However, this simple reconfiguration to provide the analogue self test has problems, namely:

Mixed analogue/digital system test 449

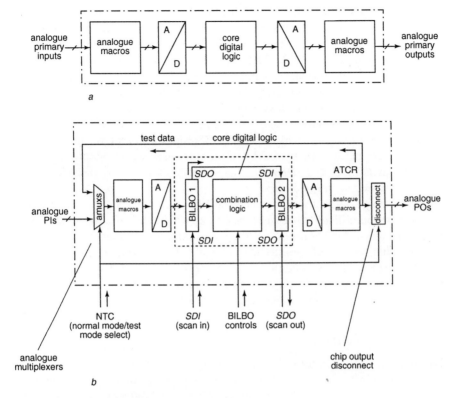

Figure 8.16   Built-in self-test considerations
  a basic chip architecture, with perimeter analogue macros; clock and other digital PIs and POs have been omitted for clarity
  b the basic concept of using the digital self-test BILBO registers to provide a self-test loop around the analogue macros

- the number of output signals from the analogue macros may not match the number of inputs on the analogue macros;
- the analogue test signals produced by BILBO 2 and the following D-to-A converter may not constitute a satisfactory test for the output analogue macros;

and, similarly,

- the analogue test response of the output analogue macros may not constitute satisfactory test data for the input analogue macros.

If BILBO 2 is run as a normal maximum-length pseudorandom test generator, then some analogue response will undoubtedly be captured by BILBO 1 in its MISR mode, and gross faults around this loop will be detected, but this is an abnormal functional test of the analogue macros which may or may not be adequate. It is certainly not a parametric test which seeks to check on acceptable analogue circuit tolerances.

The published literature has considered some of these aspects. One proposal[42] is to modify the test vector generation produced by BILBO 2 so that it is not the conventional maximum-length test vector circuit with a sequence length of $2^n - 1$. The inclusion of the all-zeros state of the sequence may be desirable, see Chapter 5, Figure 5.29, plus the addition of a pattern manipulator circuit which can recode the normal pseudorandom vectors before their D-to-A conversion. A simple five-bit generator is shown in Figure 8.17. An alternative suggested strategy is to feed the BILBO 2 register with scan-in data patterns supplied by an off-chip source, the resultant overall test response being observable by simultaneously scanning out the data captured in BILBO 1 if desired. This strategy may be compatable with a boundary-scan chip implementation.

In reviewing this and similar analogue BIST strategies, every mixed-signal design must be individually considered to ascertain the effectiveness of such functional tests. Hard faults only rather than soft faults, see Section 7.3 of the previous chapter, would seem to be the possible fault coverage, and hence the closeness of any of the analogue macro tolerances to a critical value may not be detectable. Further comments and other similar proposals for built-in analogue self test may be found in the published literature[29,30,32,42,43].

8.4.3 Built-in analogue test generators

Rather than utilise pseudorandom or other autonomous digital test vector sequences followed by D-to-A conversion to exercise the analogue elements during test, consideration has been given to on-chip generation of conventional analogue test signals. This can in theory provide more satisfactory real-life test stimuli than the above methods for the analogue macros, ideally covering (some) soft faults as well as hard faults.

The basic test strategy for testing the analogue macros is shown in Figure 8.18a. The normal-mode analogue inputs are replaced by the on-chip signals generated by some appropriate signal processing circuit, which provides repeatable test data and which ideally can be synchronised to other requirements in the overall self test procedure. Minimum silicon area is also an important consideration in the provision of this test resource.

However, the a.c. response from the analogue macros under test has to be detected on-chip if the whole IC is to be a BIST design, and this depends upon which of the architectures shown in Figure 8.15b is present. Assuming input analogue macros with core digital logic, then the A-to-D and BILBO techniques shown in Figure 8.18b are possible; if the analogue macros are fully buried then some strategy as in Figure 8.18c becomes appropriate. It is clear that the provision of an on-chip signal source to test the analogue macros is only part of the problem; indeed, the greatest difficulty may be the satisfactory capture of the a.c. response data, this representing the greatest practical weakness in the present state of the art.

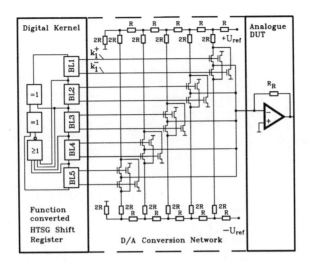

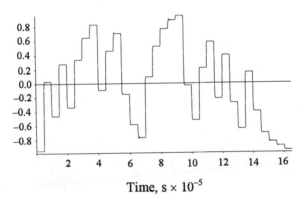

Figure 8.17 Example five-bit modified autonomous test vector generator for built-in analogue self test, with the D-to-A output test stimulus (Acknowledgement, Reference 41)

A number of proposals have been published for the design of built-in analogue test signal generators[44–48]. In general, circuits which require resistance and inductance are inappropriate, and hence capacitance must be the principal reactive element if required, which combined with operational amplifiers provides the necessary frequency selectivity. The following design constraints have been listed by Roberts and Lu[44] as being desirable in the formulation and design of built-in analogue signal sources:

- the source should be capable of generating high-precision a.c. signals with a quality higher than that of the analogue macros under test;

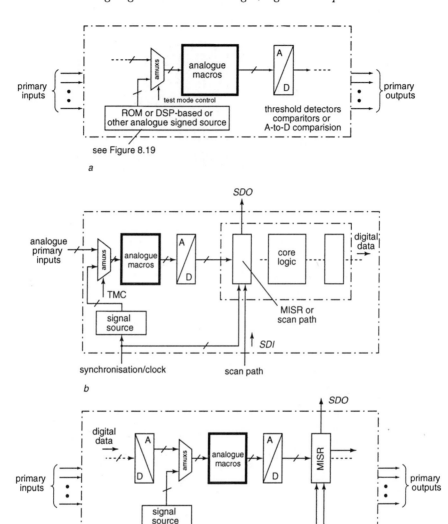

Figure 8.18 On-chip autonomous analogue test stimuli
  a basic needs
  b primary input analogue macros
  c buried analogue macros

- the source should allow complete digital programmability;
- the source should ideally be capable of simultaneously generating multiple tones;
- the source should be insensitive to process variations, with no external timing or calibration being necessary;
- as previously noted, the source should occupy minimum silicon area, ideally not exceeding, say, 10 % of the total mixed-signal chip area.

One obvious way of generating an on-chip analogue waveform is to use a read-only memory (ROM) store, loaded with data which when read out and D-to-A converted approximates to the required a.c. test waveform. An output filter to remove the quantisation noise will then give the a.c. test stimulus, see Figure 8.19a. In theory any waveform can be produced, the limitation being the amount of data required in the ROM and the frequency of the sample and conversion process.

However, it is possible to incorporate some form of feedback into the digital data store, with the object of making the output frequency more readily digitally controllable compared with the simple fixed memory readout of Figure 8.19a. These signal processing methods have been termed direct digital frequency synthesis (DDFS). Details of the evolution of DDFS circuits may be found published[49–53], many circuits giving excellent peformance if no constraints on memory size and/or filter performance are imposed. For example, Nicholas and Samueli[53] generate an adjustable d.c. to 75 MHz spectrally pure sinusoidal output using a 150 MHz sample rate, and employing 2^{15} 12-bit data words stored in a nine-bit coarse ROM plus a three-bit fine ROM. The general outline schematic is shown in Figure 8.19b.

However, the silicon area requirements of many of these published circuits tends to be high, possibly higher than vendors or OEMs will readily accept for built-in self test, and therefore there is continued research into alternative ways of producing on-chip test signal generation, with minimisation of silicon area being a design criteria. Continuous-time analogue signal generation for built-in self test purposes does not appear to have been pursued or published to any noticable extent, although full-custom ICs produced by specialist design houses may incorporate such means. The increasing use of BiCMOS technology, which allows very significant analogue/digital macros to be realised, may modify this status, but the difficulty of exactly synchronising any continuous-time signal generation to a scan chain or other digital means of response observation remains. Notice that if we were not attempting to provide the test generation on-chip, then there are a number of standalone commercial ICs available which provide precision function generation, including sine, square, triangular, sawtooth and other waveforms, with frequency ranges from below 1 Hz to many tens of MHz, and with further provision for sweep frequency and frequency modulation of the output, see for example Reference 54. These and other off-the-shelf ICs such as analogue switches and analogue multiplexers are potentially very useful for on-board

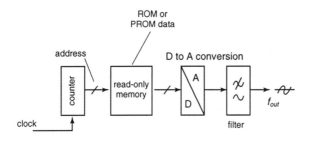

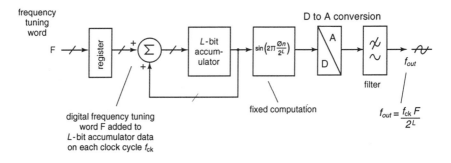

Figure 8.19 *Digital generation of analogue waveforms*
a simple fixed stored data generation
b a basic DDFS architecture

test of PCB system assemblies, but do not help us in our present consideration of fully on-chip test resources.

The broad choice of on-chip analogue test generation is therefore summarised in Figure 8.20, with the continuous-time analogue solution having application difficulties. Here we will look a little further at the third possibility, namely the use of some form of DSP, as this is likely to be the area of increasing significance in mixed-signal IC test strategies of the future. For readers who may wish to be reminded of digital signal processing theory, there are many excellent available texts including the following, and reference should be made to them and other sources for further information and refreshment[23,45,52,55–62].

One of a number of published proposals for on-chip DSP signal generation is the use of delta-sigma modulation, which it is claimed offers efficient silicon area solutions[44,48]. Figure 8.21*a* shows the block schematic of an oversampled ΔΣ-based function generator, the closed loop circuit within the the dotted

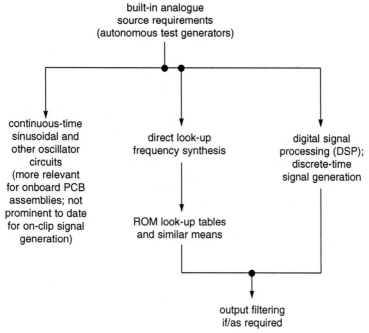

Figure 8.20   The three general possibilities for generating on-chip analogue test signals

rectangle being a second-order digital resonator producing an n-bit digital sine wave at the delta-sigma modulator input. This data may be passed through a D-to-A converter followed by a filter to remove the quantisation noise, to give a final sinusoidal analogue output, but preferably the rapidly switching output bit stream from the modulator is lowpass filtered to provide this analogue signal. The circuits for delta-sigma modulators are shown in Figures 8.21b and c, the second-order version being employed in the published details of Figure 8.21a. Z^{-1} is a unit delay register, the quantiser is a threshold detection circuit whose output is logic 1 if its n-bit input is above a given binary number and is otherwise logic 0, and the D/D converter gives an n-bit output 1 1 1... 1 1 if its input is logic 1, otherwise its output is 0 0 0 ... 0 0.

Further theoretical and practical details of this generator may be found in Reference 44, including the possibilities of multitone and triangular/trapezoidal/square-wave signal generation. However, there is still a very high gate count in the schematic of Figure 8.21 when all the register stages, full-adder circuits and other elements are totalled, and therefore the silicon area is likely to be considerable even if it is claimed to be less than earlier DDFS solutions—a continuous-time sinusoidal oscillator is likely to require the smallest silicon area, but then the advantages of any digital control of its output is not so readily available. A further problem with the $\Delta\Sigma$ modulator

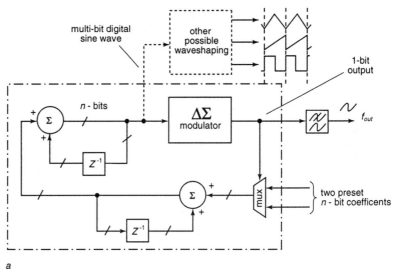
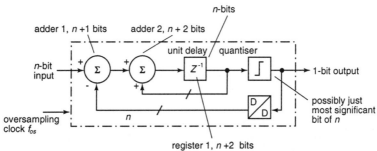
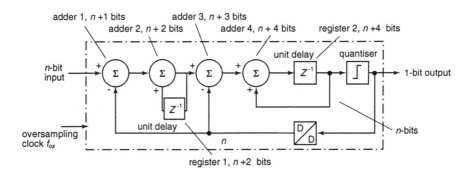

Figure 8.21 Sinusoidal waveform generation using delta-sigma modulation[43]
  a schematic omitting the clocks for clarity
  b a first-order $\Delta\Sigma$ modulator with oversampling clock f_{os}
  c a second-order $\Delta\Sigma$ modulator with oversampling clock f_{os}

is that the analogue output frequency f_{out} is only a small fraction of the oversampling frequency f_{os}, possibly $f_{out} < 10^{-3} f_{os}$, and therefore the maximum possible output frequency f_{out} is severely limited.

Other possibilities for on-chip analogue signal generation include the following:

(i) If the digital logic core itself contains a microprocessor or some other digital processing resource, then it may be relevant to consider reprogramming this resource to generate a.c. test stimuli. Simple tone generation using sine/cosine approximation algorithms and other mathematical functions may be generated; adaptive IIR filters and FIR filters and Fourier and other transforms may also be executed[61], which may be relevant for the analysis of the analogue response data.

(ii) The special signal processing test strategies such as referenced in the previous chapter, including impulse testing, correlation techniques and other concepts, are also relevant to consider as BIST test resources, possibly very much more so than many of the other alternatives. In this context we must certainly consider built-in voltage and current monitoring[32,63,64], which with some form of correlation may provide successful on-chip test strategies—Figure 8.22 illustrates on-chip voltage and current transients when filter macros are subjected to an impulse test, the resulting waveform signatures being correlated with known healthy responses to give go/no-go results.

(iii) Finally, developments are still taking place in continuous-time amplifier and oscillator design; in particular operational transconductance amplifiers (OTAs) for the generation of sinusoidal signals are receiving on-going attention, which may have relevance for future on-chip mixed-signal design and test[65-67].

However, because of the diversity of analogue circuits and the almost unique nature of every mixed-signal IC or PCB assembly, no one test strategy for all analogue and mixed analogue/digital circuits is ever likely to emerge. Rather, a menu of accepted test strategies is likely to be formulated. Hence, the circuit or system designer should become conversant with what may be possible, which must include a much greater familiarity with digital signal processing theory and practice than has previously been the case for digital designers.

8.5 Boundary scan and IEEE standard 1149.4

Moving on from the practical uncertainties of built-in self test for mixed analogue/digital circuits to an area which is receiving international consideration, it will be recalled from Section 5.4 of Chapter 5 that IEEE standard 1149.1, 1991, has been established to provide a recognised means of

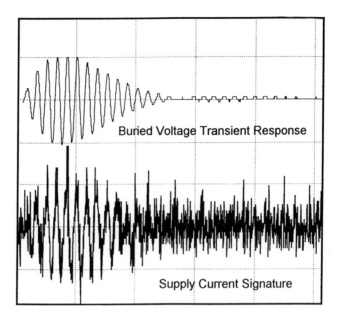

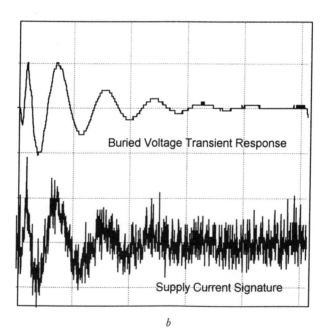

Figure 8.22 Normalised voltage and current signatures for buried analogue macros when subject to transient response tests
a supervisory audio tone filter
b receiver bandpass filter (Acknowledgement, Reference 63)

providing test access to digital VLSI ICs by the provision of boundary scan cells[68,69]. As well as providing test access into and out from the ICs, an important objective of standard 1149.1 was also to allow the interconnections between ICs to be checked on final OEM system assemblies, the increased pin count per IC combined with new forms of packaging making the old-established bed of nails test probing techniques increasingly impossible to implement. Figures 5.18 to 5.22 of Chapter 5 illustrate the main concepts involved in standard 1149.1.

International working parties are currently finalising a corresponding boundary scan standard for analogue macros, this being IEEE standard 1149.4. It was originally hoped that a single universal standard covering both analogue and digital macros would be possible, but partly because the digital standard was considered first, being the most urgent need of the late 1980s, and partly because it became evident that the boundary scan requirements for digital and analogue macros would not have a great deal of practical commonality, standard 1149.1 was established for the digital boundary scan requirements, with 1149.4 now being finalised as the corresponding analogue standard.

Perhaps even more important than in the digital-only case of 1149.1, the analogue boundary scan is very heavily concerned with the interconnect between macros. This is particularly necessary because there are discrete components still required on certain analogue I/Os for timing or other purposes which cannot readily be incorporated on-chip. An interconnect situation such as shown in Figure 8.23 therefore arises which the analogue boundary scan standard has to accommodate. The boundary-scan cells shown in this Figure are:

    ABSC = analogue boundary scan cells

    DBSC = digital boundary scan cells

the latter being exactly as detailed in Figure 5.18. We will have no need to refer to the DBSCs in detail in the following discussions.

If it was readily possible and convenient to probe each interconnect line, then the correct presence of all passsive components outside the IC I/Os could be confirmed by the analogue measurement techniques introduced in Chapter 1[70]. The ideal characteristics of operational amplifiers, namely infinite input impedance and virtual earth at the input terminals, allows component measurements to be undertaken as in Figure 8.24; however, the principle pursued in standard 1149.4 is generally to apply the interconnect stimulus via the output pin of one IC and collect the test result at the input pin of a following IC, as is done for the simple continuity test in the digital case. Therefore, the analogue boundary scan cells are required to provide this analogue controllability and observability, via an overall scan-in/scan-out test structure.

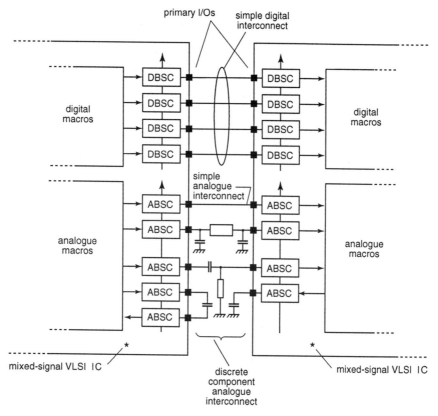

Figure 8.23 The interconnect situations which may be encountered in mixed analogue/digital circuits and systems

It will be apparent that to test for interconnect impedences two test connections need to be made to each component, plus some means of disconnecting or disabling any paths which may be in parallel with the component being tested. Also, the interconnect impedances may be, indeed in many cases will be, frequency sensitive, and hence magnitude and phase data may be involved. There is also the need to detect bridging faults which may occur between any two adjacent analogue or digital lines, and this can become more difficult if one or more of the interconnect lines has any series component such as shown in Figures 8.23 and 8.24. The actual detection of bridging faults can be done with digital data, as in the digital-only situation of standard 1149.1, but clearly the presence of a series component in an analogue interconnect line requires that the bridging tests shall be carried out either side of such a component. Bridging across the component itself is not difficult to detect, since this will show as a faulty impedance in the component test. However, overall a considerable complexity is beginning to build up if all these possibilities are to be covered in the test standard.

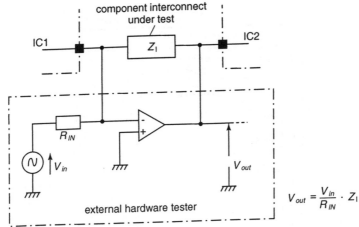

Figure 8.24 External hardware measurements of interconnect components; additional complexities may be present in order not to measure paths which may effectively be in parallel with the component Z_I

The basic boundary scan structure proposed to satisfy these requirements is shown in Figure 8.25. Notice that the analogue and the digital boundary scan cells both have the standard 1149.1 TAP controller (see Figures 5.19 and 5.21) for the input/output digital test data and the test control, but in addition we now have four test connections to every analogue boundary scan cell, namely:

AT1 = analogue test bus 1
AT2 = analogue test bus 2
+V = positive voltage supply bus
G = ground potential supply bus

Notice that these four ABSC connections are buses and not serial scan paths like the TDI/TDO digital scan path, with AT1 and AT2 providing the two-wire connections required for analogue tests such as in Figure 8.24, and +V and G providing test voltages, see below.

The proposed arrangement of each ABSC circuit is shown in Figure 8.26a. The five switches S1 to S5 provide the following facilities:

S1 = core disconnect, to isolate the analogue core signal from the boundary scan circuit
S2 = to connect a positive d.c. voltage (possibly but not necessarily V_{DD}) into the boundary scan circuit
S3 = to connect a zero (ground) voltage (possibly but not necessarily V_{SS}) into the boundary scan circuit
S4 = to connect the primary I/O pin to one of the two test buses
S5 = to connect the primary I/O pin to the other of the two test buses.

462 VLSI testing: digital and mixed analogue/digital techniques

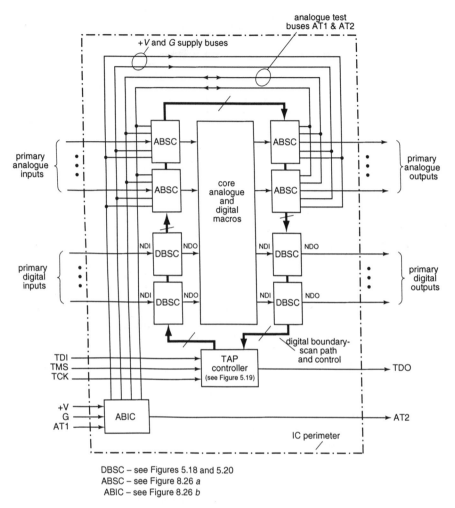

DBSC – see Figures 5.18 and 5.20
ABSC – see Figure 8.26 a
ABIC – see Figure 8.26 b

Figure 8.25 Outline schematic for the on-chip analogue boundary scan proposals of IEEE standard 1149.4

Notice that the latter two switches enable any of the analogue I/Os to be connected to either or both of the test buses, thus providing full flexibility to access any or all of the I/Os and hence interconnections to the IC.

The analogue bus interface circuit ABIC is also a switching matrix similar to the ABSC circuit. As shown in Figure 8.26b, it has provision for switching each internal analogue test bus to either or both the test bus I/Os, together with switches to connect the two d.c. voltage inputs to either or both test bus lines. If the particular IC is not being tested, then the I/Os AT1 and AT2 of ABIC can be directly connected by switches S3 and S8 or S4 and S7, with the switches S1, S2, S5 and S6 in the ABSC circuits being open. There may also be provision in both the ABICs and ABSCs for:

Mixed analogue/digital system test 463

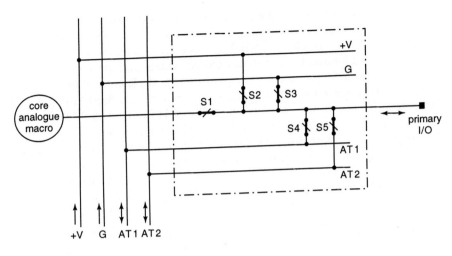

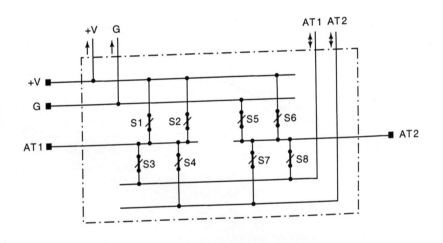

Figure 8.26 The basic provisional switching resources of standard 1149.4
  a the analogue boundary scan cell (ABSC)
  b the analogue bus interface circuit (ABIC)

(i) generating specific values of +V and G within the cells if required for special circuit applications, for example if specific threshold voltages are required;
(ii) the provision of an operational amplifier comparator circuit within the cells to allow monitoring of actual voltage responses during test.

We will not pursue these possibilities here, but instead will review the facilities which the switches provide in test mode.

Control of the switches in the ABSC circuits and in ABIC is a moderately complex requirement; certain combinations such as S2 and S3 simultaneous closed in an ABSC are clearly forbidden, but there remains a considerable number of combinations of switches closed which may be required for a complete IC boundary-scan test. One published possible control structure for each ABS cell is shown in Figure 8.27, which uses the standard 1149.1 digital controls Shift_DR, Clock_DR, Update_DR and Mode provided by the TAP controller, see Figures 5.20 and 5.21. Appropriate control data is scanned into the latches L1 to L3 of this circuit in order to set the necessary switches for each test, the interconnection tests for example being considered in the continuing discussion. The control provided by the latches L1 to L3 when loaded into L4 to L6 is given in Table 8.1 below.

Table 8.1 *The switch permutations with the control structure shown in Figure 8.27*

Mode input	Latched control data			Resultant state of switches (C = closed, O = open)					Resultant connections
	Q_4	Q_5	Q_6	S1	S2	S3	S4	S5	
0	0	0	0	C	O	O	O	O	normal mode*
0	0	0	1	C	O	O	O	C	core, I/O, AT2
0	0	1	0	C	O	O	O	O	normal mode
0	0	1	1	C	O	O	O	C	core, I/O, AT2
0	1	0	0	C	O	O	C	O	core, I/O, AT1
0	1	0	1	C	O	O	C	C	core, I/O, AT1, AT2
0	1	1	0	C	O	O	O	O	normal mode
0	1	1	1	C	O	O	O	C	core, I/O, AT2
1	0	0	0	O	O	O	O	O	fully isolated
1	0	0	1	O	O	O	O	C	I/O, AT2
1	0	1	0	O	C	O	O	O	I/O, +V
1	0	1	1	O	C	O	O	C	I/O, +V, AT2
1	1	0	0	O	O	O	C	O	I/O, AT1
1	1	0	1	O	O	O	C	C	I/O, AT1, AT2,
1	1	1	0	O	O	C	O	O	I/O, G
1	1	1	1	O	O	C	O	C	I/O, G, AT2

*normal mode = core to I/O only

With these switching provisions available in the analogue boundary scan cells, interconnection testing between ICs can readily be undertaken. For a simple interconnection path between two ICs not involving any interconnect components, +V or G may be connected to one I/O and the result monitored by connecting the second I/O to AT1 or AT2. It may also be possible to set up a complete test path through all such simple IC connections in series by appropriate connections within each boundary scan cell, all core macros

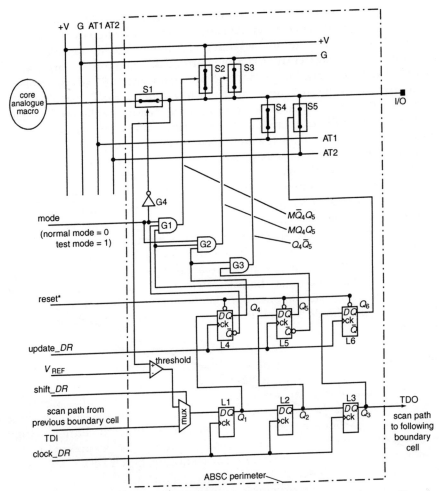

Figure 8.27 *A proposed control structure for the analogue boundary scan cells using the normal TAP controller, cf. the digital boundary scan control shown in Figure 5.20b*

being isolated from this test path by the mode input signal being held at logic 0. The appropriate control data for this series test would need to be determined and then scanned in under the control of the individual TAP controllers.

To test for the correct presence and value of interconnect components, the operational amplifier technique shown in Figure 8.24 can be employed, but a simpler alternative may be used which avoids having to feed any operational amplifier connections through the various boundary scan switches, the losses through which may have to be taken into account if accurate interconnect measurements are required. The test circuit for a single series interconnect impedance is shown in Figure 8.28a. A known constant current is fed into one side of the interconnect via, say, AT1, the other end being connected to G (zero volts). The voltage at the source end is monitored via AT2, which knowing the constant current is a measure of the component impedance. Provided that the series impedance of the series switch S5 is negligible in comparison with the high impedance of the voltmeter, as will usually be the case, then the meter reading will be an accurate measure of the input voltage to Z_I, but there will be an error in this impedance due to the small but finite impedance of the switch S3. If a more accurate measurement of Z_I is required then a difference voltage measurement as shown in Figure 8.28b can be undertaken. Should the AT2 buses of IC1 and IC2 not be separate, then two individual voltage readings, the first from IC1 with S5 of IC2 open and the second from IC2 with S5 of IC1 open, will be necessary, and the difference between them then calculated.

More complex interconnection assemblies such as two-port networks involving two or more analogue boundary scan circuits on each IC can be tested by appropriate multiple measurements, from which branch currents and branch impedances may be calculated. Simple network theory will be involved in these calculations, but the flexibility of the switches in the boundary cells should ensure that appropriate measurements are always possible.

However, although the first objective of standard 1149.4 was to formulate a means for testing the interconnect between analogue macros in a mixed-signal VLSI IC, there is also the desirability of being able to scan in test data for testing the on-chip analogue macros themselves. This must take as its starting point that the I/Os of the analogue macros are accessible or made accessible for such test purposes, and not completely buried within surrounding circuitry. This partitioning will generally be present in the majority of mixed-signal circuits, or can be arranged as a DFT requirement during the design phase.

A number of possible structures have been discussed in test conferences, but a final standard has yet to be announced. All proposals use the AT1 and AT2 bus lines to feed in analogue test signals and feed out analogue response data, but further switching beyond that contained in the boundary scan cells clearly becomes necessary in order to access all the potentially required internal macro nodes.

One outline proposal which indicates the general test strategy is shown in Figure 8.29a. Each analogue macro can be isolated from its following neighbour, and test signals fed into it via AT1 and output signals fed out via AT2. To separate the individual internal macros, a further design of control cell is necessary to control the switches S1, S2 and S3; the first and the last analogue macros in a series cascade can use the normal ABSC cells for analogue input and output data, respectively.

Mixed analogue/digital system test 467

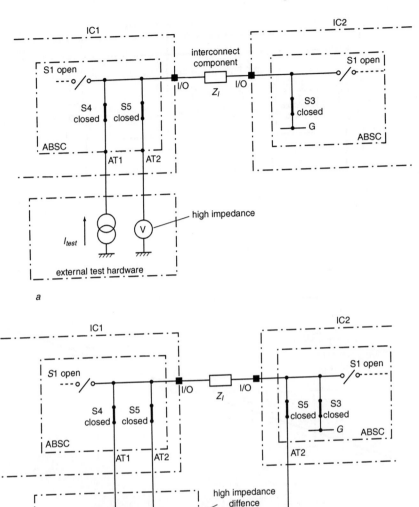

Figure 8.28   Analogue measurements for interconnect components
   a simple single component between two I/Os, not taking into consideration the small impedance of S3
   b compensation for the impedance of S3; the difference between two individual single-input voltage measurements may be necessary instead of this single difference-voltage measurement

468 VLSI testing: digital and mixed analogue/digital techniques

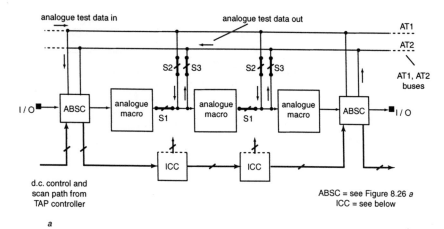

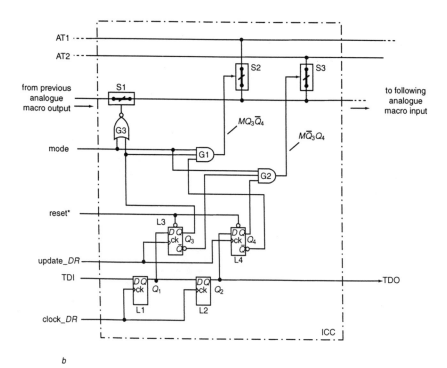

Figure 8.29   *A strategy for scan test of the internal analogue macros*
   a general concept to allow injection of analogue test signals into a
      macro via AT1 bus and extraction of response data via AT2 bus
   b possible circuit for the internal control cells (ICCs)

A possible circuit for the internal control cells (ICCs) is given in Figure 8.29*b*. The mode input is not essential and may be deleted to simplify the circuit. In comparison with the six latches in the ABSC cell, four latches are now required, controlling the three switches S1 to S3 as shown in Table 8.2. To test, say, the centre macro illustrated in Figure 8.29a, switches S1 and S3 preceeding the macro would be opened and switches S1 and S3 after the macro would be closed, thus establishing a test path for external test stimuli fed in on test bus AT1 through the macro to output on bus AT2.

Table 8.2 *The switch permutations with the control structure shown in Figure 8.29b*

Mode input	Latched control data		Resultant state of switches (C = closed, O = open)			Resultant connections
	Q_3	Q_4	S1	S2	S3	
0	0	0	C	O	O	normal mode
0	0	1	C	O	O	normal mode
0	1	0	C	O	O	normal mode
0	1	1	C	O	O	normal mode
1	0	0	C	O	O	normal mode
1	0	1	C	O	C	data to AT2
1	1	0	O	C	O	data from AT1
1	1	1	O	O	O	fully isolated *

*Not useful

The inclusion of this scan-test resource adds considerable silicon overhead to the circuit. Additionally, there could be a degrading of the normal circuit performance due to the presence of S1 and the additional stray circuit capacitances, as well as the imperfections in S2, S3 and the remaining AT1, AT2 bus connections in applying and monitoring the test mode data. The switch S1 of Figure 8.29b may be an active operational amplifier circuit in order to minimise any degrading of the normal circuit performance, but this will add to the silicon area penalty which has to be paid.

The impact on circuit performance of all the switches in the analogue boundary cells and (if used) the internal control cells is a critical consideration; also recall from Figure 8.14b that if the switches are simple CMOS n-channel/p-channel FET pairs the peak to peak analogue voltages must be less than the d.c. supply voltages used to control the switches, which if voltages come down from 5 V to 3 V or less to reduce VLSI power dissipation may cause yet further problems. Thus, although there is a very strong case to be made for analogue boundary scan facilities to confirm the IC interconnections, justification for scan testing of the internal analogue macros remains an open question. On the other hand, built-in self test for the analogue macros has been seen in previous pages to also be an expensive strategy, and hence the choice between analogue BIST and analogue self test remains an open question, one which has not been fully compared and quantified in real-life applications.

The final publication of IEEE standard 1149.4 will include the recommendation of the working party considering all these aspects. Differential I/Os requiring four analogue buses AT1 to AT4 may also be included as a possible means to improve the analogue testing performance, but this again will impact upon the silicon area requirements. However, the basic concepts introduced here should be present in the final documentation, although the circuits which have been shown here may not be exactly as finally documented.

Further information on all the developments and discussions to date may be found in conference proceedings and other publications[32,71–78]. Circuit details for what has been termed pin electronics, that is the many methods of measurement of component and interconnect impedance, including the essential guarding principles to ensure that only the one component or path of interest is being measured, may be found in Bateson[70].

8.6 Chapter summary

If at the end of this chapter the reader is left with a feeling of uncertainty concerning the practice of mixed-signal VLSI testing, this is possibly a correct picture of the present state of the art. In spite of considerable efforts from the mid 1980s onwards, when mixed-signal ICs began to grow from manageable MSI size towards LSI and VLSI complexities, the subject still remains one of continuing development in both design techniques and testing strategies and, most importantly, in appropriate hardware and software CAD resources to assist in this area. Mixed-signal design for testability is therefore still considered to be the most challenging area of present-day microelectronic circuit and system design.

As we have seen, it is invariably necessary to partition the complete analogue/digital circuit so as to give access to the analogue I/Os. The broad DFT choice between:

(i) some form of built-in self test using on-chip autonomous continuous-time analogue sources or some appropriate digital signal processing (DSP) source;
(ii) some form of scan-in test using external test signal generators and response measuring instrumentation;

is one where there is no obvious winner, and largely depends in the end upon:

- the precise nature of the mixed-signal system;
- whether hardware resources are readily available to provide relevant off-chip test stimuli and response measurement;
- an analysis of the silicon area penalties for the different DFT choices;
- the expertise and inclination of the chip or system designer(s).

What is clear from all these considerations is that there will never be one best solution covering all the permutations of analogue and digital macros that can be present in real-life systems. The possible test options must be considered early in the design cycle so as to determine as far as possible the costs and benefits of the DFT choices in the final design.

External analogue and increasingly DSP instrumentation provides the most ready means of testing all the detailed parameters of an analogue core or individual analogue macros. Modern instruments linked by the IEEE standard 488.1/1978 instrumentation bus (otherwise known as the general-purpose instrumentation bus, GPIB, or the Hewlett-Packard instrumentation bus, HPIB) will be prominent in many current test resources[79-85], providing the vendor or OEM with fully flexible means of test for all the analogue requirements. However, when detailed parametric testing is not required then on-chip or custom test resources may become appropriate.

The IEEE standard 1149.1 covering boundary scan for digital circuits is now well established, providing a means for the interconnection testing and scan testing of digital macros. The publication of standard 1149.4 will establish the corresponding standard for the analogue macros, but as we have seen may involve a considerable silicon area penalty to implement in an IC design. However, if it does become accepted in its final form, then the pros and cons between BIST and scan test for mixed-signal circuits may become more readily quantifiable.

Finally, in all the preceeding technical discussions on test, no mention has been made of the specific testing requirements for military, avionic or other specialised application areas of microelectronics. In general these will be very heavily concerned with environmental and life testing rather than just one-off tests for functional faults with an immediate pass/fail result. There are procedures which the OEM must follow for specific customers, following for example the various USA military specifications and standards such as MIL-STD-883D and others[86-89]. This is beyond our concern in this text, but is an essential part of the mechanical, electrical and electronic system test for many specialist areas.

8.7 References

1 SOIN, R., MALOBERTI, F., and FRANCA, J. (Eds.): 'Analogue/digital ASICs: circuit techniques, design tools and applications' (Peter Peregrinus, 1991)
2 TRONTELJ, J., TRONTELJ, L., and SHENTON, G.: 'Analog digital ASIC design' (McGraw-Hill, 1989)
3 HURST, S. L.: 'VLSI custom microelectronics: digital, analog and mixed signal' (Marcel-Dekker, 1998)
4 PLETERSEK, T., TRONTELJ, J., TRONTELJ, L., JONES, I., and SHENTON, G.: 'High-performance designs with CMOS analogue standard cells', *IEEE J. Solid-State Circuits*, 1986, **SC-21**, pp. 215–222
5 DEDIC, I.J., KING, M. J., VOGT, A.W., and MALLISON, N.: 'High-performance converters on CMOS', *J. Semicust. ICs*, 1989, **7**, September, pp. 40–44

6 SPARKES, R.G., and GROSS, W.: 'Recent developments and trends in bipolar analog arrays', *Proc. IEEE*, 1987, **75**, pp. 807–815
7 'Quickchip 2 designers' guide'. Tektronix, Inc., OR
8 'Semicustom linear array brochure'. AT & T Technologies, PA
9 'RLA and RFA series linear array design manual'. Raytheon Corp., CA
10 HORNUNG, R., BONNEAU, M., and WAYMEL, B.: 'A versitile VLSI design system for combining gate array and standard circuits on the same chip'. Proceedings of IEEE conference on *Custom IC*, 1987, pp. 245–247
11 'Mixed signal ASIC handbook'. GEC Plessey Semiconductors, UK
12 TORMEY, T.: 'Mixed-signal simulation eases system integration', *Comput. Des.*, 1989, **28** (9), pp. 103–106
13 MEYER, E.: 'ASIC users shift analog on-chip', *Comput. Des.*, 1990, **29** (5), pp. 67–72
14 LAMBINET, P.: 'Trends in mixed-signal ASIC partitioning', *Electron. Prod. Des.*, 1993, **14** (10), pp. 37–41
15 FAWCETT, B.K.: 'Reconfigurable FPGAs in test equipment', *Electron. Prod. Des.*, 1996, **17** (2), pp. 20–26
16 SHEPHERD, P.R.: 'Integrated circuit design, fabrication and test' (Macmillan, 1996)
17 ARMSTRONG, J.R., and GRAY, F.G.: 'Structured logic design with VHDL' (Prentice Hall, 1993)
18 GU, R.X., SHARAF, K.M., and ELMASRY, M.I.: 'High-performance digital VLSI design' (Kluwer, 1996)
19 GRAY, P., and MEYER, R.: 'Analysis and design of analog integrated circuits' (Wiley, 1993)
20 GOYAL, R.: 'High-frequency analog integrated circuit design' (Wiley, 1995)
21 VERGHESE, N.K., SCHMERBECK, T.J., and ALLSTOT, D.J.: 'Simulation techniques and solutions for mixed-signal coupling in integrated circuits' (Kluwer, 1995)
22 PUCKNELL, D.A., and ESHRAGHIAN, K.: 'Basic VLSI design: systems and circuits' (Prentice Hall, 1988)
23 GEIGER, R.L., ALLEN, P.E., and STRADER, N.R.: 'VLSI design techniques for analog and digital circuits' (McGraw-Hill, 1990)
24 ISMAIL, M., and FRANCA, J. (Eds.): 'Introduction to analog VLSI design' (Kluwer, 1990)
25 BEENKER, F.P.M., and EERDEWICK, K.J.E.: 'Macro testing: unifying IC and board test'. Proceedings of IEEE conference on *Design and test*, 1986, pp. 26–32
26 PRILIK, R., VAN HORN, J., and LEET, D.: 'The loophole in logic test: mixed-signal ASICs'. Proceedings of IEEE conference on *Custom integrated*, 1988, pp. 1–5
27 O'LEARY, P.: 'Practical aspects of mixed analogue and digital design', *in* [1], pp. 213–238
28 SHEPHERD, P.R.: 'Testing mixed analogue and digital circuits', *in* [1], pp. 334–359
29 FASSANG, P.P., MULLINGS, D., and WONG, T.: 'Design for testability for mixed analog/digital ASICs'. Proceedings of IEEE conference on *Custom integrated*, 1988, pp. 16.5.1–16.5.4
30 WAGNER, K.D., and WILLIAMS, T.W.: 'Design for testability of mixed signal integrated circuits'. Proceedings of IEEE international conference on *Test*, 1988, pp. 823–828
31 DUHAMEL, P., and RAULT, J.-C.: 'Automatic test generation techniques for analog circuits and systems: a review', *IEEE Trans.*, 1979, **CAS-26**, pp. 411–440
32 HARVEY, R.J.A., BRATT, A.H., and DOREY, A.P.: 'A review of testing methods for mixed-signal ICs', *Microelectronics J.*, 1993, **24**, pp. 663–674

33 BUTLER, I.C., TAYLOR, D., and PRITCHARD, T.I.: 'The effect of response quantisation on the accuracy of transient response testing', *IEE Proc., Circuits Devices Syst.*, 1995, **142**, pp. 334–338
34 TAYLOR, D., EVANS, P.S.A., and PRITCHARD, T.I. 'Testing of mixed-signal systems using dynamic stimuli', *Electron. Lett.*, 1993, **29**, pp. 811–813
35 BINNS, R.J., TAYLOR, D., and PRITCHARD, T.I.: 'Testing linear macros in mixed-signal systems using transient response testing and dynamic supply current monitoring', *Electron. Lett.*, 1994, **30**, pp. 1216–1217
36 ECKERSALL, K.R.,TAYLOR, G.E., BANNISTER, B.R., and BELL, I.M.: 'Testing an analogue circuit using a complimentary signal set'. Proceedings of IEE colloquium on *Testing mixed signal circuits*, London, May 1992
37 SAVIR, J., and BARDELL, P.H.: 'Built-in self test: milestones and challenges', *VLSI Des.*, 1993, **1** (1), pp. 23–44
38 *Test synthesis seminar*, digest of papers, IEEE Computer Society International Test Conference, Altoona, PA, 1994
39 WEY, C.-L., JIANG, B.L., and WIERZBA, G.M.: 'Built-in self test for analog circuits'. Proceedings of IEEE 31st midwest symposium on *Circuits and systems*, 1988, pp. 862–865
40 WEY, C.-L., JIANG, B.L., and WIERZBA, G.M.: 'Built-in self test (BIST) design of large-scale analogue circuit networks'. Proceedings of IEEE international symposium on *Circuits and systems*, 1989, pp. 2048–2051
41 WEY, C.-L.: 'Built-in self test (BIST) structure for analog circuit fault diagnosis', *IEEE Trans.*, 1990, **IM-39**, pp. 517–521
42 OHLETZ, M.J.: 'Hybrid built-in self test (HBIST) for analogue/digital integrated circuits'. Proceedings of European conference on *Test*, 1991, pp. 307–315
43 OHARA, H., NGO, H.X., ARMSTRONG, M.J., RAHIM, C.F., and GRAY, P.R.: 'A CMOS programmable self-calibrating 13-bit channel data acquisition peripheral', *IEEE J. Solid-State Circuits*, 1987, **SC-22**, pp. 930–938
44 ROBERTS, G.W., and LU, A.K.: 'Analog signal generation for built-in self test of mixed-signal integrated circuits' (Kluwer, 1995)
45 MAHONEY, M.: 'DSP-based testing of analog and mixed-signal circuits' (IEEE Computer Society Press, 1987)
46 TONER, M., and ROBERTS, G.W.: 'A BIST scheme for a SNR test of a sigma-delta ADC'. Proceedings of IEEE international conference on *Test*, 1993, pp. 805–814
47 BURST, L., and TSEY, M.-S.: 'Mixing signals and voltages on a chip', *IEEE Spectr.*, 1993, **30**, pp. 40–43
48 LU, A.K., ROBERTS, G.W., and JOHNS, D.: 'A high quality analog oscillator using oversampling D/A conversion techniques', *IEEE Trans.*, 1994, **CAS-41/2**, pp. 437–444
49 TIERNEY, J., RADAR, C.M., and GOLD, B.: 'A digital frequency synthesiser', *IEEE Trans.*, 1971, **AU-19**, pp. 48–57
50 'Direct digital synthesiser handbook'. Stanford Telecom, Inc., CA, 1990
51 NEW, B.: 'Complex digital waveform generation'. Application note XAPP.008.002, Xilinx, Inc. Programmable Logic Data Book, 1994
52 MANASSEWITSCH, V.: 'Frequency synthesizers, theory and design' (Wiley, 1989)
53 NICHOLAS, H.T., and SAMUELI, H.: 'A 150 MHz direct digital frequency synthesiser in 1.25 μm CMOS with –90 dB spurious performance', *IEEE J. Solid-State Circuits*, 1991, **26**, pp. 1959–1969
54 'New releases data book, vol. IV'. Maxim Integrated Products, CA, 1995
55 OPPENHEIM, A.V., and SCHAFER, R.W.: 'Discrete-time signal processing' (Prentice Hall, 1989)
56 MITRA, S.K., and KAISER, J.F.: 'Handbook for digital signal processing' (Wiley, 1993)

57 LYNN, P.A., and FUERST, W.: 'Introductory digital signal processing with computer applications' (Wiley, 1996)
58 FLIEGE, N.J.: 'Multirate digital signal processing' (Wiley, 1994)
59 JONES, N.B., and WATSON, J.D. McK. (Eds.): 'Digital signal processing: principles, devices and applications' (IEEE Peter Peregrinus, 1990)
60 ROBERTS, R.A., and MULLIS, C.T.: 'Digital signal processing' (Addison-Wesley, 1987)
61 'ADSP-21000 family applications handbook, Vol. 1'. Analog Devices, Inc., MA, 1994
62 CANDY, J.C., and TEMES, J.: (Eds.): 'Oversampling delta-sigma converters' (IEEE Press, 1991)
63 NIGH, P., and MALY, W.: 'Test generation for current tests', *IEEE Des. and Test comput.*, 1987, **7** (1), pp. 26–38
64 BINNS, R.J., TAYLOR, D., and PRITCHARD, T.I.: 'Generating, capturing and processing supply current signatures from analogue macros in mixed-signal ASICs', *Microelectron. J.*, 1996, **27**, pp. 723–729
65 SENANI, R.: 'New electronically tuneable OAT-C sinusoidal oscillators', *Electron. Lett.*, 1989, **25**, pp. 286, 287
66 KHAN, I.A., AHMED, M.T., and MINHAJ, N.: 'Tuneable OTA-based multi-phase sinusoidal oscillators', *Int. J. Electron.*, 1992, **72**, pp. 443–450
67 HOU, C.-L., WU, J.-S., HWANG, J., and LIN, H.-C.: 'OTA-based even phase sinusoidal oscillators', *Microelectron. J.*, 1997, **28**, pp. 49–54
68 'Standard 1149.1, 1990, standard test access port and boundary scan architecture'. Institute of Electrical and Electronics Engineers, NY, 1990
69 Supplement to IEEE standard 1149.1, 1149.1(b). Institute of Electrical and Electronics Engineers, NY, 1994
70 BATESON, J.: 'In-circuit testing' (Van Nostrand Reinholt, 1985)
71 'P1149.4 mixed signal test bus framework proposals'. Proceedings of IEEE international conference on *Test*, 1992, pp. 554–557
72 WILKINS, B.R.: 'A structure for board-level mixed-signal testability'. Proceedings of IEEE international conference on *Test*, 1993, pp. 556–557
73 PARKER, K.P., McDERMID, J., and ORESO, S.: 'Structure and methodology for analogue testability bus'. Proceedings of IEEE international conference on *Test*, 1994, pp. 309–322
74 WILKINS, B.R., ORESO, S., and SUPARJO, B.S.: 'Towards a mixed-signal testability standard P.1149.4'. Proceedings of European conference on *Test*, 1993, pp. 58–64
75 WILKINS, B.R.: 'Stretching the boundaries: mixed-signals and P1149.4', IEE colloquium on *Systems design for testability*, digest no. 1995/083, 1995, pp. 5.1–5.10
76 LEE, N.-C.: 'A hierarchical analog test bus framework for testing mixed signal integrated circuits and printed circuit boards', *J. Electron. Test.: Theory and Appl.*, 1993, **4**, pp. 361–368
77 PARKER, K.: 'Standards-based designing for testability'. Proceedings of *NEPCON West '95*, 1995, pp. 1921–1929
78 LEE, K.J., LEE, T.-P., WEN, R.-C., and LIN, Z.-T.: 'Analogue boundary scan architecture for d.c. and a.c. testing', *Electron. Lett.*, 1996, **32**, pp. 704, 705
79 KULARATNA, N.: 'Modern electronic test and measuring instruments' (Institution of Electrical Engineers, 1996)
80 FLUKE, T. M.: 'Instrumentation', *IEEE Spectr.*, 1989, **26** (1), pp. 50–53
81 ANSI/IEEE standard 488.1/1987, 'Standard interface for programmable instrumentation'.
82 ANSI/IEEE standard 488.2/1987, 'Standard codes, formats, protocols and common commands'

83 CARISTI, A. J.: 'IEEE 488 general purpose instrumentation bus manual' (Academic Press, 1989)
84 'Tutorial description of the Hewlett-Packard interface bus'. Hewlett-Packard, CA, 1984
85 LEIBSON, S. H. 'IEEE 488.2 products are now appearing', *EDN*, 1991, **25**, April, pp. 91–99
86 MIL-STD-883D, 1995, 'Test methods and procedures for microelectronics'
87 MIL-STD-2165A, 1993, 'Testability program for systems and equipments'
88 MIL-STD-38510J, 1993, 'General specification for microcircuits'
89 MIL-ST-55110E, 1993, 'Specification for rigid wiring boards'

Chapter 9
The economics of test and final overall summary

9.1 Introduction

There is a considerable amount of published information concerning the relative design and fabrication costs of the competing methods of producing microelectronic circuits for OEM use, the broad choice being between standard off-the-shelf parts and some form of customised IC. However, there is less information available concerning the financial pros and cons of test, and particularly the design for testability (DFT) methodologies such as considered in previous chapters of this text. We have in these previous pages raised the question of the silicon area penalties which can be involved, see for example Figures 5.11 and 5.38, but so far we have not specifically looked at actual or relative financial costs.

We will, therefore, in this final chapter attempt to bring into the picture these financial matters. We will not be able to cite any absolute costs in $, $k or $M, since such detail is continually changing and in most cases depends upon the internal financial cost structures of individual suppliers, vendors and OEMs; however, we can consider the many factors that may be involved, together with the possible equations and financial models which may help in quantifying the cost of a product involving VLSI technology.

The overall importance of costing does not need emphasising, since it is the responsibility of every vendor and OEM not to make a financial loss across their range of products. Microelectronic costs, however, may be difficult to assess, and may be based upon previous experience rather than on precise financial data relating to a new product. Within the theme of this text, namely testing, it would clearly be cheaper for an IC vendor not to test production wafers and finished ICs, and for an OEM not to test a final product; on the other hand it is clearly equally unacceptable to subject every manufactured product to lengthy exhaustive life tests. This is the dilemma facing every vendor, what and how much testing is optimum for the particular product; as broadly indicated in Figure 9.1, too little and the end user will be upset and

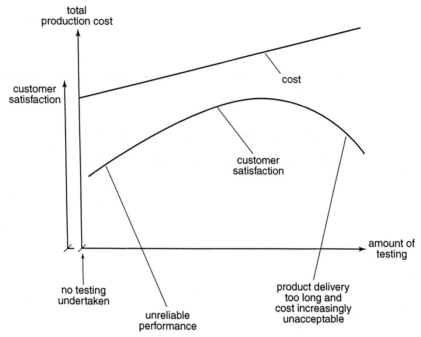

Figure 9.1   Too much or too little testing?

custom will be lost; too much and risk being uncompetitive in the market place. Testing costs have, however, been reported as being as high as 30 to 40 % of VLSI production costs, although no detailed breakdown of this figure is known.

9.2 A comparison of design methodologies and costs

Before considering the specific cost implications of testing and DFT, let us briefly review the other factors which vendors and OEMs encounter, on top of which we will later consider the testing considerations.

The design of every product involves nonrecurring engineering (NRE) costs, including designers' time and overheads, and other factors which are ideally completely paid for by the time the product reaches the production line. Recall from Figures 6.18, 7.1 and 8.4 that in addition to standard off-the-shelf parts there are many variants of customised ICs available, ranging from field-programmable logic devices through to full hand-crafted custom design, each with their own market advantages and costs. In general the lower the production volume required and the lower the technical requirements, then the more relevant will be standard parts and PLDs. The OEM therefore may have to consider costs such as listed in Table 9.1. Some of these costs may not arise, and others, such as training, may not be considered by some companies to be NRE expenditure directly costed to a particular new product. Nevertheless, they are costs which must be considered somewhere within the company finances. Note that external PLD design costs should not arise, since the whole objective of PLDs is to enable the OEM to undertake all design activities in house so as to give personal control over the design and prototype stages.

Table 9.1 *The IC NRE costs which may be incurred by the OEM with the three main categories of design styles. The individual item costs may vary greatly between the three styles*

NRE item	standard off-the-shelf IC	PLD	custom IC
In-house NRE costs:			
Purchase or hire of CAD tools*	–	✓	✓
Designer retraining	–	✓	✓
Design time and overheads	✓	✓	✓
Special assembly and test resources	✓	✓	✓
Purchase of PLD programmer	–	✓	–
Outside NRE costs:			
Subcontract some or all of the IC design work	–	–	✓ ⎫
Cost of mask making	–	–	✓ ⎬ †
Cost of custom jigs to interface with standard IC tester	–	–	✓ ⎭
Iteration costs	(low)	(low)	✓

* We exclude here any in-house CAD tools which are for general-purpose use, such as PCB layout CAD
† Possibly a single lump sum NRE cost to the OEM

To illustrate how NRE costs should broadly be considered at an early product planning stage, let us define the following:

E_1 = the total IC design NRE costs involved in the chosen design style

E_2 = the remaining system design NRE costs for the rest of the product, i.e. mechanical design, PCB design, tooling, etc.

A = per-unit cost to the OEM of the standard ICs, PLDs or custom IC

B = per-unit cost of the remaining product components

C = per-unit assembly and test costs

Then for a production run of N units the total OEM costs are:

$$E_1 + E_2 + N(A + B + C)$$

giving a per-unit total system cost of

$$PUC_{system} = \frac{E_1 + E_2}{N} + (A + B + C) \tag{9.1}$$

Clearly, if E_1 is large and/or N is small, then the IC design cost will have a profound effect on per-unit cost, and the OEM should consider a design style which reduces this factor as much as possible. On the other hand, NRE costs become much less significant for large volume requirements, since $E_1 + E_2$ can be amortised over a much larger production N, with $A + B$ now becoming the significant parameters. (An IC vendor's sale of standard parts is the extreme example of this where, for example, the original design costs and also VLSI tester costs of $M may be spread over millions of production circuits, adding only a very small sum to the per-unit selling cost.)

In theory this simple equation, eqn. 9.1, can indicate which is the most economical design style for a given product and production quantity; the often insuperable difficulty, however, is to obtain or estimate accurate numbers to put into the equation for the alternative design styles, which may involve different potential IC vendors, different external design houses and other aspects. Overall the best economic decision will generally be as shown in Figure 6.18, but the precise break-even points between the categories will be almost impossible to quantify. Other nonfinancial matters such as long-term availability, commercial product security, company image, etc.[1-3] can also influence a final choice.

To fill out the main parameters involved in eqn. 9.1, more detailed tabulations such as shown in Table 9.2 may be compiled. The values given in this Table are illustrative only, and must not be taken as accurate or up to date, but they serve to illustrate the broad financial considerations which an OEM may address. The cost of DFT and IC and system test, which we will consider in greater detail in the following pages, will be noted as being but a part of the overall picture; indeed, DFT and test strategies may be more important for the intangible factors such as reliability and customer satisfaction than for purely financial reasons.

The figures given in Table 9.2 should be appreciated as being simplistic, not involving all the factors of a real-life situation. In particular, the per-unit cost of custom ICs is highly dependent upon the production quantities required, and hence the per-unit production cost is nonlinearly related to N. None the less, if the OEM can obtain estimates for all NRE and per-unit component costs, then effective forecasting and comparison of per-unit system costs may be possible. Very detailed cost analyses of custom circuits have been published, particularly by Fey and Paraskevopoulos[4-7], involving gate counts and production quantities but not specifically any DFT or other test strategy, but there remains the problem of finding accurate data to enter into all these financial equations.

New in-house CAD costs and designer learning (retraining) costs have not been built into Table 9.2. Hardware and software CAD costs may be very

Table 9.2 Example OEM costs for a small digital product, ignoring any capital equipment costs for new in-house CAD resources, etc. Design style 1 = using off-the-shelf ICs, 2 = using PLDs, 3 = full-custom, giving maximum performance and minimum silicon area, 4 = standard-cell IC, 5 = gate-array IC. All costs in $k

Nonrecurring OEM costs item:		Design style			
	1	2	3	4	5
(i) OEM circuit design + PCB and remaining hardware design costs	20	15	10	10	10
(ii) Custom or other outside design costs to the OEM	–	–	150	20	5
(iii) OEM and end-user documentation	4	3	2	2	2
Total NRE Costs	24	18	162	32	27
Per-unit OEM production costs item:		*	*	*	
(i) ICs plus their PCB assembly	0.3	0.2	0.1	0.1	0.1
(ii) Other hardware and assembly costs	0.7	0.5	0.4	0.4	0.4
(iii) Unit test	0.3	0.2	0.1	0.1	0.1
Total per-unit cost (PUC)	1.3	0.9	0.6	0.6	0.6

* Very dependent upon volume, see text

appreciable, and also have the unfortunate feature that systems tend to become obsolete within possibly three years of purchase. This is particularly so with mixed-signal CAD software, which is still in the throes of rapid development and obsolescence. CAD resources for programmable logic devices are perhaps the most easy to cost, since complete hardware/software packages are commercially available from vendors; a problem that may arise, however, is that the CAD resource for the programmable products of vendor x may not be compatable with the products of vendor y, and therefore more than one software package may become necessary if more than one type of PLD is adopted. Hardware and software maintenance will also add at least ten per cent of the capital cost of the CAD resources for each year of use.

A broad generalisation of CAD costs with which the OEM may be involved is shown in Figure 9.2, to which software costs and hardware and software maintenance must be added. In total at least twice the initial hardware costs may be involved in establishing and running a particular in-house CAD resource, the cost of which has to be amortised over production sales in, say, three years' activity.

Design time, correction time and learning time are also factors which the OEM has to consider. Figure 9.3 illustrates the time factors which are involved in the design of systems using differing microelectronic design styles, which must translate in some complex way into financial considerations. Notice also

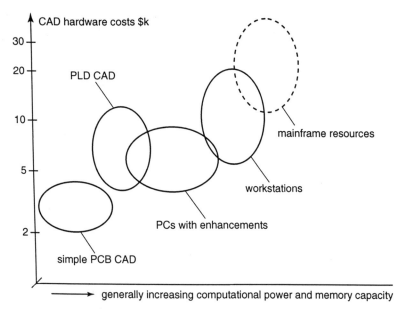

Figure 9.2 A general picture of CAD hardware costs to which software costs must be added. The computing power of newer hardware continually increases but overall costs remain very roughly constant

that the more sophisticated the microelectronic content, for example hand-crafted full-custom ICs, then in general the greater risk of requiring a design iteration at the prototype stage, and the greater the cost and time then necessary to implement this iteration. On the other hand, for simple applications, the adoption of PLDs allows the quickest and cheapest design iteration, with no external agency being involved outside the OEM's organisation.

Learning time (or retraining time) for OEM designers encountering a new CAD design environment for the first time is a further factor which is difficult to quantify financially, and is generally lost in company overheads. In broad terms it takes perhaps two years for a new OEM designer to become conversant and up to speed with CAD tools involving IC design activities, of which time a possible 25 % is spent in nonprofitable activities. With a salary plus company overhead of, say, $150k per designer per annum, this implies a company learning cost of around $75k per new designer. For new/updated CAD tools the subsequent learning times will be considerably less, costing, say, $10–$15k per designer per new system.

These design iteration times and learning/retraining times have a further crucial effect on final market sales. Most microelectronic-based products have a sales life of only about two to three years at the most before being superseded by improved or more innovative products. Therefore any delay in availability involves the risk of losing a considerable share of the market; a six-

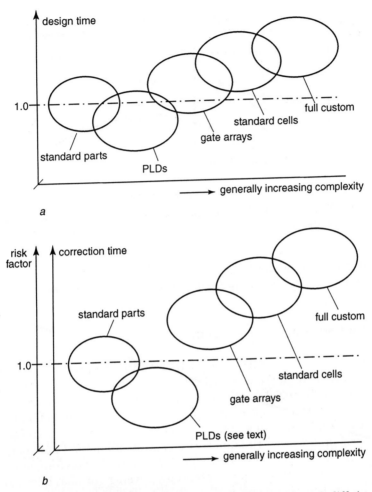

Figure 9.3   Relative comparison of design and correction times with differing design styles. Exact quantification is product, company and vendor dependent
 a relative design to prototype times, taking off the shelf standard parts as the norm
 b similarly, relative risk and correction times

month unforeseen delay in production and sales has been estimated to cost at least 30 % loss in total sales of a given product design. Figure 9.4 shows the usual representation of a microelectronic product life cycle and profit revenue; the general effect of being late to market will leave the right-hand part of this picture relatively unchanged, but the ramp up to maturity will be depressed thus making total sales and total profits corresponding smaller [1,8].

These interlocking financial factors may be found more fully discussed in Reference 1. Further general discussions on cost may also be found in Needham[8] but without a great deal of quantification; PCB assemblies will be

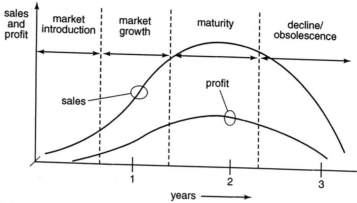

Figure 9.4   *Typical product life cycle and company profits for a microelectronic-based product or system, with obsolescence rapidly approaching after two to three years' life*

found discussed in Bennetts[9] and in Bateson[10]; other financial items will also be found in several of the individual author chapters in Di Giacomo[11].

9.3 The issues in test economics

So far we have seen that the costs of DFT and IC and system test must be present somewhere in the financial calculations, but we have not specifically addressed such financial costs. Indeed, the costs directly associated with test is not a subject which appears very prominently in published literature; the special issue on VLSI testing of the journal 'VLSI Design', for example[12], has a great deal on test strategies with very extensive and helpful References, but contains no financial data or cost equations. We will, therefore, attempt to give here and in the following section some coverage of what information is available, particularly on cost modelling and on-going work in this area, but inevitably precise financial figures will not be possible.

A problem with the economics of testing is that it has so many and sometimes conflicting aspects. There is the short-term aspect—how cheaply can we make the product and still sell it—and the long-term aspect of lifetime costs embracing subsequent maintenance and repair. Factors including mean time between failure (MTBF) and mean time to repair (MTTR), see Section 4.6.2 of Chapter 4[13,14], will become involved when lifetime costs are being considered. One frequently quoted maxim is the rule of tens law, which states that if the cost of detecting a fault at the component level is, say, $1, then to detect the same fault at the assembled PCB level would be $10, at completed system level would be $100 and in service with the end-user would be $1000. This is a classic rule to be borne in mind—that the later a fault is detected in the production and life of an equipment then the more costly it will be to correct the fault. The exact ratios of 1:10:100:1000 are, perhaps, oversimplistic, but on the other hand a recent study which suggested ratios of about 10:1:10:100, indicating that board-level testing is the least costly test level, may not be universally valid[15].

If we concentrate our considerations here primarily on VLSI testing costs, since complete system costs are very individualistic, clearly the more exhaustive the IC testing the fewer the problems that should be encountered in product assembly and later testing. However, if no DFT provision is built into the IC design, then exhaustive testing can become prohibitively time consuming and therefore expensive, but if DFT is built in then there will be the cost penalties of increased chip area and potentially lower yield to be considered. These factors are broadly illustrated by Figure 9.5, which indicates that there must be some optimum test overhead for a given VLSI design. Notice that to build in increased controllability and observability or BIST involves not only increased silicon area but also more IC pins, and therefore a higher package cost to add to the overall picture.

Because of the difficulties of accurately costing any form of DFT, it has been argued that it is not economically viable to include any design for testability features in a new chip design. Also, until recently the chip design time was unduly and indeterminately influenced by attempting to build in such features, resulting in delayed time to market and loss of total sales (see Figure 9.4), but this has to a large degree been overcome by the increasing availablity of relevant CAD software which supports DFT. There was, for example, a strong post-experience reaction to BILBO by some IC vendors in the late 1980s, that it involved an unnecessary high expenditure and also some loss of state of the art circuit performance. This reaction was to some degree fostered by the increasing yield and decreasing defect density of production wafers, which to some extent made the comprehensive BILBO test strategy not as necessary as previously anticipated. However, with the continuing increase in size and complexity of VLSI designs, and in spite of continuing improvements in yield, it is now accepted that some DFT strategy must be incorporated in complex chips. This acceptance is reinforced by:

(i) increasing knowledge of DFT theory and practice by circuit designers and the availability of relevant CAD support;
(ii) increasing demand by OEM system designers for boundary scan and other means to aid their PCB and system testing difficulties;
(iii) the publication of IEEE standard 1149.1 for digital scan test and the (forthcoming) standard 1149.4 for analogue scan test, which establish agreed international standards for all circuit and system designs;
(iv) the increased acceptance by all IC vendors that they must accept some responsibility for easing the use and testing of their products by their customers after the products have left their hands—lifetime costs are as much a part of their professional responsibility as that of the OEM and end-user.

486 VLSI testing: digital and mixed analogue/digital techniques

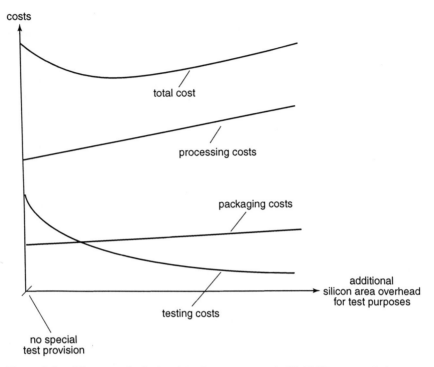

Figure 9.5 *The general relationships between per-unit VLSI IC costs and the amount of built-in DFT to ease the testing problem*

The latter is expressed very aptly by Turino[16], who states that '... what becomes clear is that trade-offs at chip level are too myopic. It is thus important to consider where the chip resides in the food chain of electronic product and use'. Further introductory comments, including the terminology 'over the wall mentality' to highlight the dangers of being too parochial, may be found in Dislis *et al.*[17].

A flow diagram which may form the basis of cost modelling such as will be discussed in the following section is shown in Figure 9.6. This diagram implies that to incorporate some form of DFT increases the design time, but reduces the test time and, equally important, possibly the testing hardware. This will generally be the case, particularly if there is a learning curve situation where the chip designer has to learn to use new CAD support or other DFT software. The very broad financial balance sheet with the availability of increasingly relevant CAD software support is therefore as follows:

(i) For:
- insertion of scan path, boundary scan, etc. into the chip design can be automatic;

- test generation data and execution is simplified and/or made more automatic.

(ii) Against:

- increased die size and therefore slightly reduced yield;
- increased package size;
- possible performance degradation.

The complete economic model for the IC should therefore ideally encompass the following factors:

- design time costs;
- material costs;
- fabrication and packaging costs;
- test;
- time to market/% market capture;
- perceived increased quality to customer;
- perceived innovative factors;
- capital/depreciation/maintenance costs of CAD and test resources;
- other overheads, profit margins, etc.

The first four of the above nine factors directly involve the IC chip design; fabrication and packaging will also involve the wafer production line yield. The remaining five factors are more intangible and company specific. The wafer production-line yield now has its own yield management software to model and cost this area[18], but this very specialist topic is outside our concern here.

Further general discussions on the cost and cost effects of test and DFT, including philosophical considerations, may be found in the published literature[1-3,16,17,19-28]. Further specific discussions may be found covering the following areas:

- the finances of the IC testers themselves may be found in References 10, 21 and 23, and emphasised in 28;
- the effect of boundary scan on the economics of printed-circuit boards may be found discussed and quantified in Reference 29;
- cost of ownership by the OEM of automatic equipment for the assembly of ICs on printed-circuit boards, including very detailed cost of ownership financial modelling, may be found in Reference 30;
- the implications of the use of multichip modules (MCMs) by the OEM in production equipment may be found discussed in Reference 31.

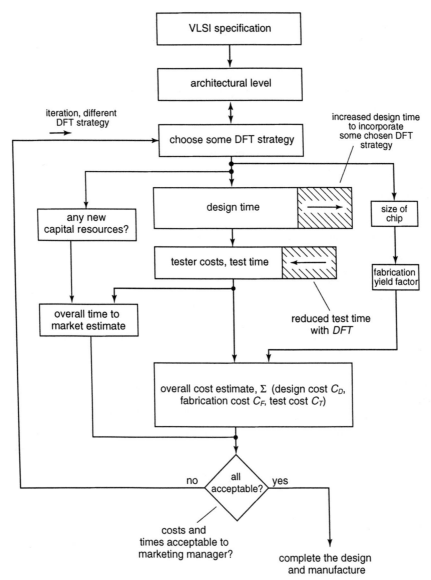

Figure 9.6 *Flow diagram showing the basis upon which quantitative cost estimates may be made. Qualitative considerations such as customer acceptance etc. are, however, additional aspects which may have to be considered*

It may be noted that in this final cited reference an essential element is stated to be the provision of standard 1149.1 boundary scan on all chips, and also known-good chips before assembly wherever possible, thus allowing the boundary scan to be an effective interconnection test and repair means for the final product.

9.4 Economic models

9.4.1 Custom ICs

The earliest economic models for integrated circuit costs were largely concerned with custom ICs, where it was desirable to show how the cost of, say, a gate-array design compared with other possible design methods. No detailed consideration of testing costs was done in these early financial equations, and certainly very little consideration of the possible minimisation of such costs in the final IC design. The choice of design style from a broad technical point of view has been published[1-3], but no detailed financial models or spreadsheets were built into these techno-economic discussions and tabulations.

The earliest studies involving the cost of test were concerned with purely combinational networks and the expense of automatic test pattern generation. The work of Goel[22] was one of the first studies of ATPG costs, and involved investigations into the relationship between the number of test vectors and the resultant fault coverage, and hence cost versus fault coverage. No built-in test or scan test was involved in these early considerations—they may therefore be regarded as the costs of not considering any form of DFT. Varma et al.[32] built upon this work, and showed the relative advantages of DFT methods where final system tests and field tests are concerned, but again combinational networks only were generally considered. The very extensive publications of Fey and Paraskevopoulos[4-7] were entirely based on gate count, with gates per IC being a common parameter in their financial developments, and therefore their work could be considererd to be relevant for comparing just the silicon cost of an IC with and without DFT if detailed gate counts are evaluated. The acual cost of test remained outside these figures.

A later work which produced a cost model and supporting software for costing different custom IC design strategies was that of Edward[33], but again no direct means of comparison of a design with and without DFT was built in; each possibility would have to be separately evaluated, effectively at gate level and without any detailed time to test and cost to test parameters.

One of the early publications which specifically incorporated a testing cost parameter was that of Dislis et al.[34]. Here three different sizes of gate array were considered with and without DFT in each of them. Scan test was assumed to require between 5 % and 15 % silicon area on a chip, and self test between 20 % and 30 %, costs being normalised to the nonDFT case. Figure 9.7 shows the published results for a 5000-gate gate array with production quantities between 1k and 1000k per circuit. An interesting result of these estimations indicated that when production quantities were large, then the increased cost of the extra silicon with DFT outweighed other factors such as easier test, the crossover point being in the tens of thousands of circuits in the case of 5000-gate and 10 000-gate ICs.

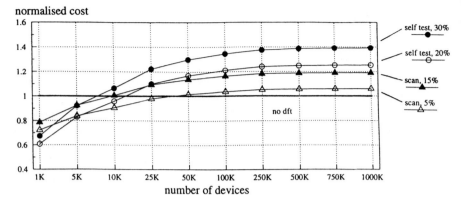

Figure 9.7 The results of Dislis et al. giving the total cost per IC of incorporating DFT into a 5000-gate gate array (Acknowledgements, based on Reference 34)

9.4.2 The general VLSI case

Turning, therefore, to the more general case rather than specifically custom ICs, and considering the requirements necessary to define an economics model that can lead to a CAD test strategy planning tool for use at the initial design phase of an IC.

Ideally this economics model should consider the lifetime finances of the product[16], which therefore implies OEM and in-service costs as well as the initial IC design and production costs. However, unless the IC vendor is also the OEM, which implies a large high-tech. organisation which will in any case probably do very detailed company cost studies, it is more realistic to consider an economics model which covers the IC design, production and test aspects only, so that (at least) an estimation of the optimum test strategy for the new circuit itself can be made.

The hierarchy of costs for a new production IC fall under three headings, namely:

(i) NRE silicon design costs, which include labour and CAD hardware and software costs;
(ii) NRE test program generation costs, which include labour and CAD and/or ATE hardware and software costs;
(iii) production costs, which include mask costs, labour, wafer fabrication, probing, scribing, bonding and packaging, final test and any equipment costs.

Notice that it is not possible to draw a clear distinction between all the NRE costs, since the considerations of test are a part of the chip design. However, we will analyse these three categories a little more below before merging them into one flow diagram covering NRE costs and a second flow diagram covering per-unit manufacturing costs.

(i) Silicon design costs

The silicon design costs are dependent upon (a) the circuit size, which may be measured in gate count, I/O count and routing area, and (b) the circuit originality and complexity, encompassing, for example, what library of existing macro designs may already be available. The cost model to cover all possible aspects may be expanded to a very large number of parameters, a design model by Ambler *et al.*, for example, listing 23 individual parameters which may be incorporated in this design equation[20]. Different authorities may however add or subtract to those specifically listed in published material, but all will embrace the three main costs of in-house design time and labour costs, CAD costs and any outside costs. Note that a recalculation of this and the following costs is necessary for each DFT strategy that may be considered.

(ii) Test program generation costs

The test program generation costs depend upon the circuit complexity, the fault coverage required and the form of the final test, that is whether scan test, BIST or even no DFT at all is built into the design.

For mainly combinational logic blocks with no DFT, then automatic test pattern generation (ATPG) using appropriate CAD resources may be appropriate; for even smaller circuits, manual test pattern generation with simulation to assess the fault coverage may be used. See the discussions in Section 3.2 of Chapter 3. However, the test pattern generation time to acheive near 100 % fault coverage may be very high[22], see Figure 3.13, with corresponding high labour and CAD costs unless suitable partitioning to ease controllability and observability is built into the design. For all complex VLSI circuits the design team should therefore consider the available DFT techniques discussed in the earlier chapters of this text, which include:

- the use of pseudorandom test vectors, with or without test response signature analysis;
- online self test with information or hardware redundancy;
- scan-path test;
- boundary scan to IEEE standards 1149.1 and 1149.4;
- offline built-in self test (BIST) for both analogue and digital cores;
- I_{DDQ} testing if CMOS technology is being used, with possible on-chip current sensors.

Again, it is the range of possibilities which complicates matters, and emphasises the need for some available test strategy planning software resource to help evaluate and compare all the financial options and implications. Ideally, the chip design team should be able to justify an acceptable test strategy for a new design, and determine not only the NRE costs involved in the design and test data but also the time to test which will form part of the following per-unit production costs.

(iii) Production costs

In considering production costs, the size of chip and the production line yield are crucial factors, together with the required production volume, packaging and test.

Yield has been considered in Chapter 1 of this text, and was shown to be related to the efficiency of test when the defect level after test is under consideration. However, there are two yield factors which should be considered, namely the yield factor of fault-free chips on the wafer itself before scribing, bonding and packaging, and the yield factor of the final packaged ICs after completion of all the manufacturing steps. Both of these factors should be considered when any calculations are being made to determine the number of wafers that need to be produced in order to provide a required number of final working ICs. Notice that DFT will entail more pins per IC and potentially impair the yield factors.

The expense of an initial wafer probe test must be borne, since to package faulty chips is not cost effective. Exactly what tests are done at wafer test stage is not clear cut; probe testing of the wafer drop-ins, see Figure 1.1, is always part of the production line test in order to ensure that all fabrication steps have been properly implemented, but to agree upon the minimum number of functional tests per chip before scribing is difficult. These tests may be the same as the tests made on the final packaged circuits, or a subset of the latter which will serve to detect an acceptably high percentage of possible circuit faults. Either way there is a wafer-stage testing cost and a final package-stage testing cost to be considered.

The mask costs for making the ICs are much more readily determined. The majority of masks used in production are made by outside mask-making companies, and hence the cost is a simple quotation exercise. The size rather than the detailed geometric complexities per mask largely determines the cost. If direct write-on-wafer rather than photolithography is used to produce the wafer patterns, then again such costs are reasonably straightforward to calculate, basically being machine time cost per hour × number of hours required to write all the individual levels of fabrication. Note that the same design data in magnetic tape or disk form is used for either mask making or for E-beam direct write-on-wafer, and therefore there should be little difference in the NRE design costs whichever wafer fabrication method is used. There may be some data translation involved to give the accepted data standard for mask making or E-beam machine control[1], but this data translation will be entirely automatic with no human intervention to cause potential errors or omissions.

Thus, in total, there are material costs, labour costs, testing costs and capital costs to be summed in determining the per-unit price of IC manufacture. This may be expanded into a very lengthy list of individual items, particularly as the hourly cost of the variety of process and test hardware may vary widely depending upon capital cost, utilisation and life expectancy. Some sixty or more separate parameters have been identified in published information relating to IC production costing, but again each design team or organisation will have its own choice of parameters to include or exclude.

Having introduced in this short analysis the three main cost categories, the NRE design and test costs may be combined in the flow diagram shown in Figure 9.8, and the production costs may similarly be summarised by the flow diagram of Figure 9.9. These two diagrams indicate the basis of any overall planning resource which may be developed to quantify the economics of VLSI design and test. The total NRE cost C_D of Figure 9.8 and the per-unit manufacturing cost C_T/N of Figure 9.9 give a total per-unit IC cost of:

$$PUC_{IC} = \left\{ \frac{C_D}{N} + \frac{C_T}{N} \right\} \qquad (9.2)$$

which is similar to the per-unit system cost given in eqn. 9.1.

The capital equipment costs involved in the above computations will be an appreciable part of the total cost when added to a specific design and production activity. The CAD hardware and software, the production equipment and all the required test resources have to be covered, although it may be company policy to sum the total running, maintenance and depreciation costs over a given period, together with all other in-house overheads, and allocate an appropriate percentage of this grand total to individual designs. Indeed, it may be possible to derive the approximate cost of a new circuit by considering all the company costs, labour, manufacturing and test, and arrive at an estimated per-unit cost as follows:

- total company salary costs per year, $\$C_S$;
- total capital equipment expenditure per year, including maintenance, repair and depreciation, $\$C_E$;
- total buildings, maintenance, marketing and all other company costs per year, $\$C_M$;
- estimated percentage of the total company manpower expended in the design and production of N units over x months, m %;
- estimated percentage of the total company manufacturing resources used in the production of N units over y months, e %;
- company final profit margin, k %;

Therefore, the estimated per-unit production cost is:

$$PUC_{IC} = \frac{1}{N} \left\{ \left[\frac{1}{12} C_S \times x \times \frac{m}{100} \right] + \left[\frac{1}{12} (C_E + C_M) \times y \times \frac{e}{100} \right] \right\} \qquad (9.3)$$

and per-unit selling cost to an OEM is this estimate $\times (1 + k)$. The accuracy of this rather coarse estimate depends heavily upon the estimated design and production times x and y, with DFT considerations impacting strongly upon both of these parameters. We will not pursue this form of overall costing any further here.

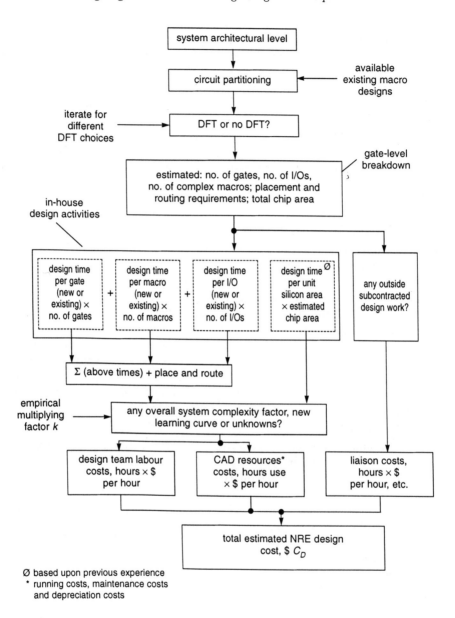

Figure 9.8 *Expansion of the IC NRE design costs C_D. Company overheads and profit margins may be applied to the total estimated design cost, or to the final overall design, production and test cost*

Other tangible and intangible factors may also be considered, such as time to market which may involve additional manpower and/or equipment availability, perceived market sales, guarantees and replacement parts, and

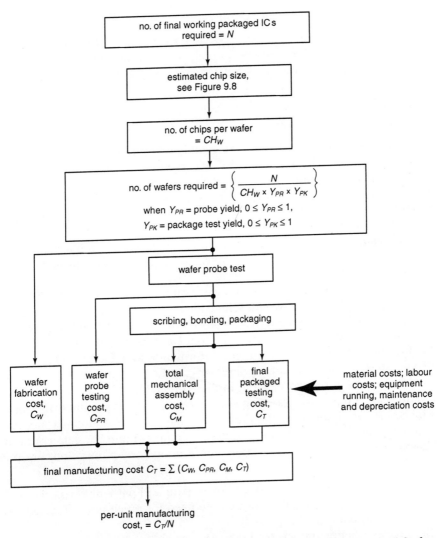

Figure 9.9   Expansion of the per-unit IC fabrication and testing cost. Each of the four boxes C_W, C_{PR}, C_M and C_T may be further expanded to cover the several activities and capital equipment costs involved

other sales-driven factors. The actual selling cost has, therefore, to be an appropriate multiple of the estimated per-unit manufacturing cost to cover company profitability and possibly expansion. Most of these factors, and other considerations such as sensitivity analyses of cost equations to variations in individual parameter values, have been discussed in the published literature[17,24–26,34–36].

9.4.3 Test strategy planning resources

Research into a number of knowledge-based systems incorporating the above broad cost considerations so as to provide a test strategy planning resource has been published. The proposed systems fall into two main categories, namely:

- single test strategy systems;
- overlay test strategy systems.

In the former, one specific test strategy only is covered, such as built-in self test or scan test. The cost of BIST, for example, may be analysed with different partitions and/or circuit details so as to find some optimum acceptable test coverage; with scan-path testing, partial scan or other variations may be considered, again with the objective of determining some optimum advantage. With the overlay test strategy systems, different test strategies applied to the same basic design are considered, with the objective of giving a comparison between them which will lead to an educated choice of the particular test strategy to adopt.

The general structure of a test strategy planner is shown in Figure 9.10. In practice the middle boxes of this schematic may take different forms, but all require (i) the normal design data to be entered in, say, HDL or netlist form, (ii) the knowledge-based information on DFT methodologies which will modify the circuit details and (iii) cost information which will enable the cost of the design with an appropriate DFT strategy to be estimated. The knowledge base is the key element in these procedures.

Published single test strategy planning systems include many early works which perhaps did not consider the actual financial costs as much as later test strategy planners, but instead considered circuit complexity, gate count or silicon area as an overall cost parameter. Among early publications are the following:

(i) Trischler, 1983, who considered scan path and particularly partial scan as a DFT strategy, and indicated the need for much more detailed test strategy planning[37];
(ii) Plessey Semiconductors, 1986, who considered in some depth the introduction of BIST into VLSI circuits [20,38];
(iii) Agrawal *et al.*, 1987, who also considered partial scan for a possible optimum incorporation of DFT[39];
(iv) Racal-Redac, 1989, with their industrial development of INTELLIGEN test pattern software to incorporate scan cells automatically and optimally into a circuit design[40];
(v) Gundlach and Müller-Glaser, 1990, who considered the automatic placement of test points in scan-test circuits, with an emphasis upon minimising the number of required test points[41],

and others[17,21]. These developments may be considered as forerunners of later, more embracing economic models.

The economics of test and final overall summary 497

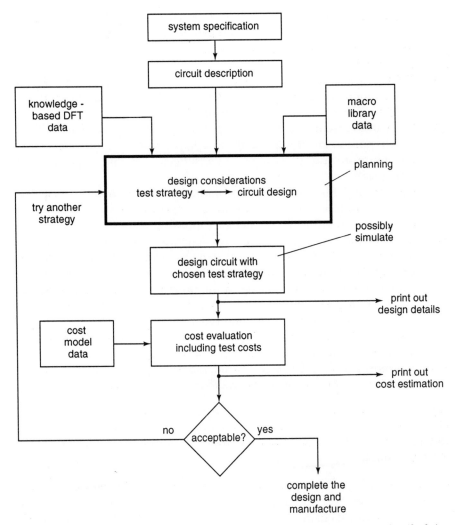

Figure 9.10 The general schematic of test strategy planners, exact details being variable

Several early considerations of optimum DFT strategies were in the specific area of programmable logic arrays. Zhu and Breuer [42] considered a large number of different alternative DFT methods for PLA realisations; similar considerations were also made by Dislis *et al.*[17] and others. However, more general publications not confined to PLAs include the following

(i) TDES (testable design expert system) developed at the University of Southern California, 1985, which used a RTL description of the circuit to partition it into testable macros, from which a test plan and an optimal test strategy were suggested[43];

(ii) Jones and Baker, 1986, who developed a knowledge-based system for determining an optimum test strategy to improve the testability of a design by using BIST, including chosen constraints in overheads and test time[44];

(iii) TIGER (testability insertion guidance expert) system, 1989, which was a development of TDES using a more sophisticated partitioning and weighting system to indicate optimality[45];

(iv) Fung and Hirschhorn, 1986, who published a system called TESTPERT which combined design and testability considerations in mainly telecommunications circuits, choosing a possible mix of DFT strategies[46];

(v) Gebotys and Elmasry, 1988, who published a system called CATREE which considered test cost, silicon area and circuit delay penalties as parameters to influence the design synthesis process[47];

(vi) ECOtest and ECOvbs software developments at Brunel University, UK, of the early 1990s, which were detailed economic models for test strategy planning including design time, test time and other possible parameters to give an estimate of the overall finances of a new design incorporating any chosen DFT strategy. PCBs and field maintenance were also considered in these developments[17,48,49];

(vii) Davis, 1994, who considered the economic modelling of PCB testing, including the need for lifetime cost considerations[50];

(viii) Sandborn et al., 1994, who considered the economics of multichip modules (MCMs) as an alternative to traditional PCB assemblies[51].

Details of these research and development activities may be found in the cited references, with further general discussions and considerations in References 17, 21 and 36. Note two essential points in all these financial models, namely

- there are no new or novel test strategies or concepts introduced in these developments over and above those which have been considered in the earlier technical chapters of this text; the principles and practice of BIST, boundary scan, self test and other methodologies remain the structures which are considered in all these economic models;
- the actual costs generated by cost equations, however complex, are only as accurate as the data entered into the equations.

A company with a great deal of experience in costing new products may be able to make reasonably accurate cost estimates, but there is always the possibility of new unknowns, such as the learning time for new CAD, CAM or ATE resources, which will always make forecasting an art rather than an exact science. Therefore, most authorities at present consider that a test economics model should be used to provide a comparison between different possible test strategies at the initial design stage, rather than as a means to provide exact design, manufacturing and test costs. The fact that an economics model has a very large number of cost parameters does not necessarily make it more accurate than one with fewer parameters, and possibly heightens the need for sensitivity analysis to be applied to the cost equations[52].

In all the published information so far available on test modelling and test strategy planning, the vast majority has been concerned with digital rather than analogue or mixed-signal circuits. A mention of the latter consideration may be found in the publications of Ambler *et al.*[17,24], but the difficulties of analogue testing, which as we have seen in previous chapters involves parametric as well as functional testing, makes this a very difficult financial area to model. In principle the test models developed for digital circuits are applicable to analogue and mixed analogue/digital ICs, no completely new cost principles being likely, but the precise tests and test hardware requirements will of course be specific to this task. The advent of the boundary scan standard IEEE standard 1149.4 will undoubtedly be reflected in future analogue cost modelling, in exactly the same way that the digital standard 1149.1 is already reflected in digital cost estimates.

In summary, therefore, this area of test economics and test strategy planning is still evolving, and is becoming of increasing significance to both IC and OEM designers. The pressures of time to market, market share and company reputation and profitability, together with the increasing complexity of VLSI circuits, will all impact upon the economics of both design and test. Further financial aspects may be found discussed in References 7, 24, 25 and 36, and the further references contained therein.

9.5 Final overall summary

9.5.1 The past and present

The last two decades have seen an unsurpassed escalation in the consideration of microelectronic testing, both from the point of view of the IC circuits themselves and now increasingly from the point of view of systems testing. This has come about due to the growth in fabrication capabilities, and the increasing power of computer-aided design resources which can utilise this growth for new circuit and system designs.

The increased complexity of VLSI circuits has made it necessary to consider carefully how the final packaged IC can be tested, how exhaustive a test procedure is economically appropriate, whether external ATE resources or built-in self test is preferable, and other interlocking aspects of the design and test procedures. From the system designer's point of view, with more capability being present within a single chip and with a single IC package now containing upwards of hundrds of I/Os within extremely small dimensions, the assembled board-level testing requirements have passed the possible limits of bed of nails probing of the I/Os. This and the increasing use of multichip-module assemblies are therefore driving the demand for boundary scan as the only present practical means of providing this controllability and observability of closely-packed I/Os.

As we have seen in the earlier chapters of this text, the testing of digital logic networks occupied almost all the early test research and development activities, with combinational logic networks rather than sequential networks being the primary focus of attention. In this work, a very high concentration on fault modelling and automatic test pattern generation was pursued, but even so the increase in gate count per circuit has tended to outstrip the useful capabilities of ATPG programs, necessitating unacceptably long computer run times to achieve an acceptable fault coverage from the generated test vectors. The stuck-at fault model, which was the basis of virtually all ATPG programs, has proved to be an excellent model for such programs, but the volume of computation remains the problem with the increased circuit size and complexity. The fundamentals of this work, however, remain entirely valid and should continue to be understood.

The testing of sequential networks has not had such a concentrated consideration. Basically, the problem of sequential testing with its inherent feedback and memory structures is that it is not amenable to solution by any fault modelling unless most (or possibly all) the feedback loops are broken and the circuit decomposed to gate level. It will be recalled that for a sequential network with n primary inputs and s internal storage circuits (latches or flip-flops), a fully-exhaustive functional test to exercise every possible combination of the logic would require 2^{n+s} input test vectors, an impossibly large number to consider as n and s increase. This numbers problem in both combinational and sequential logic has been the prime factor in making design for testability an essential design consideration.

Testing based upon fault modelling has therefore largely been superseded by DFT strategies, which involve some form of partitioning of a complete circuit plus offline scan test or built-in self test, or online (concurrent) self test, or other functional means of test. This has been the subject matter of the central chapters of this text, and may broadly be summarised as shown in Figure 9.11. In addition to these general DFT strategies we must add the considerable research and development which has been undertaken on design methods for RAM, ROM and other strongly-structured circuits, as covered in Chapter 6.

Work in all these areas continues, with boundary scan and scan test perhaps receiving the most attention for digital networks. A phase in the late 1980s when, due to improved fabrication and yield, the need for built-in self test with its silicon area penalties was questioned by several authorities, has now largely passed, although there are still voices which do not allow complacency in an unthinking adoption of DFT[53]. Whether current critisisms are meant as a spur to improve testing strategies or whether they are a real opposition to the excesses of DFT is difficult to judge.

The economics of test and final overall summary   501

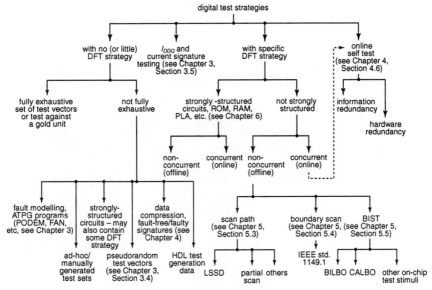

Figure 9.11   The broad summary of digital circuit testing methodologies. Note, online self test may be regarded as a particular form of DFT, but has been shown here separately since it is primarily used in specific applications

Following the coverage of the very considerable volume of work undertaken into digital logic testing and DFT, the problems of analogue and mixed analogue/digital testing were finally considered in the later chapters of this text. As was seen, the numbers problem of digital test is now largely replaced by the problem of the wide diversity of parametric tests which may be involved, and which is further compounded by the variety of circuits, applications and depth of testing required. Figure 7.1 illustrated the hierarchy of possible OEM analogue requirements, with Figure 7.5 illustrating the three-dimensional range of possible circuits.

Figure 9.12 attempts to summarise the possible analogue circuit test strategies. However, every new circuit and system will have its own unique requirements, ranging from relatively straightforward testing by commercial instrumentation through to situations with difficult controllability and observability of buried macros requiring built-in analogue self test or analogue scan test. The out of tolerance testing specification for almost all systems will be unique.

If full access to the I/Os of buried analogue macros in mixed-signal circuits is made available at the design stage, then the analogue tests will not be dissimilar to that of an analogue-only system. This has been the general situation to date, the chip or system designer providing the necessary controllability and observability of the analogue I/Os to allow this to take place. Indeed, this is still likely to remain in the future, unless particular

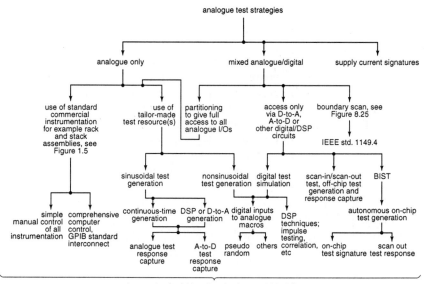

Figure 9.12 The broad summary of analogue and mixed analogue/digital test strategies. Both functional and parametric tests may be required

circumstances dictate otherwise. However, the increasing interest in and significance of digital signal processing and the use of digital signals to test analogue macros, as illustrated in Chapters 7 and 8, is the most significant development for mixed analogue/digital circuit testing. Impulse response, correlation techniques and the mathematics of digital signal processing which have largely been established in the data transmission (communications) field are now becoming necessary subjects for the IC designer to know, in order that test strategies might be correctly understood and evaluated.

In contrast to all these fault-model based and functional test strategies, there is also the completely dissimilar strategy of current monitoring. As shown in Chapter 3, current monitoring was developed as a means of test for CMOS circuits, the low quiescent current of CMOS being impaired if certain short-circuit and bridging faults are present in the chip fabrication. However, the current signature for such circuits has been shown to be more powerful than originally anticipated, see Figure 3.30, and developments to produce on-chip current monitoring circuits promise to be particularly valuable. Recent work continues with investigations into current monitoring for analogue circuits, thus making current signatures a potentially universal tool for digital, analogue and mixed analogue/digital testing[54,55]. Further comments on current and voltage monitoring will be made in the following concluding section.

Computer-aided design hardware and software resources have not been analysed in depth in this text, largely because the detailed software programs are usually vendor-specific and in any case are usually obsolete within a three-year period. The most noticable exception to this is SPICE and SPICE-based derivatives, which have established an international acceptance as the simulation program for microelectronic devices. Continued improvements in process simulation, in device simulation and in circuit simulation will undoubtedly continue[56]. A pre-eminent review of the significance and future of CMOS will be found in Reference 57.

The present status of both digital and analogue microelectronic testing therefore has a very secure foundation. All the fundamental concepts and basic economic considerations of test have now been researched, but practice will continue to evolve in order to try to keep pace with the still-increasing complexity of VLSI circuit and system design. The IEEE standards for boundary scan and also for the test of distributed systems[58-63] will continue to be a significant factor in the design and test area.

9.5.2 The future

The classic publication of Gordon Moore in 1979 which has become known as Moore's Law, and which forecast the doubling of on-chip circuit densities every eighteen months or so[1,64], has remained valid up to now, but the question of how much longer this rate of increase of complexity can be financially sustained has recently been raised.

The reason for this questioning is not so much the approaching limits of possible fabrication, but more the escalating cost of newer state of the art deep submicron fabrication equipment. This dilemma was highlighted by Maly[65]; this and other financial considerations such as testing has resulted in a proposal by the Semiconductor Industry Association (SIA) to document a number of milestones which might be achieved by the semiconductor industry over the next fifteen years[66-68]. This is the 'National technology roadmap for semiconductors', known as the SIA Roadmap for short, and may be read as the future industrial device fabrication and economic targets.

A summary of some key figures taken from the SIA Roadmap report is given in Table 9.3. This may be seen as a continuation of Moore's Law, but with a slightly reduced gradient in comparison with the past. Thomas and others have considered the implications of these projected figures on other VLSI parameters, such as shown in Table 9.4 which illustrates the incredible complexity of interconnect that may be present if the summary details of Table 9.3 are realised[67,69]; six or more levels of metal interconnect may be required in these next-generation VLSI circuits. (Note that not all the values cited in Table 9.4 may be simultaneously present.)

Table 9.3 *SIA Roadmap figures for VLSI fabrication limits, illustrating decreasing feature size and increasing maximum possible die size*

Year	min. feature size µm	max. die size mm²	Target data max. no. of I/Os	max. operating frequency, MHz	transistor density, transistors per mm²
1998	0.25	300	512	450	7×10^4
2001	0.18	360	512	600	1×10^5
2004	0.13	430	512	800	2×10^5
2007	0.10	520	800	1000	5×10^5
2010	0.07	630	1024	1100	9×10^5

Table 9.4 *Some suggested possible figures based upon the SIA Roadmap data (Acknowledgement, based on References 67 and 69)*

Year	Possible no. of transistors per I/O	Possible no. of transistors per die	Possible interconnect density, m/cm² per level	Possible max. interconnect length per circuit, m
1988	25×10^3	21×10^6	50	840
2001	35×10^3	47×10^6	70	2100
2004	45×10^3	107×10^6	105	4100
2007	85×10^3	290×10^6	125	6300
2010	140×10^3	558×10^6	155	10000

It is clear that if these projected increases in VLSI capability take place, which will mean the realisation of systems on a chip and greatly increased memory capacity in memory-only chips, then testing and DFT will become crucial to the success of the products. It is inconceivable that no such provisions will be made in the name of saving the silicon area overhead which may be necessary for some appropriate DFT strategy; scan test, BIST or newer developments must be an essential part of these VLSI and ULSI circuits, unless some completely new method of production testing not based upon directly monitoring test voltages or currents in the circuit under test can be developed—see Figure 9.13. In all cases, however, the ATE and other off-chip test resources must also be considered—can their performance and throughput match the requirements of these newer circuits, and overall can the market justify the cost of not only the new fabrication lines but also these test resources?

It may well be that (i) rate of learning, (ii) capital costs and (iii) test, will control the growth of VLSI capability rather than the possible manufacturing limits, although from the past history of the semiconductor industry there is no evidence to support this view, the market to date always absorbing all increases in IC fabrication complexity. Memory should continue to be the driving force in this evolution, since the demand from product and system designers for ever more memory will never be satisfied provided that the cost of bytes per cent remains acceptable.

Looking, therefore, at possible developments and emphases in the near future in testing, we may anticipate research and development activities in the following areas, some of which will prove valuable, others not so. All will, however, add to our knowledge and perhaps generate testing concepts not yet envisaged.

(i) Failure mechanisms and defect-oriented testing

Continuing work is undoubtedly required to analyse the types of faults and identify the failure mechanisms which are present in microelectronic circuits, particularly in deep submicron VLSI fabrication[70,71]. Three-dimensional fault models may become necessary[73]. Such studies and quantification may point the way to defining more directly realistic test methods for the detection of such faults, rather than considering functionally-based test strategies. Memory failure mechanisms will be particularly important in such developments, especially where high-density memory is on chip with processors, DSP and other digital or mixed analogue/digital circuitry [71-74].

(ii) Three-dimensional considerations

At present all silicon fabrication processes are planar, the circuit basically being two-dimensional with multilayer polysilicon and metal interconnect. This means that the main bulk of the circuit, the substrate, is doing nothing except physically carrying the working top layers. It may therefore be possible to consider more truly three-dimensional structures, which may allow controllability and observability considerations to be incorporated. This is clearly the province of the device physicist rather than the chip or circuit designer, but the design and test problems of the latter may provide a spur to the consideration of new forms of device technology and fabrication. Optoelectronics (optronics) may be significant in this and other contexts[75,76].

(iii) Multichip module developments

In parallel with the increasing amount of circuitry on a single chip must go the increasing interest in and development of multichip modules (MCMs), which may be a preferred way of partitioning a complex system for test purposes rather than a single system-on-a-chip IC[77]. This must therefore be one of the active continuing research and development areas, but whose exact future place cannot be accurately forecast.

(iv) E-beam and other diagnostic tools

Electrical measurements currently represent the accepted means of IC and system testing, virtually all developments in test strategies assuming this without question. However, there are other possibilities, including:

- electron beam (E-beam) techniques;
- thermal imaging techniques;
- ion beam techniques;
- scanning tunnelling microscopy (STM);
- atomic force microscopy (MFM);
- infrared microscopy;

and other physical means of detecting electrical activity. Introductory comments on these together with a very extensive reference listing may be found in Soden and Anderson[78].

Electron beam techniques using a scanning electron microscope (SEM) basically employ a beam of primary electrons, the probe, which is powerful enough to generate sufficient secondary electrons for detection purposes, but gentle enough not to affect the working of the IC under test. The probe may be focused to about 0.1 μm, and may be scanned across the whole surface of a VLSI circuit. By adjusting the E-beam energy, different detection capabilities ranging from physical defects in the IC fabrication to working waveform measurements at given points of the circuit can be achieved.

E-beam probe techniques are already well established for physical failure location in known-faulty circuits, being used by IC manufacturers for fabrication defect analysis. Figure 9.13 illustrates a commercial E-beam probe station[79], which may be integrated with the CAD design software to define the layout geometry of the IC under test. Waveforms of the working circuit can also be generated and displayed in real time, or captured in memory and compared with known fault-free response waveforms.

The potential use of E-beam and SEM technology for IC testing therefore is well established; the electron probe with its good focusing makes it applicable for submicron testing where mechanical probing is no longer possible, and the very low electron beam intensities already possible, down as low as a few tens of electrons per pulse, make its use on very low power VLSI circuits a practical proposition. Whether this capability can be advanced so as to be used as a viable production line testing resource as well as a prototype circuit and fault location test tool remains to be seen.

Thermal imaging, another of the possible measuring techniques listed above, is a further candidate for VLSI testing, and is receiving noticeable attention. Temperature distributions and thermal gradients in working ICs can be monitored; both two-dimensional and three-dimensional thermographic imaging has been reported[80,81]. However, with the latest versions of Xilinx® field-programmable gate arrays which use antifuse technology in their dedication, it is reported that the state of the antifuses cannot so far be detected by thermal means when the circuit is powered up[82]. Hence, it is not yet clear whether thermal means can be used for production testing of packaged ICs, or (as now) mainly for producing thermal contour maps of new designs in order to identify hot spots or hot areas which may give rise to

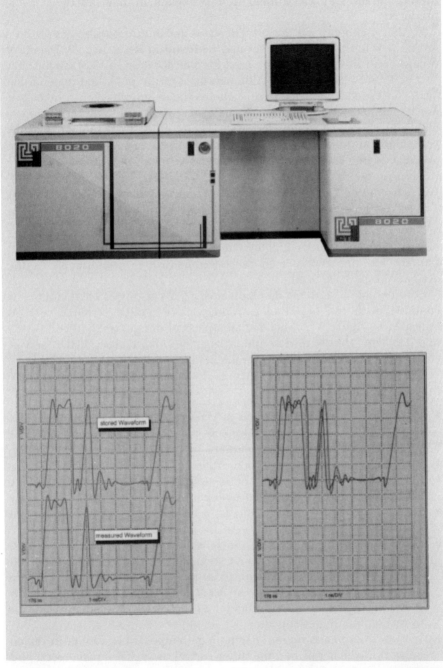

Figure 9.13 An E-beam scanning electron microscope resource, with the capability of measuring and comparing on-chip working voltage waveforms (Details courtesy Avantest/Integrated Circuit Engineering GmbH, Germany)

thermal stress and electromigration problems in new microelectronic products.

Further comments on the present status and future possibilities of these several new tools and techniques may be found in Reference 78. Details of voltage contrast which has been used for fault location in VLSI circuits may also be found published[83-87]. Although most of the published material has been concerned with digital circuits, there is no reason why their content may not be applied advantageously to analogue circuits; the waveform capture and comparison capability shown in Figure 9.13, for example, may be particularly appropriate for analogue and mixed analogue/digital testing, but once again cost may be a critical factor.

(v) CAD resources

It has been cynically suggested that all the troubles of IC testing are a result of shortcomings in the available IC design software. This is an unfair observation, but it does highlight the problem that the available design software has not always been as comprehensive or user-friendly as the designer would have liked.

Considerable continuing developments in CAD resources will undoubtedly continue, with the object of providing an integrated resource such as indicated in Figure 9.14, covering all aspects of design, test and cost within one database. Hardware description languages, particularly VHDL and its commercial derivatives, will be prominent in the top hierarchy of design data.

Specific areas of continuing CAD research and development may include the following:

- mixed-level fault simulation in a VHDL environment, leading to test vector generation which provides known fault coverage[88];
- automatic test data generation for analogue circuits[89];
- improved analogue and mixed analogue/digital design tools[90];
- automatic incorporation of IEEE standards 1149.1 and 1149.4 boundary scan requirements[91], and other built-in test strategies;
- power dissipation and automatic generation of I_{DDQ} data for test purposes[92];
- newer design methodologies, such as asynchronous logic design to improve maximum performance capabilities, intelligent sensors and systems on a chip, all of which will impact upon methods of test. Multi-valued logic will continue to be an academic research area, but is unlikely to become a practical proposition unless completely new and novel means of realisation are developed;
- increased emphasis upon delay testing (temporal behaviour) in digital networks, which will become more critical as digital circuit speed and density increase, interconnect delays becoming as if not more significant than gate delays;
- formal design methodologies which give correct by design circuit

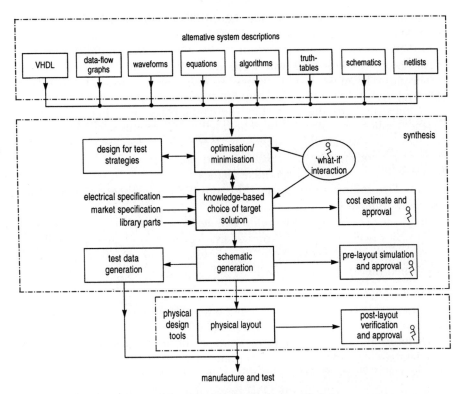

Figure 9.14 Future integrated CAD resources to link all aspects of design, test, cost and manufacture, using knowledge-based data to optimise design and test strategies, with minimum human intervention except for 'what-if' and decision making

realisations, but which have not to date specifically covered any test strategies,

and finally, and possibly the most important:

- developments in digital signal processing (DSP) strategies for analogue and mixed analogue/digital test[93–95], and also fuzzy logic expert system tools to handle test predictions and test data[94,96–98].

The biggest problem in all these disparate CAD developments may remain the difficulties of integrating them into a single, comprehensive system, with seamless interfaces between all the individual software packages and hardware tools.

Therefore, in conclusion we may say with confidence that testing will remain one of the key factors, if not the key factor, in future VLSI and ULSI products. Design, testing, manufacture and in-service needs will become more closely interrelated[99], with the experts and researchers of each area having to look increasingly outside their own specialisation if the benefits of their knowledge and the capabilities of fabrication lines are to be fully exploited.

All these developments will form the subject matter of many future publications. From the testing point of view, the major sources of information will remain:

- Journal of Electronic Testing: Theory and Application (JETTA)
- EEE Transactions on Computer Aided Design
- IEEE Design & Test of Computers
- IEEE Journal of Solid-State Circuits

plus the Proceedings of the many IEEE and IEE conferences, workshops and symposia on IC design and test.

Finally, it may be useful to reference again several of the available textbooks which deal with specific areas of test in greater fundamental detail than it has been possible to cover here. This should reinforce the basic and largely established theory of this still evolving and expanding area[14,17,56,70,100–106].

9.6 References

1. HURST, S.L., 'VLSI custom microelectronics: digital, analog and mixed signal' (Marcel-Dekker, 1998)
2. 'Microelectronics matters'. The Open University, Microelectronics for Industry PT504M, UK, 1987
3. 'Microelectronic decisions'. The Open University, Microelectronics for Industry, PT505MED, 1988
4. FEY, C.F., and PARASKEVOPOULOS, D.E.: 'Economic aspects of technology selection: level of integration, design productivity and development schedules', *in* [11], pp. 25.3–25.27
5. FEY, C.F., and PARASKEVOPOULOS, D.E.: 'A techno-economic assessment of application-specific circuits: current status and future trends', *Proc. IEEE*, 1987, **75**, pp. 829–841
6. FEY, C.F., and PARASKEVOPOULOS, D.E.: 'Studies in LSI technology economics: a comparison of product costs using MSI, gate arrays, standard cells and full custom VLSI', *IEEE J. Solid-State Circuits*, 1986, **SC-21**, pp. 297–303
7. PARASKEVOPOULOS, D.E., and FEY, C.F.: 'Studies in LSI technology applications: design schedules for application-specific integrated circuits', *IEEE J. Solid-State Circuits*, 1987, **SC-22**, pp. 223–229
8. NEEDHAM, W.D.: 'Designer's guide to testable ASIC devices' (Van Nostrand Reinhold, 1991)
9. BENNETTS, R.G.: 'Introduction to digital board testing' (Crane Russak, 1982)
10. BATESON, J.: 'In-circuit testing' (Van Nostrand Reinhold, 1985)
11. DI GIACOMO, J. (Ed.): 'VLSI handbook' (McGraw-Hill, 1989)
12. Special issue on VLSI Testing, *VLSI Des.*, 1993, **1** (1)
13. FUQUA, N.B.: 'Reliability engineering for electronic design' (Marcel Dekker, 1987)
14. O'CONNER, P.D.T.: 'Practical reliability engineering' (Wiley, 1991)
15. DAVIS, B.: 'Economic modeling of board test strategies'. Proceedings of 2nd international workshop on the *Economics of design and test*, Austin, TX, May 1993

16 TURINO, J.: 'Lifetime cost implications of chip-level DFT'. Digest of test synthesis seminar, IEEE international conference on *Test*, 1994, pp. 2.2.1–2.2.4
17 DISLIS, C., DICK, J.H., DEAR, I.D., and AMBLER, A.P.: 'Test economics and design for testability' (Ellis Horwood, 1995)
18 HACKEROTT, M.: 'Yield management', *Eur. Semicond.*, 1996, 18 (4), pp. 76–80
19 TRONTELJ, J., TRONTELJ, L., and SHENTON, G.: 'Analog digital ASIC design' (McGraw-Hill, 1989)
20 AMBLER, A.P., PARASKEVA, M., BURROWS, D.F., KNIGHT, W.L., and DEAR, I.D.: 'Economically viable automatic insertion of self test features for custom VLSI'. Proceedings of IEEE international conference on *Test*, 1986, pp. 232–243
21 DAVIS, B.: 'The economics of automatic testing' (McGraw-Hill, 1994)
22 GOEL, P.: 'Test generation costs, analysis and projections'. Proceedings of IEEE conference on *Design automation*, 1980, pp. 77–84
23 HOUSTON, R.E.: 'An analysis of ATE testing costs'. Proceedings of IEEE international conference on *Test*, 1983, pp. 396–411
24 AMBLER, A.P., ABADIR, M., and SASTRY, S. (Eds): 'Economics of design and test for electronic circuits and systems' (Ellis Horwood, 1992) (extended version of papers presented at the first international workshop on the *Economics of Test*, Austin, TX, 1992)
25 TURINO, J.: 'Design to test: a definitive guide for electronic design, manufacture and service' (Van Nostrand Reinhold, 1990)
26 DISLIS, C., DEAR, I.D., and AMBLER, A.P.: 'The economics of chip level testing and DFT'. Digest of test synthesis seminar, IEEE international conference on *Test*, 1994, pp. 2.1.1–2.1.8
27 DESENA, A.: 'Solving the ASIC test dilemma', *ASIC and EDA: Technologies for System Design*, 1992, 1 (1), pp. 54–57
28 RUSSELL, G., and SAYERS, I.L.: 'Advanced simulation and test methodologies for VLSI' (Van Nostrand Reinhold, 1989)
29 BLEEKER, H., VAN DEN EIJNDEN, P., and DE JONG, F.: 'Boundary-scan test: a practical approach' (Kluwer, 1993)
30 HEGARTY, C., and MEIER, F.: 'COO analysis for automatic semiconductor assembly', supplement to 'Packaging, assembly and test', *Solid State Techn. and Euro. Semicond.*, 1996, 4 (3), pp. S10–S14
31 STOREY, T.: 'Multichip module testing in a foundry environment, supplement to 'Packaging, assembly and test', *Solid State Techn. and Euro. Semicond.*, 1996, 4 (3), pp. S16–S20
32 VARMA, P., AMBLER, A.P., and BAKER, K.: 'An analysis of the economics of self-test'. Proceedings of IEEE international conference on *Test*, 1994, pp. 20–30
33 EDWARD, L.N.M.: 'USIC cost simulation: a new solution to an old problem', *J. Semicust. ICs*, 1990, 8 (2), pp. 3–12
34 DISLIS, C., DEAR, I.D., LAU, S.C., and AMBLER, A.P.: 'Cost analysis and test method environments'. Proceedings of IEEE international conference on *Test*, 1989, pp. 875–883
35 DAVIS, B.: 'The economics of design and test', *in* [24]
36 'Economics of electronic design, manufacture and test', special issue of *J. Electron. Test. Theory and Appl.*, 1994, 5 (2/3), pp. 127–312
37 TRISCHLER, E.: 'Testability analysis and incomplete scan path'. Proceedings of IEEE international conference on *Computer aided design*, 1983, pp. 38,39
38 PARASKEVA, M., BURROWS, D.F., and KNIGHT, D.F.: 'A new structure for VLSI self-test: the structured test register (STR)', *Electron. Lett.*, 1985, 21, pp. 856, 857
39 AGRAWAL, V.D., CHENG, K.T., JOHNSON, D.D., and LIN, T.: 'A complete solution to the partial scan problem'. Proceedings of IEEE international conference on *Test*, 1987, pp. 44–51

40 'INTELLIGEN manual'. Racal-Redac Microelectronics, UK, 1989
41 GUNDLACH, H.H.S., and MÜLLER-GLASER, K.D.: 'On automatic test point insertion in sequential circuits'. Proceedings of IEEE international conference on *Test*, 1990, pp. 1072–1079
42 ZHU, X., and BREUER, M.A.: 'Analysis of testable PLA designs', *IEEE Des. Test Comput.*, 1988, **5** (4), pp. 14–28
43 ABADIR, M.S., and BREUER, M.A.: 'A knowledge based system for designing VLSI chips', *IEEE Des. Test Comput.*, 1985, **2** (4), pp. 56–56
44 JONES, N.A., and BAKER, K.: 'An intelligent knowledge based system tool for high level BIST design'. Proceedings of IEEE international conference on *Test*, 1986, pp. 743–749
45 ABADIR, M.S.: 'TIGER: testability insertion guidance expert system'. Proceedings of IEEE international conference on *Computer aided design*, 1989, pp. 562–565
46 FUNG, H.S., and HIRSCHHORN, S.: 'An automatic DFT system for the SILK silicon compiler', *IEEE Des. and Test Comput.*, 1986, **3** (1), pp. 45–57
47 GEBOTYS, C.H., and ELMASRY, M.I.: 'Integration of algorithmic VLSI synthesis with testability incorporation'. Proceedings of IEEE conference on *Custom integrated circuits*, 1988, pp. 2.4.1–2.4.4
48 DISLIS, C., DICK, J., and AMBLER, A.P.: 'An economics based test strategy planner for VLSI design'. Proceedings of European conference on *Test*, 1991, pp. 437–446
49 DEAR, I.D., DISLIS, C.D., and AMBLER, A.P.: 'Test strategy planning using economic analysis', *in* [36], pp. 11–29
50 DAVIS, B.: 'Economic modelling of board test strategies', *in* [36], pp. 157–170
51 SANDBORN, P.A., GHOSH, R., DRAKE, K., ABADIR, M., BAL, L., and PARIKH, A.: 'Multichip systems trade-off analysis tool', *in* [36], pp. 207–218
52 DICK, J.H., TRISCHLER, E., DISLIS, C.D., and AMBLER, A.P.: 'Sensitivity analysis in economics based test strategy planner', *in* [36], pp. 239–252
53 THOMPSON, K.M.: 'Intel and the myths of test'. Keynote address, Proceedings of IEEE international conference on *Test*, 1995
54 ZWOLINSKI, M., CHALK, C., and WILKINS, B.R.: 'Analogue fault modelling and simulation for supply current monitoring'. Proceedings of IEEE European conference on *Design and Test*, 1996, pp. 547–552
55 COBLEY, R.A.: 'Approaches to on-chip testing of mixed-signal macros in ASICs'. Proceedings of IEEE international conference on *Test*, 1996, pp. 553–557
56 CAREY, G.F., RICHARDSON, W.B., REED, C.S., and MULVANEY, B.: 'Circuit device and process simulation mathematical and numerical aspects' (Wiley, 1996)
57 WILKES, M.V.: 'The CMOS end-point and related topics in computing', *Comput. Control Eng. J.*, 1996, **7** (2), pp. 101–107
58 IEEE standard 1149.1, 1990 'Standard test access port and boundary scan architecture'
59 Supplement 1149.1b to IEEE standard 1149.1, 1994
60 IEEE standard P.1149.4, 1992 'Mixed signal test bus framework proposals'
61 IEEE standard P.1149.5, 1994 'Standard module test and maintenance bus'
62 SU, C., and TING, Y.-T.: 'Decentralized BIST for 1149.1 and P.1149.5 based interconnects'. Proceedings of IEEE European conference on *Design and Test*, 1996, pp. 120–125
63 'Boundary scan tutorial and BSDL reference guide'. Report PN.E1017–90001, Hewlett-Packard, CA, 1990
64 MOORE, G.: 'VLSI: some fundamental changes', *IEEE Spectr.*, 1979, **16** (4), pp. 30–37
65 MALY, W.: 'Cost of silicon viewed from a VLSI design perspective'. Proceedings of IEEE conference on *Design automation*, 1994, pp. 135–142

The economics of test and final overall summary 513

66 'The national technology roadmap for semiconductors'. Semiconductor Industry Association, CA, 1994
67 MALY, W.: 'Future of testing: re-integration of design, testing and manufacture'. Keynote address, Proceedings of IEEE Eurpoean conference on *Design and Test*, 1996
68 SPLINTER, M.: 'Semiconductor manufacturing comes of age'. Keynote address, Proceedings of SEMI conference on *Advanced semiconductor manufacturing*, MA, November 1995
69 THOMAS, M.E.: 'Review of NTRS interconnect roadmap'. Proceedings of SEMI conference on *Advanced semiconductor manufacturing*, November 1995
70 AMERAESKERA, A.: 'Failure mechanisms in semiconductor devices' (Wiley, 1996)
71 SACHDEV, M.: 'Defect oriented testing for CMOS circuits'. PhD thesis, Brunel University, UK, 1996
72 KHARE, J., and MALY, W.: 'Inductive contamination analysis with SRAM applications'. Proceedings of IEEE international conference on *Test*, 1995, pp. 552–560
73 SACHDEV, M., and VERSTRAELEU, M.: 'Development of a fault model and test algorithms for embedded DRAMs'. Proceedings of IEEE international conference on *Test*, pp. 815–824
74 KUIJSTERMANS, F.C.M., SACHDEV, M., and THIJSSEN, L.: 'Defect oriented test methodology for complex mixed-signal circuits'. Proceedings of IEEE European conference on *Design and test*, 1995, pp. 18–23
75 MÄRZ, R.: 'Integrated optics: design and modeling' (Artech House, 1995)
76 TOCCI, C.S., and CAULFIELD, H.J. (Eds.): 'Optical interconnection: foundations and applications' (Artech House, 1994)
77 SHERWANI, N.: 'Introduction to multichip modules' (Wiley, 1996)
78 SODEN, J.M., and ANDERSON, R.E.: 'IC failure analysis: techniques and tools for quality and reliablity improvement', *Proc. IEEE*, 1993, **81**, pp. 703–715
79 'E-beam probing: E-beam station 8000'. Integrated Circuit Testing GmbH, Mùchen, Germany, 1992
80 GUI, X., DEW, S.K., and BRETT, M.J.: '3-dimensional thermal analysis of high-density triple-layer interconnect structures in very large scale integrated circuits', *J. Vac. Sci. Tech., Part A*, 1994, **12** (1), pp. 59–62
81 LEE, H.: 'Thermal analysis of integrated circuit chips using thermographic imaging techniques', *IEEE Trans. Instrum. Meas.*, 1994, **IM-43**, pp. 824–829
82 'Data sheets XC8100 FPGA Family'. Xilinx, Inc., CA, 1995, also on WEBLINK at http://www.xilinx.com
83 CONRAD, D., RUSSELL, J.D., LAURENT, J., SKAF, A., and COURTOIS, B.: 'Multiple adjacent image processing for automatic failure location using electron-beam testing', *Microelectron. Reliab.*, 1992, **32**, pp. 1615–1620
84 PICKARD, B., and MINGUEZ, S.D.: 'Test methods in voltage contrast mode using liquid crystals', *Microelectron. Reliab.*, 1992, pp. 1605–1613
85 GIRARD, P., RENAUD, J.F., SODINI, D., COLLIN, J.P., QUESTEL, J.P., and LEBRUM, S.: 'The stroboscope phase selective voltage contrast—a fast signal processing for electron-beam testing of ICs', *Microelectron. Engin.*, 1992, **16**, pp. 475–480
86 DINNIS, A.R.: 'Voltage contrast maps using the time-dispersive electron spectrometer', *Microelectron. Eng.*, 1995, **23**, pp. 523–526
87 CHIM, K.: 'A novel correlation scheme for quantitive voltage contrast measurements using low extraction fields in the scanning electron microscope', *Meas. Sc. Tech.*, 1995, **6**, pp.488–495
88 LEUNG, E., RHODES, T., and TSAI, T.-S.: 'VHDL integrated test development', *Electron. Prod. Des.*, 1995, **16** (10), pp. 53–56

89 MIR, S., and COURTOIS, B.: 'Automatic test generation for maximal diagnosis of linear analogue circuits'. Proceedings of IEEE European conference on *Design and Test*, 1996, pp. 254–258
90 HUISING, J.H., VAN DE PLASSCHE, R.J., and SANSEN, W.M.C.: 'Analog circuit design' (Kluwer, 1995)
91 MAUNDER, C.M., and TULLOSS, R.E.: 'The test access port and boundary scan architecture' (IEEE Computer Society Press, 1994)
92 RUNYON, S.: 'Tool automatically generates I_{ddq} vectors', *Electron. Engin. Times*, 1996, January 29, reprinted by Systems Science, Inc., CA, 1996
93 MAHONY, M.: 'DSP-based testing of analog and mixed-signal circuits' (IEEE Computer Society Press, 1987)
94 MOHAMED, F., MARZOUKI, M., and TAUATI, M.H.: 'FLAMES: a fuzzy logic ATMS and model-based expert system for analog diagnosis'. Proceedings of IEEE European conference on *Design and Test*, 1996, pp. 259–263
95 DA SILVA, J.M., and SILVA MATOS, J.: 'Evaluation of cross-correlation for mixed current/voltage testing of analogue and mixed-signal circuits'. Proceedings of IEEE European conference on *Design and Test*, 1996, pp.264–268
96 FARES, M., and KAMINSKA, B.: 'Fuzzy optimization models for analog test decisions', *in* [36], pp. 299–305
97 SCHNEIDER, M., KANDEL, A., LANGHOLZ, G., and CHEW, G.: 'Fuzzy expert system tools' (Wiley, 1996)
98 PATYRA, M.J. (Ed.): 'Fuzzy logic: implementations and applications' (Wiley, 1996)
99 MALY, W.: 'Cost of silicon from VLSI perspective'. Proceedings of IEEE conference on *Design automation*, 1994, pp. 135–142
100 VAN DE GOOR, A.J.: 'Testing semiconductor memories: theory and practice' (Wiley, 1991)
101 ABRAMOVICI, M., BREUER, M.A., and FRIEDMAN, A.D.: 'Digital systems testing and testable design' (Computer Science Press, 1990)
102 YARMOLIK, V.N.: 'Fault diagnosis of digital circuits' (Wiley, 1990)
103 SALEH, R., JOU, R.-S., and NEWTON, A.R.: 'Mixed-mode simulation and analog multilevel simulation' (Kluwer, 1994)
104 SCHWARZ, A.F., 'Handbook of VLSI chip design and expert systems' (Academic Press, 1993)
105 PERRY, W.: 'Effective methods for software testing' (IEEE Computer Society Press, 1995)
106 BARDELL, P.H., McANNEY, W.H., and SAVIR, J.: 'Built-in test for VLSI: pseudorandom techniques' (Wiley, 1987)

Appendix A
Primitive polynomials for $n \leq 100$

The following gives the primitive polynomials with the least number of terms which may be used to generate an autonomous maximum length pseudo-random sequence (an M-sequence) from an n-stage linear feedback shift register (LFSR). Alternatives are possible in many cases, particularly as n increases.

Recall from Chapter 3 that the primitive polynomial has the form:

$$1 + a_1 x^1 + a_2 x^2 \ldots a_n x^n$$

where $a_i \in \{0, 1\}$. The polynomial:

$$1 + x^3 + x^4 + x^7 + x^{12}$$

represents a generating polynomial for $n = 12$. For convenience in the following table we will merely record the powers of x in reverse order from the above, for example:

    12   7   4   3   0

This indicates a 12-stage LFSR with taps at $n = 12$, 7, 4 and 3 exclusive-ORed back to the serial input of the LFSR. The length of the shift register n will therefore always be the first (left-hand) number in this table.

Additional comments are given following the table.

Table A1 Minimum primitive polynomials for $n \leq 100$

1	0				51	16	15	1	0
2	1	0			52	3	0		
3	1	0			53	16	15	1	0
4	1	0			54	37	36	1	0
5	2	0			55	24	0		
6	1	0			56	22	21	1	0
7	1	0			57	7	0		

Table A1 continued

8	6	5	1	0	58	19	0		
9	4	0			59	22	21	1	0
10	3	0			60	1	0		
11	2	0			61	16	15	1	0
12	7	4	3	0	62	57	56	1	0
13	4	3	1	0	63	1	0		
14	12	11	1	0	64	4	3	1	0
15	1	0			65	18	0		
16	5	3	2	0	66	10	9	1	0
17	3	0			67	10	9	1	0
18	7	0			68	9	0		
19	6	5	1	0	69	29	27	2	0
20	3	0			70	16	15	1	0
21	2	0			71	6	0		
22	1	0			72	53	47	6	0
23	5	0			73	25	0		
24	4	3	1	0	74	16	15	1	0
25	3	0			75	11	10	1	0
26	8	7	1	0	76	36	35	1	0
27	8	7	1	0	77	31	30	1	0
28	3	0			78	20	19	1	0
29	2	0			79	9	0		
30	16	15	1	0	80	38	37	1	0
31	3	0			81	4	0		
32	28	27	1	0	82	38	35	3	0
33	13	0			83	46	45	1	0
34	15	14	1	0	84	13	0		
35	2	0			85	28	27	1	0
36	11	0			86	13	12	1	0
37	12	10	2	0	87	13	0		
38	6	5	1	0	88	72	71	1	0
39	4	0			89	38	0		
40	21	19	2	0	90	19	18	1	0
41	3	0			91	84	83	1	0
42	23	22	1	0	92	13	12	1	0
43	6	5	1	0	93	2	0		
44	27	26	1	0	94	21	0		
45	4	3	1	0	95	11	0		
46	21	20	1	0	96	49	47	2	0
47	5	0			97	6	0		
48	28	27	1	0	98	11	0		
49	9	0			99	47	45	2	0
50	27	26	1	0	100	37	0		

The first list of 168 primitive polynomials was published by Stahnke in 1973[1]. The list of $n \leq 300$ may be found in Bardell, et al.[2], and for $n = 301$ to 500 in Bardell[3]. A listing of the primitive trinomials for $n \leq 100$, that is those with only three terms in the generating polynomial, thus requiring only one two-input exclusive-OR gate, may be found in Bardell[4]. The same data as given here may also be found in other publications. Further information may be found in References 2, 5–7. Some comments on the above tabulation:

1. It will be seen that for $n \leq 100$ never more than four taps on the LFSR are required to give an M-sequence. This also hold for $n \leq 500$, as shown in Reference 3. It is not known whether this also applies for all $n > 500$. Three taps are never necessary.
2. There are no trinomial polynomials for $n = $ any power of 2 greater than 2^3. All require four feedback taps, and hence need three two-input exclusive-OR gates in their hardware realisation.
3. The primitive polynomials listed in the above table are not the only ones possible. See Chapter 3, page 87. A complete table for all the irreducible primitive polynomials up to $n = 34$ is given in Peterson[5].
4. Notice that if an n-stage maximum-length LFSR requires to be increased to a longer M-sequence, it almost always necessary to change the taps on the earlier stages. For example, to change from $n = 12$ to $n = 13$ requires the taps on stages 7, 4 and 3 to be changed to 4, 3 and 1, with the nth stage tap of course moved to the $n+1$th stage. There are a few exceptions, for example $n = 2$, 3 and 4, 6 and 7, 26 and 27, and others, but in general it is not possible to modify the length of a LFSR M-sequence without some feedback modification to the stages before the final stage. This is in contrast to a CA maximum-length pseudorandom sequence generator, see Section 3.4.3.2, where connections between earlier stages of the CA generator can remain unaltered if n is increased.

References

1 STAHNKE, W.: 'Primitive binary polynomials', *Math. Comput.*, 1973, **27**, pp. 977–980
2 BARDELL, P.H., McANNEY, W.H. and SAVIR, J.: 'Built-in test for VLSI: pseudorandom techniques' (Wiley, 1987)
3 BARDELL, P.H.: 'Primitive polynomials of degree 301 through 500', *J. Electron. Test., Theory Appl.*, 1992, **3**, pp. 175, 176
4 BARDELL, P.H.: 'Design considerations for parallel pseudorandom pattern generators', *J. Electron. Test., Theory Appl.*, 1990, **1**, pp.73-87
5 PETERSON, W.W.: 'Error correcting codes' (Wiley, 1961)
6 LAWIS, T.G., and PAYNE, W.H.: 'Generalised feedback shift register pseudo random number generator', *J. ACM*, 1973, **20**, pp. 456-468
7 WILLETT, M.: 'Characteristics of M-sequences', *Math. Comput.*, 1976, **30**, pp. 306-311

Appendix B
Minimum cost maximum length cellular automata for $n \leq 100$

The following gives the positions in n-stage autonomous cellular automata of the type 150 cells necessary to produce a maximum-length sequence of $2^n - 1$ states. All remaining cells in the n-stage string of cells are the simpler type 90 cells. See Figure 3.21 of Chapter 3.

Only the type 150 cells are listed below. The entry 12 3 7, for example, represents a 12-stage CA, with a type 150 cell in stages 3 and 7. All remaining stages 1, 2, 4, 5, 6, 8, 9, 10, 11 and 12 are type 90 cells.

The method of generating this data may be found in References 1, 2. The listings below are not the only possible minimal cost (fewest number of type 150 cells) realisations, and alternatives are possible. One disadvantage with a minimal cost realisation is that if the number of stages, n, requires to be increased, or decreased, then the positions of the type 90 and 150 cells in the string of cells will almost always require changing. This may not be necessary with a nonminimal realisation, where it may be possible to add or take away stages in the string without altering the other 90 and 150 cells. Conversely, the most costly realisation would be to standardise on type 150 cells throughout, using them as type 90 circuits where required by disconnecting the Q_k exclusive-OR signal.

Table A2 Minimal cost cellular automata for $n \leq 100$

No. of stages and position of type 150 cells			No. of stages and position of type 150 cells			No. of stages and position of type 150 cells		
1	1		51	1		101	1	20
2	1		52	2	29	102	33	
3	1		53	1		103	15	
4	1	3	54	9		104	2	40
5	1		55	17		105	1	
6	1		56	4	14	106	30	
7	3		57	9		107	19	

Table A2 continued

8	2	3	58	17		108	1	35
9	1		59	4	15	109	1	4
10	2	7	60	2	38	110	13	
11	1		61	1	10	111	27	
12	3	7	62	5		112	2	5
13	5		63	31		113	1	
14	1		64	3	5	114	22	
15	3		65	1		115	41	
16	1	15	66	1	19	116	16	
17	5		67	15		117	33	
18	1	17	68	8		118	30	
19	3		69	1		119	1	
20	2	3	70	1	37	120	3	73
21	1	10	71	17		121	45	
22	5		72	6	55	122	14	
23	1		73	9		123	51	
24	8	12	74	1		124	21	
25	9		75	7		125	13	
26	1		76	2	22	126	40	
27	1	20	77	3	44	127	15	
28	3		78	1	41	128	1	29
29	1		79	9		129	49	
30	1		80	1	71	130	1	27
31	11		81	1		131	1	
32	1	15	82	1	69	132	18	
33	1		83	1		133	1	4
34	1	19	84	36		134	26	
35	1		85	1	46	135	1	
36	6		86	1		136	1	97
37	9		87	13		137	1	132
38	7		88	5		138	28	
39	1		89	1		139	11	
40	8		90	1		140	8	
41	1		91	15		141	25	
42	19		92	3	71	142	5	
43	3		93	33		143	35	
44	4	26	94	42		144	13	
45	9		95	1		145	1	46
46	2	10	96	6		146	1	
47	13		97	1	82	147	1	136
48	15		98	8		148	8	
49	1	10	99	13		149	1	108
50	11		100	1	67	150	2	102

References

1 ZHANG, S., MILLER, D.M., and MUZIO, J.C.: 'Determination of minimal cost one-dimensional linear hybrid cellular automata', *Electron. Lett.*, 1991, **27**, pp.1625-1627
2 SERRA, M., SLATER, T., MUZIO, J.C., and MILLER, D.M.: 'The analysis of one-dimensional linear cellular automata and their aliasing properties', *IEEE Trans.* 1990, **CAD-9**, pp. 767-778

Appendix C
Fabrication and yield

In Chapter 1 it was shown that the defect level, *DL*, after test was given by the theoretical relationship

$$DL = \{1 - Y^{(1-FC)}\} \times 100\,\%$$

Both the 'goodness' of the fabrication process, the yield, *Y*, and the effectiveness of the testing procedure, *FC*, are involved in this result.

Modelling of the number of good die on a wafer, which is principally dependent upon die size and process goodness and not upon circuit complexity, has been extensively studied, since with accurate modelling the yield, and hence the cost, of new circuits may be forecast. Also, once the detailed modelling parameters have been determined, the quality of the production lines can be maintained and possibly improved. However, the available modelling theory and available parameter values usually lag the latest production process, and as a result the yield of most production lines has historically tended to be higher than that predicted by modelling theory. (This is also true of reliability predictions for most products, where actual reliability, except for catastrophic occurrences, tends to be somewhat higher than predicted.)

Wafer yield *Y*, where *Y* has some value between 0 and 1, is normally considered to be a function of (i) the density of defects per unit area of the wafer, D_0, (ii) the area of the die, *A*, and (iii) an empirical parameter, α, which is specific to a particular fabrication process or model being used. Many equations have been proposed for *Y*, all of which have a statistical basis. Many variants try to take into account factors such as the defect density, D_0, not being uniform across the whole wafer, being higher towards the edges, but all predict that the yield, *Y*, decreases as die sizes increase.

One of the more simpler modelling equations which has been used is:

$$Y = \left\{ \frac{1 - e^{D_0 A}}{D_0 A} \right\} \times 100\,\%$$

This equation does not consider any variations in the value of D_0 with die size, or variations across the wafer surface. The difficulty of finding a realistic value for D_0, which will vary with process efficiency and wafer size, is the problem. Vendors rarely release their own in-house figures of yield, so most published data is inevitably obsolete and pessimistic. Some early 1980 published values for D_0 are:

$$D_0 = 10^5 \text{ defects per m}^2 \text{ for bipolar technology}$$

and

$$D_0 = 5 \times 10^4 \text{ defects per m}^2 \text{ for MOS technology}$$

These figures give the yield characteristics shown in Figure A.1. Other equations may be found published[1].

Although the general characteristics for wafer yield shown in Figure A.1 remain valid, the values for the yield of present day process lines are possibly one or two orders of magnitude better than these early figures. The

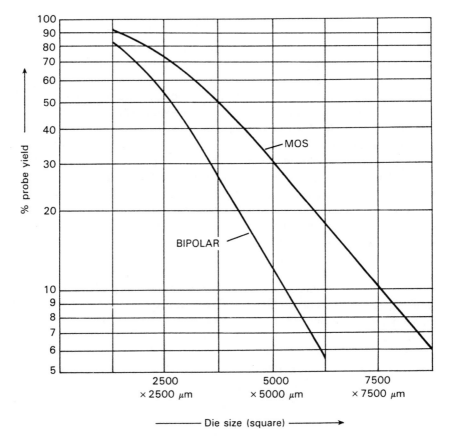

Figure A.1   Statistical good die yield from a 6 inch (15 cm) wafer as a function of die size

underlying principles of resulting yield, however, remain. Further information may be found published[1-6].

References

1 MOORE, W.R., MALY, W., and STROJAW, A.: 'Yield modelling and defect tolerance in VLSI' (Adam Hilger, 1988)
2 CHRISTOU, A.: 'Integrating reliability into microelectronic manufacturing' (Addison-Wesley, 1988)
3 JAEGER, R.C.: 'Introduction to microelectronic fabrication' (Addison-Wesley, 1988)
4 SZE, S.M. (Ed.): 'VLSI technology' (McGraw-Hill, 1983)
5 BAKOGLU, H.B.: 'Circuits, interconnects and packaging for VLSI' (Addison-Wesley, 1990)
6 HURST, S.L.: 'VLSI custom microelectronics: digital, analog and mixed-signal' (Marcel-Dekker, 1998)

Index

AC faults *see* delay faults
acceptance tests 1
accumulator syndrome testing 128, 191
ad-hoc DFT strategies 201, 371
algorithmic state machine (ASM) 238
algorithmic (or automatic) test pattern generation (ATPG) 48
aliasing 127, 143, 248, 257, 268
analogue arrays 381
analogue multiplexers (AMUXs) 382, 384, 439, 454
analogue scan path 442
analogue shift register 443, 445
analogue testing 9, 381, 427, 436, 450, 502
analogue-to-digital conversion 403
analogue-to-digital converters (ADCs) 403, 419
appearance fault 300
application-specific ICs (ASICs) 2, *327* (*see also* user-specific ICs (USICs))
application-specific standard parts (ASSPs) 429
arithmetic logic unit (ALU) 353, 358
arithmetic operations 165
arithmetic spectral coefficients 130, 139, 192
artificial intelligence (AI) 110
ASICs (*see* also USICs) 2, 327
assembly related faults 39
asynchronous operation 206
asynchronous reset 72
autocorrelation 79, 396
automatic test pattern generation (ATPG) 43, 47, 97, 111
autonomous test vector generation 74, 90, 94, 142, 245, 260, 265
AXE 358

back-to-back testing 404
backward trace 58, 61
balanced structure scan test (BALLAST) 226
'bed-of-nails' 13, 382, 499
Berger codes 161, 184
Bessel filters 400
BIC circuits 402
bidirectional I/Os 211
binary counters 74, 346
binary decision diagrams (BDDs) 194
bonding failures 39
bonding pads 8
Boolean difference 51, 62
boundary scan 230, 350, 457
boundary scan description language BSDL 288
bridging faults 25, 30, 33, 332, 450
built-in analogue test generators 446
built-in current (BIC) testing 104, 402
built-in digital circuit observer (BIDCO) 260
built-in logic block observation (BILBO) 241, 280, 447
Butterworth filters 400

CAD future 508
calculus of D-cubes 62
CAMELOT 23, 25
canonic expansion 131, 132, 133
capital equipment 6, 10, 242, 391, 395, 481
CATA 286

528 Index

catastropic faults 385
Cauer filters 400
CEBS BIST methodology 280
cellular arrays 178, 297, 362
cellular automata (CAs) 90, 265, 272
cellular automata logic block observation (CALBO) 241, 264
central processing unit (CPU) 352
characteristic polynomial 85, 146, 253
Chebyshev filters 400
chip-related faults 39
CMOS memory faults 31, 95, 206
CMOS technology 31, 95
coefficient test signatures 138
coherent sampling 409
combinational bridging faults 30
companion matrix 94
component fault location 43
component redundancy 169, 187, 193
compression *see* input compression, output compression
computer-aided design (CAD) 288, 480, 498, 508
computer-aided manufacture (CAM) 498
COMTRAC 358
concurrent built-in logic block observation (CBILBO) 260
concurrent fault simulation 45
concurrent testing *see* on-line self-test
confidence limits 37
consistancy operation 58
controllability 20, 232, 305, 382, 436
controllability factor 23
controllability transfer factor (CTF) 23
COPRA 358
co-prime rule 284
correction time 483
correlation 79, 138, 271, 396
cost *see* economics
crosspoint fault model 299
CSTP test methodology 280
C-testability 364
cumulative input polynomial 255
current monitoring 95, 350, 457, 491
current probes 106
C.vmp 358

D-algorithm *see* Roth's D-algorithm
DC fault dictionary 400
D-drive 56, 59
de Bruijn counter 260, 262
deductive fault simulation 45
deep sub-micron geometry 332
defect density 524

defect level 4, 525
delay faults 106
delay fault testing 105, 508
delta-sigma modulation 454
dependant faults 385
design checks 1
design-for-test, design-for-testability 11, 201
design time 459, 483
deviation faults 385
DFT guidelines 204
diagnostic reasoning 109
diagonal patterns 334
die size 523
digital modelling 386, 402
digital signal processing (DSP) 349, 392, 440
digital testing 501
digital-to-analogue conversion 415
digital-to-analogue converters (DACs) 414
D-intersection 59
direct digital frequency synthesis (DDFS) 453
disappearance fault 300
distinguishing input sequence 71
divisor polynomial 146, 253
don't care conditions 205
double error correction (DEC) 340, 358
dual (of a function) 99
dual-path sensitisation 50

E-beam lithography 492
E-beam diagnostics 505
economics of test 477, 484, 488
economics modelling 489
edge-triggered (cf. level-triggered) 212
emitter-coupled logic (ECL) 330
environmental faults 36, 385
environmental testing 188, 358
equally-likely fault assumption 127
erasable programmable logic devices (EPLDs) 323
error address 159
error correction 154, 340
error detecting codes 156, 159
error detection 154
error polynomial 150, 256
error probabilty 37
Eulerian cycle 97
even parity 154, 159
exhaustive test pattern generation 3, 73
ex situ self-test 276, 279
fabrication checks 1, 525

failure rate λ 171, 188, 342
fail safe 357
fail soft 355, 359
FAN 66, 298
fast Fourier transform (FFT) 393, 411, 417
fault coverage FC (*see also* test coverage) 4, 69, 205, 288, 523
fault detection (cf. fault location) 43
fault detection efficiency *see* fault coverage
fault diagnosis 108
fault dictionary 400
fault location (cf. fault detection) 43
fault masking 126, 143, 248, 257, 268
fault modelling, fault models 19, 25, 385
fault simulation 44, 45
fault tolerant, fault tolerance 355, 357
ferroelectric random access memory (FRAM) 330, 336
FET transmission gates 35
field-programmable gate arrays (FPGAs) 320, 324
filters 398
finances *see* economics of test
financial models, financial modelling 489, 496
firm configurable 362
flip-flops 211, 265
folding (of PLA structures) 320
forward driving 50
forward propagating *see* forward driving
forward trace *see D*-drive
four-wire bus *see* JTAG four-wire bus
FTMP 358
FTBBC 358
FTSC computer 358
full custom 381
fully exhaustive testing 4, 8
functional K-maps 54, 314, 336, 402
functional testing 14, 19

gate level 38
gain error 406
galloping patterns 334
Galois field of GF(2) 81
GALPAT 334
generating polynomial 87
general-purpose instrumentation bus (GPIB) 9, 390
Gibb's dyadic differentiator 55
glued simulator 434
Grey code 74
growth fault 300

guided probe testing 109

Hadamard transform 127, 134
Hamming codes 157, 358
Hamming distance 156
hard configurable 362
hard faults (cf. soft faults) 332, 400, 450
hardware description language (HDL) 286
hardware redundancy (*see also* component redundancy) 119, 169, 193
HILDO test methodology 276
histogram testing 410
homing input sequence 71, 205
HP instrumentation bus (HPIB) 9, 390
hybrid built-in self-test (HBIST) 447
hybrid construction 382, 505
hybrid modular redundancy 174
hypercube 97, 156

in-circuit testing 13
I_{DDQ} testing *see* current monitoring
IDT 286
IEEE Std. 1076 289
IEEE Std. 1149.1 230, 289, 485, 499
IEEE Std. 1149.4 457, 485, 499
IEEE Std. 1149.5 289
implicit testing *see* on-line self-test
impulse response 396
information redundancy 151, 193
initialisation 71
input compaction *see* input compression
input compression 120
input test vector *see* test vector
input zones 122
instrumentation (*see also* rack-and-stack) 9, 386
in situ self-test 276
integral nonlinearity 406
intermittent faults (*see also* soft faults) 35, 385
isomorphism 93, 268
iterative arrays 362
I-testability 365

Joint Test Action Group (JTAG) 231
JTAG four-wire bus 232

Karnaugh map 301, 303, 402
k out of $2k$ codes 177, 181

L1/L2 circuit 213
L1/L2* circuit 214

530   *Index*

latches (*see also* flip-flops) 211, 265
Latin square 341
lead-zirconate-titanate (PZT) 336
learning time, learning curve 480, 482
level-sensitive (cf. edge-triggered) 204, 212
level-sensitive scan design (LSSD) 211
lifetime costs 496
life cycle 484
linear feedback shift register (LFSR) 76, 142, 281, 286, 447
LOCUST test methodology 277
logic automated stimulus and response (LASAR) 63
logical decomposition 402

macro test 288
majority gates 178, 188
manual test pattern generation 48
marching patterns 334
Marcov model 37
masking redundancy 169
maximal-length Berger code 185
maximum length pseudo-random sequence *see* M-sequence
Mealy model 71, 206
mean time between failure (MTBF) 172
mean time to repair (MTTR) 172
megacells 352
memory 297
'memory' faults *see* CMOS memory faults
metallisation failures 39
microprocessors 297, 347
MIL specification 188, 358, 471
minimal cost cellular automata 519
minterms 329
mixed analogue/digital test 11, 427, 435, 441, 502
mixed-mode circuits 428, 430
modular redundancy 170, 174, 362
modulo-2 arithmetic 81
Moore model 71, 206
m out of n codes 161, 163, 176
M-sequence 76, 90
multiple autonomous test vector generators 281
multi-chip modules (MCMs) 487, 500, 505
multiple input signature register (MISR) 153, 192, 245, 250, 285, 448
multiplexers (*see* also analogue multiplexers) 202, 209, 437

near-neighbour patterns 334
negative reconvergence 50

non-coherent sampling 409
non-exhaustive test pattern generation 44, 74
non-monotonicity error 416
non-recurrent engineering (NRE) 479, 490
non-separable codes *see m* out of *n* codes
normal mode or normal operating mode 119

observability 20, 232, 305, 382, 436
observability factor 24
observability transfer factor (OTF) 24
odd parity 154
off-chip current measurements 103
off-line built-in self-test 153, 241, 306, 333, 441
offset error 406
on-chip current measurement 103
on-chip test generators 75, 90, 450
one-hot state assignment 189
on-line self-checking *see* on-line self-test
on-line self-test 119, 153, 311, 337
ones count *see* syndrome
operationally related faults 39
operational transconductance amplifier (OTA) 457
orthogonal Latin squares *see* Latin squares
orthogonal transforms 134
output compaction *see* output compression
output compression 124
output signature *see* MISR

package costs 486
packaging (of ICs) 485
parallel fault simulation 45
parallel reconvergence 50
parametric tests 8, 386
parity checks 154, 162, 176
partial scan 225
partitioning 11, 72, 201, 210, 266, 284, 339, 441, 436
path oriented decision making (PODEM) 63, 66
path sensitising 50
pattern sensitive faults 26, 332, 334
peripheral interconnect bus (PI-bus) 241
per-unit cost (PUC) 480, 493
photolithography defects 39
physical fault location 39
pins (on IC package) 8
placement and routing 210, 218, 274

Index 531

PLURIBUS 358
polynomials 81, 145, 253
polynomial expansion 81
polynomial multiplication and division 81, 145, 254
positive reconvergence 50
primary inputs 20, 72, 208
primary outputs 20, 72, 208
primitive D-cubes of failure 56
primitive polynomials 86, 515
PRIME 358
printed-circuit board (PCB) testing 13, 260
probability 4, 36, 143, 285, 411
probabilty matrices 37
probe (*see also* 'bed-of-nails') 109
probing, E-beam 506
probing, mechanical 109, 111, 142, 382
production tests 1
production yield 4
product test 1
programmable array logic (PALs) 323
programmable gate arrays (PGAs) 320
programmable logic arrays (PLAs) 162, 298, 302, 306, 311, 314
programmable logic devices (PLDs) 298, 323
programmable read-only memory (PROM) 329
propagating (or propagation) D-cubes 57

quantisation 404
quantisation error 404

rack-and-stack instrumentation 9, 390, 429
Rademacher transform 135
random-access memory (RAM) 331, 337, 345
random-access scan (RAS) 218, 223
random path sensitising (RAPS) 70
read-only memory (ROM) 162, 329
read/write cycle 331
reciprocal polynomial 87
reconvergence *see* parallel reconvergence
redundancy *see* hardware redundancy and information redundancy
Reed-Muller coefficients 130, 139, 192
reliability $R(t)$ 172, 343
residue (in polynomial division) 145, 254

residue codes 163
retraining time *see* learning time
rollback 359
Roth's D-algorithm 55, 63
RTD test methodolgy 280
rule-of-tens law 484

S^3 BIST methodology 276
sample-and-hold 443
SASP BIST methodology 277
scanning electron microscope (SEM 506
scan path design 206, 218, 220, 230, 288, 441
scan path shift register 207
scan-set 218, 221
scan test, scan-path test 12, 350, 439
SCOAP 23, 25, 205
screening tests 187
secondary inputs 208
secondary outputs 208
self checking *see* self-test
self checking checkers 175, 312
self configurable 362
self-dual logic function 368
self-test 119, 201, 314, 345
sensitising 22
sensitivity (cost) analysis 495
separate-mode scan test 441
sequential bridging faults 30
sequential circuits 70, 188, 362
serial bit stream 12
serial shadow register (SSR) 354
S-graph 350
Sharp operator 302
shrinkage fault 300
short-circuit fault modelling 95
shuffle exchange 366
SIA roadmap 503
SIFT 358
signatures *see* test signatures
signature analyser 143, 245
silicon overheads 217, 274, 310, 442
simulation-based diagnostics 109
simultaneous scan test 209
single error correction (SEC) 156, 340, 358
sneak software conditions 15
soft-configurable 362
soft faults (cf. hard faults) 332, 344, 450
software testing 15
space compression (cf. time compression) 125

532 *Index*

spectral techniques, spectral coefficients 130, 135, 139, 192
spectrum 135
STAR computer 357
state transition matrix 91
static redundancy 169
storage *see* memory
STR BIST methodology 277
STRATUS 358
stuck-at-0 (s-a-0) 26, 56
stuck-at-1 (s-a-1) 26, 56
stuck-at faults 25, 44
STUMPS BIST methodology 279
SUPERCAT 286
switch-level fault modelling 102
synchronising input sequence 71, 205
syndrome 125, 191, 305
systems-on-a-chip 504
systems intregration software 509
systolic arrays 367

TANDAM 358
TDES 286, 497
TEST/80 23
test access port (TAP) 232, 238, 464
test coverage 44
test economics *see* economics of test
test mode (cf. normal mode) 119
test patterns *see* test vectors
test pattern generation (TPG) (*see also* automatic test pattern generation (ATPG)) 43, 47
test resources 6, 205, 230, 390, 429, 450, 461
test set 7
test strategy planning 496
test signatures 119, 141, 244, 265
test systems, test equipment, test hardware 390, 393
test vectors 7
thermal imaging 506
thick film 382
thin film 382

tiles 320
time compression (cf. space compression) 125
TMEAS 23
totally self-checking (TSC) checkers 175
TPLA 301
TRAM 335
transmission gate *see* FET transmission gate
transient fault 36
transition count testing 129, 192
transistor length L 100
transistor width W 100
tridiagonal matrix 92
trinomials 517
true polynomial divider 146
two-dimensional codes 160
two-rail checkers 176, 312
type 90 function 90, 264
type 150 function 90, 264
type A LFSR 88, 146, 251, 274
type B LFSR 88, 146, 251

user-specific ICs (USICs) 2, 327, 430
unreliability $U(t)$ 172

vanishing fault 300
VHDL 286
VICTOR 23, 25
voting circuits 171

wafer probe 2, 3, 492
walking patterns 334
WALKPAT 334
watchdog processor 361
weighted BIST 280, 355
weighted built-in logic block observation (WBILBO) 281
weighted random patterns 281
weighted syndrome testing 305

yield 4, 523
yield management software 487